Forests for a Better Future: Sustainability, Innovation and Interdisciplinarity

Forests for a Better Future: Sustainability, Innovation and Interdisciplinarity

Editors

Angela Lo Monaco
Cate Macinnis-Ng
Om P. Rajora

MDPI • Basel • Beijing • Wuhan • Barcelona • Belgrade • Manchester • Tokyo • Cluj • Tianjin

Editors

Angela Lo Monaco	Cate Macinnis-Ng	Om P. Rajora
University of Tuscia	University of Auckland	University of New Brunswick
Italy	New Zealand	Canada

Editorial Office
MDPI
St. Alban-Anlage 66
4052 Basel, Switzerland

This is a reprint of articles from the Special Issue published online in the open access journal *Forests* (ISSN 1999-4907) (available at: https://www.mdpi.com/journal/forests/special_issues/IECF2020).

For citation purposes, cite each article independently as indicated on the article page online and as indicated below:

LastName, A.A.; LastName, B.B.; LastName, C.C. Article Title. *Journal Name* **Year**, *Volume Number*, Page Range.

ISBN 978-3-0365-4863-0 (Hbk)
ISBN 978-3-0365-4864-7 (PDF)

Cover image courtesy of Prof. Dr. Om P. Rajora

Contents

About the Editors

Angela Lo Monaco

Angela Lo Monaco is an Associate Professor of Wood Technology and Forest Logging at the University of Tuscia (Viterbo, Italy), where she is also Head of the Wood Science and Technology Laboratory. She is currently the Editor-in-Chief of the Section "Wood Science" of *Forests*. Angela Lo Monaco's research interests mainly concern wood science and technology both for the environment as well as for cultural heritage. Trees and timber are taken into consideration not only as a source of materials and byproducts but also as sensors of anthropic and climatic events. Her research interests focus on the properties of wood in artefacts of historical and artistic interest, with a special emphasis on the characteristics of wood degraded by environmental factors, in which the color and chemical characteristics of surfaces are used as monitoring systems. The impacts of forest logging on residual trees and seedlings, as well as the influence of human activities on the quantity and quality of dead wood in a forest, are some of the topics of her studies.

Cate Macinnis-Ng

Cate Macinnis-Ng is an Associate Professor and Rutherford Discovery Fellow in the School of Biological Sciences at the University of Auckland. She is also a PI in Te Pūnaha Matatini—New Zealand's Centre for Research Excellence in Complex Systems. Cate is a tree ecophysiologist with an interest in plant–climate intersections. Her main research is focused on a field-based throughfall exclusion experiment which examines the effects of drought on the southern conifer, kauri (*Agathis australis*). Cate is the immediate past president of the New Zealand Ecological Society.

Om P. Rajora

Prof. Dr. Om P. Rajora is a professor of forest genetics and genomics at the University of New Brunswick, Canada, where he held the prestigious Senior Canada Research Chair in Forest and Conservation Genomics and Biotechnology. He is the Coordinator of the International Union of Forest Research Organizations (UFRO) unit 2.04.01 "Population, Ecological and Conservation Genetics". Prof. Dr. Rajora has served as an Associate Editor, Academic Editor, or Editorial Board Member of several international journals, and is currently an Associate Editor for *Botany* and *Frontiers in Conservation Science-Conservation Genomics* and an Academic Editor for *Forests*. Prof. Dr. Rajora is an internationally eminent scholar with outstanding recognition and reputation. Over the past four decades, Prof. Dr. Rajora has conducted highly original, cutting-edge research in forest tree genetics and genomics, addressing basic and applied aspects, resulting in ground-breaking discoveries and new insights in the population, conservation, evolutionary and ecological genetics and genomics, and structural and functional genomics of forest trees. He has served on many national and international science boards and committees and has received many international and national awards, honors, and distinctions. Prof. Dr. Om Rajora has developed a pioneering and internationally renowned book series on Population Genomics. He is the Editor-in-Chief of this book series and editor or co-editor of about 10 volumes in this series, being published by Springer *Nature*.

Preface to "Forests for a Better Future: Sustainability, Innovation and Interdisciplinarity"

This book is the compilation of the selected articles published in the Special Issue "Forests for a Better Future: Sustainability, Innovation, Interdisciplinarity—Selected Papers from the 1st International Electronic Conference on Forests (IECF2020)". The articles/chapters in this book highlight the role of research in innovation and sustainability in the forest sector. It is well known that the sustainability of forests is essential for the provisioning of the goods and services forest ecosystems offer for the sustenance of life on this planet. This requires innovative and interdisciplinary forest research and forestry practices. The articles in this book deal with these critical issues. The contributions included fall within the broad thematic areas of forest science and cover crucial topics such as biocontrol, forest fire mapping, harvesting and logging practices, quantitative and qualitative assessments of forest products, urban forests, and wood treatments—topics which have also been approached from an interdisciplinary perspective. The authors are researchers from both universities and research institutes. This book is suitable for researchers, students, and practitioners in the forest sector. The contributions also have practical applications and implications, as they deal with the ecological and economic importance of forests and new technologies for the conservation, monitoring, and improvement of services and forest value. The editors are grateful to the authors and reviewers who have helped us significantly improve the quality of these publications. We would like to thank Leah Lu and other *Forests* assistant editors and staff for their invaluable support in this endeavor.

Angela Lo Monaco, Cate Macinnis-Ng, and Om P. Rajora
Editors

Editorial

Forests for a Better Future: Sustainability, Innovation and Interdisciplinarity

Angela Lo Monaco [1],*, Cate Macinnis-Ng [2] and Om P. Rajora [3]

1   Department of Agricultural and Forest Sciences, University of Tuscia, 01100 Viterbo, Italy
2   School of Biological Sciences and Te Pūnaha Matatini, The University of Auckland, Private Bag 92019, Auckland 1142, New Zealand; c.macinnis-ng@auckland.ac.nz
3   Faculty of Forestry and Environmental Management, University of New Brunswick, Fredericton, NB E3B 5A3, Canada; om.rajora@unb.ca
*   Correspondence: lomonaco@unitus.it

Citation: Lo Monaco, A.; Macinnis-Ng, C.; Rajora, O.P. Forests for a Better Future: Sustainability, Innovation and Interdisciplinarity. *Forests* **2022**, *13*, 941. https://doi.org/10.3390/f13060941

Received: 7 June 2022
Accepted: 13 June 2022
Published: 16 June 2022

Publisher's Note: MDPI stays neutral with regard to jurisdictional claims in published maps and institutional affiliations.

Forests offer a solution to climate change through carbon storage and providing ecosystem services and sustainable products. While reducing greenhouse gas emissions are required to address the climate change crisis, restoring and expanding forests increase terrestrial carbon sinks. Therefore, it is critical to manage our forests sustainably. This requires innovative and interdisciplinary research and practices. To address this topical and critical issue, we organized the first International Electronic Conference on Forests (IECF) held in November 2020 under the theme of 'Forests for a Better Future: Sustainability, Innovation, Interdisciplinarity'. The main topics of the Conference included Forest Ecology, Management and Restoration, Forest Genetics, Ecophysiology and Biology, Forests and Urban Forests Sustainability, Forest Inventory, Quantitative Methods and Remote Sensing, Wood Science, Production Chain and Fuelwood, Forest Opertaions and Engineering, and Fire Risks and Other Natural hazards.

While face-to-face conferences provide opportunties to meet, network and socialise, the IECF first edition provided the chance to share recent work when the global pandemic prevented travel and large gatherings. Online conferences have the added benefit of being climate friendly and accessible to all. Participation in the event was free of charge so it was an excellent opportunity for research students and early career scientists to build their connections and profiles with exposure that would otherwise not have been possible in 2020. We thank all contributors for taking part at the confernce. Conference materials are available at the conference website https://iecf2020.sciforum.net/conference/IECF2020 (accessed on 24 May 2022).

A Special Issue was developed and has been published consisting of 20 selected papers presented at the conference. We are delighted to present this book which is a compilation of these selected published articles. The authors are from many countries (Argentina, Germany, Greece, Italy, Japan, Korea, Poland, Portugal, Russia, Serbia, and Spain), from both universities and reseach intitutions. The articles covered many forest trees species (*Abies sachalinensis*, *Alnus hirsuta*, *Betula ermanii*, *Fraxinus mandshurica*, *Lomatia hirsuta*, *Picea abies*, *Picea glehnii*, *Picea jezoensis* var. *microsperma*, *Pinus pumila*, *Pinus sylvestris*, *Populus* spp., *Prunus laurocerasus*, *Quercus serrata*, *Robinia pseudoacacia*, *Salix* spp., *Thuja occidentalis*, *Ulmus davidiana* var. *japonica* and *Triplochiton scleroxylon*).

The articles present research results addressing a wide range of topics relevant to improving forestry practices and sustainability, including forest ecology biocontrol, wood treatments, urban forests, mapping of fire spread and harvesting and logging practices. The articles present topical examples of world-class research activities.

Here, we outline key research activities and highlights of the publications included in this book. Forests were studied as urban ecosystems aimed at recreation, therapy and health for humans [1,2] but also as natural environments placed in urban contexts frequented by wild animals. Urbanization has led to severe habitat fragmentation and loss, and has brought humans and wildlife in close proximity, affecting both [3]. The stand dynamics of natural forests associated with the mortality of trees was investigated, estimating survival probabilities using non-parametric methods [4]. Computer vision on forest composition [5] was used to assess the tree types and distribution, helping improve timber stock calculation. Sentinel-2 satellite images from the European Space Agency and modelling allowed the assessment of cumulative risks posed to habitats and were also used in monitoring high-ignition-risk areas, starting from the concept that firefighting success depends on both fire ignition prevention and early ignition detection [6]. Changes in the temperature regime of post-fire and post-technogenic cryogenic soils of Central Siberia were investigated using remote sensing data and numerical simulation results [7]. The pollution stress in Scots pine stands, assessed dendrochronologically, was quantified as disturbances of incremental dynamics and long-term strong reduction in growth [8]. The environmental impact generated by the production of thermal energy was assessed for a poplar short-rotation coppice, considering the entire life cycle [9]. Kühn et al. [10] evaluated the possibility of improving quality wood production in coppice stands of *Lomatia hirsute*. Identification of early bark beetle infestation (e.g., *Ips typographus* and *Pityogenes chalcographus*) was performed by drone-based monoterpene detection testing semiconductor gas sensor arrays under artificial and real-life field conditions [11]. *Robinia pseudoacacia* plantations for the rehabilitation of post-mining lignite areas in northwestern Greece were evaluated in terms of above-ground biomass and deadwood, indicating low biomass accumulation in bruised trees and deadwood, due to their young age (5–30 years) [12]. Blanco and Blanco [13] developed simple volume equations for hybrid poplar plantations in the Órbigo river basin (NW Spain), providing a simple but acceptably accurate tool to empower aging rural owners when estimating standing volume by themselves. The feasibility of a harvesting system for small-diameter trees as forest biomass was investigated in Japan [14]. Environmental sustainability was emphasized by both environmentally friendly forest products for disease control [15] and modified wood with low environmental impact [16]. *Pinus sylvestris* bark represents a rich source of active compounds for products with high added value, such as those that have antifungal, antibacterial, and antioxidant properties. Karličić et al. [15] estimated the antifungal potential of *P. sylvestris* bark through its chemical (water extracts) and biological components (*Trichoderma* spp. isolated from the bark). A reliable prediction of the proportion of standing Scots pine bark is therefore also important in this sense but also for classical use (estimating timber volume without bark) [17]. The influence of log crookedness and taper on stack volume was simulated using a 3D model, which provided compelling results for buyers who can estimate the log content more efficiently than a quick visual assessment [18]. The delivery of forest products from the forest or plantation to the final processing sites can be one of the most cost-effective operations in the forest sector. Finally, Trzciński and Tymendorf [19] and Sperandio et al. [20] deal, respectively, in the transport of large-sized pine log to the sawmill and of biomass products to small-scale plants for energy production.

In summary, the articles included in this book highlight the ecological and economic importance of forests and new technologies for conserving, monitoring and improving forest services and value.

Our thanks go to the authors of the papers for their contributions to the conference and the Special Issue. We also thank the reviewers.

Conflicts of Interest: The authors declare no conflict of interest.

References

1. Wajchman-Świtalska, S.; Zajadacz, A.; Lubarska, A. Recreation and Therapy in Urban Forests—The Potential Use of Sensory Garden Solutions. *Forests* **2021**, *12*, 1402. [CrossRef]
2. Kim, J.; Park, D.-B.; Seo, J.I. Exploring the Relationship between Forest Structure and Health. *Forests* **2020**, *11*, 1264. [CrossRef]
3. Jasińska, K.; Jackowiak, M.; Gryz, J.; Bijak, S.; Szyc, K.; Krauze-Gryz, D. Habitat-Related Differences in Winter Presence and Spring–Summer Activity of Roe Deer in Warsaw. *Forests* **2021**, *12*, 970. [CrossRef]

4. Wijenayake, P.; Hiroshima, T. Age-Based Survival Analysis of Coniferous and Broad-Leaved Trees: A Case Study of Preserved Forests in Northern Japan. *Forests* **2021**, *12*, 1014. [CrossRef]
5. Illarionova, S.; Trekin, A.; Ignatiev, V.; Oseledets, I. Tree Species Mapping on Sentinel-2 Satellite Imagery with Weakly Supervised Classification and Object-Wise Sampling. *Forests* **2021**, *12*, 1413. [CrossRef]
6. Santos, L.; Lopes, V.; Baptista, C. MDIR Monthly Ignition Risk Maps, an Integrated Open-Source Strategy for Wildfire Prevention. *Forests* **2022**, *13*, 408. [CrossRef]
7. Ponomareva, T.; Litvintsv, K.; Finnikov, K.; Yakimov, N.; Sentyabov, A.; Ponomarev, E. Soil Temperature in Disturbed Ecosystems of Central Siberia: Remote Sensing Data and Numerical Simulation. *Forests* **2021**, *12*, 994. [CrossRef]
8. Chojnacka-Ożga, L.; Ożga, W. The Impact of Air Pollution on the Growth of Scots Pine Stands in Poland on the Basis of Dendrochronological Analyses. *Forests* **2021**, *12*, 1421. [CrossRef]
9. Sperandio, G.; Suardi, A.; Acampora, A.; Civitarese, V. Environmental Sustainability of Heat Produced by Poplar Short-Rotation Coppice (SRC) Woody Biomass. *Forests* **2021**, *12*, 878. [CrossRef]
10. Kühn, H.; Loguercio, G.; Caselli, M.; Thren, M. Growth and Potential of *Lomatia hirsuta* Forests from Stump Shoots in the Valley of El Manso/Patagonia/Argentina. *Forests* **2021**, *12*, 923. [CrossRef]
11. Paczkowski, S.; Datta, P.; Irion, H.; Paczkowska, M.; Habert, T.; Pelz, S.; Jaeger, D. Evaluation of Early Bark Beetle Infestation Localization by Drone-Based Monoterpene Detection. *Forests* **2021**, *12*, 228. [CrossRef]
12. Spyroglou, G.; Fotelli, M.; Nanos, N.; Radoglou, K. Assessing Black Locust Biomass Accumulation in Restoration Plantations. *Forests* **2021**, *12*, 1477. [CrossRef]
13. Blanco, R.; Blanco, J.A. Empowering Forest Owners with Simple Volume Equations for Poplar Plantations in the Órbigo River Basin (NW Spain). *Forests* **2021**, *12*, 124. [CrossRef]
14. Yoshioka, T.; Tomioka, T.; Nitami, T. Feasibility of a Harvesting System for Small-Diameter Trees as Unutilized Forest Biomass in Japan. *Forests* **2021**, *12*, 74. [CrossRef]
15. Karličić, V.; Zlatković, M.; Jovičić-Petrović, J.; Nikolić, M.P.; Orlović, S.; Raičević, V. *Trichoderma* spp. from Pine Bark and Pine Bark Extracts: Potent Biocontrol Agents against *Botryosphaeriaceae*. *Forests* **2021**, *12*, 1731. [CrossRef]
16. Gennari, E.; Picchio, R.; Monaco, A.L. Industrial Heat Treatment of Wood: Study of Induced Effects on Ayous Wood (*Triplochiton scleroxylon* K. Schum). *Forests* **2021**, *12*, 730. [CrossRef]
17. Wilms, F.; Duppel, N.; Cremer, T.; Berendt, F. Bark Thickness and Heights of the Bark Transition Area of Scots Pine. *Forests* **2021**, *12*, 1386. [CrossRef]
18. de Miguel-Díez, F.; Tolosana-Esteban, E.; Purfürst, T.; Cremer, T. Analysis of the Influence That Parameters Crookedness and Taper Have on Stack Volume by Using a 3D-Simulation Model of Wood Stacks. *Forests* **2021**, *12*, 238. [CrossRef]
19. Trzciński, G.; Tymendorf, Ł. Transport Work for the Supply of Pine Sawlogs to the Sawmill. *Forests* **2020**, *11*, 1340. [CrossRef]
20. Sperandio, G.; Acampora, A.; Civitarese, V.; Bajocco, S.; Bascietto, M. Transport Cost Estimation Model of the Agroforestry Biomass in a Small-Scale Energy Chain. *Forests* **2021**, *12*, 158. [CrossRef]

 *forests*

Article

Exploring the Relationship between Forest Structure and Health

Jinki Kim [1,*], **Duk-Byeong Park** [2] **and Jung Il Seo** [3,*]

[1] Department of Landscape Architecture, Kongju National University, 182 Shinkwan-dong, Gongju 32588, Chungcheongnam-do, Korea

[2] Department of Regional Development, Kongju National University, 182 Shinkwan-dong, Gongju 32588, Chungcheongnam-do, Korea; parkdb84@kongju.ac.kr

[3] Department of Forest Resources, Kongju National University, 182 Shinkwan-dong, Gongju 32588, Chungcheongnam-do, Korea

* Correspondence: jkkim12@kongju.ac.kr (J.K.); jungil.seo@kongju.ac.kr (J.I.S.); Tel.: +82-41-330-1447 (J.K.); +82-41-330-1302 (J.I.S.)

Received: 18 November 2020; Accepted: 26 November 2020; Published: 27 November 2020

Abstract: There is abundant evidence that green space in urban neighborhood is associated with physical activity and it is well known that physical activity contributes to human health. Physical activity fosters normal growth and development, can reduce the risk of chronic diseases, and can make people feel better and function better. Evidences also show that exposure to natural places can lead to positive mental health outcomes, whether a view of nature from a window, being within natural places, or exercising in these environments. The study aims to identify the factors of forest structure and socioeconomic characteristics influencing adults' physical activity and health. A sample of 148,754 respondents from the Korea Community Health Survey, conducted in 2016, was analyzed. Measures included frequency of physical activity, stress, depression, and landscape metrics of forest patch. Hierarchical multiple regression analysis, controlling for socio-demographic characteristics, revealed that larger forest patches and the more irregular shapes were associated with more physical activity. The study also showed that the shape of forest patch and slope were associated with less mental health complaints, whereas composition related landscape metrics were not.

Keywords: physical activity; mental health; landscape metrics; hierarchical multiple regression

---

1. Introduction

There is abundant evidence that green space in urban neighborhoods is associated with physical activity [1–3]. Positive associations with increased levels of physical activity were reported for the amount of green space close to home [1], the distance to the nearest green space [4,5], the size of the nearest green space [5,6], and the presence of certain features [6]. Schipperijn et al. (2013) showed that urban green space can facilitate a wide range of free or low cost activities and the availability of urban green space is one of the environmental factors that is frequently linked to increased levels of physical activity [3].

It is well known that physical activity contributes to human health. In addition to fostering normal growth and development, it can reduce the risk of chronic diseases and conditions [7,8] such as depression [9], anxiety [10], and stress [11], and also can make people feel better, function better, and sleep better [12]. Evidence also shows that exposure to natural places can enhance psychophysiological recovery [13], and lead to positive mental health outcomes [14]. Contact with natural environments can similarly promote restoration from stress and mental fatigue [14,15].

The European Commission has devoted specific attention to nature-based solutions and has defined the concept within the range of ecosystem-based approaches [16]. Nature-based solutions aim to help societies address a variety of environmental, social and economic challenges in sustainable ways. The concept builds on and supports other closely related concepts, such as ecosystem services, ecosystem-based adaptation/mitigation, and green and blue infrastructure. They all recognize the importance of nature and require a systemic approach to environmental change based on an understanding of the structure and functioning of ecosystems, including human actions and their consequences [17].

Forest ecosystem provide recreational services whether it be hiking, birdwatching, hunting, or taking in the scenic beauty and forests can generate recreational revenue through use- or entry-fees and ecotourism [18]. Agimass (2018) investigated the choice of forest sites for recreation [19] and Birch et al. (2014) describe that forests provide direct net income by charging visitors to access the picnic areas [20]. There is strong evidence that forests ecosystem can affect human health, well-being, and public health as well. Hunter et al. (2019) describe the relationship between duration of a nature experience, and changes in two physiological biomarkers of stress-salivary cortisol and alpha-amylase. They reported the efficiency of a nature pill per time expended was greatest between 20 and 30 min [21]. Focusing on forests, research reported that forest bathing, is associated with clinical therapeutic effects for the human physiological and psychological systems [22,23].

There are studies, on the other hand, that reported no association between green space and physical activity [24–26] or physical activity and health [27]. The research regarding physical activity as a mechanism explaining the relationship between green space and health has produced mixed results. Lachowycz and Jones (2011) identified 50 studies examining the relationship between green space and physical activity. Among them, only twenty studies (40%) reported a positive association between green space and physical activity. Two reported a negative association and 28 found no evidence for a relationship [28]. These mixed results were found at both a regional and national level.

Recent models of public health and environment reflect the fact that individual characteristics and preferences are active within the context of socio-economic, political, cultural and environmental factors that operate at different scales from the household and community to wider geographic levels [29]. Research since the 1990s using ecological models of behavior has increasingly emphasized the need to consider the physical environment more carefully in such studies [30,31]. This approach is of particular interest to those responsible for planning and designing the environment. A number of approaches to environment-behavior research have developed versions of this ecological model, reflecting similar understandings of the transactional nature of the relationship between person and place [32,33].

Within this context, public health policy has generally adopted a model of the relationship between environment and health that reflects Bronfenbrenner's human ecology theory where the individual is located within nested ecological systems [34]. Bronfenbrenner's work underlines how the individual can exert an influence over his or her environment and, at the same time, how the environment exerts an influence on the individual. Barton and Grant (2006) presented a conceptual model of the settlement as ecosystem in its context. The settlement ecosystem health map shows the relationship between health and the physical, social, and cultural environment (Figure 1). In this map, people are at the center of the built and natural environments, reflecting not only the focus on health, but also sustainable development activities. The human settlement is set within its natural environment and the global ecosystem on which it ultimately depends [29].

On a landscape scale, the conceptual models produced by Forman for the kinds of landscape patterns are necessary to support different groups of animal species—concepts such as the landscape mosaic of patch, corridor and matrix—which offer a basis for making planning decisions that minimize habitat loss [35,36]. During the past few decades, landscape pattern metrics have been employed for analyzing the composition and configuration of landscape structure and spatial pattern [37]. Composition refers to features associated with the variety and abundance of patch types within the landscape, but without considering the spatial character, placement, or location of patches within the mosaic. The principal metrics are number of categories, proportions, and diversity.

Spatial configuration refers to the spatial character and arrangement, position, or orientation of patches within the class or landscape [38]. Some aspects of configuration are measures of the spatial character of the patches themselves, even though the aggregation may be across patches at the class or landscape level.

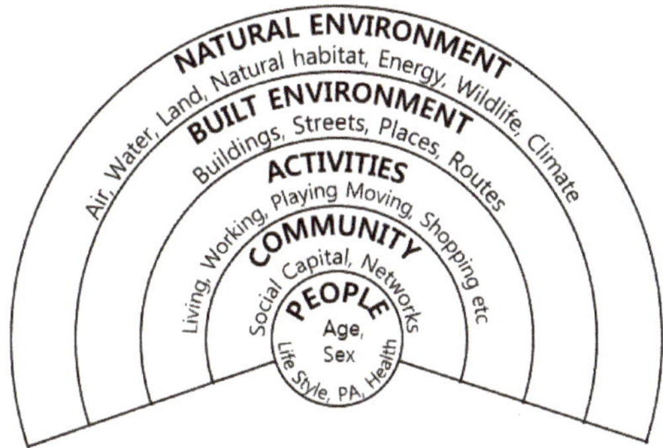

Figure 1. The relationship between health and environment (adapted from models on the determinants of health by Barton and Grant (2006)).

To date, it has been largely unclear how many green areas, and what size of natural elements are needed to achieve certain health benefits, and which type of green space will most benefit the health of local residents. The purpose of this study is to investigate the relationship between landscape structure, physical activity, and health and identify the factors of landscape structure and socioeconomic characteristics influencing adults' activity and health. Forest features such as size, shape, edge, number, slope, and elevation are discussed on the premise that they exert an influence on promoting physical activity and improving health.

2. Materials and Methods

2.1. Data

Data used in this study included the Korea Community Health Survey (KCHS), Land Use Land Cover (LULC), a digital elevation model (DEM), and geographic information system (GIS) files including administrative boundary (Figure 2). The KCHS data were collected from Korea Centers for Disease Control and Prevention in 2016. The centers conduct a nationwide survey to provide data for planning, monitoring, and evaluating community health promotion and disease prevention programs.

The survey was designed to establish a basis for implementing health services by producing community health statistics that establish and evaluate the community health care plan. It also tried to integrate evaluation indices for the health services of local governments by standardizing survey indices and procedures comparable among communities.

This survey was conducted by 253 community health centers, 35 community universities, and 1500 interviewers [39]. Among them, the data collected from 166 community health centers were used for this study. Seven metropolitan areas were excluded since the land cover conditions of those areas are not comparable for the calculation of landscape structure due to the severe urbanization.

LULC and DEM data were downloaded from the National Environmental Information Network System [40] and used to calculate landscape structure indices and topographic conditions respectively. LULC data consist of water, developed areas, barren, wetlands, forest, croplands, grasslands, and so forth. Among them, forest class was selected and exported in Grid file format in ArcGIS 10.1. The Grid file was converted to GeoTiff format for the calculation of landscape metrics in FRAGSTATS 4.2.1.

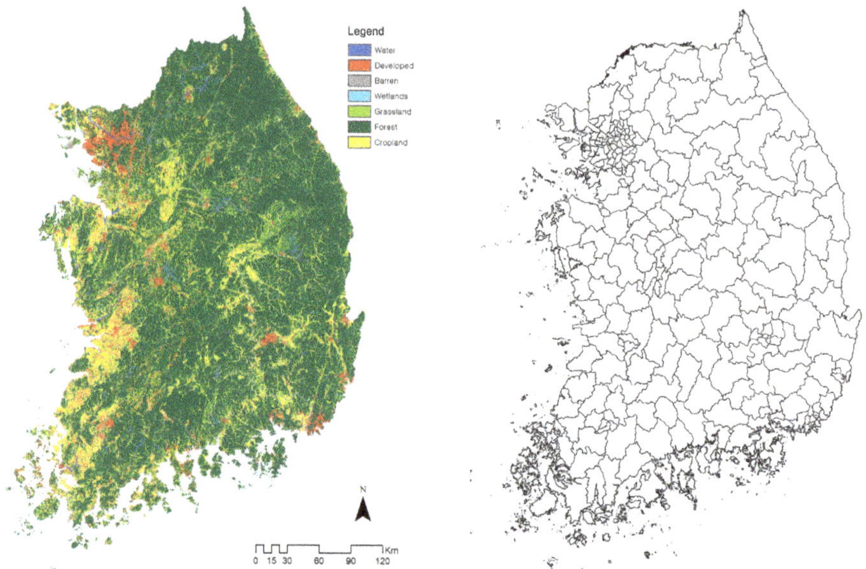

Figure 2. Land Use Land Cover map (**left**) and County boundary (**right**).

DEM data were used to calculate local elevation and slope. The DEM was clipped with the county boundary. The nearest-neighborhood resampling method was used to resample the created DEM in 30 m resolution. A spatial analysis was conducted to calculate elevation and slope.

2.2. Study Population

The target population was defined as adults aged over 19 years and who live in the jurisdiction of one community health center. The registered population data were obtained from the Ministry of Public Administration and Security. The data include gender, age, and population structure as well as the stratification of the surveyed population.

The sample size in each community health center was selected so that the main health index in each community health center has ±3% desired sampling error with a 95% confidence level. Nine hundred people on average were selected in each community health center and the total population was 148,754.

The KCHS collected information of many aspects of physical activity including running (jogging), tracking, biking, swimming, soccer, basketball, jump rope, squash, tennis, and other types of exercise. The metrics were level of physical activity in a typical day, weekly frequency of participation in vigorous physical activities for at least 10 min, and weekly frequency of participation in walking exercise for at least 10 min. The duration of these activities was included in this survey. Mental health status was measured by a question on subjective stress recognition and experience of depression.

The KCHS collected demographic and health-related information from participants. This included gender, age, education, smoking status, drinking, and oral and mental health. The metric of income had eight categories from low income of <$1000/month in a household to high income of >$5000/month.

2.3. Landscape Structure

The landscape structure was quantified with a number of landscape metrics. In this study, quantifying the landscape structure involved the use of statistics that described the landscape configuration and composition. Landscape metrics include Class Area (CA), Percent of Landscape (PLAND), Number of Patches (NP), Largest Patch Index (LPI), Edge Density (ED), and Mean Shape Index (MSI) (Table 1).

Table 1. Landscape metrics and description.

Metrics	Formula	Description
Class Area (ha)	$CA = \sum_{j=1}^{n} a_{ij}(\frac{1}{10,000})$	• CA equals the sum of the areas (m^2) of all patches of the corresponding patch type
Percent of Landscape (%)	$PLAND = P_i \dfrac{\sum_{j=1}^{n} a_{ij}}{A}(100)$	• %LAND equals the percentage the landscape comprised of the corresponding patch type.
Number of Patches	$NP = n_1$	• NP equals the number of patches of the corresponding patch type (class)
Largest Patch Index	$LPI = \dfrac{\max\limits_{j=1}^{n} a_{ij}}{A}(100)$	• LPI equals the percentage of the landscape comprised by the largest patch
Edge Density (m/ha)	$ED = \dfrac{\sum_{k=1}^{m'} eik}{A}(10,000)$	• ED equals the sum of the lengths (m) of all edge segments involving the corresponding patch type, divided by the total landscape area (m^2), multiplied by 10,000
Mean Shape Index	$MSI = \dfrac{\sum_{j=1}^{n}\left(\frac{0.25pij}{\sqrt{aij}}\right)}{ni}$	• MSI equals the average shape index of patches of the corresponding patch type. When all patches are circular (vector) or square (raster) MSI equals 1.
Slope		• Slope equals the average slope (%) of each county
Elevation		• Elevation equals the average elevation of each county

2.4. Statistical Analysis

The data were analyzed using SPSS 24. Descriptive characteristics of the study participants were tabulated. A hierarchical multiple regression analysis, controlling for socio-demographic characteristics, was used to investigate associations between landscape structure and physical activity and health. Figure 3 represents the research framework using landscape metrics, geographic information, and community health survey in this study.

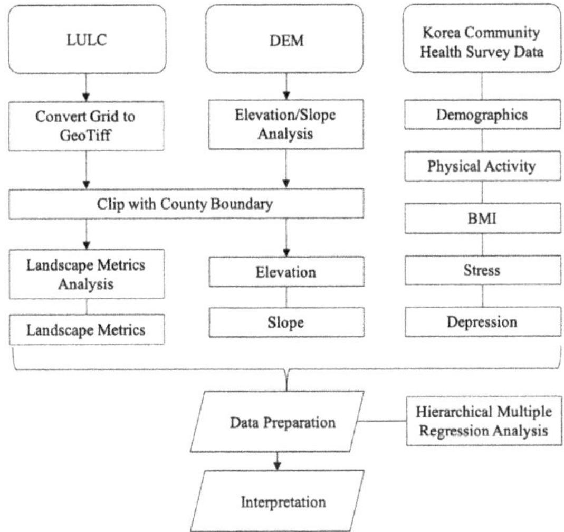

Figure 3. Framework of data used for geographic information and community health survey.

3. Results

We performed a hierarchical regression analysis to determine whether landscape structure influenced adults' physical activity and health beyond age, gender, education level, and household income. Eight predictors of landscape structure indices were used in the equation because of their statistically significant correlations with physical activity and health. Descriptive statistics and regression results are as follows.

3.1. Demographics

Of the 148,754 samples, 54.9% (81,665) were male and 45.1% (67,089) were female. According to the 2016 Korean Statistical Information Service data, the percentage of males and females in Korea is 50.12% and 49.88% respectively [41]. The average age of the participants was 54.57 years old and the oldest was 105. The highest participation age range was ages 50 to 59 (20.3%) and the lowest was over 80 (7.2%). The highest number of respondents for degree of education was secondary (30.8%) and the lowest degree of education held by the participants was graduate (1.5%). As for the monthly income, median income (ranging from $1000 to $4000) accounted for 50.3% of the participants and low income (<$1000/month in a household) occupied 24.1% (Table 2).

Table 2. Socioeconomic status.

Variables	N	Percent
Age Goup (Years)		
19–29	13,992	9.4
30–39	18,376	12.4
40–49	25,591	17.2
50–59	30,143	20.3
60–69	26,279	17.7
70–79	23,611	15.9
80+	10,762	7.2
Sex		
male	67,089	45.1
female	81,665	54.9
Education		
No school	12,881	9.7
Elementary	30,821	23.1
Subsecondary	17,159	12.9
Secondary	40,974	30.8
2–3 years college	15,497	11.7
College	13,644	10.3
Graduate	1967	1.5
Income		
1 (lowest)	14,611	9.9
2	20,985	14.2
3	25,233	17.1
4	26,628	18.1
5	22,318	15.1
6	15,240	10.3
7	8819	6.0
8 (highest)	13,535	9.2

3.2. Physical Activity and Health

Exercise experts measure activity in metabolic equivalents (METs), which is defined as the energy it takes to sit quietly. For the average adult, this is about one calorie per every one kilogram of body weight per hour. Vigorous-intensity physical activities (VPA) burn more than 6METs and include hiking, jogging, shoveling, bicycling fast, basketball, soccer games, and participating in a strenuous fitness class [42]. As for the frequency of VPA, 77.7% of respondents did not report vigorous physical activities. However, 3.6% respondents spent more than 10 min every day for physical activity (Table 3).

Table 3. Physical activity and health status.

Variables	N	Percent
Physical Activity		
0 (lowest)	115,333	77.7
1	6705	4.5
2	6330	4.3
3	6430	4.3
4	2664	1.8
5	4033	2.7
6	1606	1.1
7 (highest)	5391	3.6
Body Mass Index		
underweight	6890	4.9
normal	94,706	67.6
overweight	34,241	24.4
obese	4211	3.0
Stress		
1 (highest)	5343	3.6
2	31,054	20.9
3	76,346	51.3
4 (lowest)	35,925	24.2
Depression		
1 (highest)	1645	1.1
2	18,810	12.6
3 (lowest)	128,260	86.2

Regarding Body Mass Index (BMI), a measure of body fat based on height and weight that applies to adult men and women, 67.6% of respondents were normal, 24.4% of respondents were overweight, and 4.9% and 3% of respondents reported underweight and obese respectively. As for the health, while 24.2% of respondents did not feel stress at all, almost half of respondents felt stress a little, 20.9% felt stress considerably, and 3.6% felt strongly. In addition, 86.2% of respondents did not feel any depression at all, and 12.6% suffered from depression a little and 1.1% suffered from depression strongly.

3.3. Landscape Metrics and Topography

Table 4 indicates average values of landscape metrics and geographical characteristics. The average area of forest (CA) is 34,456.92 ha and the average proportion of the class (PLAND) is 58.88%. The average values of number of forest patches (NP), the largest forest percentage (LPI), edge density (ED), and mean shape of forest class (MSI) are 1576.14, 35.40%, 48.90 m/ha, and 1.23 respectively. The average slope of the study area is 19.93% and elevation is 294.67 m (Table 4).

Table 4. Descriptive Statistics.

Metric Name	Acronym	Units	Mean	Std. Dev.	Range
Class Area	CA	ha	34,456.92	30,894.44	CA > 0, without limit
Percent of Landscape	PLAND	%	58.88	20.77	0 < PLAND ≤ 100
Number of Patches	NP	none	1,576.14	1,237.77	NP ≥ 1, without limit
Large Patch Index	LPI	%	35.40	24.74	0 < LPI ≤ 100
Edge Density	ED	m/ha	48.90	15.67	ED ≥ 0, without limit
Mean Shape Index	MSI	none	1.23	0.062	MSI ≥ 1, without limit
Slope	Slope	%	19.93	8.94	-
Elevation	Elevation	m	294.67	190.95	-

3.4. Associations between Landscape Structure and Physical Activity and Health

Associations between landscape structure and overall physical activity, Body Mass Index, stress, and depression were observed in the hierarchical multiple regression analysis. Table 5 contains the standardized regression coefficients (*ßeta*), R^2, and change R^2 (ΔR^2) in each health indicator. The Durbin-Watson statistics is a test for autocorrelation in the residuals from a statistical regression analysis. The Durbin-Watson statistic had a value from 1.655 to 1.911. This means that there is no autocorrelation detected in the sample.

Table 5. Hierarchical multiple regression analysis of health indicators by socioeconomic and landscape structure (N = 148,754).

	Physical Activity (No. of Day)		Body Mass Index		Stress [1]		Depression [2]	
	ßeta		*ßeta*		*ßeta*		*ßeta*	
	Step 1	Step 2	Step 1	Step 2	Step 1	Step 2	Step 1	Step 2
Age	−0.100 **	−0.107 **	0.013 **	0.011 **	0.242 **	0.235 **	−0.043 **	−0.047 **
Gender	−0.117 **	−0.118 **	−0.164 **	−0.163 **	−0.033 **	−0.033 **	−0.088 **	−0.087 **
Education	−0.030 **	−0.030 **	−0.009 *	−0.010 *	0.050 **	0.045 **	0.065 **	0.064 **
Income	0.036 **	0.036 **	0.039 **	0.042 **	0.003	0.012 **	0.098 **	0.104 **
CA		−0.029 **		0.006		0.002		0.015 **
PLAND		0.016		0.009		−0.002		0.000
NP		0.025 **		0.007 *		0.006		0.005
LPI		0.023 **		−0.012 *		−0.005		−0.031 **
ED		−0.011 **		−0.009 *		−0.020 **		−0.019 **
MSI		0.044 **		0.008		0.019 **		0.026 **
Slope		−0.015		−0.023 *		0.036 **		0.068 **
Elevation		0.010		0.020		−0.001		−0.030 **
R^2	0.025	0.027	0.027	0.028	0.044	0.046	0.043	0.045
ΔR^2	-	0.002	-	0.001		0.002		0.002
F	838.851	307.072	897.410	302.759	1532.116	300.721	1472.619	518.769
Durbin-Watson		1.655		1.911		1.827		1.787

* $p < 0.05$, ** $p < 0.01$, variable input: enter, (1) 1 = more stressful 4 = less stressful, (2) 1 = very depressed 3 = not depressed. CA: Class Area, PLAND: Percentage of Landscape, NP: Number of Patch, LPI: Largest Patch Index, ED: Edge Density, MSI: Mean Shape Index.

3.4.1. Landscape Structure and Physical Activity

In Step 1, age, gender education, and income were forced into the equation, $R^2 = 0.025$, $F (4, 131852) = 838.851$, $p < 0.01$. The four predictors used in Step 1 accounted for 2.5% of the variance in physical activity. In Step 2, physical activity including socio-economic variables was plugged into the equation $R^2 = 0.027$, $F (12, 131844) = 307.072$, $p < 0.01$. Two point seven percent of the variance in physical activity was accounted for after Step 2. Comparison of the two steps also indicated that the change in R^2 was statistically significant ($\Delta R^2 = 0.002$, $p < 0.05$).

Significant associations between physical activity and landscape metrics in class area (CA), number of patches (NP), largest patch index (LPI), edge density (ED), and mean shape index (MSI) were observed. Regression results showed that number of patches (NP), largest patch index (LPI), and mean shape index (MSI) were positively associated with physical activity while class area (CA) and edge density (ED) were negatively associated (Table 5).

3.4.2. Landscape Structure and BMI

The regression analysis showed that individual factors contributed significantly to the regression model, $R^2 = 0.027$, F (4, 125330) = 897.410, $p < 0.01$, and accounted for 2.7% of the variation in Body Mass Index in Step 1. In Step 2, Body Mass Index was plugged into the equation $R^2 = 0.028$, F (12, 125322) = 302.759, $p < 0.01$. Two-point eight percent of the variance in Body Mass Index was accounted for after Step 2. Comparison of the two steps also indicated that the change in R^2 was statistically significant ($\Delta R^2 = 0.001$, $p < 0.05$).

The findings also indicated that there were significant associations with number of patches (NP), largest patch index (LPI), edge density (ED), and slope. A regression analysis revealed that BMI was positively associated with number of patches (NP), whereas it was negatively associated with largest patch index (LPI), edge density (ED), and slope (Table 5).

3.4.3. Landscape Structure and Mental Health

In Step 1, age, gender education, and income were plugged into the equation $R^2 = 0.044$, F (4, 131852) = 1532.116, $p < 0.01$. Four-point four percent of the variance in stress was accounted for by the four predictors used in Step 1. In Step 2, stress was plugged into the equation, $R^2 = 0.046$, F (12, 131844) = 531.833, $p < 0.01$. Four-point six percent of the variance in stress was accounted for after Step 2. Comparison of the two steps also indicated that the change in R^2 was statistically significant ($\Delta R^2 = 0.002$, $p < 0.05$).

In the case of depression, the analysis revealed that individual factors contributed significantly to the regression model, $R^2 = 0.043$, F (4, 131814) = 1472.619, $p < 0.01$, and accounted for 4.3% of the variation in Step 1. In Step 2, depression was plugged into the equation $R^2 = 0.045$, F (12, 131806) = 518.769, $p < 0.01$ Four-point five percent of the variance in depression was accounted for after Step 2. Comparison of the two steps also indicated that the change in R^2 was statistically significant ($\Delta R^2 = 0.002$, $p < 0.05$).

Stress level was associated with edge density (ED), mean shape index (MSI), and slope significantly, whereas depression was associated with class area (CA), largest patch index (LPI), edge density (ED), mean shape index (MSI), slope, and elevation. Among these associations, regression results showed that stress was positively associated with MSI and slope while depression was associated with class area (CA), mean shape index (MSI), and slope (Table 5).

4. Discussion

Green space is increasingly considered as an important factor in relation to physical activity and mental health. Studies revealed that the amount of green space close to home and the distance and size of the nearest green space are positively associated with physical activity and health. This study examined the relationship between landscape structure, physical activity, and health with visually or psychologically sensible metrics focused on size, shape, number and edge of the forest patch.

Findings of this study show that the size, shape, and number of forest patches are positively associated with physical activity, whereas only the shape index is positively associated with mental health. Demographic characteristics in this study revealed that older people reported less physical activity and women demonstrated less physical activity than men. It was also found that those who have a higher education level reported less physical activity, whereas higher income was related to more physical activity (Table 6).

Table 6. The relationships among landscape structures, PA, and health indicators.

	Model	PA	BMI	Stress	Depression
1	Age	−	+	+	−
	Sex	−	−	−	−
	Education	−	−	+	+
	Income	+	+		+
2	CA	−			+
	PLAND				
	NP	+	+		
	LPI	+	−		−
	ED	−	−	−	−
	MSI	+		+	+
	Slope		−	+	+
	Elevation				−

CA: Class Area, PLAND: Percentage of Landscape, NP: Number of Patch, LPI: Largest Patch Index, ED: Edge Density, MSI: Mean Shape Index.

4.1. Landscape Structure and Physical Activity

Hierarchical multiple regression analyses revealed that the largest forest patch and more irregular forest shape were positively related to physical activity. These results support the findings of previous studies that there are relationships between green spaces and physical activity [24,25,43]. The percentage of the largest forest patch was positively related among area related landscape metrics in this study. The total forest area, however, was negatively associated with physical activity and percentage of forest was not significantly related. It seems that only large forest area promotes respondents to perform frequent physical activity in the community.

The analysis also revealed that the number of forest patches was positively associated with physical activity. This result supports the findings of Kaczynski (2009). He examined how the number and total size of neighborhood parks within 1 km of participants' homes, as well as distance to the closest park, were related with physical activity. The study indicated that living near more parks and parkland showed more positive relationships with activity. A positive association was found with physical activity that may be more strongly linked to park or forest [1]. Based on this context, it appears that a greater number of forests is associated with better accessibility. However, landscape composition, which is an abundance of forest patches or proportion of green space, is less likely to be associated with physical activity.

A negative association was found with edge density (ED) and a positive association was found with mean shape index (MSI). Edge density is edge length on a per unit area basis, which facilitates comparison among landscapes of varying size. The forest edge effect results primarily from differences in wind and light intensity and quality reaching a forest patch, which alter microclimate and disturbance rates [44,45]. Edge index in a landscape is very critical information in the study of fragmentation and the total amount of edge is directly related to the degree of spatial heterogeneity. Spatial heterogeneity plays a critical role in determining ecological function. However, the results showed that it was negatively associated with physical activity (Table 6).

4.2. Landscape Structure and Health

The results showed that shape related landscape indices were associated with fewer mental health complaints and better body mass indices. Area related indices such as class area (CA), percentage of landscape (PLAND), number of patches (NP), and largest patch index (LPI) did not respectively show a significant relationship with stress, whereas mean shape index (MSI) and slope were positively associated with depression and edge density (ED) was associated negatively. With regard to depression, proportion and number of forest patches did not show a significant relationship with depression.

However, class area (CA), mean shape index (MSI), and slope were positively associated with depression while largest patch index (LPI), edge density (ED), and elevation were associated negatively (Table 6).

Several studies have shown that people who have access to green areas have less stress, and those who visit green areas more are less stressed than those who visit green areas less frequently [46,47]. Previous studies also indicated that human perception of landscapes is correlated with health and stress reduction [13], increased neighborhood satisfaction [48], and landscape structure. In this study, shape of the forest and slope positively influenced stress and depression. Since the shape and slope of forest are important elements of human perception of landscapes, the results support previous findings in relation to human perception and mental health. On the other hand, elevation negatively influenced to depression. As the elevation increases, people are more prone to depression. This result also supports previous findings [49,50] of the positive association between high altitude and the risk of depression, suicide.

5. Conclusions

This study identified the factors of landscape structure and socioeconomic characteristics influencing physical activity and health among adults who live in the jurisdiction of community health centers. A hierarchical multiple regression analysis was used to examine associations among these factors.

The study revealed that larger forest patches and more irregular forest shapes were associated with more physical activity, whereas total area of forest patches was negatively associated with physical activity. Landscape metrics, such as largest patch index (LPI), mean shape index (MSI), and class area (CA) were associated with physical activity with a significance level of 0.01.

Despite the importance of community-level contextual effects in health, few studies have focused on comprehensively understanding their aspects. The present study revealed that the shape of forest patches and slope are associated with fewer mental health complaints, whereas composition related landscape metrics, i.e., number or proportion of forest patches, were not associated with stress and depression. Mean shape index (MSI) and slope were positively associated with stress with a significance level of 0.01 while number of patch (NP) and percentage of landscape (PLAND) metrics were not.

This study offers a conceptual basis and a methodological framework for landscape ecological planning considerations for promoting physical activity and improving health. The results confirmed the usefulness of landscape metrics for estimating their relationships with physical activity and health, and deriving more quantitative evidence.

Understanding how elements of landscape structure influence people's physical activity and health is important for landscape architects, planners, and policy makers for planning and designing a healthy environment. For further studies, community capacity such as social capital as well as landscape structures should be considered as an independent variable for community-level contextual effects in public health.

Author Contributions: J.K. and D.-B.P. conceived the investigation, analyzed data and wrote the paper. J.K. acquired funding, realized literature review, analyzed GIS data sets, and performed images and tables production; J.I.S. provided revision of the written paper and analysis of data and was involved in project administration. All authors have read and agreed to the published version of the manuscript.

Funding: This work was supported by the research grant of the Kongju National University in 2019 (2019-0278-01).

Conflicts of Interest: The authors declare no conflict of interest.

References

1. Kaczynski, A.T.; Potwarka, L.R.; Smale, B.J.A.; Havitz, M.E. Association of Parkland Proximity with Neighborhood and Park-based Physical Activity: Variations by Gender and Age. *Leis. Sci.* **2009**, *31*, 174–191. [CrossRef]
2. McMorris, O.; Villeneuve, P.J.; Su, J.; Jerrett, M. Urban greenness and physical activity in a national survey of Canadians. *Environ. Res.* **2015**, *137*, 94–100. [CrossRef] [PubMed]
3. Schipperijn, J.; Bentsen, P.; Troelsen, J.; Toftager, M.; Stigsdotter, U.K. Associations between physical activity and characteristics of urban green space. *Urban For. Urban Green.* **2013**, *12*, 109–116. [CrossRef]
4. Foster, C.; Hillsdon, M.; Thorogood, M. Environmental perceptions and walking in English adults. *J. Epidemiol. Community Heal.* **2004**, *58*, 924–928. [CrossRef] [PubMed]
5. Giles-corti, B.; Broomhall, M.H.; Knuiman, M.; Collins, C.; Douglas, K.; Ng, K.; Lange, A.; Hon, B.A.; Donovan, R.J. Increasing Walking; How Important Is Distance To, Attractiveness, and Size of Public Open Space? *Am. J. Prev. Med.* **2005**, *28*, 169–176. [CrossRef]
6. Kaczynski, A.T.; Potwarka, L.R.; Saelens, B.E. Association of Park Size, Distance, and Features With Physical Activity in Neighborhood Parks. *Am. J. Public Health* **2008**, *98*, 1451–1456. [CrossRef]
7. Warburton, D.E.R.; Nicol, C.W.; Bredin, S.S.D. Health benefits of physical activity: The evidence. *Can. Med. Assoc. J.* **2006**, *174*, 801–809. [CrossRef] [PubMed]
8. Sallis, J.F.; Floyd, M.F.; Rodrı, D.A.; Saelens, B.E. Role of Built Environments in Physical Activity, Obesity, and Cardiovascular Disease. *Circulation* **2012**, *125*, 729–737. [CrossRef]
9. Rethorst, C.D.; Wipfli, B.M.; Landers, D.M. The Antidepressive Effects of Exercise. *Sports Med.* **2009**, *39*, 491–511. [CrossRef]
10. Mackay, G.J.; Neill, J.T. The effect of "green exercise" on state anxiety and the role of exercise duration, intensity, and greenness: A quasi-experimental study. *Psychol. Sport Exerc.* **2010**, *11*, 238–245. [CrossRef]
11. Barton, J.; Pretty, J. What is the best dose of nature and green exercise for improving mental health? A multi-study analysis. *Environ. Sci. Technol.* **2010**. [CrossRef] [PubMed]
12. How much Physical Activity Do You Need? Available online: https://www.cdc.gov/physicalactivity/basics/index.htm (accessed on 13 January 2018).
13. Ulrich, R.S. View through a Window May Influence Recovery from Surgery. *Science* **1984**, *224*, 420–421. [CrossRef] [PubMed]
14. Hartig, T.; Evans, G.W.; Jamner, L.D.; Davis, D.S.; Tommy, G. Tracking restoration in natural and urban field settings. *J. Environ. Psychol.* **2003**, *23*, 109–123. [CrossRef]
15. Van den Berg, A.E.; Jorgensen, A.; Wilson, E.R. Evaluating restoration in urban green spaces: Does setting type make a difference? *Landsc. Urban Plan.* **2014**, *127*, 173–181. [CrossRef]
16. Faivre, N.; Fritz, M.; Freitas, T.; de Boissezon, B.; Vandewoestijne, S. Nature-Based Solutions in the EU: Innovating with nature to address social, economic and environmental challenges. *Environ. Res.* **2017**, *159*, 509–518. [CrossRef]
17. European Commission. *Towards an EU Research and Innovation policy agenda for Nature-Based Solutions & Re-Naturing Cities: Final Report of the Horizon 2020 Expert Group on "Nature-Based Solutions and Re-Naturing Cities"*; Publications Office of the European Union: Brussels, Belgium, 2015.
18. Jenkins, M.; Schaap, B. *Forest Ecosystem Services—Background Analytical Study 1*; United Nations Forum on Forests, 2018; Available online: https://www.un.org/esa/forests/wp-content/uploads/2018/05/UNFF13_BkgdStudy_ForestsEcoServices.pdf (accessed on 20 October 2019).
19. Agimass, F.; Lundhede, T.; Panduro, T.E.; Jacobsen, J.B. The choice of forest site for recreation: A revealed preference analysis using spatial data. *Ecosyst. Serv.* **2018**, *31*, 445–454. [CrossRef]
20. Birch, J.C.; Thapa, I.; Balmford, A.; Bradbury, R.B.; Brown, C.; Butchart, S.H.M.; Gurung, H.; Hughes, F.M.R.; Mulligan, M.; Pandeya, B.; et al. What benefits do community forests provide, and to whom? A rapid assessment of ecosystem services from a Himalayan forest, Nepal. *Ecosyst. Serv.* **2014**, *8*, 118–127. [CrossRef]
21. Hunter, M.C.R.; Gillespie, B.W.; Chen, S.Y.P. Urban nature experiences reduce stress in the context of daily life based on salivary biomarkers. *Front. Psychol.* **2019**, *10*, 1–16. [CrossRef]
22. Hansen, R.; Rall, E.; Chapman, E.; Rolf, W.; Pauleit, S. Urban Green Infrastructure Planning: A Guide for Practitioners. GREEN SURGE; 2017. Available online: https://www.researchgate.net/publication/319967102_Urban_Green_Infrastructure_Planning_A_Guide_for_Practitioners (accessed on 27 September 2019).

23. Song, C.; Ikei, H.; Miyazaki, Y. Physiological effects of nature therapy: A review of the research in Japan. *Int. J. Environ. Res. Public Health* **2016**, *13*. 781. [CrossRef]

24. Jones, A.; Hillsdon, M.; Coombes, E. Greenspace access, use, and physical activity: Understanding the effects of area deprivation. *Prev. Med.* **2009**, *49*, 500–505. [CrossRef]

25. De Vries, S.; van Dillen, S.M.E.; Groenewegen, P.P.; Spreeuwenberg, P. Streetscape greenery and health: Stress, social cohesion and physical activity as mediators. *Soc. Sci. Med.* **2013**, *94*, 26–33. [CrossRef] [PubMed]

26. Sugiyama, T.; Giles-Corti, B.; Summers, J.; du Toit, L.; Leslie, E.; Owen, N. Initiating and maintaining recreational walking: A longitudinal study on the influence of neighborhood green space. *Prev. Med.* **2013**, *57*, 178–182. [CrossRef] [PubMed]

27. Maas, J.; Verheij, R.A.; Spreeuwenberg, P.; Groenewegen, P.P. Physical activity as a possible mechanism behind the relationship between green space and health: A multilevel analysis. *BMC Public Health* **2008**, *8*, 1–13. [CrossRef] [PubMed]

28. Lachowycz, K.; Jones, A.P. Greenspace and obesity: A systematic review of the evidence. *Obes. Rev.* **2011**, e183–e189. [CrossRef]

29. Barton, H.; Grant, M. A health map for the local human habitat. *J. R. Soc. Promot. Public Heal.* **2006**, *126*, 252–261. [CrossRef]

30. Bull, F.; Giles-corti, B.; Wood, L. Active landscapes: The methodological challenges in developing the evidence on urban environments and physical activity. In *Innovative Approaches to Researching Landscape and Health: Open Space: People Space 2*; Thompson, C.W., Aspinall, P., Bell, S., Eds.; Routledge: Abingdon, UK, 2010; pp. 97–119.

31. Sallis, J.F. Measuring Physical Activity Environments. A Brief History. *Am. J. Prev. Med.* **2009**, *36*, S86–S92. [CrossRef]

32. Ittelson, W.H. Environmental perception and urban experience. *Environ. Behav.* **1978**, *10*, 193–213. [CrossRef]

33. Thompson, C.W. Activity, exercise and the planning and design of outdoor spaces. *J. Environ. Psychol.* **2013**, *34*, 79–96. [CrossRef]

34. Bronfenbrenner, U. *Making Human Beings Human; Bioecological Perspectives on Human Development*; Bronfenbrenner, U., Ed.; Sage Publications, Inc.: Thousand Oaks, CA, USA, 2005.

35. Forman, R.T.T. *Land Mosaics: The Ecology of Landscapes and Regions*; Cambridge University Press: New York, NY, USA, 1995; ISBN 0521474620.

36. Dramstad, W.E.; Olson, J.D.; Forman, R.T.T. *Landscape Ecology Principles in Landscape Architecture and Land-Use Planning*; Island Press: Washington, DC, USA, 1996.

37. Forman, R.T.T.; Godron, M. *Landscape Ecology*; Wiley: New York, NY, USA, 1986; ISBN 0471870374.

38. McGarigal, K.; Marks, B.J. *FRAGSTATS: Spatial Pattern Analysis Program. for Quantifying Landscape Structure*; Department of Agriculture, Forest Service, Pacific Northwest Research Station: Portland, OR, USA, 1995.

39. Kang, Y.W.; Ko, Y.S.; Kim, Y.J.; Sung, K.M.; Kim, H.J.; Choi, H.Y.; Sung, C.; Jeong, E. Korea Community Health Survey Data Profiles. *Osong Public Heal. Res. Perspect.* **2015**, *6*, 211–217. [CrossRef]

40. NEINS National Environmental Information Network System. Available online: http://www.neins.go.kr/GIS/MNU01/doc01a.asp (accessed on 11 November 2018).

41. KOSTAT Population Census. Available online: http://kostat.go.kr/portal/eng/pressReleases/8/7/index.board?bmode=read&bSeq=&aSeq=370993&pageNo=1&rowNum=10&navCount=10&currPg=&sTarget=title&sTxt (accessed on 27 September 2018).

42. Olson, R.D.; Piercy, K.L.; Troiano, R.P.; Ballard, R.M.; Fulton, J.E.; Galuska, D.A.; Pfohl, S.Y. *Physical Activity Guidelines for Americans*, 2nd ed.; U.S. Department of Health and Human Services: Washington, DC, USA, 2018.

43. Akpinar, A. Landscape Structure and Health Indicators: How Are They Related? In Proceedings of the International Congress on Landscape Ecology, Antalya, Turkey, 23–25 October 2014.

44. Chen, J.; Franklin, J.F.; Spies, T.A. Vegetation Responses to Edge Environments in Old-Grwoth Douglas-Fir Forests. *Ecol. Appl.* **1992**, *2*, 387–396. [CrossRef]

45. Harper, K.A.; Burton, P.J.; Chen, J.; Saunders, S.C.; Roberts, D.; Jaiteh, M.S.; Essen, P.A. Edge Influence on Forest Structure and Composition in Fragmented Landscapes. *Conserv. Biol.* **2005**, *19*, 768–782. [CrossRef]

46. Nielsen, T.S.; Hansen, K.B. Do green areas affect health? Results from a Danish survey on the use of green areas and health indicators. *Health Place* **2007**, *13*, 839–850. [CrossRef] [PubMed]

47. Stigsdotter, U.K.; Grahn, P. Stressed individuals' preferences for activities and environmental characteristics in green spaces. *Urban. For. Urban. Green.* **2011**, *10*, 295–304. [CrossRef]

48. Kaplan, R. The Nature of the View from Home: Psychological Benefits. *Environ. Behav.* **2001**, *33*, 507–542. [CrossRef]
49. Brenner, B.; Cheng, D.; Clark, S.; Camargo, C.A. Positive association between altitude and suicide in 2584 U.S. counties. *High Alt. Med. Biol.* **2011**, *12*, 31–35. [CrossRef]
50. Ha, H.; Tu, W. An ecological study on the spatially varying relationship between county-level suicide rates and altitude in the United States. *Int. J. Environ. Res. Public Health* **2018**, *15*. 671. [CrossRef]

Publisher's Note: MDPI stays neutral with regard to jurisdictional claims in published maps and institutional affiliations.

Article

Recreation and Therapy in Urban Forests—The Potential Use of Sensory Garden Solutions

Sandra Wajchman-Świtalska [1,*], Alina Zajadacz [2] and Anna Lubarska [2]

[1] Department of Forestry Management, Faculty of Forestry and Wood Technology, University of Life Sciences in Poznań, Wojska Polskiego St. 71C, 60-625 Poznan, Poland

[2] Faculty of Geographical and Geological Sciences, Adam Mickiewicz University in Poznan, Boguslawa Krygowskiego St. 10, 61-680 Poznan, Poland; alina@amu.edu.pl (A.Z.); anna.lubarska@amu.edu.pl (A.L.)

* Correspondence: sandra.switalska@up.poznan.pl

Abstract: Urban forests are not only woodlands or groups of trees, but also individual trees, street trees, trees in parks, trees in derelict corners, and gardens. All of which are located in urban and peri-urban areas and diversify the landscape and provide a wide range of social benefits. Sensory gardens play a specific therapeutic and preventive role. Designing such gardens as a recreational infrastructure element can successfully enrich urban forests. Following the principles of universal design may provide enjoyment for all city-dwellers, with special attention given to the needs of individuals with disabilities. We studied 15 gardens and one sensory path located in various regions in Poland. The inventory was carried out on the basis of the features considered important in spatial orientation by blind and partially sighted people. The results showed that the solutions used were only partly adequate for the needs of selected users. We found neither tactile walking surface indicators (e.g., communication lines and terrain), spatial models, nor applications in mobile devices. However, these could be useful for all visitors. We confirmed that although problems with the use of forest tourist space are dependent on the type of disability, by implementing the idea of universal design for all elements of recreational infrastructure, forests may be accessible for all users.

Keywords: urban forests; forest therapy; urban environment; sensory gardens; wellbeing; social inclusion; recreational development; universal design; urban green areas; therapeutic space

Citation: Wajchman-Świtalska, S.; Zajadacz, A.; Lubarska, A. Recreation and Therapy in Urban Forests—The Potential Use of Sensory Garden Solutions. *Forests* **2021**, *12*, 1402. https://doi.org/10.3390/f12101402

Academic Editor: Idalia Kasprzyk

Received: 14 August 2021
Accepted: 12 October 2021
Published: 14 October 2021

Publisher's Note: MDPI stays neutral with regard to jurisdictional claims in published maps and institutional affiliations.

1. Introduction

Forests, due to their relatively large area and free accessibility, are most often used in recreational activities [1]. Research has shown that physical activity in the natural environment is preferable to physical activity in a closed space in terms of the feeling of relaxation, well-being, and the reduction of stress and aggression [2]. However, undertaking tourist and physical activities by people with disability (PwD) is much more complicated than in the case of non-disabled people. For disabled people, participating in tourism and recreation activities presents real barriers—environmental and interactive— which make it difficult [3]. These barriers result directly from the type of disability or are indirectly related to it (e.g., overprotection of parents or guardians and inadequate education), and they are primarily internal barriers. The real barriers include a lack of knowledge, health problems, social inefficiency, and physical and mental dependence. Among the environmental barriers are attitude, architecture, ecology, transport, and laws and regulations. Interactive barriers include non-adaptation of the ability to challenge and communication barriers.

Urban forests are defined as networks or systems comprising all woodlands or groups of trees. They include also individual trees located in urban and peri-urban areas, street trees, trees in parks, trees in derelict corners, and gardens [4]. The main objective of urban forests is to meet the recreational needs of people and to contribute to the ecological and physical structure of the city [5].

In general, gardens are a stimulating sensory environment for recreation, education, and therapy outdoors [6]. They are especially conducive to sensory gardens, defined as a self-contained area that is focused on a variety of sensory experiences [7,8]. Such an area, if designed, maintained, and managed well, offers a positive resource that caters for a variety of needs, from education to recreation. It promotes health and well-being, giving the individual the sense of mental and physical well-being [9–11]. With the sensory element (hard and soft landscaping, colors, and textures) as the key factor for designing these gardens, its role is to encourage users to touch, smell, and actively experience the garden with all their senses [12].

Experience during the COVID-19 pandemic has shown, in the context of widely understood well-being, the importance of health and the therapeutic use of direct social contacts, relaxation in the natural environment and in open space, especially among city dwellers [13]. Long-term social isolation, enforced work, and online learning intensify the need for regeneration in a different, multisensory natural environment. Both the experiences from the lockdown period and the observed change in recreational activities in the direction of tourism and individual recreation, based on nature, indicate an increase in the importance of green areas, including forests, as areas with a tourism and recreational function. The frequency of visits to natural areas during the COVID-19 social distancing restrictions was found to have increased in different parts of the world [14–16]. These trends, in combination with social needs, but also with the principles of sustainable development of the natural environment, imply the necessity to properly arrange selected forest enclaves as universally designed recreation places.

In connection with the above, the aim of our study is to show that designing sensory gardens as one of the many elements of forest recreational development is a form of diversifying the infrastructure that can enrich urban forests, including the aspect of the enjoyment of all citizens (those with disability and those without any). The key question is as follows: which features of universal design are of elementary importance from the point of view of PwD (e.g., in terms of spatial orientation, information, and sense of security) and should constitute a basis for designing recreational zones in forest so that they are more accessible to all.

2. Materials and Methods

2.1. Study Area

The study covered 15 gardens and one sensory path. These objects are located in various regions in Poland, in cities, rural areas, and areas of natural value, including national parks (Figure 1).

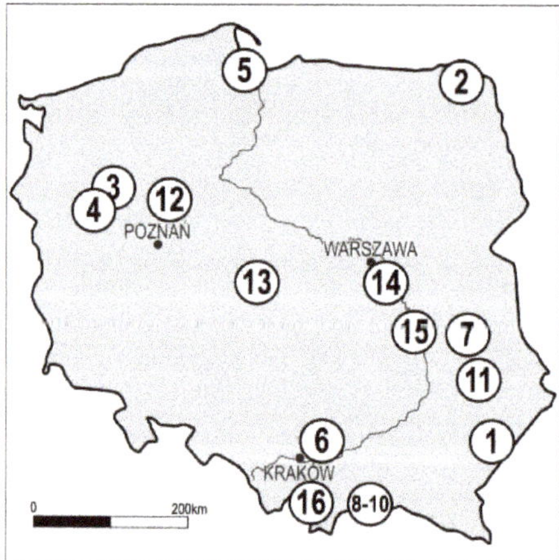

Figure 1. The location of the study area. 1. Bolestraszyce, 2. Bród Nowy, 3. Bucharzewo, 4. Chorzępowo, 5. Gdańsk-Oliwa, 6. Kraków, 7. Lublin, 8. Muszyna (Ogrody Biblijne), 9. Muszyna (Ogrody Magiczne), 10. Muszyna (Ogrody Sensoryczne), 11. Osmolice, 12. Owińska, 13. Poddębice, 14. Powsin, 15. Trzcianki, 16. Zawoja (Babiogorski National Park).

Table 1. Details of the location and type of sensory gardens included in the study.

Location of the Sensory Garden	Type of Environment	Setting	Coordinates
1. Bolestraszyce	Rural	Arboretum	49.81767470425935 22.85963205188485
2. Bród Nowy	Rural	Educational and recreational school garden	54.13310633220314 22.87824015575021
3. Bucharzewo	Rural	Forest educational garden	52.67929426341419 16.09764799617132
4. Chorzępowo	Rural	Forest educational garden	52.69849904750158 16.09713467457603
5. Gdańsk-Oliwa	Urban	City park	54.414561388875526 18.56869532665728
6. Kraków	Urban	Commercial science park	50.07044044210445 19.997914018768107
7. Lublin	Urban	University botany garden	51.26596084591114 22.51659339812219
8. Muszyna (Ogrody Biblijne)	Urban	Commercial educational and spiritual garden	49.35895057351808 20.901974361122278
9. Muszyna (Ogrody Magiczne)	Urban	City garden	49.345923233427236 20.884421943659394
10. Muszyna (Ogrody Sensoryczne)	Urban	City park	49.347567900990136 20.88853191824345
11. Osmolice	Rural	Private garden	51.579692111665814 22.06933798840454
12. Owińska	Rural	Educational and recreational school garden	52.511276130254345 16.97413330331303
13. Poddębice	Urban	City park	51.88970856853506 18.952417769156217
14. Powsin	Urban	Botany garden of a scientific institution	52.10628212831032 21.095847051626784
15. Trzcianki	Rural	Commercial theme park	51.36551912934566 21.910141838449825
16. Zawoja	Rural	National park education area	49.6120128648685 19.518040540235628

2.2. Study Design

The research was carried out between June and August 2018. In the garden inventory, the following assessment criteria presented in the study of Jakubowski et al. were used [17]: scents, clear path layout, diversified surface of path surface, advice from others, waypoints described in Braille, convex plans of communication routes, audible information, tactile walking surface indicator, spatial models, and applications on mobile devices. The research was conducted according to a standardized form elaborated by Jakubowski et al. [17].

The results of the presented research narrowed-down to the adaptation of sensory gardens to the needs of visually impaired people are presented in the studies by Zajadacz and Lubarska [18–20].

This study will discuss the arrangement of forest recreational areas in a broader context, taking into account the needs of PwD.

3. Results

3.1. Universal Design in Sensory Gardens—Potential Applications in Forests

The inventory was carried out on the basis of the features considered important in spatial orientation by blind and partially-sighted people. The results of the field research showed the most commonly used solutions that facilitate relaxation, as well as independent spatial orientation, for blind and partially-sighted people, which include places identified by their smell, clear layout of paths, different path surfaces, and tips from other people (Figures 2–5). Unfortunately, other amenities considered essential were less frequently represented, and such as tactile walking surface indicators, spatial models (e.g., communication lines and terrain), and applications on mobile devices did not occur in the gardens studied (Table 2). Thus, it can be concluded that in sensory gardens, despite many amenities in terms of the uniqueness of their arrangement, many possibilities for their better adaptation to the needs of people with disabilities have not been used. The features presented in Table 2, based on the recommendations of the PWD community [17], are therefore potential applications in sensory gardens and forests.

(a)   (b)

Figure 2. Scents and sounds, (**a**) Poddębice (Fot. A. Lubarska), (**b**) Owińska (Fot. A. Zajadacz).

(a)   (b)

Figure 3. Clear path layout, (**a**) Owińska, and (**b**) facilities tailored to the needs of the blind, Chorzępowo (fot. A. Zajadacz).

Figure 4. *Cont.*

Figure 4. Diversified surface of path surface, Owińska and Kraków (fot. A. Zajadacz).

Figure 5. Waypoints described in Braille, Owińska (Fot. A. Zajadacz).

Table 2. Elements in the sensory garden to facilitate spatial orientation.

Feature	Sensory Gardens and Paths																$\sum$/%
	a															b	
	1	2	3	4	5	6	7	8	9	10	11	12	13	14	15	16	
1. Scents (Fot.1)	x	x	x	x	x	x	x	x	x	x	x	x	x	x	x	x	16/100
2. Clear path layout (Fot. 2)	x	x	x	x	x	x	x	x	x	-	-	-	x	x	x	x	13/81
3. Diversified surface of path surface (Fot. 3)	x	x	x	x	-	x	-	x	-	x	x	-	x	x	x	-	11/69
4. Advice from others	x	x	x	x	x	x	x	x	-	-	-	x	-	-	-	x	10/63
5. Waypoints described in Braille (Fot.4)	x	x	-	-	-	-	-	-	-	-	-	x	-	-	x	-	x	5/31
6. Convex plans of communication routes	x	-	x	x	-	-	-	-	-	-	-	-	-	-	x	-	-	4/25
7. Audible information	x	x	-	-	-	-	-	-	-	-	-	-	-	-	-	-	x	3/19
8. Tactile walking surface indicators	-	-	-	-	-	-	-	-	-	-	-	-	-	-	-	-	0/0
9. Spatial models	-	-	-	-	-	-	-	-	-	-	-	-	-	-	-	-	0/0
10. Applications in mobile devices	-	-	-	-	-	-	-	-	-	-	-	-	-	-	-	-	0/0

(a): Sensory gardens: 1. Bucharzewo, 2. Owińska, 3. Zawoja, 4. Bolestraszyce, 5. Osmolice, 6. Trzcianki, 7. Bród Nowy; 8. Kraków, 9. Gdańsk, 10. Lublin, 11. Muszyna Ogród Zmysłów, 12. Muszyna Ogród Biblijny, 13. Muszyna Ogród Magiczny, 14. Poddębice, 15. Powsin (Warszawa). (b): Sensory path in a village: 16. Chorzępowo; x, the element is present; (-), the element is not present. Source: Field inventory results, July–August 2018 [20].

3.2. Recreational Facilities in Forests—Desk Research of Applied Practices

In Poland, there are guidelines commissioned by the General Directorate of State Forests presenting practical tips on how to shape the space of recreational and leisure facilities and their individual elements in order to make them accessible to people with reduced motor skills [17]. Table 3 presents technical aspects of selected elements of the recreational forest management infrastructure. The selection of the presented elements and their parameters was based on the most frequently indicated obstacles [21] in the use of forests by PwD.

Table 3. Technical aspects of selected elements of the recreational forest management infrastructure (based on [22]).

Infrastructure	Basic Parameters	Additional Info
One-way paths	• Minimum width of 90 cm • Recommended to be not less than 150 cm (such a width allows for safe passing of wheelchairs)	• It is allowed to reduce the width of the track to 80 cm • When it is unavoidable, e.g., due to existing rock formations—closure on very short sections, no longer than 60 cm
Bidirectional paths	• Minimum width of 150 cm	-
Longitudinal slope of routes	• Should not exceed 5%	• From 5 to 8%—if the section is no longer than 50 m (in the case of existing paths); additionally, at the beginning and end of such a section there should be a flat surface free from other terrain obstacles, which is a resting place • From 8 to 10%—if the section is no longer than 9 m; additionally, at the beginning and end of such a section there should be a flat surface without any other terrain obstacles, which is a resting place • From 10 to 12.5%—if the distance is no longer than 3 m; additionally, at the beginning and end of such a section there should be a flat surface without any other terrain obstacles, which is a resting place

Table 3. *Cont.*

Infrastructure	Basic Parameters	Additional Info
Cross slope of routes	• Should not exceed 3%	• On most paved surfaces water drainage is effective at a slope of 2% • Over 3% is acceptable only in situations where it is necessary due to the drainage of water from the surface of the path • It may never exceed 5%
Avoidance spots on routes (without infrastructure)	• Minimum width of 180 cm (including the track width) • Minimum length of 240 cm	• Be created on routes less than 150 cm wide every 300 m • Have a cross slope of the surface not exceeding 3%
Stairs	• Excluded or limited to the absolute minimum • Create an alternative path in their vicinity	• There should be no more than 10 steps in one flight of stairs • The minimum width of the flight of stairs should be at least 150 cm • The tread depth must be at least 150 cm, the greater it is, the more comfortable it will be for disabled people to move • Stairs with a tread depth of less than 90 cm are an insurmountable architectural barrier • The maximum height of one riser should not exceed 5 cm • If it is not possible to obtain such a height, it is permissible to increase the height of the steps to a maximum of 10 cm, while limiting their number in one run to 5, providing a handrail and an alternative route
Handrails	• The basic handrail should be 90 cm high • The second additional handrail should be 75 cm above the surface of the stairs	• Be installed near dangerous places, e.g., at viewpoints and on rock outcrops • Have fender rails • The diameter of the grab part of the handrail should be 3.5−5 cm

4. Discussion

4.1. Sensory Gardens—A Creative Solution for People Not Only with Disabilities

Gardens are an integral part of urban forests [4]. The main point in trying to define the concept of "sensory garden" is to find out what makes it different from other gardens. The main difference in the sensory garden is that all its elements, "hard" and "soft" (plants, shapes, colors, and textures) must be carefully selected and designed to provide maximum sensory stimulation. Therefore, the sensory garden is defined as follows:

- "A stand-alone area that focuses on a variety of sensory experiences " (sensory trust) [7];
- "A composition designed so that extra-visual stimuli are used on purpose and at a greater intensity than usual" [23];
- A garden in which the influence of plants and other elements on certain senses is particularly emphasized (the garden of color, sound, smell, touch, and taste) [24];
- A garden, which refers to the idea that the garden can stimulate the senses (e.g., sight, taste, hearing, smell and touch) [25].

The definitions of a sensory garden assume that a garden of this type [18]:

- Must be designed and created in a process with a set purpose;
- Is a closed whole, separate from the surrounding space;
- Stimulates all human senses;
- Focuses on non-visual experiences;
- Has vegetation, but also other elements, both natural and anthropogenic.

Nowadays, creating a "specially dedicated" or "closed" space for people with disabilities is not a desirable strategy. For many years, there has been a visible tendency to integrate the amenities necessary for PwD into the space in such a way that they are an integral, natural part of it. Our study confirmed that sensory gardens, created as "gardens for the blind", are currently designed for everyone interested in resting in a given location [26]. Designed in the rural or urban environment as a part of, e.g., arboretum, educational and recreational school gardens, city parks, city gardens, private gardens, or university botany gardens, are widely accessible recreational spaces.

The intensity of natural stimuli in sensory gardens is also conducive to therapy. Garden therapy—or hortitherapy—is a form of therapy "that uses plants to improve the physical and mental condition of a person" [27]. Hortitherapy can be performed in the following ways [27]:

- Passive—by staying in a given place and experiencing the stimulation of the senses, by listening to birds singing, and feeling smells, wind, sunlight;
- Active—by performing physical work in the garden related to the maintenance of the garden: picking fruit and flowers, etc. The outdoor activities that most people like, such as sunbathing, games and fun, and walks, are interpreted differently. According to Hagedorn [28], this is passive use of the garden, while Gonzalez and Kirkevold [25] consider it an active form of using the space.

Sensory gardens play a therapeutic and preventive role. These include, for example, parks equipped with memory gyms—i.e., memory training devices, intended mainly for elderly people with signs of dementia, but if skillfully used, they can also be an attraction for other user groups. They can be equipped with "moto-sensory paths, i.e., a place intended to serve the elderly and people who experience limitations in mobility as a result of illness or injury. The proposal of the various available solutions is to help maintain or increase the mobility of the beneficiaries [29]. Moto-sensory pathways are used for therapy [29], as follows:

- "As a place of passive therapy—walking, watching nature, and being in green space, as well as aromatherapy, especially to stimulate immunity and control stress.
- As a place of application of other types of therapy loosely related to the garden— physical therapy, kinesiotherapy, and psychotherapy.
- As a place of active hortitherapy, using gardening for healing purposes, for treating depression, addictions, etc., as well as for activating elderly people with cognitive disorders that often occur at this age. It also works well for people with intellectual disabilities and emotional disorders."

At the same time, activities like walking, watching nature, being in green space, aromatherapy, etc. do not exclusively represent the needs of PwD. They are the needs of all society. Considering the elements mentioned above within the study area, our results show that the solutions used were only partly adequate for the needs of selected users. The most frequently present features were scents, clear path layouts, and a diversified surface of the paths. We found neither tactile walking surface indicators, spatial models, nor applications on mobile devices that could be useful for increasing users' mobility.

4.2. Barriers in Tourism and Recreation of People with Disabilities

Many barriers to tourism and recreation for people with disabilities have been identified in the literature. In one of the classifications, they were divided into the following groups: urban, architectural, and social barriers; barriers of communication; lack of tourist equipment; high costs of participation in various forms of tourism; and insufficient information on the tourist needs of people with disabilities [30]. The low level of affluence, low income in relation to the costs of tourism trips, and the commercial attitude of travel agencies mean that the main barriers reported by the disabled are economic ones, relegating the remaining architectural and urban barriers to the background [31,32].

A broad overview of the classifications of barriers and constraints to undertaking tourism and recreation developed so far has been presented, among others, by Lubarska [33]. Of particular note is the new approach to barriers (according to the title of the study: "Re-conceptualizing") of Darcy and McKercher. The four-tier model they propose defines a hierarchy of barriers, with lower tiers referring to basic barriers. If there is a barrier at this level, the tourist trip may no longer take place. The first level concerns barriers that everyone can encounter, with or without a disability. Tier 2 barriers are in some ways common to people with all types of disabilities. At this level, there are barriers that are less relevant for people without disabilities, e.g., the information barrier, which, with some disabilities, cannot be overcome without the support of others. The two higher tiers relate to specific disabilities. At tier 3, there are barriers that are typical for certain disabilities, e.g., the limited possibility of written contact (instant messaging or e-mail) with the tourist facility and the focus on telephone contact, which is a major obstacle for deaf people. Here, the individual adaptation of the venue to the needs of people with specific disabilities becomes important. Adaptation to the needs of a specific disability is not always sufficient at tier 4, where individual limitations come into play. However, even tier 4 barriers can sometimes be offset by solutions targeted at the individual tourist [34].

Our study revealed that most of architectural barriers are easy to eliminate by implementing guidelines, as commissioned by the Polish General Directorate of State Forests, relating to specific technical aspects of the infrastructure. While social barriers or high costs of participation in various forms of tourism and recreation in general may concern everyone, undoubtedly insufficient information on tourist needs for people with disabilities (in general or within selected areas) is a barrier to the effective development of recreational and leisure facilities.

4.3. Therapeutic Functions of Forests, Green Areas and Sensory Gardens

The health benefits of contact with nature have been widely researched [35]. Studies show that the less green there is in the neighbourhood, the higher level of cortisol, the "stress hormone", in the blood of residents [36]. The most frequently mentioned feelings resulting from the experience of nature are freedom, unity with nature and with one's self, luck, and happiness [37]. There is a wide range of research on the effects of forest-based therapies on stress (or depression) [38–43]. The therapeutic properties of various plant communities have an impact on specific different medical aspects, including disinfection, blood pressure lowering, anti-asthma, or immune-boosting [44]. Factors in the forest environment that may provide health benefits include the aroma of plants, light intensity, humidity, wind, temperature, and oxygen concentrations [45]. Increasing outdoor recreation can be considered beneficial both on an individual level and to society as a whole [46]. For people living in large and dense cities, urban green space plays an important social integrative role [47,48]. The health benefits of contact with nature are described, among others, by the "Nature Therapy Theory" [49,50]. One example of this form of therapy is forest bathing, also known as shinrin-yoku, a practice that combines a series of outdoor exercises and tasks based on mindfully using all five senses [51,52]. There are also "therapeutic landscapes", which are places that, for various reasons, can have a beneficial effect on health and well-being [53]. Moreover, a positive impact on behaviour and interpersonal self-improvement has "wilderness therapy" [54]. This therapy combines experiential education, as well as individual and group therapy with adventure-based therapy in a wildlife environment. Research has shown that physical activity in the natural environment is preferable to physical activity in a closed space in terms of the feeling of relaxation, well-being, and the reduction of stress and aggression [2].

An extensive systematic review of the benefits of urban parks shows that all of the above benefits have been confirmed by numerous scientific studies, including the fact that the creation of urban green spaces promotes physical activity and social integration among residents and visitors, and thus contributes not only to physical, but also mental health [55]. The benefits for the latter are particularly evident in the long term, and a positive association was observed regardless of whether the green-space was in an urban or rural area [56]. Our study showed that the number of sensory gardens located in Polish towns/cities and villages is comparable (7 and 8).

Accordingly, sensory gardens in cities or rural areas can also play the role of restorative and therapeutic gardens. They are places that offer conditions favorable to maintaining the body's internal balance, which have an effect on emotional and psychological wellbeing [57].

4.4. Recreational Development of Forests and Green Areas

Generally, the technical management of forests and other green areas for recreation consists of designing the recreational facilities, arranging them in a way that ensures natural comfort of rest, and minimizing the conflicts caused by this procedure in the naturalness of the environment. Furthermore, there is the need to individualize the design procedure and to use natural local materials. The infrastructure necessary for recreation consist of technical construction objects, linear objects, and surface objects. Finally, as urban forests belong to the city where they are located, and have a characteristic of public urban green space, these areas should also reflect the socio-cultural characteristics of the city [58].

Practicing tourism and recreation by PwD is impeded by a number of barriers [3,30–32]. Problems with the use of forest tourist spaces are dependent on the type of disability. Wheelchair users report problems such as stairs (steep stairs, no landings, and no handrails); obstacles and obstructions on the path; slippery, gravel, and sandy surfaces; narrow trails and passages; missing or incorrect information; and crowds. For visually impaired people, obstacles and difficulties on the path are slippery surfaces, insufficient lighting, and crowds. In contrast, elderly people recognize the lack of stairs and landings, slippery surfaces, and crowds as a problem [22].

Note, however, that the definition of disability is very wide. According to Polish legislation, disabled people are "those whose physical, mental, or mental fitness permanently or temporarily hinders, restricts, or prevents everyday life, study, work, and performing social roles, in accordance with legal and customary norms" [59]. This means that people with limited mobility may not only be wheelchair users, but also people with difficulties in moving independently without assistive devices (e.g., crutches, canes, and walkers), people with sight and/or hearing and/or voice impairment; people suffering from multiple disorders of a physical and mental nature (e.g., asthma or heart problems, and personality disorders), elderly people (over 60 years of age); and pregnant women and minors (under the age of 5).

Regarding sensory gardens, despite being directed towards specific users and having a certain specialization, overall, they are areas of activity for users of all age groups—children, youth, adults, and seniors [60]. Therefore, the preparation of recreational infrastructure in forests and other green areas, including sensory gardens, in order to enable a comfortable rest should take into account the possibilities and limitations of people with different psychophysical abilities and should implement the principle of "design for all". While spending time outdoors is in itself beneficial to health (including public health), the fact that the space is designed by professionals knowledgeable about accessibility and friendliness of the outdoors greatly enhances the value of that space [61]. There is a positive link between the number and quality of leisure facilities and the well-being of residents in the area. The better the infrastructure and the more walking/hiking paths and active recreation routes, the healthier and more satisfied visitors are, including health "at-risk" groups [62]. However, again, the benefits are not limited to this group, but apply to all visitors.

In addition to the importance of urban parks being accessible to all, it should be attractive to visitors. Some of the characteristics of a green space that influence the willingness to visit are location, including ease of access, condition of the access road, etc.; facilities, the ability to use the space for recreational purposes; and atmosphere, which is created by factors that are difficult to grasp [63]. Sensory gardens are special spaces in this respect, as they are prepared and designed precisely with a view to attractiveness, and in terms of accessibility, they go beyond the usual level. As for the atmosphere, it is helped by the aforementioned multisensoriality, which characterizes this space. Hence, it follows that sensory gardens are beneficial for urban forests as they can positively influence the motivation to visit green areas, make the stay more attractive, and allow groups requiring increased accessibility, as mentioned before (the elderly, the weakened or injured, or families with children), for both active and passive therapy.

5. Conclusions

The location of forests within or near the administrative boundaries of cities promotes intensive recreational use of the forest environment. Thus, these areas can be an excellent places for city-dwellers to provide health support. The provision of urban green spaces and their associated benefits are considered "a key ingredient for city sustainability" [37]. The existence of urban green areas benefits citizens and increases welfare levels [64]. Unfortunately, the potential of natural settings in contributing to the quality of working and housing environments, which could enhance the health and well-being of residents, is not fully considered in the current trend of building compact cities [65]. What is more, urban forests, besides contributing to human well-being positively, affect the private value of properties [66].

The therapeutic function of forests is based on the influence of plants (hortitherapy) and contributes to the improvement of the physical (somatic) and mental condition of a person. This therapy can be both passive and active. Forest zones, which perform recreational functions, in addition to influencing the individual well-being of a given person, thanks to universal design aimed at eliminating barriers to the use of public space, can foster the process of social inclusion of people who, for many reasons, encounter difficulties in using the open space of public places. This applies especially to PwD, but also to the elderly and families with children—social groups for which rehabilitation, therapeutic, and preventive activities are of particular importance.

Technical aspects of selected elements of the recreational forest management infrastructure, presented in the article, in conjunction with the results of research on the adaptation of sensory gardens to the needs of people with visual impairments, provide many tips for the universal design of forest recreational zones. Their adaptation requires taking into account the local specificity of the natural environment, which varies depending on the type of geographical environment. An open question, which has so far been an unsolved research problem, is still the definition of proportions and finding a balance in the arrangement of forest recreation zones between the "artificiality" of their development (adaptation to the needs of many user groups) and the "naturalness" of environmental conditions with a wide therapeutic impact.

Author Contributions: Conceptualization, S.W.-Ś., A.Z. and A.L.; methodology, S.W.-Ś., A.Z. and A.L.; software, S.W.-Ś., A.Z. and A.L.; validation, S.W.-Ś., A.Z., and A.L.; formal analysis, S.W.-Ś., A.Z. and A.L.; investigation, S.W.-Ś., A.Z. and A.L.; resources, S.W.-Ś., A.Z. and A.L.; writing—original draft preparation, S.W.-Ś., A.Z. and A.L.; writing—review and editing, S.W-Ś., A.Z. and A.L.; visualization, S.W.-Ś., A.Z. and A.L.; supervision, S.W.-Ś., A.Z. and A.L.; project administration, S.W.-Ś., A.Z. and A.L; funding acquisition, S.W.-Ś. and A.Z. All authors have read and agreed to the published version of the manuscript.

Funding: The research was co-financed within the framework of the Ministry of Science and Higher Education program as "Regional Initiative Excellence" in years 2009–2022, Project No. 005/RID/2018/2019 and Project "GEO +: high-quality doctoral study program implemented at the Faculty of Geographical and Geological Sciences of the Adam Mickiewicz University in Poznan no. POWR.03.02.00-00-I039/16" co-financed by the European Social Fund under the Knowledge Education Development Operational Program (PO WER).

Institutional Review Board Statement: Not applicable.

Informed Consent Statement: Not applicable.

Data Availability Statement: Data sharing not applicable.

Conflicts of Interest: The authors declare no conflict of interest. The funders had no role in the design of the study; in the collection, analyses, or interpretation of data; in the writing of the manuscript; or in the decision to publish the results.

References

1. Mazurek-Kusiak, A. Charakterystyka popytu na rekreację konną w polskich lasach. *Sylwan* **2018**, *162*, 785–792. [CrossRef]
2. Thompson Coon, J.; Boddy, K.; Stein, K.; Whear, R.; Barton, J.; Depledge, M.H. Does Participating in Physical Activity in Outdoor Natural Environments Have a Greater Effect on Physical and Mental Wellbeing than Physical Activity Indoors? A Systematic Review. *Environ. Sci. Technol.* **2011**, *45*, 1761–1772. [CrossRef] [PubMed]
3. Smith, R.W. Leisure of disabled tourist. Barriers to participation. *Ann. Tour. Res.* **1987**, *14*, 376–389. [CrossRef]
4. Urban and Peri-Urban Forestry. Available online: http://www.fao.org/forestry/urbanforestry/87025/en/ (accessed on 26 June 2021).
5. Van Elegem, B.; Embo, T.; Muys, B.; Lust, N. A methodology to select the best locations for new urban forests using multicriteria analysis. *For. Int. J. For. Res.* **2002**, *75*, 13–23. [CrossRef]
6. Spring, J.A. Design of evidence-based gardens and garden therapy for neurodisability in Scandinavia: Data from 14 sites. *Neurodegener. Dis. Manag.* **2016**, *6*, 87–98. [CrossRef] [PubMed]
7. Sensory Trust. Sensory Garden Design Advice. 2003. Available online: http://www.sensorytrust.org.uk/information/factsheets/sensory-garden-4.html (accessed on 25 May 2018).
8. Wajchman-Świtalska, S.; Zajadacz, A.; Lubarska, A. Therapeutic functions of forests and green areas with regard to the universal potential of sensory gardens. *Environ. Sci. Proc.* **2021**, *3*, 8. [CrossRef]
9. Hussein, H.; Omar, Z.; Ishak, S.A. Sensory garden for an inclusive society. *Asian J. Behav. Stud.* **2016**, *1*, 33–43. [CrossRef]
10. Souter-Brown, G.; Hinckson, E.; Duncan, S. Effects of a sensory garden on workplace wellbeing: A randomised control trial. *Landsc. Urban Plan.* **2021**, *207*, 103997. [CrossRef]
11. Kucks, A.; Hughes, H. Creating a Sensory Garden for Early Years Learners: Participatory Designing for Student Wellbeing. In *School Spaces for Student Wellbeing and Learning*; Hughes, H., Franz, J., Willis, J., Eds.; Springer: Singapore, 2019. [CrossRef]
12. Hussein, H.; Abidin, N.M.N.Z.; Omar, Z. Engaging Research and Practice in Creating for Outdoor Multi-sensory Environments: Facing Future Challenges. *Procedia—Soc. Behav. Sci.* **2013**, *105*, 536–546. [CrossRef]
13. Zajadacz, A. Changes in leisure time in the large cities in Poland caused by the COVID-19 pandemic: The types of activities and the amount of leisure time. In *Seria Turystyka i Rekreacja—Studia i Prace*; Zajadacz, A., Ed.; Bogucki Wydawnictwo Naukowe: Poznań, Poland, 2021; Volume 23, p. 136.
14. Grima, N.; Corcoran, W.; Hill-James, C.; Langton, B.; Sommer, H.; Fisher, B. The importance of urban natural areas and urban ecosystem services during the COVID-19 pandemic. *PLoS ONE* **2020**, *15*, e0243344. [CrossRef]
15. Venter, Z.; Barton, D.; Gundersen, V.; Figari, H.; Nowell, M. Urban Nature in a Time of Crisis: Recreational Use of Green Space Increases during the COVID-19 Outbreak in Oslo, Norway. Preprint. Available online: https://osf.io/preprints/socarxiv/kbdum/ (accessed on 30 September 2021).
16. Mackenzie, S.H.; Goodnow, J. Adventure in the Age of COVID-19: Embracing Microadventures and Locavism in a Post-Pandemic World. *Leis. Sci.* **2020**, *43*, 62–69. [CrossRef]
17. Jakubowski, M.; Szczepańska, M.; Ogonowska-Chrobrowska, H. *Ogrody i Ścieżki Zmysłów w Procesie Rekreacji i Edukacji Przyrodniczo Leśnej Osób Niewidzących i Niedowidzących*; Specjalny Ośrodek Szkolno-Wychowawczy dla Dzieci Niewidomych w Owińskach: Owińskach, Poland, 2018.
18. Zajadacz, A.; Lubarska, A. *Ogrody Sensoryczne Jako Uniwersalne Miejsca Rekreacji Dostosowane do Potrzeb Osób Niewidomych w Kontekście Relacji Człowiek-Środowisko*; Boguski Wydawnictwo Naukowe: Poznań, Poland, 2020; p. 180.
19. Zajadacz, A.; Lubarska, A. Sensory gardens in the context of promoting well-being of people with visual impairments in the outdoor sites. *Int. J. Spa Wellness* **2019**, *2*, 3–17. [CrossRef]
20. Zajadacz, A.; Lubarska, A. Sensory gardens as places for outdoor recreation adapted to the needs of people with visual impairments. *Studia Perieget.* **2020**, *30*, 25–43. [CrossRef]
21. Nowacka, W.Ł. Projektowanie leśnej przestrzeni turystycznej z punktu widzenia niepełnosprawnego użytkownika. *Studia I Mater. Cent. Edukac. Przyr.-Leśnej* **2010**, *26*, 30–39.
22. Kacprzyk, W. *Turystyka w Lasach Państwowych. Tom I. Las bez Barier—Obiekty Terenowe*; Ośrodek Rozwojowo-Wdrożeniowy Lasów Państwowych w Bedoniu: Nowy Bedoń, Poland, 2015; p. 95.
23. Pawłowska, K. Ogród sensoryczny. *Pr. Kom. Kraj. Kult.* **2008**, *9*, 143–152.
24. Latkowska, M.J.; Miernik, M. Therapeutic gardens—Places of passive and active green therapy. *Archit. Czas. Tech.* **2012**, *8A*, 245–250.
25. Gonzalez, M.T.; Kirkevold, M. Clinical use of sensory gardens and outdoor environments in Norwegian nursing homes: A cross-sectional e-mail survey. *Issues Ment. Health Nurs.* **2015**, *36*, 35–43. [CrossRef]
26. Hussein, H. Sensory gardens. *Access Des.* **2009**, *118*, 13–17.
27. Zawiślak, G. Hortiterapia jako narzędzie wpływające na poprawę zdrowia psychicznego i fizycznego człowieka. *Ann. UMSC* **2015**, *25*, 21–31.

28. Hagedorn, R. Environment and opportunity: The potential of horticulture for enriching the life of disabled people. *Clin. Rehabil.* **1988**, *2*, 249–251. [CrossRef]
29. Haupt, P.; Skalna, B.; Rekuć, M.; Mikołajska, I.; Furlaga, Z.; Kusińska, E.; Gajewski, Ł. Modelowa Koncepcja Ścieżki Moto-sensorycznej. 2019. Available online: https://www.rops.krakow.pl/pliki/MIIS/Innowacje_za__aczniki/Ko__cowa_wersja_koncepcji_projektu_architektonicznego.pdf (accessed on 1 June 2021).
30. Łobożewicz, T. Wpływ turystyki i rekreacji na przywracanie sprawności psychofizycznej osób o specjalnych potrzebach. In *Postęp w Turystyce na Rzecz Osób o Specjalnych Potrzebach*; Ślężyński, J., Petryński, W., Eds.; Polskie Stowarzyszenie Osób Niepełnosprawnych: Kraków, Poland, 1995; p. 46.
31. Zajadacz, A. *Turystyka Osób Niesłyszących—Ujęcie Geograficzne*; Bogucki Wydawnictwo Naukowe: Poznań, Poland, 2012; p. 227.
32. Tabęcki, R. Ograniczenia i perspektywy rozwoju turystyki osób niepełnosprawnych w Polsce i w wybranych krajach europejskich. In *Krajoznawstwo i Turystyka Osób Niepełnosprawnych*; Midura, F., Żbikowski, J., Eds.; Wydawnictwo PWSZ im. Papieża Jana Pawła II: Biała Podlaska, Poland, 2005; p. 125.
33. Lubarska, A. Przegląd klasyfikacji barier i ograniczeń dla turystyki osób z niepełnosprawnością. In *Seria Turystyka i Rekreacja Studia i Prace*; Młynarczyk, Z., Zajadacz, A., Eds.; Bogucki Wydawnictwo Naukowe: Poznań, Poland, 2018; pp. 57–72.
34. McKercher, B.; Darcy, S. Re-conceptualizing barriers to travel by people with disabilities. *Tour. Manag. Perspect.* **2018**, *26*, 59–66. [CrossRef]
35. Doimo, I.; Masiero, M.; Gatto, P. Forest and wellbeing: Bridging medical and forest research for effective forest-based initiatives. *Forests* **2020**, *11*, 791. [CrossRef]
36. Roe, J.J.; Thompson, C.W.; Aspinall, P.A.; Brewer, M.J.; Duff, E.I.; Miller, D.; Mitchell, R.; Clow, A. Green Space and Stress: Evidence from Cortisol Measures in Deprived Urban Communities. *Int. J. Environ. Res. Public Health* **2013**, *10*, 4086–4103. [CrossRef]
37. Chiesura, A. The role of urban parks for the sustainable city. *Landsc. Urban Plan.* **2004**, *68*, 129–138. [CrossRef]
38. Antonelli, M.; Barbieri, G.; Donelli, D. Effects of forest bathing (shinrin-yoku) on levels of cortisol as a stress biomarker: A systematic review and meta-analysis. *Int. J. Biometeorol.* **2019**, *63*, 1117–1134. [CrossRef] [PubMed]
39. Kotera, Y.; Richardson, M.; Sheffield, D. Effects of Shinrin-Yoku (Forest Bathing) and Nature Therapy on Mental Health: A Systematic Review and Meta-analysis. *Int. J. Ment. Health Addict.* **2020**. [CrossRef]
40. Farrow, M.R.; Washburn, K.A. Review of Field Experiments on the Effect of Forest Bathing on Anxiety and Heart Rate Variability. *Glob. Adv. Health Med.* **2019**. [CrossRef]
41. Lee, I.; Bang, K.S.; Kim, S.; Choi, H.; Lee, B.; Song, M. Effect of Forest Program on Atopic Dermatitis in Children—A Systematic Review. *J. Korean Inst. For. Recreat.* **2016**, *20*, 1–13. [CrossRef]
42. Lee, I.; Choi, H.; Bang, K.S.; Kim, S.; Song, M.; Lee, B. Effects of forest therapy on depressive symptoms among adults: A systematic review. *Int. J. Environ. Res. Public Health* **2017**, *14*, 321. [CrossRef]
43. Rosa, C.D.; Larson, L.R.; Collado, S.; Profice, C.C. Forest therapy can prevent and treat depression: Evidence from meta-analyses. *Urban For. Urban Green.* **2020**, *57*, 126943. [CrossRef]
44. Krzymowska-Kostrowicka, A. *Geoekologia Turystyki i Wypoczynku*; Wydaw. Naukowe PWN: Warsaw, Poland, 1997; pp. 1–239.
45. Loureiro, G.; Rabaça, M.A.; Blanco, B.; Andrade, S.; Chieira, C.; Pereira, C. Urban versus rural environment–any differences in aeroallergens sensitization in an allergic population of Cova da Beira, Portugal? *Eur. Ann. Allergy Clin. Immunol.* **2005**, *37*, 187–193. [PubMed]
46. Eriksson, L.; Nordlund, A. How is setting preference related to intention to engage in forest recreation activities? *Urban For. Urban Green.* **2013**, *12*, 481–489. [CrossRef]
47. Dwyer, J.; Schroeder, H.; Gobster, P. The significance of urban trees and forests: Toward a deeper understanding of values. *J. Arboric.* **1991**, *17*, 276–284.
48. Germann-Chiari, C.; Seeland, K. Are urban green spaces optimally distributed to act as places for social integration? Results of a geographical information system (GIS) approach for urban forestry research. *For. Policy Econ.* **2004**, *6*, 3–13. [CrossRef]
49. Miyazaki, Y.; Park, B.J.; Lee, J. Nature therapy, in designing our future. In *Local Perspectives on Bioproduction, Ecosystems and Humanity*; Osaki, M., Braimoh, A., Nakagami, K., Eds.; United Nations University Press: Tokyo, Japan, 2011; Volume 2011, pp. 407–412.
50. Song, C.; Ikei, H.; Kobayashi, M.; Miura, T.; Li, Q.; Kagawa, T.; Kumeda, S.; Imai, M.; Miyazaki, Y. Effects of viewing forest landscape on middle-aged hypertensive men. *Urban For. Urban Green.* **2017**, *21*, 247–252. [CrossRef]
51. Hansen, M.M.; Jones, R.; Tocchini, K. Shinrin-yoku (forest bathing) and nature therapy: A state of the art review. *Int. J. Environ. Res. Public Health* **2017**, *14*, 851. [CrossRef]
52. Wen, Y.; Yan, Q.; Pan, Y.; Gu, X.; Liu, Y. Medical empirical research on forest bathing (Shinrin-yoku): A systematic review. *Environ. Health Prev. Med.* **2019**, *24*, 70. [CrossRef]
53. Bell, S.L.; Foley, R.; Houghton, F.; Maddrell, A.; Williams, A.M. From therapeutic landscapes to healthy spaces, places and practices: A scoping review. *Soc. Sci. Med.* **2018**, *196*, 123–130. [CrossRef]
54. Fernee, C.R.; Gabrielsen, L.E. Wilderness therapy. In *Outdoor Therapies: An Introduction to Practices, Possibilities, and Critical Perspectives*, 1st ed.; Harper, N.J., Dobud, W.W., Eds.; Routledge: New York, NY, USA, 2020; pp. 69–80.

55. Konijnendijk, C.C.; van den Bosch, M.; Annerstedt, M.; Nielsen, A.B.; Maruthaveeran, S. Benefits of Urban Parks—A Systematic Review; A Report for IFPRA; Copenhagen & Alnarp. 2013. Available online: https://www.theparksalliance.org/benefits-of-urban-parks-a-systematic-review-a-report-for-ifpra-published-in-january-2013/ (accessed on 10 June 2021).

56. Coldwell, D.F.; Evans, K.L. Visits to Urban Green-Space and the Countryside Associate with Different Components of Mental Well-Being and Are Better Predictors than Perceived or Actual Local Urbanisation Intensity. *Landsc. Urban Plan.* **2018**, *175*, 114–122. [CrossRef]

57. Westphal, J.M. Hype, Hyperbole, and Health: Therapeutic site design. In *Urban Lifestyles: Spaces, Places People*; Benson, J.F., Rowe, M.H., Eds.; Brookfield, A.A. Balkema: Rotterdam, The Netherlands, 2000.

58. Murat, K. Factors affecting the planning and management of urban forests: A case study of Istanbul. *Urban For. Urban Green.* **2020**, *54*, 126739. [CrossRef]

59. Uchwała Sejmu Rzeczypospolitej Polskiej z dnia 1 Sierpnia 1997 r. Karta Praw Osób Niepełnosprawnych (M.P .Nr 50, poz. 475). Available online: http://isap.sejm.gov.pl/isap.nsf/DocDetails.xsp?id=WMP19970500475 (accessed on 10 June 2021).

60. Krzeptowska-Moszkowicz, I.; Moszkowicz, Ł.; Porada, K. Evolution of the Concept of Sensory Gardens in the Generally Accessible Space of a Large City: Analysis of Multiple Cases from Kraków (Poland) Using the Therapeutic Space Attribute Rating Method. *Sustainability* **2021**, *13*, 5904. [CrossRef]

61. Ward Thompson, C. Activity, Exercise and the Planning and Design of Outdoor Spaces. *J. Environ. Psychol.* **2013**, *34*, 79–96. [CrossRef]

62. Rosenberger, R.S.; Bergerson, T.R.; Kline, J.D. Macrolink-ages between Health and Outdoor Recreation: The Role of Parks and Recreation Providers. *J. Park Recreat. Admi.* **2009**, *27*, 8–20.

63. Irvine, K.; Warber, S.; Devine-Wright, P.; Gaston, K. Understanding Urban Green Space as a Health Resource: A Qualitative Comparison of Visit Motivation and Derived Effects among Park Users in Sheffield, UK. *Int. J. Environ. Res. Public Health* **2013**, *10*, 417–442. [CrossRef] [PubMed]

64. Kolimenakis, A.; Solomou, A.D.; Proutsos, N.; Avramidou, E.V.; Korakaki, E.; Karetsos, G.; Maroulis, G.; Papagiannis, E.; Tsagkari, K. The Socioeconomic Welfare of Urban Green Areas and Parks: A Literature Review of Available Evidence. *Sustainability* **2021**, *13*, 7863. [CrossRef]

65. Tsunetsugu, Y.; Lee, J.; Park, B.; Tyrväinen, L.; Kagawa, T.; Miyazaki, Y. Physiological and psychological effects of viewing urban forest landscapes assessed by multiple measurements. *Landsc. Urban Plan.* **2013**, *113*, 90–93. [CrossRef]

66. Bonilla-Duarte, S.; Gómez-Valenzuela, V.; Vargas-de la Mora, A.L.; García-García, A. Urban Forest Sustainability in Residential Areas in the City of Santo Domingo. *Forests* **2021**, *12*, 884. [CrossRef]

Article

Habitat-Related Differences in Winter Presence and Spring—Summer Activity of Roe Deer in Warsaw

Karolina D. Jasińska [1,*], Mateusz Jackowiak [1,2], Jakub Gryz [3], Szymon Bijak [4], Katarzyna Szyc [4] and Dagny Krauze-Gryz [1]

1. Department of Forest Zoology and Wildlife Management, Institute of Forest Sciences, Warsaw University of Life Sciences, Nowoursynowska 159, 02-776 Warsaw, Poland; mateusz.jackowiak@ios.gov.pl (M.J.); dagny_krauze_gryz@sggw.edu.pl (D.K.-G.)
2. Central Laboratory for Environmental Analysis—CentLab, Institute of Environmental Protection—National Research Institute, Krucza 5/11D, 00-548 Warsaw, Poland
3. Department of Forest Ecology, Forest Research Institute, Sękocin Stary, Braci Leśnej 3, 05-090 Raszyn, Poland; j.gryz@ibles.waw.pl
4. Department of Forest Management Planning, Dendrometry and Forest Economics, Institute of Forest Sciences, Warsaw University of Life Sciences—SGGW, Nowoursynowska 159, 02-776 Warsaw, Poland; szymon_bijak@sggw.edu.pl (S.B.); katarzyna_szyc@sggw.edu.pl (K.S.)
* Correspondence: karolina_jasinska@sggw.edu.pl

Abstract: Preliminary research conducted in Warsaw in the 1970s and 2000s showed that roe deer (*Capreolus capreolus*) stayed in forest habitat and avoided anthropogenic areas. Activity and exploration patterns of animals are shaped by indices of anthropogenic disturbances, elevated in large cities. The aims of the study were (1) to compare the presence of roe deer in natural and anthropogenic habitats of Warsaw during three periods: 1976–1978, 2005–2008 and 2017–2021, based on snow tracking on transect routes (681.2 km in total), and (2) to describe the presence and activity of roe deer in relation to human disturbances in selected urban forests in its reproductive period (March–August), based on camera trap survey (2019–2020, 859 observations, 5317 trap-days in total). The number of tracks was higher in natural habitat during all three periods, with the highest value in 2017–2021 (9.85/km/24 h). The peak of roe deer activity was recorded at dusk, and it changed with moon phases between spring and summer. Landscape connectivity and level of light pollution did not affect the activity pattern of roe deer. Our research showed that roe deer inhabiting urban areas avoided human presence by using well-covered habitats and being active in periods when the level of human disturbance was lower.

Keywords: *Capreolus capreolus*; ungulate; urban forests; human disturbances; daily activity; moon phases

Citation: Jasińska, K.D.; Jackowiak, M.; Gryz, J.; Bijak, S.; Szyc, K.; Krauze-Gryz, D. Habitat-Related Differences in Winter Presence and Spring—Summer Activity of Roe Deer in Warsaw. *Forests* **2021**, *12*, 970. https://doi.org/10.3390/f12080970

Academic Editors: Angela Lo Monaco, Cate Macinnis-Ng and Om P. Rajora

Received: 28 June 2021
Accepted: 19 July 2021
Published: 22 July 2021

Publisher's Note: MDPI stays neutral with regard to jurisdictional claims in published maps and institutional affiliations.

1. Introduction

Urbanization is considered a global threat to biodiversity [1]. The development of urban areas has increased worldwide over the last 30 years, and is expected to triple between 2000 and 2030, with an increase of world urban population to nearly 5 billion by that time [2]. Urban areas are characterized by high human density, large areas of impervious surfaces and built infrastructure [1,3–6]. That causes profound and ongoing changes in environment, i.e., abiotic environmental conditions (e.g., pollution) and to landscape structure [7]. Urban-associated landscape changes result mainly in habitat loss, fragmentation and reduced size and connectivity of landscape patches [8–12]. By limiting the movement of most species, the richness of biodiversity declines in the dense core of built-up urban areas [13].

In the present world, areas of undisturbed wilderness are rapidly decreasing, compelling wild mammals to integrate into urban environments. In recent decades, the number of studies on wildlife functioning in urban areas has rapidly increased, most of them concerning mesopredators (e.g., red fox *Vulpes vulpes*, badger *Meles meles*, stone marten *Martes foina*) [14–17], rather than ungulates (with the exception of wild boar *Sus scrofa*) [18]. Some research showed that human presence and activities accumulating in urban areas result in disturbances perceived by animals as analogous to the presence of natural predators [19–21], or can even exceed the effect of predation risk [22]. To deal with anthropogenic stressors, animals may shift their activity to more sheltered habitats, darker nights or become more nocturnal [23,24]. On the other hand, some of the papers showed that animals maintain their natural rhythms [25].

One of the most numerous ungulates in Poland is the roe deer *Capreolus capreolus*, whose population increased extensively in recent decades [26]. Roe deer, like other ungulates, inhabits mainly woodland and open habitats, often utilizing the ecotone between forests and agricultural areas [27–29]. However due to overabundant population in natural habitats, roe deer can inhabit suboptimal, human-transformed landscape [30,31]. Roe deer has recently been observed in urban areas, inhabiting mainly suburbs [31,32], where higher share of well-covered habitats enables roe deer to hide from human and human-related predators, i.e., dogs (*Canis familiaris*) [8].

Snow tracking conducted in Warsaw in the 1970s showed the presence of roe deer only in urban forests, especially in peripheral zones of the city, while the snow tracking data from the 2000s showed incidental presence of roe deer in anthropogenic habitat (urban parks), closer to the city center (Goszczyński J., unpubl. data). This suggests a possible gradual increase of roe deer in suboptimal, human-modified habitats, despite high sensitivity to human disturbances. Therefore, the aims of the research were (1) to compare the presence and abundance of roe deer in natural and anthropogenic habitats, based on historical (Goszczyński J., unpubl. data) and present snow tracking data collected along transect routes in various habitats in Warsaw; and (2) to describe changes in the presence and activity of roe deer in relation to human disturbances in selected urban forests, as the most often selected habitat. We hypothesized that in the last few years the abundance of roe deer in the anthropogenic habitat has increased. Nevertheless, urban forests remain the main habitat for roe deer, so with high level of human disturbances in urban forests, roe deer will have to adapt behaviorally to avoid contact with human and human-related predators. This will be shown by (1) observing higher daily activities at nights because human activity is then lowest, (2) higher activity patterns during dark nights compared to bright nights using moon phases and (3) observing a lower frequency of occurrence in forest complexes with higher level of light pollution.

2. Materials and Methods

2.1. Study Area

Warsaw (52°13′47″ N, 21°00′42″ E), the capital of Poland, is the largest (517 km²) and most populous (3437 inhabitants/km²) city in the country [33]. It is situated at an altitude of 113 m above sea level in a temperate zone, with an annual rainfall of about 500 mm and an average temperature of 7.7 °C [34]. The Vistula River flows throughout Warsaw and divides city into two parts. Warsaw is characterized by a high proportion of green areas [33,35]: urban forests, parks, botanical and zoological gardens, squares, cemeteries, allotments, home gardens, residential and roadside vegetation, and natural riparian forests (Nature 2000 protected). Urban forests constitute ca. 15% of the city area. Urban forests are located mainly in peripheral districts of the city.

2.2. Data Collection

We used present and historical snow tracking data on transect routes to describe the temporal changes in the roe deer abundance in different habitats in Warsaw. Snow tracking was conducted during three different periods: (1) 1976–1978 (two winter seasons) done by Professor J. Goszczyński (unpubl. data), (2) 2005–2008 (three winter seasons) and (3) 2017–2021 (four winter seasons). Snow tracking was conducted in winter months (December to February) when the snow cover depth exceeded 1 cm. The number of tracks was noted per 100 m of tracking route. Snow footprints were easily trampled in areas with high dog and human activity, so we started snow tracking as early as 12 h after the snowfall (but only if it stopped snowing in the afternoon at the latest). Simultaneously, snow tracking was done up to two days after snowfall in areas less penetrated by humans. In all cases, the number of snow registered tracks was recalculated per 24 h after a snowfall.

Tracking routes were distributed randomly throughout Warsaw, in different types of habitats (e.g., forests, agriculture areas, urban parks and cemeteries, built-up areas). The length of transect routes differed between habitats. For further calculations we distinguished only between two types of habitats—natural (forests and open areas) and anthropogenic (urban area, other green areas: e.g., cemeteries, parks). In total, 681.2 km of snow tracking on transect routes was done (Table 1). Results of snow tracking allowed for showing the relative index of abundance of roe deer in natural and anthropogenic habitats in Warsaw, defined as number of tracks/1 km of transect routes/24 h.

Table 1. The length of snow tracking transect routes to determine roe deer abundance in natural and anthropogenic habitats of Warsaw during three research periods (1976–1978, 2005–2008, 2017–2021).

Research Period	Length of Transect Routes in Habitats [km]	
	Natural	Anthropogenic
1976–1978	96.2	79.5
2005–2008	39.0	180.8
2017–2021	146.0	139.7
in total	281.2	400.0

To determine the activity patterns of roe deer, we used camera traps set randomly in selected urban forests (Figure 1) in spring and summer of 2019 and 2020, assuming that urban forests are the main habitat of roe deer. Moreover, setting camera traps in open areas, urban parks and cemeteries, or built-up areas may result in their theft or damage.

Camera traps were located in eleven different forest areas, six on the northeastern side of Warsaw (on the right bank of the Vistula River) (Henryków and Dąbrówka Forest Park, Bródno Forest Park, Utrata Forest, Sobieski Forest, Olszynka Grochowska Nature Reserve and Kawęczyn Nature Reserve), and five on the west bank of Vistula River (Młociny Forest, Bielany Forest, Morysin Nature Reserve, Natolin Forest Nature Reserve, Kabaty Forest Nature Reserve). The forest complexes investigated differed in size (45–917 ha) (Table A1). During the exposition period, the camera trap was regularly inspected at 1–1.5 month intervals. During the inspection, the batteries and memory card were changed, and the camera trap locations were changed to minimize the risk of recording the same individuals.

We used several types of camera traps (Reconyx HyperFire: PC90, PC800, PC850, PC900, RECONYX, Inc., Holmen, WI, USA; Ltl Acorn 6210 MC, Zhuhai Ltl Acorn Electronics Co., Ltd., Zhuhai, Guangdong, China; Browning Spec Ops Advantage, Browning Trail Cameras, Morgan, UT, USA) that differed slightly in the way of records acquisition. Reconyx camera traps took a series of three photos, at one-second interval, while Acorn and Browning devices took a single photo at one-second interval. The camera traps were set on trees, 30 cm above the ground to register adult and juvenile individuals of roe deer. We did not use any attractant to lure animals.

Camera traps were set in 27 (2019) and 34 (2020) locations in February and disassembled mostly late summer/early autumn. For the analysis, we used data from March to August, i.e., roe deer reproductive period (as defined by implantation in females, antler growth through roe bucks territorialism to rut) [37]. In total, data for 5317 camera trap-days were collected.

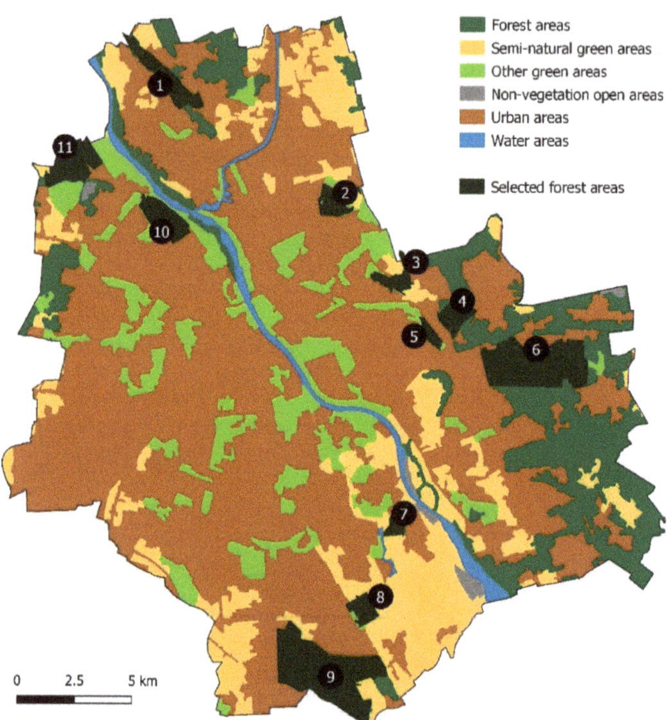

Figure 1. Distribution of urban forests in Warsaw where camera traps were set in the years 2019–2020 to study roe deer activity patterns. The numbers show selected forest areas: 1. Henryków and Dąbrówka Forest Park, 2. Bródno Forest Park, 3. Utrata Forest, 4. Kawęczyn Nature Reserve, 5. Olszynka Grochowska Nature Reserve, 6. Sobieski Forest (incl. Jan III Sobieski Nature Reserve), 7. Morysin Nature Reserve, 8. Natolin Forest Nature Reserve, 9. Kabaty Forest Nature Reserve, 10. Bielany Forest and 11. Młociny Forest. Types of land cover were taken from CORINE Land Cover [36]. Six types of land cover were distinguished: forest areas (all natural wooded areas), semi-natural green areas (arable land, pastures, heterogeneous agricultural areas, shrub and/or herbaceous associations), other green areas (artificial, non-agricultural vegetated areas and permanent crops), non-vegetation open areas (open spaces with little or no vegetation, mine, dump and construction sites), urban areas (urban fabric, industrial, commercial and transport units) and waters (water bodies).

We recorded each roe deer appearing in the images without distinguishing between individuals. A new observation was considered if a minimum of 15 min elapsed between subsequent photos or series of photos showing an animal/animals. This rule was abandoned only when an animal in the photo was different in age, sex, body condition and antler development, indicating clearly that the animal in the photo was a different individual than the one previously registered. A group of different individuals appearing in one picture or a series of pictures was also recorded as a single observation. Camera traps recorded date of the observation, time (24 h record) in Central European Time (CET), and a moon phase.

We analyzed activity of roe deer in months, seasons (spring: March–May, summer: June–August) and within the 24 h period. We used the time of sunrise and sunset for Warsaw in the years 2019–2020 [38] converted to CET and defined diurnal period as the time between sunrise and sunset, and night as the time between 1 h after sunset and 1 h before sunrise. We defined the crepuscular periods as one hour before sunrise (dawn) and one hour after sunset (dusk) [39]. We analyzed the activity of roe deer in eight moon phases: new moon, waxing crescent, first quarter, waxing gibbous, full moon, waning gibbous, last quarter and waning crescent. As the moonlight may influence the behavior of prey species, we divided moon phases into dark (new moon, waning crescent, waxing crescent) and bright nights (full moon, waning gibbous, waxing gibbous). The activity of roe deer in moon phases and in dark and bright nights was analyzed for nocturnal period (time of a day when there was no sunlight: dawn, dusk and night) [40].

Finally, we analyzed the impact of human disturbances, defined as light pollution and landscape fragmentation, on roe deer frequency of occurrence (defined as N observations per 100 camera trap days for a given forest complex) in urban forests under study. To investigate influence of human disturbances on roe deer occurrence, we adopted level of light pollution based on Visible Infrared Imaging Radiometer Suite (VIIRS) after [41] for localization of each camera trap and averaged the values for each forest complex (Table A2). We assumed that landscape connectivity is provided by green spaces in the city: wooded areas, shrubs and green open spaces (arable lands and remaining green spaces). Therefore, we analyzed the level of isolation of urban forests under study. We set a 250 m buffer zone around each forest, in which the shares of forest areas, semi-natural green areas (arable land, pastures, heterogeneous agricultural areas, shrub and/or herbaceous associations), other green areas (artificial, non-agricultural vegetated areas, permanent crops), non-vegetation open areas (open spaces with little or no vegetation, mine, dump and construction sites), urban areas (urban fabric, industrial, commercial and transport units) and waters (water bodies) were calculated (Table A3). The information and data on topographic objects were taken from CORINE Land Cover published in 2021 [36], which presented the biophysical characteristic of Earth [42]. All maps and GIS analyses were carried out in QGIS v3.10.

2.3. Statistical Analysis

All data processing, calculations and statistical analyses were performed in MS Excel spreadsheet and PAST4.03 software [43]. For observed relationships and differences, we assumed $p = 0.05$ as their significance level. Prior to the analysis, distribution of the given variable was tested for its normality with the Shapiro–Wilk test.

Density of roe deer tracks was compared using Kruskal–Wallis test followed by U Mann–Whitney post hoc comparisons. Distributions of number of roe deer observations in subsequent months in March–August period or in seasons (spring: March–May, summer: June–August) were compared using the chi-square test. The same procedure was applied for frequency of occurrence of roe deer in parts of the day or during the moon phases. To compare the activity of roe deer during the day, we calculated mean occurrence per hour of a given part of a day.

Mean number of observations of roe deer per 100 trap-days calculated for each forest complex was correlated with light pollution and share of forest areas, other green areas, other open areas, semi-natural green areas, urban areas and water in 250 m buffer zones around a given forest complex. For those analyses, Pearson correlation coefficient was used.

3. Results

3.1. Abundance of Roe Deer in Different Habitats of Warsaw Using Snow Tracking Data

The highest density of roe deer tracks (9.85/km/24 h) was recorded in natural habitats in the third period of research (2017–2021), while no roe deer were observed in anthropogenic habitat in the first period (1976–1978) (Figure 2). In general, natural habitats were characterized by a significantly higher density of roe deer tracks (U= 1,541,653, $p < 0.001$) for all analyzed periods. Various periods were characterized by significantly different roe deer abundance (H = 32.39, $p < 0.001$). Significant differences were found between the density of roe deer tracks in 1976–1978 and 2005–2008 (U = 633,900, $p < 0.001$) and between 2005–2008 and 2017–2021 (U = 16,600, $p < 0.001$).

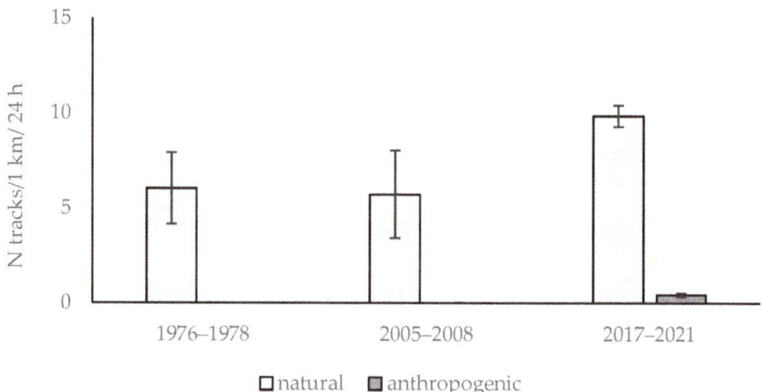

Figure 2. Mean (+SE) number of roe deer tracks in two different types of habitats in Warsaw as recorded by snow tracking along transect routes in the three study periods: 1976–1978, 2005–2008, 2017–2021.

3.2. Activity Patterns of Roe Deer in Warsaw Urban Forests Using Camera Traps Data

In total, 859 observations of roe deer were recorded in forest complexes in Warsaw during March–August for the years 2019–2020 (442 in year 2019 and 417 in 2020). There were no significant differences between the number of observations in 2019 and 2020 ($\chi2 = 0.72$, df = 1, $p > 0.05$). The total number of observations was not uniformly distributed among analyzed months ($\chi2 = 47.6$, df = 11, $p < 0.001$) and seasons ($\chi2 = 22.5$, df = 1, $p < 0.001$) with 23% of observations in April (58% observations recorded in spring) and 10% in June. Most of the observations were recorded during the day-time (67.8% of all recorded observations) (Figure 3).

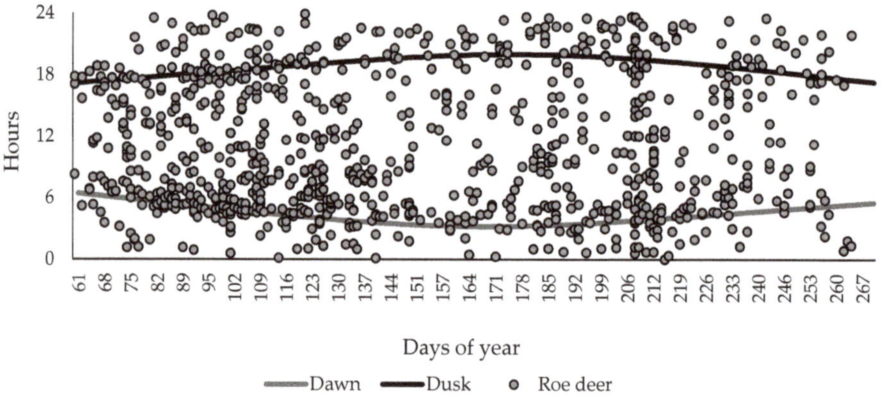

Figure 3. Temporal pattern of roe deer activity in a reproductive period (March–August) in urban forests of Warsaw in the years 2019–2020. Each dot refers to one case of roe deer presence recoded by a camera trap.

Mean number of roe deer observations per one hour did not differ between parts of the day in spring and summer ($\chi2 = 1.18$, df = 3, $p > 0.05$). The majority of the observations were noted in crepuscular time of day in spring (dawn—5.3/h, dusk—6.0/h) and summer (dawn—3.2/h, dusk—3.5/h). The lowest values were noted for nights, while in summer the numbers of observation per 1 h of day and night were similar (2.5 and 2.4 observations/h, respectively) (Figure 4). Roe deer activity during day or night was not significantly correlated with the length of day or night (r = −0.50, $p = 0.12$ and r = 0.09, $p = 0.80$, respectively).

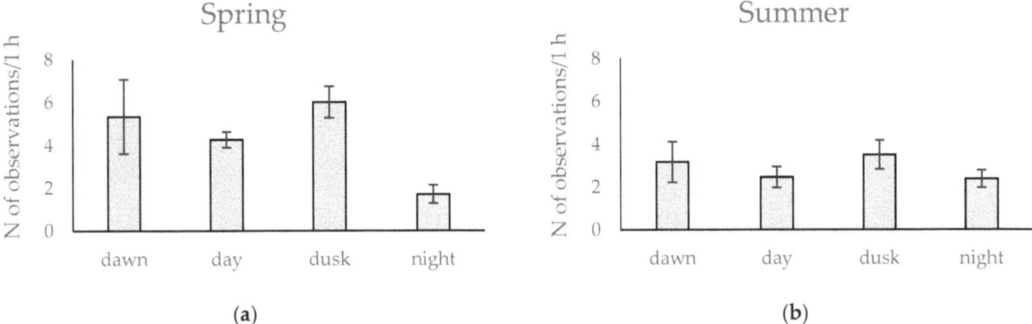

Figure 4. Mean (+SE) number of roe deer camera-trap observations per one hour during different parts of the day during (**a**) spring (March–May) and (**b**) summer (June–August) in urban forests of Warsaw in years 2019–2020.

In general, the frequency of observations in moon phases did not differ during nocturnal (night, dawn, dusk) periods ($\chi2 = 11.5$, df = 7, $p = 0.119$), but the number of roe deer observations in certain moon phases differed between spring and summer ($\chi2 = 14.60$, df = 7, $p = 0.042$) (Figure 5). The frequency of nocturnal observations at bright and dark nights was similar (51.3% and 48.7%, respectively, n = 227, $\chi2 = 0.13$, df = 1, $p = 0.721$).

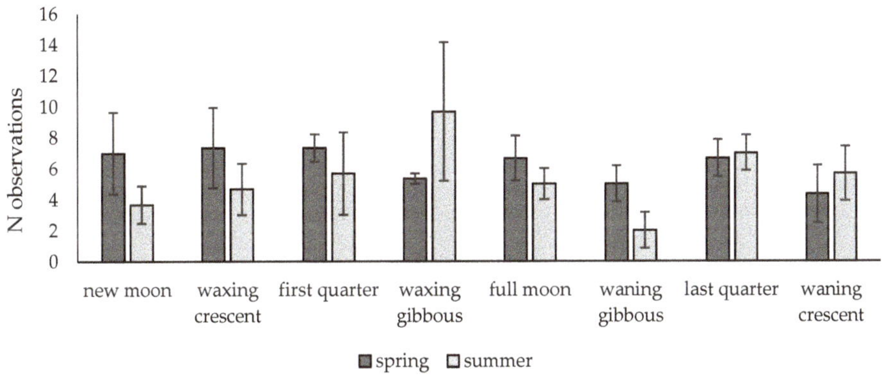

Figure 5. Number (+SE) of roe deer camera-trap observations in different moon phases during spring (March–May) and summer (June–August) in urban forests of Warsaw in years 2019–2020.

Mean numbers of roe deer recorded by individual trap per 100 trap-days were not related to the level of light pollution assigned to a given forest complex (r = 0.007, $p = 0.839$). Furthermore, in the case of studied forest complexes, we found no significant relationship between mean number of recorded animals per 100 trap-days and any of the analyzed parameters describing the impact of human disturbances on roe deer occurrence (Table A4).

4. Discussion

Ungulates are characterized by high behavioral plasticity, which allows them to inhabit heavily human-modified areas, including cities [44,45]. In these habitats, human disturbances are often considered to be analogous to predation risk [19]. This may lead to shifts in temporal and spatial activity patterns of animals [46,47], and an increase in their vigilance levels [48] or flight distances [49]. In line with the assumptions presented in the introduction, we found that human disturbances affected abundance and activity patterns of roe deer in urban areas of Warsaw. Abundance of this species was highest in more sheltered habitats (urban forests). Roe deer activity pattern varied depending on the time of day and moon phases.

Our study, based on snow tracking, showed that winter abundance of roe deer in Warsaw was highest in natural habitats (forests and open, mostly agricultural areas) in every period of research (1976–1978, 2005–2008 and 2017–2021). Ciach and Fröhlich [44] showed that winter occurrence of roe deer in urban areas was correlated with presence of open areas, providing high-quality food resources for roe deer during this season [27,50]. In turn, wooded areas provide effective camouflage for this species [23], not only in winter but all year round, therefore, roe deer may seek shelter here from humans (see [25]). Moreover, animals may buffer human disturbances by adjusting their activity in time and space [24]; in addition, roe deer may shift from open to more wooded habitats, staying hidden in forest refuges during the day, and foraging in open areas at night [23,48].

Our research showed that roe deer tracks were noticed in anthropogenic types of habitat more often in 2017–2021 than in previous study periods. This might be linked to change in various habitat types of Warsaw since 1970s, including an increasing share of built-up surfaces. Roe deer from an overabundant population inhabiting natural areas may be pushed into human-modified landscape [30]. Resulting from a high level of behavioral plasticity (e.g., [51]), roe deer can live in cities [44]. For the last two decades, Warsaw has gone through dynamic development process, which caused a decrease in the share of natural habitats. As a result, during migration or daily activities, roe deer may be forced to leave its natural habitat (forest and open areas) and penetrate more human-transformed areas. Moreover, the Vistula River flows throughout the entire city, an important ecological corridor for mammalian species, but this may also enable animals to penetrate city areas far from the source habitats, i.e., the forests [17].

Disturbance caused by humans affects selection of habitats by animals [52–54]. Proximity to trails, roads or buildings (which are associated with increased human activity), leads to avoidance of such areas [23,55]. Nevertheless, our research showed that the frequency of camera trap survey-based occurrence of roe deer in spring–summer and in certain urban forests in Warsaw was not linked to land cover around forest complexes. This is in line with findings of other studies, which claimed that human disturbances may have no significant effect on roe deer space use or vigilance, when density of human infrastructures is very high [25,48]. Urban landscape is highly heterogeneous with patches of optimal habitats surrounded by a matrix of human-transformed habitats, which most will be of low utility for the species. It may be assumed that relatively natural areas will be used by roe deer if only they offer food and shelter, and may be reached by migrating individuals. Distances from one patch of habitat to another are rather small, so with high population density in source habitats (forests located at the borders of the city and in larger suburban forests) most urban forests will be assumingly reached and inhabited.

Daily activity of animals can vary, depending on many conditions, including predation risk [56,57]. Most often, observed adjustment to living in anthropogenic landscape is shifting activity to periods of lower detectability [58]. Such shifts relate to patterns of daily activity, and influence of moon phases. In natural habitats with predators present, bimodal activity pattern is typical for roe deer [45,55,56]. Animals, living in close proximity to humans, may become more nocturnal to avoid high-risk periods, since humans are mostly active during the day [58]. Indeed, such activity pattern was observed for roe deer under human disturbance [23,45,55]. Our study showed very low activity of roe deer in urban forests during nights in the seasons that were investigated. On the one hand, this result might be linked to chosen methods. The effectiveness of camera traps during dark night hours might be lower than during day and crepuscular parts of day. On the other hand, human activity in urban areas is greatest during daylight hours [58], and despite the amicable spring and summer weather, it ends at dusk. Our results showed that activity pattern of roe deer in Warsaw was bimodal, connected with crepuscular periods of day and the higher peak occurring during dusk. Moreover, the activity of roe deer in different parts of day did not differ between spring and summer, while other authors showed monthly changes in distribution of roe deer activity in reproductive period (March–August) in natural habitat. According to Krop-Benesch et al. [37], the March and May activity of roe deer was higher at dusk, while in June–August higher activity level was observed at dawn.

Moon phases may influence animal behavior (e.g., [59]), including predator–prey interactions [60,61]. Animals being prey for predatory species shift their activity to darker nights and darker moon phases [62]. There is lack of consistency whether roe deer reacts to moonlight intensity; however, some research showed no influence of moon phases on roe deer activity [56]. Our research showed that moon phases influenced roe deer activity in urban forests of Warsaw, with the highest value obtained for waxing gibbous, one phase before the full moon, when the moonlight increases. Our findings concur with Kurt [63], who showed that roe deer may be more active during bright nights.

Even in urban areas, roe deer preferred natural to anthropogenic habitats, which was confirmed by our study that started in the 1970s in Warsaw. Roe deer inhabited mainly urban forests, which provided shelter from human disturbances, and allowed animals to be active with increasing moonlight and possibility to be detected. In turn, during the day, roe deer shifted its activity to times when the presence of humans and dogs in urban forest decreased.

5. Conclusions

Urbanization has led to severe habitat fragmentation and loss, and has brought humans and wildlife in close proximity, affecting both. In a highly heterogeneous and often hostile urban matrix, roe deer inhabits green, wooded areas, which offer shelter and exploits surrounding open spaces for feeding. Indeed, in Warsaw, winter abundance indices of roe deer snow tracks were the highest in natural habitats during all three periods of research. We found no specific spatial characteristics that would influence frequency of occurrence of roe deer within urban forests. This suggests that urban forest is a sufficient refuge regardless its surroundings and points to maintained ecological connectivity within urban matrix. Nevertheless, our research showed that roe deer inhabiting urban area avoided human presence, first by using well-covered habitats but also shifting its temporal activity. It reduced activity during the time when humans were more active, but at the same time its nocturnal (including dawn and dusk) activity was higher with increasing moonlight intensity (waxing gibbous) in the summer season. Overall, the study showed considerable plasticity of the species, which adapted to human-transformed landscape and exploited most suitable habitats, while its behavior altered to avoid human disturbance.

Author Contributions: Conceptualization, K.D.J.; methodology, K.D.J., M.J., S.B. and D.K.-G.; formal analysis, S.B., K.D.J.; investigation, K.D.J., M.J. and K.S.; resources, J.G. and M.J.; data curation, K.D.J. and M.J.; writing—original draft preparation, K.D.J., M.J., S.B. and D.K.-G.; writing—review and editing, K.D.J., M.J. and D.K.-G.; visualization, K.S. and K.D.J.; supervision, K.D.J. and D.K.-G.; project administration, K.D.J. All authors have read and agreed to the published version of the manuscript.

Funding: The article was financed by the Polish Ministry of Science and Higher Education within funds of the Institute of Forest Sciences, Warsaw University of Life Sciences (WULS), for scientific research.

Data Availability Statement: Data available on request.

Conflicts of Interest: The authors declare no conflict of interest.

Appendix A

Table A1. Characterization of urban forests where the camera traps were distributed in urban forests of Warsaw in the years 2019–2020.

Number of Urban Forest	Urban Forest	Forest Area (ha)	Number of Location	Latitude	Longitude	Dates of Camera Trap Exposition
1	Henryków i Dąbrówka Forest Park	201.37	1	52°20.349′ N	20°58.256′ E	02.03.–27.06.2019
			2	52°20.362′ N	20°58.209′ E	02.03.–06.05.2019 20.04.–13.07.2020
			3	52°21.086′ N	20°57.385′ E	02.03.–01.05.2019 17.06.–29.09.2019 25.02.–04.04.2020
			4	52°20.268′ N	20°58.275′ E	07.03.–20.04.2020
2	Bródno Forest Park	139.00	1	52°17.909′ N	21°04.015′ E	24.02.–20.05.2020
			2	52°18.037′ N	21°03.924′ E	02.03.–07.05.2019 24.02.–14.07.2020
			3	52°18.156′ N	21°03.787′ E	02.03.–07.05.2019
3	Utrata Forest	102.90	1	52°16.057′ N	21°06.698′ E	05.03.–04.06.2019
4	Kawęczyn Nature Reserve	123.18	1	52°15,249′ N	21°08,733′ E	05.03.–12.03.2019 23.02.–21.04.2020
			2	52°15,242′ N	21°08,285′ E	10.03.–21.04.2020
5	Olszynka Grochowska Nature Reserve	58.91	1	52°14.896′ N	21°07.301′ E	18.03.–30.05.2019 18.06.–22.07.2019
			2	52°14.873′ N	21°07.328′ E	12.05.–13.10.2020
6	Sobieski Forest	765.75	1	52°13.756′ N	21°10.215′ E	09.03.–01.06.2019 18.02.–07.07.2020
			2	52°13.746′ N	21°10.163′ E	09.03.–07.05.2019 21.04.–07.07.2020
			3	52°14.020′ N	21°09.664′ E	01.03.–17.07.2019
			4	52°13.941′ N	21°10.550′ E	07.05.–04.10.2019 18.02.–03.03.2020
			5	52°13.599′ N	21°10.388′ E	18.02.–01.03.2020
			6	52°13.744′ N	21°10.214′ E	08.03.–21.04.2020
7	Morysin Nature Reserve	45.04	1	52°10.508′ N	21°06.134′ E	09.03.–25.04.2019 15.07.–05.09.2019 09.02.–23.02.2020 14.09.–22.10.2020
			2	52°10.529′ N	21°06.117′ E	29.02.–11.05.2020
8	Natolin Forest Nature Reserve	104.16	1	52°08.459′ N	21°05.092′ E	11.03.–14.05.2019 22.07.–31.12.2019 10.02.–29.06.2020
			2	52°08.611′ N	21°04.223′ E	27.02.–11.03.2019 29.04.–11.06.2019 09.03.–22.04.2020
			3	52°08.421′ N	21°05.018′ E	29.04.–22.07.2019
			4	52°08.379′ N	21°05.206′ E	05.06.–07.07.2020

Table A1. *Cont.*

Number of Urban Forest	Urban Forest	Forest Area (ha)	Number of Location	Latitude	Longitude	Dates of Camera Trap Exposition
9	Kabaty Nature Reserve	918.27	1	52°06.985′ N	21°03.229′ E	03.03.–27.04.2020
			2	52°07.455′ N	21°03.054′ E	26.02.–11.04.2019 18.05.–01.09.2019 27.04.–24.10.2020
			3	52°07.753′ N	21°02.823′ E	04.03.–11.03.2019 19.05.–08.07.2019
			4	52°08.007′ N	21°02.185′ E	11.03.–07.04.2019 08.07.–15.08.2019
			5	52°07.305′ N	21°05.380′ E	17.03.–31.12.2019 07.01.–08.05.2020
			6	52°07.493′ N	21°02.105′ E	26.02.–04.03.2019
			7	52°07.592′ N	21°03.118′ E	26.02.–17.06.2019 08.07.–28.09.2019 03.03.–24.10.2020
			8	52°06.916′ N	21°03.231′ E	09.06.–24.10.2020
			9	52°07.550′ N	21°02.105′ E	14.03.–19.05.2019 24.04.–26.10.2020
			10	52°07.567′ N	21°01.944′ E	19.05.–28.09.2019 24.04.–08.06.2020
			11	52°06.878′ N	21°02.499′ E	09.03.–24.04.2020
			12	52°07.384′ N	21°05.316′ E	09.03.–24.04.2020
			13	52°06.842′ N	21°02.806′ E	09.03.–24.04.2020
			14	52°06.912′ N	21°02.416′ E	24.04.–09.06.2020
10	Bielany Forest	196.07	1	52°17.349′ N	20°58.057′ E	10.03.–22.09.2020
11	Młociny (Nowa Warszawa) Forest	293.64	1	52°18.864′ N	20°54.201′ E	02.03.–18.06.2019 02.03.–10.03.2020
			2	52°19.114′ N	20°54.025′ E	02.03.–06.05.2019

Appendix B

Table A2. Mean light pollution level (based on Visible Infrared Imaging Radiometer Suite (VIIRS)) for each urban forest in Warsaw in the years 2019–2020.

Urban Forest	Level of Light Pollution
Henryków i Dąbrówka Forest Park	10.45
Bródno Forest Park	16.38
Utrata Forest	16.00
Kawęczyn Nature Reserve	12.63
Olszynka Grochowska Nature Reserve	9.88
Sobieski Forest	24.64
Morysin Nature Reserve	6.37
Natolin Forest Nature Reserve	15.22
Kabaty Nature Reserve	5.23
Bielany Forest	17.18
Henryków i Dąbrówka Forest Park	10.45

Appendix C

Table A3. Spatial parameters levels describing landscape connectivity for urban forests in Warsaw in the years 2019–2020.

Urban Forest	% Share in 250 m Buffer of **					
	Forest Areas	Semi-Natural Green Areas	Other Green Areas	Non-Vegetation Open Areas	Urban Areas	Water
Henryków i Dąbrówka Forest Park	19.5	7.0	-	-	73.5	-
Bródno Forest Park	-	15.5	25.3	-	59.2	-
Utrata Forest	-	25.9	20.9	-	53.2	-
Kawęczyn Nature Reserve	40.9	-	-	-	59.1	-
Olszynka Grochowska Nature Reserve	-	-	21.0	-	79.0	-
Sobieski Forest	49.4	-	1.3	-	49.3	-
Morysin Nature Reserve	5.6	39.8	1.6	-	41.8	11.2
Natolin Forest Nature Reserve	-	52.3	16.2	-	31.4	-
Kabaty Nature Reserve	8.7	34.5	-	-	56.8	-
Bielany Forest	1.8	-	37.4	-	48.8	12.0
Młociny (Nowa Warszawa) Forest	23.2	9.4	29.8	1.9	35.6	-

** dashes in cells indicate lack of area type in a buffer zone.

Appendix D

Table A4. Pearson correlation coefficients between selected environmental parameters describing landscape connectivity with occurrence of roe deer (mean frequency of occurrence per 100 trap-days) in urban forests of Warsaw as recorded by camera traps in the years 2019–2020.

Analysis	Parameter	r	p
share (%) in 250 m buffer around urban forests	Forest areas	0.13	0.71
	Semi-natural green areas	0.34	0.31
	Other green areas	−0.43	0.19
	Non-vegetation open areas	−0.44	0.18
	Urban areas	−0.02	0.95
	Water	−0.22	0.52

References

1. McKinney, M.L. Urbanization as a major cause of biotic homogenization. *Biol. Conserv.* **2006**, *127*, 247–260. [CrossRef]
2. Seto, K.C.; Güneralp, B.; Hutyra, L.R. Global forecasts of urban expansion to 2030 and direct impacts on biodiversity and carbon pools. *Proc. Natl. Acad. Sci. USA* **2012**, *109*, 16083–16088. [CrossRef]
3. Prugh, L.R.; Hodges, K.E.; Sinclair, A.R.E.; Brashares, J.S. Effect of habitat area and isolation on fragmented animal populations. *Proc. Natl. Acad. Sci. USA* **2008**, *105*, 20770–20775. [CrossRef] [PubMed]
4. McDonnell, M.J.; Hahs, A.K. The use of gradient analysis studies in advancing our understanding of the ecology of urbanizing landscapes: Current status and future directions. *Landsc. Ecol.* **2008**, *23*, 1143–1155. [CrossRef]
5. Pickett, S.T.A.; Cadenasso, M.L.; Grove, J.M.; Boone, C.G.; Groffman, P.M.; Irwin, E.; Kaushal, S.S.; Marshall, V.; McGrath, B.P.; Nilon, C.H.; et al. Urban ecological systems: Scientific foundations and a decade of progress. *J. Environ. Manag.* **2011**, *92*, 331–362. [CrossRef] [PubMed]
6. Forman, R.T. *Urban Ecology: Science of Cities*; Cambridge University Press: New York, NY, USA, 2014.
7. Parris, K.M. *Ecology of Urban Environments*; Wiley-Blackwell: Chichester, UK, 2016.
8. Hewison, A.J.M.; Vincent, J.P.; Joachim, J.; Angibault, J.M.; Cargnelutti, B.; Cibien, C. The effects of woodland fragmentation and human activity on roe deer distribution in agricultural landscapes. *Can. J. Zool.* **2001**, *79*, 679–689. [CrossRef]
9. Acevedo, P.; Delibes-Mateos, M.; Escudero, M.A.; Vicente, J.; Marco, J.; Gortazar, C. Environmental constraints in the colonization sequence of roe deer (*Capreolus capreolus* Linnaeus, 1758) across the Iberian Mountains, Spain. *J. Biogeogr.* **2005**, *32*, 1671–1680. [CrossRef]
10. Syphard, A.D.; Clarke, K.C.; Franklin, J.; Regan, H.M.; McGinnis, M. Forecasts of habitat loss and fragmentation due to urban growth are sensitive to source of input data. *J. Environ. Manag.* **2011**, *92*, 1882–1893. [CrossRef]

11. Scolozzi, R.; Geneletti, D. A multi-scale qualitative approach to assess the impact of urbanization on natural habitats and their connectivity. *Environ. Impact Assess. Rev.* **2012**, *36*, 9–22. [CrossRef]
12. van Strien, M.J.; Grêt-Regamey, A. How is habitat connectivity affected by settlement and road network configurations? Results from simulating coupled habitat and human networks. *Ecol. Model.* **2016**, *342*, 186–198. [CrossRef]
13. van Strien, M.J.; Slager, C.T.J.; de Vries, B.; Gret-Regamey, A. An improved neutral landscape model for re-creating real landscapes and generating landscape series for spatial ecological simulations. *Ecol. Evol.* **2016**, *6*, 3808–3821. [CrossRef]
14. Baker, P.J.; Dowding, C.V.; Molony, S.E.; White, P.C.L.; Harris, S. Activity patterns of urban red foxes (*Vulpes vulpes*) reduce the risk of traffic-induced mortality. *Behav. Ecol.* **2007**, *35*, 716–724. [CrossRef]
15. Harris, S. Ecology of Urban Badger *Meles meles*: Distribution in Britain and Habitat Selection, Persecution, Food and Damage in the City of Bristol. *Biol. Conserv.* **1984**, *28*, 349–375. [CrossRef]
16. Herr, J.; Schley, L.; Roper, T.J. Socio-spatial organization of urban stone martens. *J. Zool.* **2009**, *277*, 54–62. [CrossRef]
17. Jackowiak, M.; Gryz, J.; Jasińska, K.; Brach, M.; Bolibok, L.; Kowal, P.; Krauze-Gryz, D. Colonization of Warsaw by the red fox *Vulpes vulpes* in the years 1976–2019. *Sci. Rep.* **2021**, *11*, 13931. [CrossRef]
18. Csókás, A.; Schally, G.; Szabó, L.; Csányi, S.; Kovács, F.; Heltai, M. Space use of wild boar (*Sus Scrofa*) in Budapest: Are they resident or transient city dwellers? *Biol. Futur.* **2020**, *71*, 39–51. [CrossRef]
19. Frid, A.; Dill, L.M. Human-caused disturbance stimuli as a form of predation risk. *Conserv. Ecol.* **2002**, *6*, 11. [CrossRef]
20. Rich, C.; Longcore, T. *Ecological Consequences of Artificial Night Lighting*; Island Press: London, UK, 2006.
21. LaPoint, S.; Balkenhol, N.; Hale, J.; Sadler, J.; van der Ree, R. Ecological connectivity research in urban areas. *Funct. Ecol.* **2015**, *29*, 868–878. [CrossRef]
22. Ciuti, S.; Northrup, J.M.; Muhly, T.B.; Simi, S.; Musiani, M.; Pitt, J.A.; Boyce, M.S. Effects of humans on behaviour of wildlife exceed those of natural predators in a landscape of fear. *PLoS ONE* **2012**, *7*, e50611.
23. Bonnot, N.; Morellet, N.; Verheyden, H. Habitat use under predation risk: Hunting, roads and human dwellings influence the spatial behavior of roe deer. *Eur. J. Wildl. Res.* **2013**, *59*, 185–193. [CrossRef]
24. Bonnot, N.; Morellet, N.; Hewison, A.J.M.; Martin, J.-L.; Benhamou, S.; Chamaillé-Jammes, S. Black-tailed deer (*Odocoileus hemionus sitkensis*) adjust habitat selection and activity rhythm to the absence of predators. *Can. J. Zool.* **2016**, *94*, 385–394. [CrossRef]
25. Wevers, J.; Fattebert, J.; Casaer, J.; Artois, T.; Beenaerts, N. Trading fear for food in the Anthropocene: How ungulates cope with human disturbance in a multi-use, suburban ecosystem. *Sci. Total Environ.* **2020**, *741*, 140369. [CrossRef] [PubMed]
26. Bragina, E.V.; Ives, A.R.; Pidgeon, A.M.; Balčiauskas, L.; Csányi, S.; Khoyetskyy, P.; Radeloff, V.C. Wildlife population changes across Eastern Europe after the collapse of socialism. *Front. Ecol. Environ.* **2018**, *16*, 77–81. [CrossRef]
27. Putman, R.J. Foraging by roe deer in agricultural areas and impact on arable crops. *J. Appl. Ecol.* **1986**, *23*, 91–99. [CrossRef]
28. McLoughlin, P.D.; Gaillard, J.-M.; Boyce, M.S.; Bonenfant, C.; Messier, F.; Duncan, P.; Delorme, D.; Van Moorter, B.; Saïd, S.; Klein, F. Lifetime reproductive success and composition of the home range in a large herbivore. *Ecology* **2007**, *88*, 3192–3201. [CrossRef]
29. Panzacchi, M.; Linnell, J.D.C.; Odden, M.; Odden, J.; Andersen, R. Habitat and roe deer fawn vulnerability to red fox predation. *J. Anim. Ecol.* **2009**, *78*, 1124–1133. [CrossRef]
30. Tinoco Torres, R.; Carvahlo, J.C.; Panzacchi, M.; Linnell, J.D.C.; Fonseca, C. Comparative use of forest habitats by roe deer and moose in a human-modified landscape in southeastern Norway during winter. *Ecol. Res.* **2011**, *26*, 781–789. [CrossRef]
31. Loro, M.; Ortega, E.; Arce, R.M.; Geneletti, D. Assessing landscape resistance to Roe deer dispersal using fuzzy set theory and multicriteria analysis: A case study in Central Spain. *Landsc. Ecol. Eng.* **2016**, *12*, 41–60. [CrossRef]
32. McCarthy, A.; Baker, A.; Rotherham, I. Urban-fringe deer management issues—A South Yorkshire case study. *Br. Wildl.* **1996**, *8*, 12–19.
33. Statistics Poland. *Statistical Yearbook of Warsaw*; Zakład Wydawnictw Statystycznych: Warsaw, Poland, 2019.
34. Climate-data.org. Available online: https://pl.climate-data.org/europa/polska/masovian-voivodeship/warszawa-4560/ (accessed on 29 October 2020).
35. Luniak, M.; Kozłowski, P.; Nowicki, W. Magpie *Pica pica* in Warsaw—Abundance, distribution and changes in its population. *Acta Ornithol.* **1997**, *32*, 77–86.
36. Büttner, G.; Kosztra, B.; Maucha, G.; Pataki, R.; Kleeschulte, S.; Hazeu, G.; Vittek, M.; Schröder, C.; Littkopf, A. Copernicus Land Monitoring Service CORINE Land Cover. User Manual. Copernicus Land Monitoring Service (CLMS). 2021. Available online: https://land.copernicus.eu/user-corner/technical-library/clc-product-user-manual?fbclid=IwAR2ZURZU-H1I5J45FSSZfGKSSHlp2AB0-i50pImfEB20ZR2iEUmFHvua7uc (accessed on 25 June 2021).
37. Krop-Benesch, A.; Berger, A.; Hofer, H.; Heurich, M. Long-term measurement of roe deer (*Capreolus capreolus*) (Mammalia: Cervidae) activity using two-axis accelerometers in GPS-collars. *Ital. J. Zool.* **2013**, *80*, 69–81. [CrossRef]
38. Dateandtime.info. Available online: https://dateandtime.info/pl/citysunrisesunset.php?id=756135&month=8&year=2019 (accessed on 29 October 2020).
39. Caravaggi, A.; Gatta, M.; Vallely, M.; Hogg, K.; Freeman, M.; Fadaei, E.; Dick, J.T.A.; Montgomery, W.I.; Reid, N.; Tosh, D.G. Seasonal and predator-prey effects on circadian activity of free-ranging mammals revealed by camera traps. *Peer J.* **2018**, *6*, e5827. [CrossRef]
40. Corsini, M.T.; Lovari, S.; Sonnino, S. Temporal activity patterns of crested porcupines *Hystvix cvistata*. *J. Zool.* **1995**, *236*, 43–54. [CrossRef]

41. lightpollutionmap.info. Available online: https://www.lightpollutionmap.info/#zoom=4.00&lat=45.8720&lon=14.5470&layers=B0FFFFFTFFFFFFFFFF (accessed on 23 June 2021).

42. Bielecka, E.; Jenerowicz, A. Intellectual structure of CORINE Land Cover research applications in Web of Science: A Europe-wide review. *Remote Sens.* **2019**, *11*, 2017. [CrossRef]

43. Hammer, Ø.; Harper, D.A.T.; Ryan, P.D. PAST: Paleontological statistics software package for education and data analysis. *Palaeontol. Electron.* **2001**, *4*, 4.

44. Ciach, M.; Fröhlich, A. Ungulates in the city: Light pollution and open habitats predict the probability of roe deer occurring in an urban environment. *Urban Ecosyst.* **2019**, *22*, 513–523. [CrossRef]

45. Bonnot, N.C.; Couriot, O.; Berger, A.; Cagnacci, F.; Ciuti, S.; de Groeve, J.E.; Gehr, B.; Heurich, M.; Kjellander, P.; Kröschel, M.; et al. Fear of the dark? Contrasting impacts of humans versus lynx on diel activity of roe deer across Europe. *J. Anim. Ecol.* **2020**, *89*, 132–145. [CrossRef]

46. Laundré, J.W.; Hernández, L.; Altendorf, K.B. Wolves, elk, and bison: Reestablishing the "landscape of fear" in Yellowstone National Park, U.S.A. *Can. J. Zool.* **2001**, *79*, 1401–1409. [CrossRef]

47. Laundré, J.W.; Hernández, L.; Ripple, W.J. The landscape of fear: Ecological implications of being afraid. *Open Ecol. J.* **2010**, *3*, 1–7. [CrossRef]

48. Benhaiem, S.; Delon, M.; Lourtet, B.; Cargnelutti, B.; Aulagnier, S.; Hewison, A.J.M.; Morellet, N.; Verheyden, H. Hunting increases vigilance levels in roe deer and modifies feeding site selection. *Anim. Behav.* **2008**, *76*, 611–618. [CrossRef]

49. Bonnot, N.C.; Hewison, A.J.M.; Morellet, N.; Gaillard, J.-M.; Debeffe, L.; Couriot, O.; Cargnelutti, B.; Chaval, Y.; Lourtet, B.; Kjellander, P.; et al. Stick or twist: Roe deer adjust their flight behaviour to the perceived trade-off between risk and reward. *Anim. Behav.* **2017**, *124*, 35–46. [CrossRef]

50. Mysterud, A.; Lian, L.-B.; Hjermann, D.Ø. Scale-dependent trade-offs in foraging by European roe deer (*Capreolus capreolus*) during winter. *Can. J. Zool.* **1999**, *77*, 1486–1493. [CrossRef]

51. Jepsen, J.U.; Topping, C.J. Modelling roe deer (*Capreolus capreolus*) in a gradient of forest fragmentation: Behavioural plasticity and choice of cover. *Can. J. Zool.* **2004**, *82*, 1528–1541. [CrossRef]

52. Lone, K.; Loe, L.E.; Gobakken, T.; Linnell, J.D.C.; Odden, J.; Remmen, J.; Mysterud, A. Living and dying in a multi-predator landscape of fear: Roe deer are squeezed by contrasting pattern of predation risk imposed by lynx and humans. *Oikos* **2014**, *123*, 641–651. [CrossRef]

53. Kays, R.; Crofoot, M.C.; Jetz, W.; Wikelski, M. Terrestrial animal tracking as an eye on life and planet. *Science* **2015**, *348*, aaa2478. [CrossRef] [PubMed]

54. Larson, C.L.; Reed, S.E.; Merenlender, A.M.; Crooks, K.R. Effects of recreation on animals revealed as widespread through a global systematic review. *PLoS ONE* **2016**, *11*, e0167259. [CrossRef]

55. Oberosler, V.; Groff, C.; Iemma, A.; Pedrini, P.; Rovero, F. The influence of human disturbance on occupancy and activity patterns of mammals in the Italian Alps from systematic camera trapping. *Mamm. Biol.* **2017**, *87*, 50–61. [CrossRef]

56. Pagon, N.; Grignolio, S.; Pipia, A.; Bongi, P.; Bertolucci, C.; Apollonio, M. Seasonal variation of activity patterns in roe deer in a temperate forested area. *Chronobiol. Int.* **2013**, *30*, 772–785. [CrossRef] [PubMed]

57. Sönnichsen, L.; Bokje, M.; Marchal, J.; Hofer, H.; Jędrzejewska, B.; Kramer-Schadt, S.; Ortmann, S. Behavioural responses of european Roe deer to temporal variation in predation risk. *Ethology* **2013**, *199*, 233–243. [CrossRef]

58. Ditchkoff, S.S.; Saalfeld, S.T.; Gibson, C.J. Animal behavior in urban ecosystems: Modifications due to human-induced stress. *Urban Ecosyst.* **2006**, *9*, 5–12. [CrossRef]

59. Kotler, B.P.; Brown, J.; Mukherjee, S.; Berger-Tal, O.; Bouskila, A. Moonlight avoidance in gerbils reveals a sophisticated interplay among time allocation, vigilance and state-dependent foraging. *Proc. R. Soc. B* **2010**, *277*, 1469–1474. [CrossRef] [PubMed]

60. Griffin, P.C.; Griffin, S.C.; Waroquiers, C.; Mills, L.S. Mortality by moonlight: Predation risk and the snowshoe hare. *Behav. Ecol.* **2005**, *16*, 938–944. [CrossRef]

61. Haddock, J.K.; Threlfall, C.G.; Law, B.; Hochuli, D.F. Light pollution at the urban forest edge negatively impacts insectivorous bats. *Biol. Conserv.* **2019**, *236*, 17–28. [CrossRef]

62. Prough, L.R.; Golden, C.D. Does moonlight increase predation risk? Meta-analysis reveals divergent responses of nocturnal mammals to lunar cycles. *J. Anim. Ecol.* **2014**, *83*, 504–514. [CrossRef] [PubMed]

63. Kurt, F. *Das Reh in der Kulturlandschaft: Sozialverhalten und Ökologie eines Anpassers Roe Deer in the Cultural Landscape: Social Behavior and Ecology of an Adaptor*; Verlag Paul Parey: Hamburg/Berlin, Germany, 1991.

 *forests*

Article

Transport Work for the Supply of Pine Sawlogs to the Sawmill

Grzegorz Trzciński * and Łukasz Tymendorf

Department of Forest Utilization, Institute of Forest Sciences, Warsaw University of Life Sciences-SGGW, 159 Nowoursynowska St., 02-776 Warsaw, Poland; luktym@gmail.com
* Correspondence: grzegorz_trzcinski@sggw.edu.pl; Tel.: +48-22-593-8128

Received: 23 November 2020; Accepted: 14 December 2020; Published: 16 December 2020

Abstract: The aim of the presented research is to characterize the scale of transport work performed on the supply of large-size pine wood to the sawmill, with indication of factors influencing structure and parameters. Analyzes were carried out for deliveries to a sawmill in northern Poland, which supplies pine sawlogs and long wood assortments. The distance of deliveries on public and forest roads was determined, as well as transport work for each type of road and the total value. The transport work was defined as a multiplication of driven kilometers with the load and the weight of the load in ton kilometers. Data on the transport distance were obtained on the basis of information from the driver, and the parameters of the transported pine sawlogs from the delivery note. Based on the collected data over a period of 12 months, the transport work was determined for selected courses. The total transport work for the 1509 analyzed deliveries was 3,447,486 ton-kilometers (tkm). The average transport work for one course amounted to 2286 tkm and was characterized by a high variability SD = 1207. The minimum value of the transport work was recorded at the level of 83 tkm, and the maximum as much as 7803 tkm. The median of the analyzed deliveries was 2220 tkm, while the first quartile Q1 = 1358, and the third quartile Q3 = 2997. With very similar cargo volumes (m^3) and cargo weight (kg) the transport distance and the total number of deliveries have a significant effect on the transport work performed with the transport of timber. Purchase of wood in seven forest districts located up to 50 km from the sawmill accounts for 30.1% of the analyzed deliveries (1509), resulting in only transport work at the level of 476,104 tkm, which is only 13.8% of the total transport work of all deliveries.

Keywords: round wood transport; wood supply chain; transport work optimization; sawlogs deliveries; sawlogs sourcing

1. Introduction

In Poland, there is one dominant round wood supplier (44.7 million m^3 in 2018 [1]): the State Forests National Forest Holding. The wood offered throughout the country is sold by 430 State Forest Districts, and the buyers are the highly dispersed and fragmented sawmill industry [2,3]. In such a situation, ensuring a properly functioning, direct wood supply chain [4,5], which is of great importance in the costs of timber harvesting [6,7], enables optimization [8,9].

In the literature on the improvement of the efficiency of wood transport to sawmills, the following issues can be distinguished:

- Cooperation of companies (the seller, sawmills and the carrier) contributing to the reduction of wood transport costs by 4–20% [10–15]. The cooperation of the companies may also contribute to reducing the demand for timber trucks [16] and the reduction of transport work [17].
- Planning of deliveries to a sawmill in the context of reducing drivers' working time and improving transport efficiency [18–20].

- Analysis of the deliveries related to timber transport and the possibility of their reduction [19–23].
- Maximizing use and increasing vehicle load capacity to improve the efficiency of timber transport [24–28].
- Transport of timber with increased load weight and its impact on the environment and infrastructure [29–33].

One of the latest analyses carried out in the transport of wood is the reduction of exhaust emissions. The latest results of Finnish research show unequivocally that the use of vehicles with a maximum permissible weight of 76 Mg in 2017 allowed to reduce the distance traveled by 4% and 0.1 Mt in reduction of CO_2 emissions in road transport [24]. Other Finnish simulations show that 76-ton trucks had a 12% lower productivity, 4% higher fuel consumption and a 6% higher transport cost compared to 84-ton vehicles [25]. Increasing the permissible weight of the transport vehicle in Finland has contributed to the reduction of CO_2 and nitrogen oxides NOx emissions and economic benefits [34]. The recently published results of research in Poland concerning the transport of large-size pine wood indicate that reducing the volume of load from 30 m^3 to 25 m^3 results in a significant, 17% increase in the quantity of deliveries, which directly lead to higher fuel consumption, increased transport costs and CO_2 emissions [35].

Aim and Scope of the Research

The aim of the presented research was to analyze the transport work performed during the delivery of large-size pine round wood to the sawmill as well as to determine the size and structure of the transport work. It was assumed that the main factors influencing the volume of transport work is the location of the single wood load (forest district, sub-forest district, stand) that determines the transport distance and the weight of the transported round wood. It was also assumed that the number of transports from individual forest districts had a significant impact on the total transport work performed on the deliveries of roundwood to the sawmill, and it may constitute the basis for optimizing transport work.

The research was related to transport work analyzes depending on the period of their realization and the distance from the State Forest District. The data of the location and related deliveries on public and forest roads were analyzed.

2. Material and Methods

The transport work in ton-kilometers was defined as the multiplication of the kilometers traveled on forest and public roads with the load and the weight of the cargo for each route and the total value for all analyzed transports. For each, State Forest Districts took into account the number of deliveries, the average weight of a load and the transport distance (minimum, average and maximum). The distance with the load was determined on the basis of the delivery note and the weight of the load from weighing the vehicles.

Large-size wood is wood with a thin end diameter of 14 cm (excluding bark), calculated in single pieces. In terms of quality and size, large-size wood is divided into four classes—A, B, C, D—and into two sub-classes—general purpose wood and special purpose wood. The large-size general purpose wood is comparable to the assortment defined as sawmill wood. Medium-size wood is wood with a minimum diameter of 5 cm and more (excluding bark), with a thick end diameter of up to 24 cm, calculated in single pieces, in pieces as groups and in stakes [36,37].

Relevant analyzes were carried out for the supply of pine sawlogs in the period from 1 April 2016 to 29 March 2017, to a sawmill in northern Poland. The transport was carried out by external companies acting on behalf of the plant. The delivery date, from the recipient's documents (sawmill) and the delivery note issued by the State Forest, allowed for the analysis for each month.

Each truck was weighed as it entered and exited the sawmill. The weight of the load was determined on the basis of weighing the truck with sawlogs and after unloading (tare) in the plant.

The place of harvesting the wood is specified on the delivery note by the State Forest, by providing the unique forest address from the information system of the State Forests (SILP).

Kilometers driven on forest and public roads with round wood were based on information from the driver when accepting the delivery. Having accurate location (from the delivery note) it was possible to verify (control) authenticity of the information and eliminate incorrect data provided by the driver. Data on the average fuel consumption in liters per 100 km for individual delivery were obtained from truck drivers. On this basis, the average combustion in L km^{-1} was calculated for one course (delivery) and for one cubic meter of transported sawlogs (total combustion for the course divided by the volume of transported wood in L·m^{-3}).

The results were analyzed statistically using the STATISTICA 12 package. The Kruskal–Wallis test was used to determine the significance of the differences. Additionally, a multiple comparison test of mean rank (Dunn's test) was performed. Analyzes were performed at the significance level of 0.05.

3. Results

In the analyzed period 280,380 m^3 of pine sawlogs was transported to the plant, with 9797 deliveries of wood from 54 forest districts. The analysis covered 1509 (13.97%) of randomly selected timber transports carried out over a period of twelve months from 40 forest districts, where a load per truck was obtained from one place, single stand. In the analyzed 1509 transports, 44,336 m^3 of round wood was delivered to the sawmill.

3.1. Characteristic of the State Forest Districts

The characteristics of Forest Districts: average distances from the sawmill, load weight and number of deliveries presented in Table 1. Most of the analyzed 129 sawlogs transports to the sawmill were carried out from the Korpele Forest District, with average distance of 34.3 km. Mileage range with load from 25 to 57 km. In the analyzed deliveries, 30.1% (454 courses) were carried out from the forest districts: Wielbark, Nidzica, Szczytno, Jedwabno, Korpele, Przasnysz and Parciaki. They are located closest to the plant, in 50 km radius (average transport distance).

Table 1. Summary of the number of deliveries, average load weight and transport distance on a public road from forest districts to a sawmill according to the average distance.

State Forest District	Number of Deliveries	Average Load Weight (t)	Deliveries with Cargo on a Public Road (km)			
			Average	SD	Minimum	Maximum
Wielbark	49	30.61	8.2	4.6	3	23
Nidzica	25	29.87	19.1	11.4	24	70
Szczytno	93	30.38	20.7	5.5	7	38
Jedwabno	58	32.00	29.6	11.4	7	47
Korpele	129	29.28	34.3	7.1	25	57
Przasnysz	64	30.37	41.5	7.6	26	64
Parciaki	36	31.81	48.5	12.4	30	81
Myszyniec	10	31.25	54.5	14.2	36	76
Spychowo	46	30.60	57.1	7.8	28	68
Olsztyn	60	30.85	57.3	11.7	37	90
Nowe Ramuki	76	29.02	61.9	10.5	40	91
Strzałowo	72	29.35	61.9	9.4	40	84
Mrągowo	14	30.87	66.4	6.8	58	84
Wipsowo	28	28.44	74.0	12.3	48	105
Olsztynek	27	30.52	75.8	9.6	64	99
Maskulińskie	61	30.67	79.4	11.2	60	120
Nowogród	79	30.23	79.7	7.1	64	98
Dwukoły	4	38.81	81.9	4.6	75	84
Jagiełek	40	29.29	82.5	8.5	63	112

Table 1. *Cont.*

State Forest District	Number of Deliveries	Average Load Weight (t)	Deliveries with Cargo on a Public Road (km)			
			Average	SD	Minimum	Maximum
Ostrołęka	41	31.89	85.9	20.9	58	132
Kudypy	48	29.85	86.0	9.9	74	123
Pułtusk	49	30.12	90.8	15.8	75	136
Pisz	119	29.64	91.3	13.5	57	116
Stare Jabłonki	21	29.96	106.6	11.2	96	138
Wichrowo	32	31.36	109.8	8.5	90	127
Miłomłyn	7	29.75	111.0	4.1	106	116
Bartoszyce	3	37.78	114.7	15.5	99	130
Drygały	84	31.79	114.8	14.5	60	160
Giżycko	71	30.18	120.9	10.9	95	150
Płaska	18	30.30	121.1	9.3	205	238
Srokowo	3	33.25	126.0	29.5	93	150
Ostrów Mazowiecka	7	33.38	132.6	10.1	118	145
Iława	21	31.22	134.3	10.4	114	160
Wyszków	2	31.15	135.0	7.1	130	140
Orneta	2	31.68	138.0	4.2	135	141
Susz	6	29.60	155.5	15.9	127	167
Łochów	4	32.53	166.3	2.5	165	170
Szczebra	1	28.30	215.0	-	215	215
Żednia	2	30.80	225.0	7.1	220	238

In the first years of operation of the plant, the source of wood supply was all forest districts in north-eastern Poland. Too much diversification of suppliers required the involvement of a significant number of transport vehicles (companies). As the sawmill developed, the pine sawlogs purchasing area was significant reduced, resulting in an average transport distance 70 km on public road.

The reduction of the purchasing radius resulted in a smaller demand for the number of trucks, and consequently reduced the number of transport companies cooperating with the plant. The cooperation was based on the principle of assigning one transport company to one State Forest Districts. It significantly improved the quality of provided transport services, execution of delivery schedules, information flow and cost reduction, and increased control throughout the wood supply chain.

3.2. Characteristics Sawlogs Deliveries and the Weight of the Load

The realization of the analyzed deliveries to the sawmill was associated with the total load deliveries (on public and forest roads) at the level of 113,728 km. The average distance of deliveries with cargo was 75.4 km and was characterized by high variability SD = 39.1 (Tables 1 and 2) in the range from 3.0 km to 274.5 km (Table 2).

Table 2. Characteristics of analyzed parameters in wood transportation to sawmill.

Measure	Mean	SD	Min	Max	Q1	Median	Q3
Mass of load (t)	30.32	2.20	21.75	38.91	28.80	30.20	31.75
Total driving with load (km)	75.4	39.1	3.0	274.5	45.2	74.0	98.0
Transport work (tkm)	2282	1207	83	7803	1358	2220	2997

Notes: SD. standard deviation; Q1. first quartile; Q3. third quartile.

The greatest number of deliveries (on a public and a forest road) was recorded in the range of 70 ÷ 80 km-186 and 80 ÷ 90 km-182 transports. In 24 cases, timber was transported over a distance of more than 200 km. Most deliveries were made within 100 km from the plant, 75.35% of transports being within this radius (Figure 1).

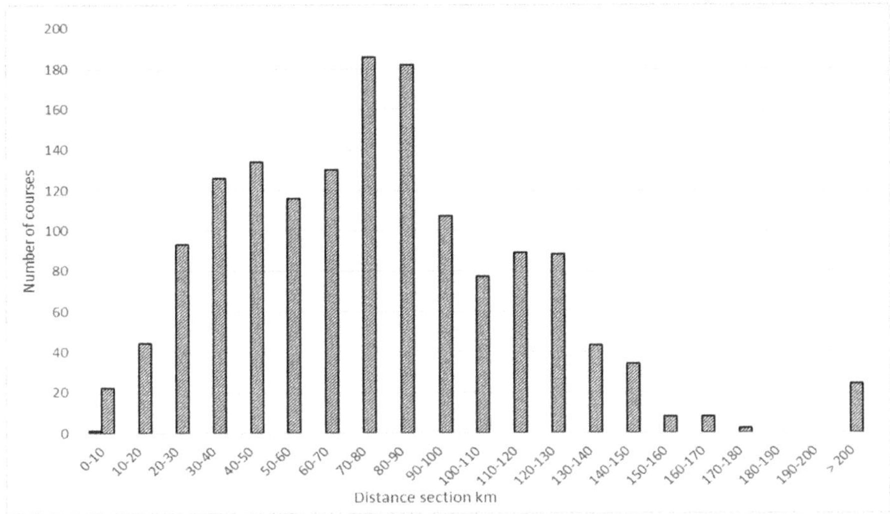

Figure 1. Number of deliveries in 10 km sections with load.

The average load weight for each delivery was 30.32 t and very similar between forest districts (SD = 2.20 and median 30.20), although the Kruskal–Wallis test showed statistically significant differences. Detailed analysis using the multiple mean rank comparison test (Dunn's test) showed that the masses of individual wood loads in transports from 18 forest districts did not differ from each other (Figure 2).

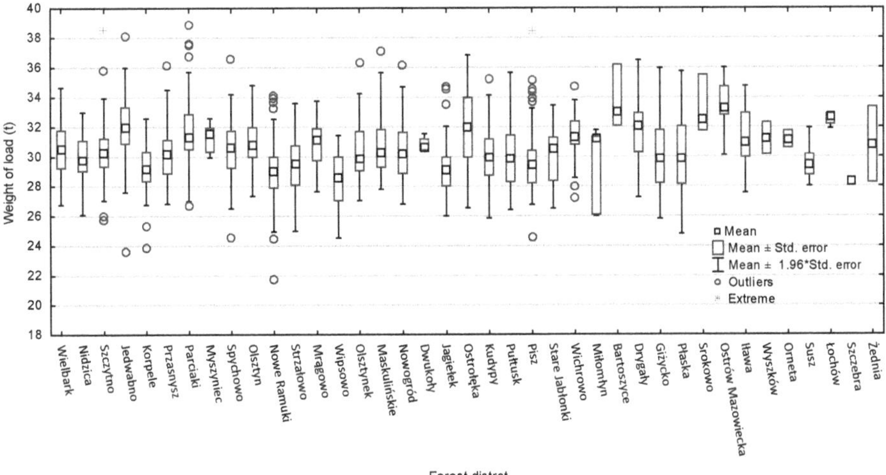

Figure 2. Weight of deliveries from individual forest districts (forest districts in the order of distance from the sawmill).

3.3. Fuel Consumption in Sawlogs Deliveries

The roundwood to the sawmill was most often delivered by Man, Scania, Volvo and Mercedes trucks with a motor power from 410 to 620 HP and a capacity of 12.0 ÷ 16.3 L. Drivers reported fuel consumption in the range from 38.0 to 87.0 L 100 km^{-1}. Calculation of fuel consumption per kilometer and summary calculations for each sawlogs delivery were made. Transport distance with load was used for all calculations. Having also registered the volume of transported wood (m^3), the fuel consumption index for the transport of 1 m^3 of wood was calculated. Average fuel consumption was at the level of 0.561 L km^{-1} with small differences in results, SD = 0.055 (Table 3). For the individual deliveries of wood to the sawmill, fuel consumption from 4.80 L to 225.94 L was recorded with an average of 77.11 L. The calculated fuel consumption index for 1 m^3 transport was on average 2.62 L·m^{-3} with the result range from 0.16 to 7.63 L·m^{-3} (Table 3).

Table 3. Fuel consumption characteristics for sawlogs delivered to a sawmill based on transport distance with load.

Measure	Mean	SD	Min	Max	Q1	Median	Q3
Fuel consumption (L·km^{-1})	0.561	0.055	0.380	0.870	0.530	0.570	0.598
Fuel consumption per 1 m^3 of delivered wood (L·m^{-3})	2.628	1.286	0.16	7.630	1.600	2.568	3.406
Fuel consumption per delivery (L)	77.116	38.099	4.800	125.940	46.170	76.160	99.456

Notes: SD. standard deviation; Q1. first quartile; Q3. third quartile.

3.4. Transport Work

The total transport work for the 1509 deliveries was 3,447,486 ton-kilometers (tkm). The average transport performance of one course amounted to 2286 tkm and was characterized by a high variability of SD = 1207. The minimum value of transport work was recorded at the level of 83 tkm, and the maximum as much as 7803 tkm. The median of the analyzed deliveries was 2220 tkm, while the first quartile Q1 = 1358 and the third quartile Q3 = 2997 (Table 2). Transport work for a single course is on a comparable level in all months, but the analysis using the Kruskal–Wallis test showed statistically significant differences ($p = 0.0000$). The comparison of the transport work for the trips carried out in individual months with the Dunn test shows statistically significant differences only when the results are compared in the months of XI to I; III; IV, VI and X as well as III from I, IV and X and the last different pair are IV from VIII, as shown in Figure 3. Statistical analyzes did not include trips in July and December due to insufficient sample. It was caused by planned breaks in the plant related to maintenance period and suspension of sawlogs deliveries.

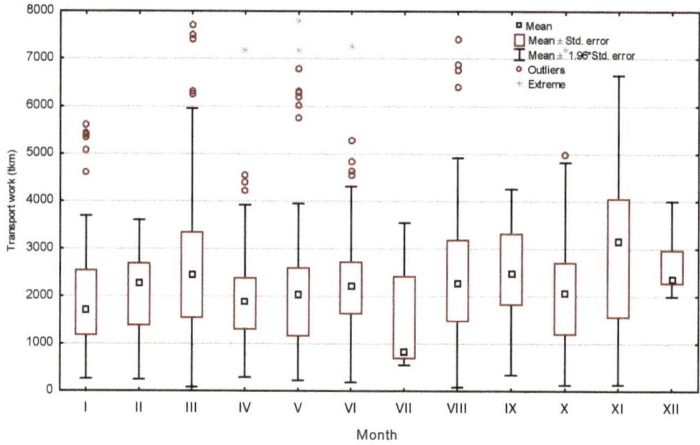

Figure 3. Characteristics of transport work for individual deliveries depending on the delivery month.

The analysis of transport work for a single course according to forest districts showed very large differences (Kruskal-Wallis test $p = 0.0000$), as shown in Figure 4. With very similar cargo volumes (m^3) and cargo weight (kg), which was shown in previous analyzes and in Table 1, the transport distance has a significant effect on the transport work performed with the transport of timber.

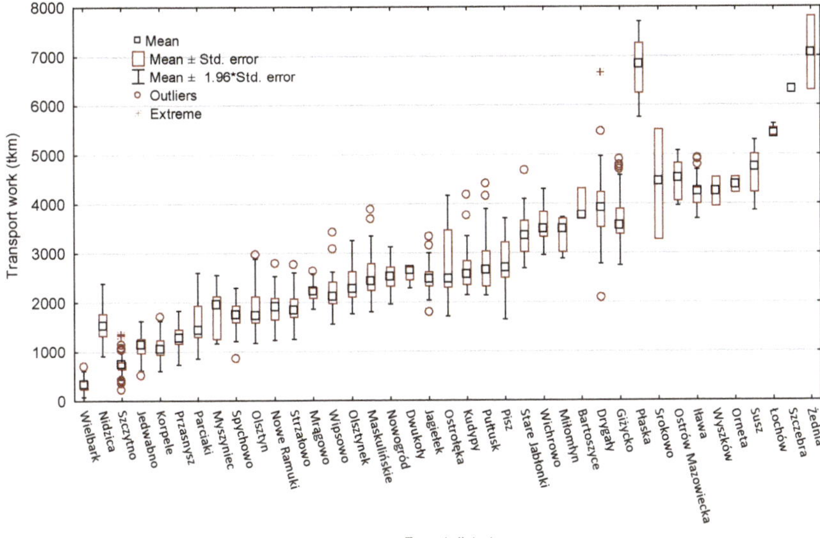

Figure 4. Characteristics of the transport work of individual deliveries from forest districts according to the transport distance.

The total transport work performed for the deliveries of wood from each forest districts depends on the distance from the sawmill and the number of deliveries made. The results presented in Figure 5 clearly show that the supply of pine sawlogs in the nearest forest districts requires much less transport work, for example in the Korpele Forest District. Purchases of round wood in remote forest districts, even with a smaller number of deliveries, are characterized by a high amount of transport work, as in the case of the Drygały or Giżycko Forest Districts. Purchase of wood in seven forest districts (Wielbark-Parciaki) up to 50 km from the sawmill (Table 1) accounts for 30.1% of the analyzed deliveries (1509), resulting in only transport work at the level of 476,104 tkm, which is only 13.8% of the total transport work of all deliveries.

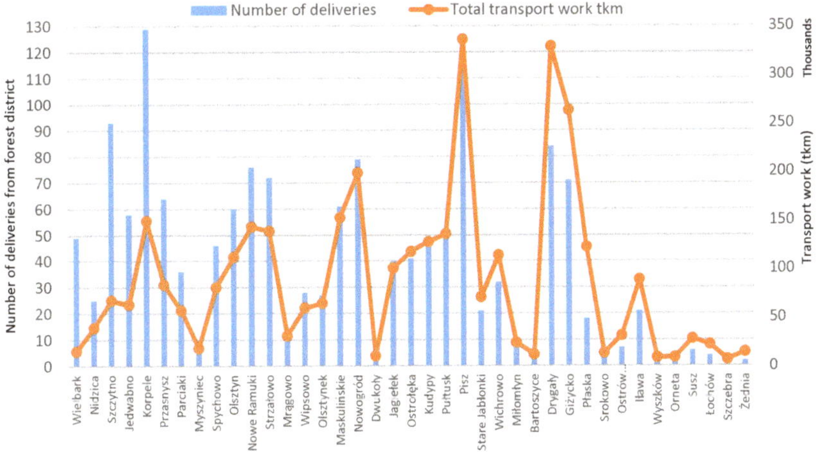

Figure 5. The completed transport work for the deliveries of timber from individual forest districts and the number of deliveries (forest districts in the order of distance from the sawmill).

4. Discussion

By making theoretical optimization by doubling the number of deliveries, made only from the closest 14 Forest Districts, with an average distance from 8.2 km to 74.0 km (in the range of min = 3.0 km to max = 105 km, Table 1), it is possible to reduce the generated transport work. Delivering the same amount of wood (at the level of 44,340 m^3) would require transport work at the level of 2,103,819 tkm, which is 39% less than in the analyzed case. By increasing the number of deliveries from the nearest forest districts threefold with an average transport distance of 8.2–57.1 km (with a range of 3.0–81.0 km), it would be possible to reduce the transport work to the level of 1,694,279 tkm, thus 51% less than in the analyzed case. At the same time, the delivery of the same wood quantity would require cooperation with only the nine closest forest districts. The sale offer of pine sawlogs by State Forest in the analyzed area is on much higher level then analyze sawmill capacity. Theoretically, this situation allows to reduce sawmill purchase circle to maximum 50 km. Harvesting and sales offer of large-size roundwood in seven forest districts (from Wielbark to Parciaki, Table 1) within 50 km from the sawmill is at the average annual level of 43,000 m^3 per State Forest District. The sales offer from 30,000 m^3 (Parciaki) up to 67,000 m^3 (Jedwabno) depends on the forest structure and is connected with 10 years Forest Management Plan. Within 50 km from the sawmill, the annual sales potential of large-size wood in seven State Forest Districts is 303,000 m^3 of wood per year. This is more than the annual demand of the analyzed sawmill. Unfortunately, the wood sales rules in Poland limits this solution. The sale of roundwood is made through an electronic restricted tender, where the State Forest sets and establishes rules for evaluating the submitted offers [38]. The presented optimization could take place in a situation of lifting the current restrictions on wood purchase. In a situation where all roundwood planned for sale by State Forest would be sold through online auctions without any restrictions, it would be possible to purchase enough quantity of sawlogs through a sawmill in the nearest State Forest Districts. On the other hand, such a solution would also have a negative impact on the possibility of selling wood by forest districts located further away from the sawmill, in the areas where there are no other significant buyers.

Based on the analyzes, the average fuel consumption was on the level 0.56 L·km^{-1}. Austrian studies show higher values at the level of 0.77 L·km^{-1} [39], which may result from a different type of assortment, road condition and different topography. With the average transport distance from the forest to the sawmill at the level of 75.4 km, with an average load volume of 29.34 m^3 [40], the amount of fuel needed to transport 1 m^3 of pine sawlogs in the conducted research is 1.44 L·m^{-3}. These results are consistent with other studies, which show the value of 0.73 L·m^{-3} at a transport distance of 31.4 km [41]. A reliable indicator for analyses and comparisons of variable loads (volume and mass) can be the mass of 1 m^3 of wood in the load, which amounted to 1033.87 kg·m^{-3}, with similar results in the population at SD = 57.87, and a median of 1031.61 kg·m^{-3} [40]. Fuel consumption per 1 m^3 of wood is primarily determined by the transport distance and the size of the load [42].

Rijkswaterstaat [43] presents calculations (estimation) that, with a reduction in transport distance of approximately 20 million kilometers per year, CO_2 emissions decrease by 0.016 Mt per year (per 1 km reduction of 0.0008 t). Assuming the estimates presented by Rijkswaterstaat [43], the reduction of the analyzed 113,728 km of deliveries by 39% (44,354 km) and 51% (58,000 km) in the first and the second analyzed delivery variant, respectively, allows to estimate the reduction of CO_2 emissions at the level of 35.5 t and 46.4 t. However, in order to fully assess the environmental impact of transport, estimation methods should be used for specific route sections, given that the amount of pollutant emissions is strongly influenced by speed, load and gradient index [44].

The calculations were presented for 1509 transports (14%) of all completed 9797 deliveries to the sawmill, so the total reduction in transport work, mainly kilometers traveled, and thus the reduction of CO_2 emissions may be much higher.

5. Conclusions

Most of the deliveries of pine sawlogs to the sawmill, 180 transports, were related to the deliveries from distance of 70 ÷ 80 and 80 ÷ 90 km. 30.1% of deliveries to the sawmill (454 courses) were made from distances of up to 50 km (average transport distance).

With very similar loads of round wood, the transport distance has a significant impact on transport work, which does not depend on the delivery date.

It is possible to optimized (reduce) the round wood deliveries, and thus transport work, as a result of a change in the structure of sawlogs sourcing by the sawmill. The annual potential of the roundwood sold in the closest State Forest Districts is higher than the sawmill capacity, but the current Wood Sales Rules limit this possibility.

Author Contributions: Conceptualization, G.T. and Ł.T.; methodology, G.T and Ł.T.; formal analysis, G.T. and Ł.T.; investigation, G.T. and Ł.T.; data curation, G.T. and Ł.T; writing—original draft preparation, G.T. and Ł.T.; writing—review and editing, Ł.T.; visualization, Ł.T.; supervision, G.T. All authors have read and agreed to the published version of the manuscript.

Funding: This research received no external funding.

Conflicts of Interest: The authors declare no conflict of interest.

References

1. GUS Forestry 2019. Available online: https://stat.gov.pl/en/topics/statistical-yearbooks/statistical-yearbooks/statistical-yearbook-of-forestry-2019,12,2.html (accessed on 2 December 2020).

2. Wnorowska, M. Raport Forestor Tartacznictwo. Sawmilling Forestor Raport. Przemysł Drzewny. *Res. Dev.* **2019**, *4*, 15–35. Available online: https://forestor.przemysldrzewny.eu/przemysl-drzewny-nr-4-2019/?v=9b7d173b068d (accessed on 20 March 2020). (In Polish)

3. Sieniawski, W. Waloryzacja Dostaw Drewna do Wybranych Segmentów Przemysłu Drzewnego. Valorization of Wood Supply to Selected Segments of Timber Industry. Ph.D. Thesis, WULS-SGGW, Warsaw, Poland, June 2012. (In Polish)

4. Cooper, M.C.; Lambert, D.M.; Pagh, J.D. Supply Chain Management: More Than a New Name for Logistics. *Int. J. Logist. Manag.* **1997**, *8*, 1–14. [CrossRef]

5. Wailgum, T. Supply Chain Management Definition and Solutions, CIO—Business Technology Leadrship. 2008. Available online: http://www.cio.com/article/40940/Supply_Chain_Management_Definition_and_Solutions (accessed on 25 March 2020).

6. Hirsch, P. Minimizing Empty Truck Loads in Round Timber Transport with Tabu Search Strategies. *Int. J. Inf. Syst. Supply Chain Manag.* **2011**, *4*, 15–41. [CrossRef]

7. Shaffer, R.M.; Stuart, W.B. A checklist for efficient log trucking. *Va. Coop. Ext.* **2005**, *420-094*, 1–5. Available online: http://hdl.handle.net/10919/54904 (accessed on 15 August 2020).

8. Greulich, F. Transportation networks in forest harvesting: Early development of the theory. In Proceedings of the International Seminar on New Roles of Plantation Forestry Requiring Appropriate Tending and Harvesting Operations, Tokyo, Japan, 29 September–5 October 2002; Available online: http://faculty.washington.edu/greulich/Documents/IUFRO2002Paper.pdf (accessed on 25 January 2020).

9. McDonald, T.P.; Haridass, K.; Valenzuela, J. Mileage savings from optimization of coordinated trucking. In Proceedings of the 2010 COFE: 33rd Annual Meeting of the Council on Forest Engineering, Auburn, Alabama, 6–9 June 2010; Available online: https://cofe.org/pdfs/COFE_2010.pdf (accessed on 25 August 2020).

10. Klocek, A. Optymalizacja przewozu drewna z miejsca jego załadunku do miejsca rozładunku. Optimizing the transport of timber from the place of its loading to the place of unloading. *Przemysł Drzewny. Res. Dev.* **2016**, *4*, 68–74. (In Polish)

11. Löfroth, C.; Svenson, G.; Rådström, L. Testing of revolutionary round wood haulage rig in Sweden. In *Raport 12/2010*; Norwegian Forest and Landscape Institute: Honne, Norway, 2010; pp. 47–48. Available online: https://nibio.brage.unit.no/nibio-xmlui/bitstream/handle/11250/2469357/SoL-Rapport-2010-12.pdf?sequence=2&isAllowed=y (accessed on 10 February 2020).

12. Frisk, M.; Göthe-Lundgren, M.; Jörnsten, K.; Rönnqvist, M. Cost allocation in collaborative forest transportation. *Eur. J. Oper. Res.* **2010**, *205*, 448–458. [CrossRef]

13. Palander, T.; Väätäinen, J. Impacts of interenterprise collaboration and backhauling on wood procurement in Finland. *Scand. J. For. Res.* **2005**, *20*, 177–183. [CrossRef]

14. Hirsch, P.; Gronalt, M. The timber transport order smoothing problem as part of the three-stage planning approach for round timber transport. *J. Appl. Oper. Res.* **2013**, *5*, 70–81.

15. Sieniawski, W.; Porter, B. Zrównoważony łańcuch dostaw drewna na przykładzie wybranego zakładu. Sustainable Wood Supply Chain based on the example of a chosen sawmill. *Studia i Materiały CEPL w Rogowie* **2012**, *14*, 254–264. Available online: http://cepl.sggw.pl/sim/pdf/sim32_pdf/sim32_Sieniawski_Porter.pdf (accessed on 20 October 2020). (In Polish).

16. Haridass, K.; Valenzuela, J.; Yucekaya, A.D.; McDonald, T. Scheduling a log transport system using simulated annealing. *Inf. Sci.* **2014**, *264*, 302–316. [CrossRef]

17. Kłapeć, B.; Tracz, W.; Janeczko, K. Optimization of the transportation of wood purchased in the State Forests units. *Sylwan* **2017**, *161*, 842–850. [CrossRef]

18. Gerasimov, Y.; Sokolov, A.; Syunev, V. Optimization of industrial and fuel wood supply chain associated with cut-to-length harvesting. *Syst. Methods Technol.* **2011**, *11*, 118–124.

19. Gerasimov, Y.; Sokolov, A.; Karjalainen, T. GIS-Based Decision-Support Program for Planning and Analyzing Short-Wood Transport in Russia. *Croat. J. For. Eng.* **2008**, *29*, 163–175.

20. Tymendorf, Ł.; Trzciński, G. Duration of stages of delivery of large–sized Scots pine wood to the sawmill. *Sylwan* **2020**, *164*, 549–559. [CrossRef]

21. Tymendorf, Ł.; Trzciński, G. Driving on forest and public roads in deliveries of large–size Scots pine wood to sawmill. *Sylwan* **2020**, *164*, 651–662. [CrossRef]

22. Czyżyk, K.; Porter, B. Geografia dostaw drewna do wybranego zakładu. Geography of wood supplies to the selected factory. *Technika Rolnicza Ogrodnicza Leśna* **2014**, *3/2014*, 2–5. (In Polish)

23. Sieniawski, W.; Trzciński, G. Analysis of large-size and medium-size wood supply. In *Raport 12/2010*; Norwegian Forest and Landscape Institute: Honne, Norway, 2010; pp. 56–57. Available online: https://nibio.brage.unit.no/nibio-xmlui/bitstream/handle/11250/2469357/SoL-Rapport-2010-12.pdf?sequence=2&isAllowed=y (accessed on 28 November 2020).

24. Liimatainen, H.; Pöllänen, M.; Nykänen, L. Impacts of increasing maximum truck weight—case Finland. *Eur. Transp. Res. Rev.* **2020**, *12*, 14. [CrossRef]

25. Väätäinen, K.; Laitila, J.; Anttila, P.; Kilpeläinen, A.; Asikainen, A. The influence of gross vehicle weight (GVW) and transport distance on timber trucking performance indicators—Discrete event simulation case study in Central Finland. *Int. J. For. Eng.* **2020**, *31*, 156–170. [CrossRef]

26. Palander, T.; Kärhä, K. Potential traffic levels after increasing the maximum vehicle weight in environmentally efficient transportation system: The Case of Finland. *J. Sustain. Dev. Energy Water Environ. Syst.* **2017**, *5*, 417–429. [CrossRef]

27. Lukason, O.; Ukrainski, K.; Varblane, U. Economic benefit of maximum truck weight regulation change for Estonian forest sektor. Veokite täismassi regulatsiooni muutmise majanduslikud mõjud eesti metsatööstuse sektorile. *Est. Discuss. Econ. Policy* **2011**, *19*. [CrossRef]

28. Trzciński, G.; Tymendorf, Ł. Timber deliveries after introduction of the normative calculators of wood density to determine the load weight. *Sylwan* **2017**, *161*, 451–459. [CrossRef]

29. McKinnon, A.C. The Economic and Environmental Benefits of Increasing Maximum Truck Weight: The British Experience. *Transp. Res. Part Transp. Environ.* **2005**, *10*, 77–95. [CrossRef]

30. Knight, I.; Newton, W.; McKinnon, A.; Palmer, A.; Barlow, T.; McCrae, I.; Dodd, M.; Couper, G.; Davies, H.; Daly, A.; et al. Longer and/or Longer and Heavier Goods Vehicles (LHVs)—A Study of the Likely Effects if Permitted in the UK: Final Report. 2008. Available online: https://www.nomegatrucks.eu/deu/service/download/trl-study.pdf (accessed on 10 February 2020).

31. Rodrigues, V.S.; Piecyk, M.; Mason, R.; Boenders, T. The longer and heavier vehicle debate: A review of empirical evidence from Germany. *Transp. Res. Part D Transp. Environ.* **2015**, *40*, 114–131. [CrossRef]

32. Pålsson, H.; Hiselius, L.W.; Wandel, S.; Khan, J.; Adell, E. Longer and heavier road freight vehicles in Sweden. *Int. J. Phys. Distrib. Logist. Manag.* **2017**, *47*, 603–622. [CrossRef]

33. Kolisoja, P.; Kalliainen, A.; Haakana, V.; Information, R. Effect of Tire Configuration on the Performance of a Low-Volume Road Exposed to Heavy Axle Loads. *Transp. Res. Rec. J. Transp. Res. Board* **2015**, *2474*, 166–173. [CrossRef]

34. Liimatainen, H.; Nykänen, L. *Impacts of Increasing Maximum Truck Weight—Case Finland*; Transport Research Centre Verne, Tampere University of Technology: Tampere, Finland, 2017; Available online: http://www.tut. fi/verne/aineisto/LiimatainenNyk%C3%A4nen.pdf (accessed on 10 January 2018).

35. Mydlarz, K.; Wieruszewski, M. Problems of Sustainable Transport of Large-Sized Roundwood. *Sustainability* **2020**, *12*, 2038. [CrossRef]

36. PN-93/D-02002. Round wood. In *Classification, Terminology and Symbols*; Polish Standardization Committee: Warsaw, Poland, 2002; p. 4. (In Polish)

37. *Zarządzenie nr 51 Dyrektora Generalnego Lasów Państwowych z dnia 30.09.2019 r. Regulation No. 51 of the General Director of the State Forests of 30.09.2019*; General Directorate of the State Forests: Warsaw, Poland, 2019. Available online: http://drewno.zilp.lasy.gov.pl/drewno/Normy/1._podzia_terminologia_i_symbole_-_ujednolicono_wg_ zarz_54-2020.pdf (accessed on 20 October 2020). (In Polish)

38. Decyzja 160 Dyrektora Generalnego LP z 12 listopada 2019 w sprawie kryteriów, parametrów i sposobu wartościowania ofert zakupu oraz regulaminów sprzedaży w Portalu Leśno-Drzewnym i aplikacji internetowej e-drewno. Decision 160 of the Director General of the State Forests of November 12, 2019 on the Criteria, Parameters and Method of Evaluating Purchase Offers As Well As Sales Regulations in the Forest-Wood Portal and the E-Timber Web Application. Available online: http://drewno.zilp.lasy.gov.pl/drewno/decyzja_nr_160_dglp_z_12.11.2019_r._w_sprawie_kryteriow_ param_i_sposobu_wart._ofert_zakupu_oraz_regulaminow_sprzedazy_w_pld_-_e-drewno.pdf (accessed on 20 October 2020). (In Polish)

39. Holzleitner, F.; Kanzian, C.; Stampfer, K. Analyzing time and fuel consumption in road transport of round wood with an onboard fleet manager. *Eur. J. For. Res.* **2010**, *130*, 293–301. [CrossRef]

40. Tymendorf, Ł.; Trzciński, G. Multi-Factorial Load Analysis of Pine Sawlogs in Transport to Sawmill. *Forests* **2020**, *11*, 366. [CrossRef]

41. Lijewski, P.; Merkisz, J.; Fuć, P.; Ziółkowski, A.; Rymaniak, Ł.; Kusiak, W. Fuel consumption and exhaust emissions in the process of mechanized timber extraction and transport. *Eur. J. For. Res.* **2017**, *136*, 153–160. [CrossRef]

42. Klvac, R.; Kolařík, J.; Volná, M.; Drápela, K. Fuel Consumption in Timber Haulage. *Croat. J. For. Eng.* **2013**, *34*, 229–240.

43. Rijkswaterstaat. Longer and Heavier Vehicles in Practice. Economic, Logistical and Social Effects. 2011. Available online: http://www.modularsystem.eu/download/facts_and_figures/3839282_longer_and_heavier_ vehicles_in_prakt.pdf. (accessed on 10 February 2020).

44. Osorio-Tejada, J.L.; Llera-Sastresa, E.; Hashim, A.H. Well-to-Wheels Approach for the Environmental Impact Assessment of Road Freight Services. *Sustainability* **2018**, *10*, 4487. [CrossRef]

Publisher's Note: MDPI stays neutral with regard to jurisdictional claims in published maps and institutional affiliations.

 *forests*

Communication

Feasibility of a Harvesting System for Small-Diameter Trees as Unutilized Forest Biomass in Japan

Takuyuki Yoshioka *, Tomoki Tomioka and Toshio Nitami

Graduate School of Agricultural and Life Sciences, The University of Tokyo, Tokyo 113-8657, Japan; tomioka@fr.a.u-tokyo.ac.jp (T.T.); nitami@fr.a.u-tokyo.ac.jp (T.N.)
* Correspondence: tyoshioka@fr.a.u-tokyo.ac.jp; Tel.: +81-3-5841-5215

Abstract: In order to secure a supply of forest biomass, as well as promote further utilization following the completion of the Feed-in-Tariff Scheme for Renewable Energy (FIT), small-diameter trees such as cleanings from young planted forests and broad-leaved trees from coppice forests are prospective resources in Japan. The goal of this study was to discuss effective methods for harvesting the small-diameter trees that are unutilized forest biomass in Japan. This study assumed a simplified model forest and conducted experiments and time studies of the harvesting of small-diameter trees with a truck-mounted multi-tree felling head. As a result, the machine used in the experiment could fell a maximum of six trees inward in a row from a forest road. However, the harvesting cost (felling, accumulating and chipping) was cheapest when the machine felled five trees inward in a row. Lengthening the maximum reach of a felling head to fell trees deeper inward in a row appeared effective in increasing the number of harvested trees. From the perspective of minimizing the harvesting cost, however, there were upper limits to the number of trees felled inward as well as to the maximum reach of a felling head. The results of a sensitivity analysis suggested the following machine improvements could be considered in future policy: increasing the moving velocity of a felling head and the maximum number of trees that can be held at a time are effective if it is possible to lengthen the maximum reach of a felling head. Meanwhile, shortening the machine's moving time among operation points is also effective if the maximum reach of a felling head cannot be lengthened.

Keywords: small-diameter tree; forest biomass; multi-tree felling head; time study; harvesting cost

Citation: Yoshioka, T.; Tomioka, T.; Nitami, T. Feasibility of a Harvesting System for Small-Diameter Trees as Unutilized Forest Biomass in Japan. *Forests* **2021**, *12*, 74. https://doi.org/10.3390/f12010074

Received: 17 December 2020
Accepted: 8 January 2021
Published: 10 January 2021

Publisher's Note: MDPI stays neutral with regard to jurisdictional clai-ms in published maps and institutio-nal affiliations.

1. Introduction

The Feed-in-Tariff Scheme for Renewable Energy (FIT) was launched in Japan in 2012 and the scheme has increased the energy utilization of forest biomass. In the case of biomass, the electric utilities have committed to buying the electricity derived from biomass at a higher price than the normal retail one for 20 years. Thus, power generation plants that accept unused forest biomass (such as thinnings and logging residues rather than wood-based materials such as mill residues and imported woods) have been built and the initiation of plant operations are progressing, in part due to the purchase price incentive [1]. As a result, 3.03 Tg of wood chips on a dry weight basis derived from thinnings and logging residues were used as energy in Japan in 2019 [2].

The use of small-diameter trees is also promising. The area covered by planted forests that have undergone final cutting and subsequent reforestation is now gradually increasing. Thus, a cleaning operation in young planted forests will be necessary 15–20 years from now, when the FIT will expire. The development of an efficient harvesting technology for small-diameter trees can thus be expected to contribute not only to securing a source of forest biomass for power generation plants but also to the continuous tending of young planted forests after regeneration.

Broad-leaved woody coppices have substantial potential. Before and during World War II, an average of 50 million m^3/y of naturally regenerated forest was felled and harvested for energy use in the form of charcoal and firewood in Japan. The annual available amount of naturally regenerated broad-leaved trees for energy is estimated to be 9 Tg/y on a dry weight basis [3]. The rich ecosystems of coppice forests were traditionally maintained by periodic cutting. Broad-leaved forests are now left unutilized and degradation is progressing. Therefore, a new approach to hardwood forest management under cyclic logging for the purpose of energy use is proposed so that the former rich ecosystems can be restored.

The authors' research group has studied technologies and systems for harvesting, transporting and chipping logging residues on steep terrain in Japan [4–10]. In the case of logging residues, the calculation of the procurement cost begins from the harvesting operation at a logging site where the limbing and bucking processes are carried out while the felling and accumulating processes must also be considered to calculate the procurement cost of small-diameter trees. Thus, in Japan, forest biomass from small-diameter trees is considered to be a resource second to that from logging residues in the Biomass Nippon Strategy [11].

Harvesting technologies for small-diameter trees have been developed and examined in North America [12–16] and Europe [17–19]. In Nordic countries, the accumulative function equipped with feller-bunchers and harvesters is utilized in harvesting small-diameter trees for bioenergy use [20–23]. For example, Belbo compared two working methods for small tree harvesting with a multi-tree felling head mounted on a farm tractor [24] and Laitila et al. examined the forwarding of whole trees after manual and mechanized felling and bunching in pre-commercial thinning [25]. Harvesting small-diameter trees has not been examined in Japan since Japanese forestry fell behind in mechanization. Nitami et al. proposed the harvesting of small-diameter trees by introducing accumulative felling and compressing machines [26] but a developed system has never been demonstrated. In the effort to clarify effective methods of harvesting such as small-diameter trees as unutilized forest biomass appropriate for Japan, this study conducted experiments and time studies in the harvesting of small-diameter trees with a truck-mounted multi-tree felling head.

2. Materials and Methods

2.1. Assumed Simplified Model Forest and Harvesting System

In this study, an effective method for harvesting small-diameter trees as unutilized forest biomass appropriate for Japan is discussed using a simplified model forest (Figure 1) in which there was a broad-leaved coppice forest on either side of a 3 m wide forest road. The stand density and biomass per unit area were assumed to be 12,000 trees/ha (growing 0.91 m apart in a reticular pattern) and 30 BDT/ha (BDT: bone-dry ton), respectively. When a felling machine harvested the coppice trees repeatedly in a clear cut way moving each operation point in turn, the number of trees felled inward in a row that minimized the harvesting cost was examined.

The following operations by two machines were assumed (Figure 2). The first machine was a chipper equipped with a multi-tree felling head. It felled and accumulated trees, which were then comminuted. The second machine, equipped with a container, followed after the first one to receive the comminuted wood chips. Such a machine as the first one shown in Figure 2 has never been in operation in Japan; thus, a harvesting experiment (felling and accumulating) was conducted in this study while the data related to the chipping operation were referred to from the previous study by the authors of this paper [27].

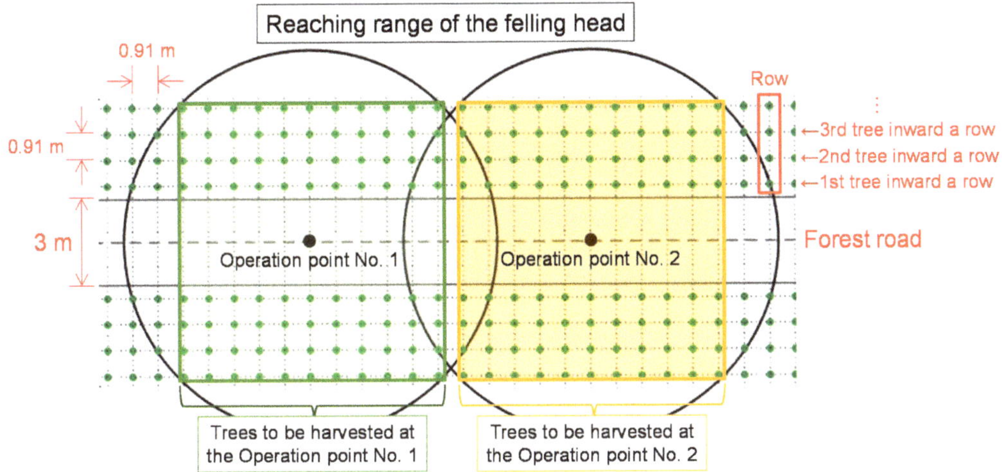

Figure 1. Assumed simplified model forest.

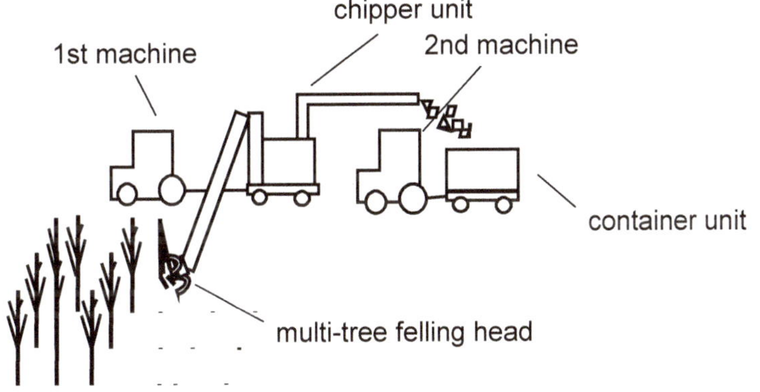

Figure 2. Assumed harvesting system.

Figure 3 contains a flow chart outlining one cycle of the harvesting operation that consists of the felling, accumulating and chipping processes. The simulation model for calculating the cycle time of harvesting operations by inputting the parameters listed in Table 1 was constructed with MATLAB (R2019a, The MathWorks, Inc., Natick, MA, USA). The productivity of harvesting could then be determined by dividing the harvest amount per cycle by the calculated cycle time as follows:

$$HP(L, n) = 3600 \times HA(L, n)/CT(L, n) \tag{1}$$

where $HP(L, n)$, $HA(L, n)$ and $CT(L, n)$ are the productivity of harvesting (BDT/h), harvest amount per cycle (BDT/cycle) and cycle time (s/cycle), respectively, when the maximum reach of a felling head is L (m) and the number of trees felled inward in a row is n. Meanwhile, this study calculated the costs taken to fell, accumulate and chip trees and considered the sum as the harvesting cost as follows:

$$HC(L, n) = MC/HP(L, n) \tag{2}$$

where $HC(L, n)$ is the harvesting cost (JPY/BDT) when the maximum reach of a felling head is L and the number of trees felled inward in a row is n and MC is the sum of the hourly costs of the two machines (JPY/h).

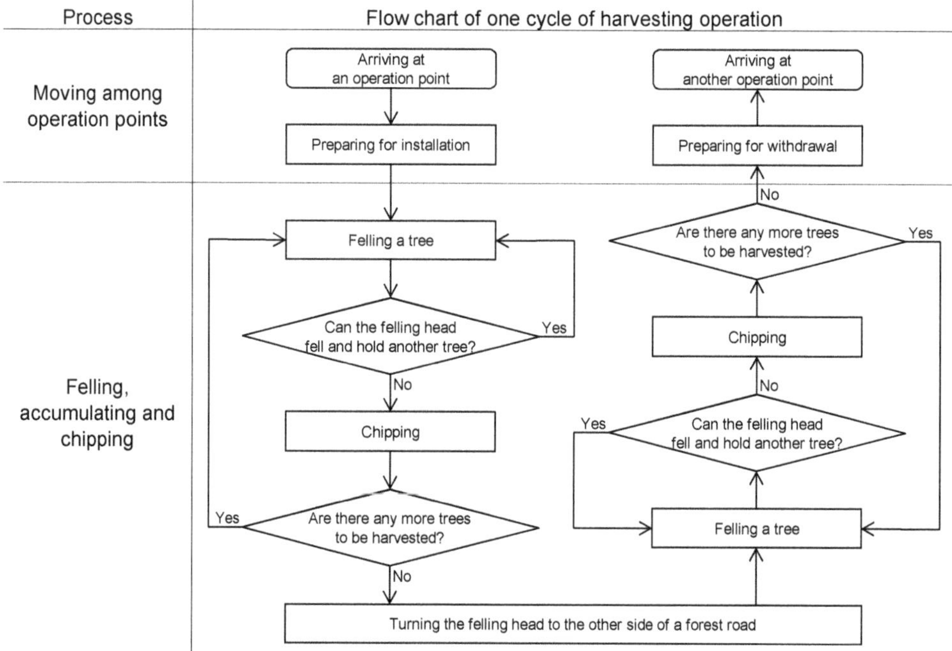

Figure 3. Flow chart of one cycle of harvesting operation.

Table 1. Parameters of the simulation model for calculating the cycle time of the harvesting operation.

Item	Parameter
Machine's moving time among operation points (s)	$a + bl$
Time for installation and withdrawal (s)	a
Machine's moving velocity (m/s)	b
Distance between adjacent two operation points (m) [1]	l
Maximum reach of the felling head (m)	L
Time for felling a tree (s)	f
Moving velocity of the felling head (m/s)	v
Maximum number of trees that can be held at a time	h
Time for chipping (s)	c

[1] Distance between adjacent two operation points, l, is determined depending on the maximum reach of a felling head, L, and the number of trees felled inward in a row, n. For example, Figure 1 shows the case of L = 6.7 m and n = 4 and the distance between the Operation points 1 and 2, which corresponds to l, is then determined to be 10 m.

2.2. Harvesting Experiment

A 10 m wide and 5 m long plot alongside a forest road was established for the harvesting experiment. The felling machine (the first one) was assumed to be located at an operation point (the center of the forest road) so as to fell all trees inside the plot. There were broad-leaved coppices of which the dominant species was konara oak (*Quercus serrata* Thunb.), once repeatedly harvested. There were 50 trees in total inside the plot, the age of the oldest tree was about 20 years old and the average diameter at ground level was 9.1 ± 3.7 cm.

The harvesting experiment was carried out with the multi-tree felling head (ENERGY WOOD GRAPPLE 300, Biojack, Finland; Table 2) used for felling and accumulating small-diameter trees in Nordic countries; the felling head was attached to a crane (LOGLIFT 61Z, Hiab, Sweden; outreach: 7.1 m, weight: 1360 kg) mounted on a log transportation truck (Figure 4). The time it took to fell and accumulate coppice trees was measured in order to collect basic data related to the parameters of the simulation model listed in Table 1.

Table 2. Technical data of the ENERGY WOOD GRAPPLE 300, Biojack [28].

Item	Technical Data
Weight	260 kg
Cutting diameter	250–300 mm
Working pressure	2.00×10^7–2.50×10^7 N/m^2 (total pressure–back pressure)
Oil flow	60–100 dm^3/min
Grapple opening	840 mm
Height in felling position	600 mm

Figure 4. Harvesting experiment with a truck-mounted multi-tree felling head.

3. Results and Discussion

3.1. Results of the Harvesting Experiment

An experiment on a felling and accumulating operation and its time study was carried out and the data necessary for calculating the cycle time were acquired (Table 3). The felling head cut a tree, of which the diameter at ground level was 20 cm, smoothly during the experiment (Figure 5). In the authors' previous study, a sugar cane harvester was used for harvesting 3- to 5-year-old willow trees (ezonokinu willow (*Salix schwerinii* E.L.Wolf.) and onoe willow (*S. sachalinensis* Fr.Schm.)) of which cultivation was aimed at short rotation forestry but it could not cut down 9 cm in diameter at ground level [29], suggesting that the felling head used in the experiment was appropriate for harvesting small-diameter trees in a broad-leaved coppice forest.

Table 3. Results of the time study.

Element Operation	Frequency	Average	Std. Dev. [1]
Time for installation	1	155 s	-
Time for withdrawal	1	115 s	-
Time for felling a tree	50	10 s	2.8 s
Moving velocity of the felling head	12	5.7 m/s	1.2 m/s
Maximum number of trees that could be held at a time	6	8.3	1.6

[1] Std. Dev.: standard deviation.

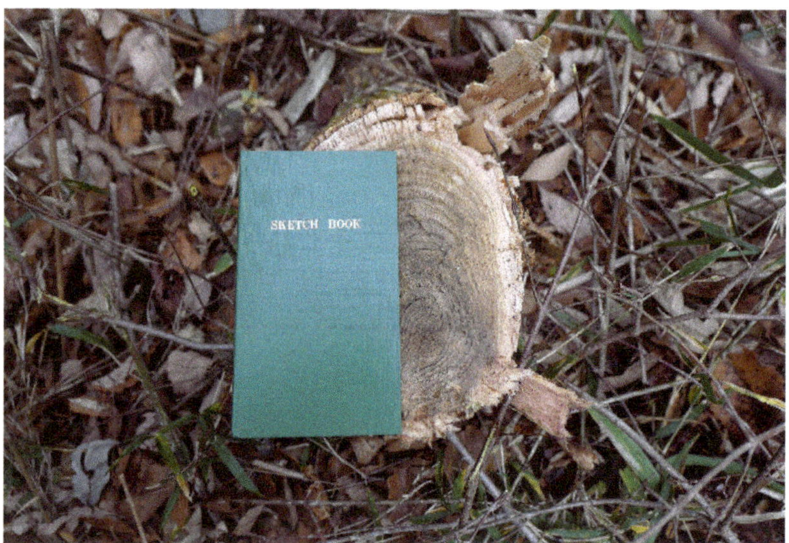

Figure 5. Cut end of a stump of which the diameter at ground level was 20 cm.

The maximum reach of the felling head used in the experiment, L, was 6.7 m, so that the machine could fell a maximum of six trees inward in a row from a 3 m wide forest road (width taken into consideration) in the model forest. With respect to the other parameters acquired from the time study, the time for installation and withdrawal, a, the time for felling a tree, f, the moving velocity of the felling head, v, and the maximum number of trees that could be held at a time, h, were set to be 270 s, 10 s, 0.57 m/s and 8, respectively, based on Table 3. The simulation model for calculating the cycle time of the harvesting operation was completed assuming that the machine's moving velocity, b, and the time for chipping, c, were 5 m/s and 10 s [27], respectively; thus, the productivity of harvesting could be calculated. Finally, the harvesting cost per BDT of small-diameter trees, $HC(L, n)$ of Equation (2), was calculated by dividing the sum of the hourly costs of the two machines (listed in Tables 4 and 5 as 12,250 JPY/h (= 7173 JPY/h for the first machine plus 5077 JPY/h for the second machine), by the harvesting productivity, $HP(L, n)$, calculated from Equation (1).

Table 4. Hourly costs of the two machines. [1.]

Item	1st Machine	2nd Machine	Note
Labor cost (JPY/h)	2000	2000	(a)
Machine cost (JPY/h) [2]	3330	1579	(b)
Fuel cost (JPY/h)	1843	1498	(c) = (d) × (e)
Hourly fuel consumption (dm^3/h)	16	13	(d)
Unit fuel price (JPY/dm^3)	115.2	115.2	(e)
Total hourly cost (JPY/h)	7173	5077	(f) = (a) + (b) + (c)

[1] The exchange rate was roughly 1 USD = 104 JPY and 1 EUR = 126 JPY in December 2020. [2] The detail of calculating the hourly costs of the two machines is listed in Table 5.

Table 5. Detail of calculating the hourly costs of the two machines.

(a) 1st Machine				
Item	Tractor	Felling Head	Chipper	Note
Price (10^6 JPY)	9.45	5.00	4.00	(a)
Hourly price (JPY/h)	900	667	533	(b) = (a) × 10^6/{(c) × (d)}
Life (y)	7	5	5	(c)
Annual operation hour (h/y)	1500	1500	1500	(d)
Hourly repair cost (JPY/h)	630	333	267	(e) = (f) × 10^3/(d)
Annual repair cost (10^3 JPY/y)	945	500	400	(f) = (a) × 10^6 × 0.1/10^3
Total hourly cost (JPY/h)	1530	1000	800	(g) = (b) + (e)

(b) 2nd machine			
Item	Tractor	Container	Note
Price (10^6 JPY)	9.45	0.30	(a)
Hourly price (JPY/h)	900	29	(b) = (a) × 10^6/{(c) × (d)}
Life (y)	7	7	(c)
Annual operation hour (h/y)	1500	1500	(d)
Hourly repair cost (JPY/h)	630	20	(e) = (f) × 10^3/(d)
Annual repair cost (10^3 JPY/y)	945	30	(f) = (a) × 10^6 × 0.1/10^3
Total hourly cost (JPY/h)	1530	49	(g) = (b) + (e)

Figure 6 shows the relationship between the number of trees felled inward in a row and the harvesting cost. The harvesting cost was cheapest when the machine felled five trees inward in a row. The following reasons are considered to explain this result: the more trees inward in a row the machine felled, the more trees were harvested at one operation point. In this case, however, the machine's total moving time markedly increased because the frequency of moving among operation points increased. Therefore, it was concluded that there was an optimum number of felled trees inward in a row that could minimize the harvesting cost.

3.2. Length of the Maximum Reach of a Felling Head

In order to increase the harvest of trees, it seemed it would be effective to lengthen the maximum reach of the felling head and fell trees deeper inward in a row; thus, the following two factors were examined in the case that the maximum reach of a felling head could be lengthened: (1) the maximum number of felled trees inward in a row that would minimize harvesting cost and (2) the minimum harvesting cost itself. With regard to factor (1), 12 trees inward in a row from a forest road was the highest possible number when the length of the maximum reach was increased to 18.2 m (Figure 7a). This meant that felling trees deeper than the twelfth one inward in a row using a longer reach felling head would not reduce the harvesting cost. Concerning factor (2), the cheapest harvesting cost of 10,658 JPY/BDT was obtained when the length of the maximum reach was 10.4 m (Figure 7b). This meant that using a felling head with a maximum reach longer than 10.4 m could not reduce the harvesting cost. These findings indicated that, from the perspective of minimizing harvesting cost, there were upper limits to the number of trees felled inward in a row from a forest road as well as a maximum reach of a felling head. This may help forest road network planning in broad-leaved coppice forests for the purpose of energy wood production and utilization when using a harvesting system for small-diameter trees such as that examined in this study.

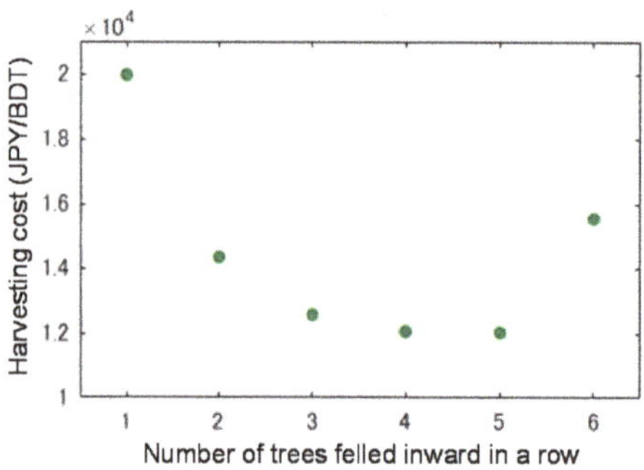

Figure 6. Relationship between the number of trees felled inward in a row and the harvesting cost.

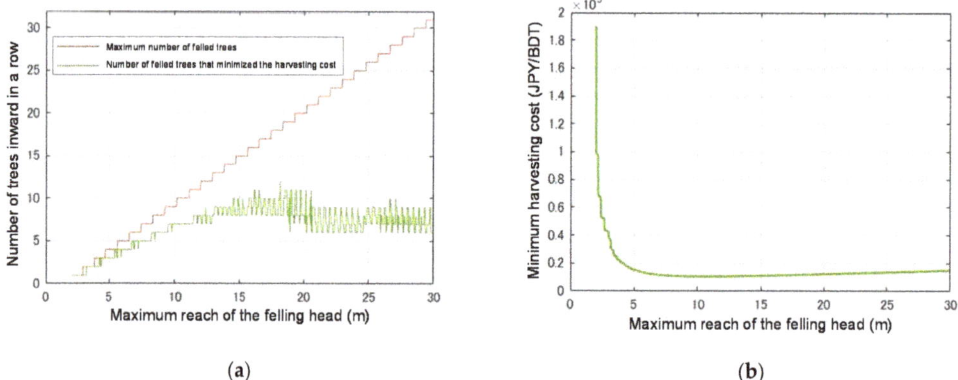

(a)

(b)

Figure 7. Two factors were examined in the case where the length of the maximum reach of a felling head was varied: (a) Maximum number of felled trees inward in a row that would minimize harvesting cost; (b) Minimum harvesting cost itself.

Although this study was a limited analysis based on various assumptions, the harvesting cost calculated in this study was more expensive than that in the U.S. [12], Italy [18] and Finland [21]. However, the general trend concerning the procurement cost of wood chips from forest biomass in Japan was identified; that is, the cost from small-diameter trees calculated in this study was more expensive than that from logging residues [30] but cheaper than that from short rotation woody crops [31].

3.3. Sensitivity Analysis

A sensitivity analysis was carried out on the results to determine how the felling machine used in the experiment might be improved. When the moving velocity of the felling head was doubled, as was the maximum number of trees that could be held at a time, the cost reduction effect increased with the longer maximum reach of the felling head used (Figure 8a,b). On the other hand, the shorter the maximum reach of the felling head used, the greater the cost reduction effect as the machine's moving time among operation points was halved (Figure 8c). Thus, the following regarding the improvements of the machine for future policy are suggested: increasing the moving velocity of a felling head and the maximum number of trees that can be held at a time is effective if it is possible to lengthen the maximum reach of a felling head. Meanwhile, shortening the machine's moving time among operation points is also effective if the maximum reach of a felling head cannot be lengthened.

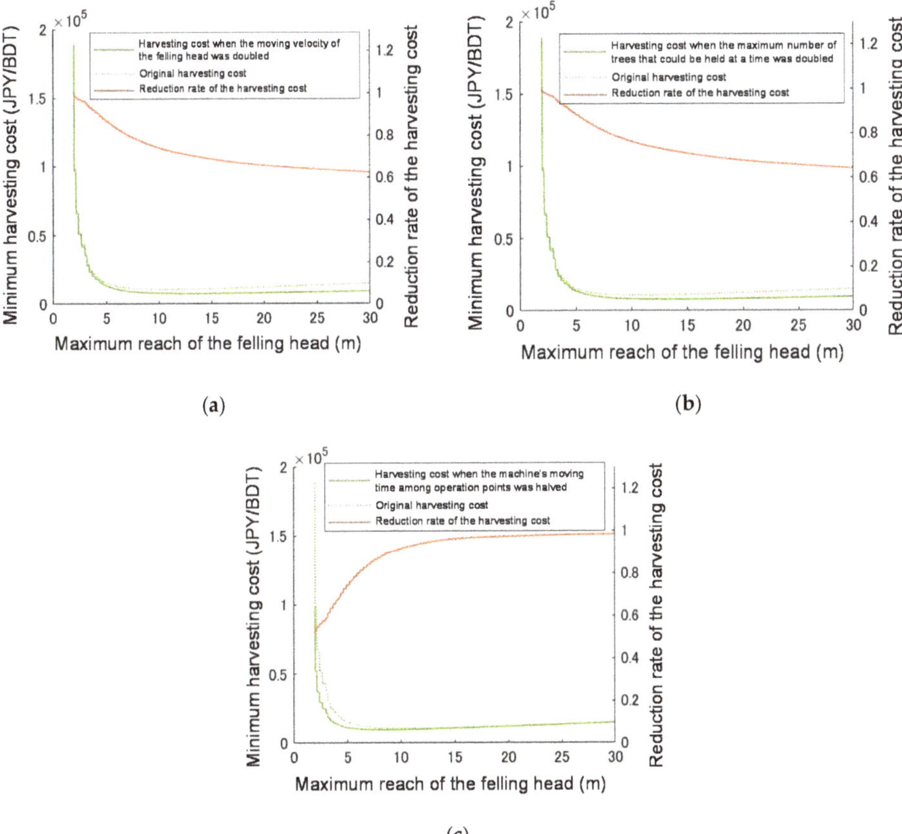

Figure 8. Results of the sensitivity analysis: (**a**) When the moving velocity of the felling head was doubled; (**b**) When the maximum number of trees that could be held at a time was doubled; (**c**) When the machine's moving time among operation points was halved.

4. Conclusions

The goal of this study was to discuss effective methods for harvesting small-diameter trees that are unutilized forest biomass in Japan. This study assumed a simplified model forest and conducted experiments and time studies of the harvesting of small-diameter trees with a truck-mounted multi-tree felling head. The findings are summarized below:

- The machine used in the experiment could fell a maximum of six trees inward in a row from a forest road. The harvesting cost was cheapest, however, when the machine felled five trees inward in a row.
- Lengthening the maximum reach of a felling head to fell trees deeper inward in a row appeared to be effective in increasing the number of harvested trees. From the

> perspective of minimizing harvesting cost, however, there were upper limits to the number of trees felled inward as well as to the maximum reach of a felling head.
> - The results of a sensitivity analysis suggested the following machine improvements could be considered in future policy: increasing the moving velocity of a felling head and the maximum number of trees that can be held at a time are effective if it is possible to lengthen the maximum reach of a felling head. Meanwhile, shortening the machine's moving time among operation points is also effective if the maximum reach of a felling head cannot be lengthened.

Author Contributions: T.Y., T.T. and T.N. conceived, designed and performed the field experiment; T.Y. and T.T. analyzed the data; T.T. contributed the simulation model; T.Y. wrote the paper. All authors have read and agreed to the published version of the manuscript.

Funding: This study was financially supported in part by JSPS KAKENHI Grant Number JP20K06121.

Institutional Review Board Statement: Not applicable.

Informed Consent Statement: Not applicable.

Data Availability Statement: The data presented in this study are available on request from the corresponding author.

Acknowledgments: The authors gratefully appreciate Kin'ichi Suzuki, the president of Amerikaya Co., Ltd. for his great cooperation in the field experiment.

Conflicts of Interest: The authors declare no conflict of interest. The funder had no role in the design of the study; in the collection, analyses or interpretation of data; in the writing of the manuscript or in the decision to publish the results.

References

1. Feed-in Tariff Scheme in Japan. Available online: https://www.meti.go.jp/english/policy/energy_environment/renewable/pdf/summary201207.pdf (accessed on 7 December 2020).
2. Survey on Woody Biomass Energy Use Trend in 2019. Available online: https://www.maff.go.jp/j/tokei/kouhyou/mokusitu_biomass/attach/pdf/index-1.pdf (accessed on 7 December 2020). (In Japanese).
3. Yoshioka, T.; Hirata, S.; Matsumura, Y.; Sakanishi, K. Woody biomass resources and conversion in Japan: The current situation and projections to 2010 and 2050. *Biomass Bioenergy* **2005**, *29*, 336–346. [CrossRef]
4. Yoshioka, T.; Iwaoka, M.; Sakai, H.; Kobayashi, K. Feasibility of a harvesting system for logging residues as unutilized forest biomass. *J. For. Res.* **2000**, *5*, 59–65. [CrossRef]
5. Yoshioka, T.; Aruga, K.; Sakai, H.; Kobayashi, H.; Nitami, T. Cost, energy and carbon dioxide (CO_2) effectiveness of a harvesting and transporting system for residual forest biomass. *J. For. Res.* **2002**, *7*, 157–163. [CrossRef]
6. Yoshioka, T.; Aruga, K.; Nitami, T.; Kobayashi, H.; Sakai, H. Energy and carbon dioxide (CO_2) balance of logging residues as alternative energy resources: System analysis based on the method of a life cycle inventory (LCI) analysis. *J. For. Res.* **2005**, *10*, 125–134. [CrossRef]
7. Yoshioka, T.; Sakai, H. Amount and availability of forest biomass as an energy resource in a mountainous region in Japan: A GIS-based analysis. *Croat. J. For. Eng.* **2005**, *26*, 59–70. Available online: https://hrcak.srce.hr/3989 (accessed on 7 December 2020).
8. Yoshioka, T.; Aruga, K.; Nitami, T.; Sakai, H.; Kobayashi, H. A case study on the costs and the fuel consumption of harvesting, transporting, and chipping chains for logging residues in Japan. *Biomass Bioenergy* **2006**, *30*, 342–348. [CrossRef]
9. Yoshioka, T.; Sakurai, R.; Aruga, K.; Nitami, T.; Sakai, H.; Kobayashi, H. Comminution of logging residues with a tub grinder: Calculation of productivity and procurement cost of wood chips. *Croat. J. For. Eng.* **2006**, *27*, 103–114. Available online: https://hrcak.srce.hr/7617 (accessed on 7 December 2020).
10. Yoshioka, T.; Sakurai, R.; Aruga, K.; Sakai, H.; Kobayashi, H.; Inoue, K. A GIS-based analysis on the relationship between the annual available amount and the procurement cost of forest biomass in a mountainous region in Japan. *Biomass Bioenergy* **2011**, *35*, 4530–4537. [CrossRef]
11. Anonymous. Appendix: Biomass-Nippon Strategy (Provisional Translation) Decided at the Cabinet Meeting, Government of Japan, December 27, 2002. *Biomass Bioenergy* **2005**, *29*, 375–398. [CrossRef]
12. Han, H.S.; Lee, H.W.; Johnson, L.R. Economic feasibility of an integrated harvesting system for small-diameter trees in southwest Idaho. *For. Prod. J.* **2004**, *54*, 21–27.
13. Pan, F.; Han, H.S.; Johnson, L.R.; Elliot, W.J. Net energy output from harvesting small-diameter trees using a mechanized system. *For. Prod. J.* **2008**, *58*, 25–30.
14. Pan, F.; Han, H.S.; Johnson, L.R.; Elliot, W.J. Production and cost of harvesting, processing, and transporting small-diameter ($\leq$ 5 inches) trees for energy. *For. Prod. J.* **2008**, *58*, 47–53.

15. Hiesl, P.; Benjamin, J.G. A multi-stem feller-buncher cycle-time model for partial harvest of small-diameter wood stands. *Int. J. For. Eng.* **2013**, *24*, 101–108. [CrossRef]
16. de Souza, D.P.L.; Gallagher, T.; Mitchell, D.; McDonald, T.; Smidt, M. Determining the effects of felling method and season of year on the regeneration of short rotation coppice. *Int. J. For. Eng.* **2016**, *27*, 53–65. [CrossRef]
17. Spinelli, R.; Cuchet, E.; Roux, P. A new feller-buncher for harvesting energy wood: Results from a European test programme. *Biomass Bioenergy* **2007**, *31*, 205–210. [CrossRef]
18. Spinelli, R.; Magagnotti, N. Comparison of two harvesting systems for the production of forest biomass from the thinning of *Picea abies* plantations. *Scand. J. For. Res.* **2010**, *25*, 69–77. [CrossRef]
19. Erber, G.; Holzleitner, F.; Kastner, M.; Stampfer, K. Effect of multi-tree handling and tree-size on harvester performance in small-diameter hardwood thinnings. *Silva Fenn.* **2016**, *50*, 1428. [CrossRef]
20. Kärhä, K.; Jouhiaho, A.; Mutikainen, A.; Mattila, S. Mechanized energy wood harvesting from early thinnings. *Int. J. For. Eng.* **2005**, *16*, 15–25. [CrossRef]
21. Laitila, J.; Heikkilä, J.; Anttila, P. Harvesting alternatives, accumulation and procurement cost of small-diameter thinning wood for fuel in Central Finland. *Silva Fenn.* **2010**, *44*, 465–480. [CrossRef]
22. Bergström, D.; Bergsten, U.; Hörnlund, T.; Nordfjell, T. Continuous felling of small diameter trees in boom-corridors with a prototype felling head. *Scand. J. For. Res.* **2012**, *27*, 474–480. [CrossRef]
23. Nuutinen, Y.; Petty, A.; Bergström, D.; Rytkönen, M.; Di Fulvio, F.; Tiihonen, I.; Lauren, A.; Dahlin, B. Quality and productivity in comminution of small-diameter tree bundles. *Int. J. For. Eng.* **2016**, *27*, 179–187. [CrossRef]
24. Belbo, H. Comparison of two working methods for small tree harvesting with a multi tree felling head mounted on farm tractor. *Silva Fenn.* **2010**, *44*, 453–464. [CrossRef]
25. Laitila, J.; Asikainen, A.; Nuutinen, Y. Forwarding of whole trees after manual and mechanized felling bunching in pre-commercial thinnings. *Int. J. For. Eng.* **2007**, *18*, 29–39. [CrossRef]
26. Nitami, T.; Iizawa, T.; Suk, S.-I.; Sakurai, R. New mechanism and efficiency of bundler for forest harvesting residuals. *Int. For. Rev.* **2010**, *7*, 266.
27. Yoshioka, T.; Sakurai, R.; Kameyama, S.; Inoue, K.; Hartsough, B. The optimum slash pile size for grinding operations: Grapple excavator and horizontal grinder operations model based on a Sierra Nevada, California survey. *Forests* **2017**, *8*, 442. [CrossRef]
28. Energy Wood Grapple Biojack 300. Available online: https://www.biojack.fi/sites/default/files/esitteet/Biojack_300_ENG_1.pdf (accessed on 7 December 2020).
29. Yoshioka, T.; Sugiura, K.; Inoue, K. Application of a sugarcane harvester for harvesting of willow trees aimed at short rotation forestry: An experimental case study in Japan. *Croat. J. For. Eng.* **2012**, *33*, 5–14. Available online: https://hrcak.srce.hr/85996 (accessed on 7 December 2020).
30. Yoshioka, T.; Kameyama, S.; Inoue, K.; Hartsough, B. A cost breakdown structure analysis on the grinding of logging residues: A comparative study of Japanese and American operations. In Proceedings of the Grand Renewable Energy 2018: International Conference and Exhibition, Yokohama, Japan, 17–22 June 2018; Anonymous, Ed.; Japan Council for Renewable Energy: Tokyo, Japan, 2018. [CrossRef]
31. Yoshioka, T.; Inoue, K.; Hartsough, B. Cost and greenhouse gas (GHG) emission analysis of a growing, harvesting, and utilizing system for willow trees aimed at short rotation forestry (SRF) in Japan. *J. Jpn. Inst. Energy* **2015**, *94*, 576–581. [CrossRef]

 *forests*

Article

Transport Cost Estimation Model of the Agroforestry Biomass in a Small-Scale Energy Chain

Giulio Sperandio [1,*], Andrea Acampora [1], Vincenzo Civitarese [1], Sofia Bajocco [2] and Marco Bascietto [1]

[1] Consiglio per la ricerca in agricoltura e l'analisi dell'economia agraria (CREA)—Centro di ricerca Ingegneria e Trasformazioni agroalimentari, Via della Pascolare 16, 00015 Roma, Italy; andrea.acampora@crea.gov.it (A.A.); vincenzo.civitarese@crea.gov.it (V.C.); marco.bascietto@crea.gov.it (M.B.)

[2] Consiglio per la ricerca in agricoltura e l'analisi dell'economia agraria (CREA)—Centro di ricerca Agricoltura e Ambiente, Via della Navicella 4, 00184 Roma, Italy; sofia.bajocco@crea.gov.it

[*] Correspondence: giulio.sperandio@crea.gov.it

Abstract: The delivery of biomass products from the production place to the point of final transformation is of fundamental importance within the constitution of energy chains based on biomass use as a renewable energy source. Transport can be one of the most economically expensive operations of the entire biomass energy production process, which limits choices in this sector, often inhibiting any expansive trends. A geographic identification, through remote sensing and photo-interpretation, of the different biomass sources was used to estimate the potential available biomass for energy in a small-scale supply chain. This study reports on the sustainability of transport costs calculated for different types of biomass sources available close a biomass power plant of a small-scale energy supply chain, located in central Italy. To calculate the transport cost referred to the identified areas we used the maximum travel time parameter. The proposed analysis allows us to highlight and visualize on the map the areas of the territory characterized by greater economic sustainability in terms of lower transport costs of residual agroforestry biomass from the collection point to the final point identified with the biomass power plant. The higher transport cost was around €40 Mg^{-1}, compared to the lowest of €12 Mg^{-1}.

Keywords: energy chain; residual biomass; isochronous rings; travel time; transport cost

Citation: Sperandio, G.; Acampora, A.; Civitarese, V.; Bajocco, S.; Bascietto, M. Transport Cost Estimation Model of the Agroforestry Biomass in a Small-Scale Energy Chain. *Forests* **2021**, *12*, 158. https://doi.org/10.3390/f12020158

Academic Editor: Raffaele Cavalli
Received: 4 January 2021
Accepted: 26 January 2021
Published: 28 January 2021

Publisher's Note: MDPI stays neutral with regard to jurisdictional claims in published maps and institutional affiliations.

1. Introduction

In recent years, the European Union has devoted growing attention to supporting and promoting actions to combat climate change that, together with the policies aimed at achieving greater energy efficiency, can foster development of a more sustainable economic-energy and environmental system [1]. All this would result in less dependence on imports, less environmental impact, and an increase in added value for rural areas. The economic and social interest for renewable energies and bioenergy can be attributed to various aspects including, mainly, the possibility of reaching energy self-supply levels more easily and, at the same time, heavily replacing the use of fossil fuels [2–4]. In this way it is possible to obtain the double effect of reducing the emissions of greenhouse gases into the atmosphere and of respecting the commitments made at world level with the ratification of various international treaties starting from the Kyoto Protocol of 1997, favoring the possibility of increasing the added value and the employment level of the local energy supply chains [5,6]. There are different types of biomass that can be used for energy purposes, both in raw form as firewood, and as woodchip or as refined material such as pellets and briquettes [7–9], which have favored the development of technologies suitable for the optimal conversion of these products into energy (heat and electricity). This product diversity corresponds to the possible activation of equally diversified energy chains to be organized in rural energy districts. The implementation of a bioenergy supply chain is mainly linked to its efficient logistics organization [10], on which most of the impacts (economic, occupational, environmental, etc.) depend, and the real convenience to the production of primary energy from agroforestry biomass. It is therefore essential to make the entire energy transformation production process sustainable when referring to the different options available, about the aspects relating to the collection, transport, and energy transformation, applying models that can improve the efficiency of the production processes and optimizing production costs for every available biomass typology [11–17].

The future challenges are aimed at improving the competitiveness of the biomass energy chain and all go in the direction of sustainable development, which involves not only the technological and production aspects, but also the governance and organization of the sector [14,18]. These aspects must inevitably be correlated to what is one of the key factors that regulate, influencing its efficiency, the entire energy transformation system, represented by the cost of transporting biomass [19]. The transport of woody biomass, in fact, significantly affects the costs, the emissions of dust and the carbon balance within the energy chain [20], with increasingly negative impacts as a function of the increase in the distance traveled and the volume of biomass transported [21,22].

The logistics of the short energy supply chain and the priority valorization of the biomass spread throughout the territory represent the base for a sustainable development of bioenergy, which can be supported by the realization of small-medium power plants, compatible with the energy demands of users in the specific area of reference [23,24].

A small territorial basin, where different sources of agroforestry biomass are available, was taken as a case study. A geographic model was built to evaluate the cost-effectiveness of the logistics of agroforestry biomass transport, in the context of the small local energy supply chain [25], taking into consideration both technical-economic data (transport, loading and unloading times and costs biomass), and data relating to the territory, such as type of biomass and its location, qualitative and quantitative availability, viability, and possible routes with respect to the transformation point.

2. Materials and Methods

2.1. Study Area and Biomass Power Plant

The study area refers to a small-scale supply chain located North-East of Rome and surrounding a biomass power plant, installed within the farm of CREA-Research Centre for engineering and agri-food processing in Monterotondo, Italy ($42°6'2.63''$ N; $12°37'37.36''$ E).

The identification of the boundaries of the potential wooden biomass supply area was based on the travel time of the trucks from any spatial point located no more than 60 min away from the biomass plant, selecting the road with shortest travel time, excluding the highways [25]. For this reason, the shape of the study area is irregular and depends on the spatial arrangement of the road network, road types and speed limits. Wider and linear roads, on which traffic can flow faster, make the study area wider than the narrower and more winding roads, on which, instead, the vehicle travels slower, reducing the study area. Five areas consisting of irregular isochronous rings were identified. The external boundaries of each ring were determined in relation to five-time bands depending on the time taken to travel from the specific biomass location to the plant. The first ring, outermost, with a travel time varying from 60 to 50 min, the second from 50 to 40 min, the third from 40 to 30 min, the fourth from 30 to 20 min and finally, the fifth ring, the closest to the biomass plant, from 20 to 0 min. The isochronous rings were calculated by applying the software R osrm package (OpenStreetMap-Based Routing Service) version 3.2.0 [26]. The furthest point of the external boundary of the area from the plant was at a linear distance of about 35 km, while the closest was about 16 km. The total area examined was 2276 km^2, distributed on each single isochronous ring in an increasing way starting from the fifth to the first.

The proposed model was constructed with reference to the specific point of energy transformation, represented by the biomass power plant used for the heating of the research Centre buildings, which are characterized by a potential volume to be heated of about 10,000 m^3. The biomass plant, equipped with a mobile grate, has a nominal heat output of

350 kW, but is also set up for micro-cogeneration by means of a steam turbine capable of generating 500 kg of steam per hour at a pressure of 12 bar, for a potential production of about 40 kWh of electricity. Currently the plant can only be used to produce thermal energy as the steam turbine is still being installed. For the building heating only, on a period of use of 130 days per year, the potential annually consumable biomass, with a water content of 35%, is around 290.1 Mg per year. When the micro-cogeneration option will be activated, then the plant could be used 24 h per day for about 300 days per year, with an annual biomass consumption estimated at 811.5 Mg.

2.2. Biomass Estimation

The potential annual supply of residual biomass from the observed area was estimated on a sampling considering the land cover area of each sampling population visualized by satellite images in Google Earth software [25]. Once the different types of biomass present in the area were identified, their quantitative estimation was subsequently carried out by applying judgment coefficients of experts of photo-interpretation. On a total of 139 observations, eight sampling populations were defined, each of which was independently sampled. In Table 1, the potential biomass range for each class and other evaluation parameters considered in the application of the biomass transport cost assessment model are reported.

Table 1. Parameters used to estimate the residual biomass potentially available from the different biomass sources.

Biomass Classes	Density	Residual Biomass Production	Frequency of Intervention	Residual Biomass Per Tree
	Tree ha^{-1}	Mg ha^{-1}	Years	Mg Tree^{-1} year^{-1}
1. Green urban area (GUA)	80	16.0–32.0	8	0.0250–0.0500
2. Sport and leisure facilities (SLF)	50	10.4–20.0	8	0.0250–0.0500
3. Vineyards (VIY)	3000–4000	2.1–3.0	1	0.0007–0.0010
4. Fruit trees and berry plantations (FBP)	400–500	2.0–3.5	1	0.0050–0.0070
5. Olive groves (OGR)	180–300	3.6–8.1	2	0.0200–0.0270
6. Complex cultivation patterns (CCP)	130–260	2.0–4.0	2	0.0100–0.0135
7. Land principally occupied by agriculture (LOA)	400–500	2.0–3.5	1	0.0050–0.0070
8. Forests (FOR)	800–1000	18.8–26.3	25	0.0009–0.0011

Based on the coverage of the tree canopy observed on the territorial map, four levels of biomass production (in Mg ha^{-1} year^{-1}) have been attributed for each of the eight types of biomass. L3, L2, L1 and L0 are the classes considered in the study and indicate, respectively, the maximum, average, minimum or absent value of the available biomass. Table 2 shows the values of these parameters.

Table 2. Biomass production levels (L) for the calculation of residual biomass available for each typological class (in Mg ha^{-1} year^{-1}).

Typology	L3	L2	L1	L0
1. Green Urban Areas (GUA)	4.00	3.00	2.00	0.00
2. Sport and Leisure Facilities (SLF)	2.50	1.90	1.30	0.00
3. Vineyards (VIY)	3.00	2.55	2.10	0.00
4. Fruit Trees and berry Plantation (FTP)	3.50	2.75	2.00	0.00
5. Olive Groves (OGR)	4.00	2.90	1.80	0.00
6. Complex Cultivation Patterns (CCP)	2.00	1.50	1.00	0.00
7. Land principally Occupied by Agriculture (LOA)	3.50	2.75	2.00	0.00
8. Forest class (FOR)	1.05	0.90	0.75	0.00

In this way, a set of data referring to each single areas of the map was constructed, identifying the type, the overall surface, the percentage of surface referring to the different biomass level potentially available, and that belonging to a specific isochronous ring.

The biomass potentially available in each sample area was then determined on the basis of the biomass estimated for each sub-area relating to each production level for each type of biomass class. The production level was determined in relation to the density of the respective typological classes. Figure 1 shows an example representation of the procedure for estimating the areas and biomass potentially available for each typological class. The biomass for each circular sample area was then determined by applying the following formula (Equation (1)):

$$Biomass\ (\mathrm{Mg}) = \sum_{k=0}^{3} \left(A_{kj} \times L_k \right) \tag{1}$$

where:

1. A—area in hectares corresponding to the level L of biomass production (Table 1);
2. k—number of areas corresponding to the same production level within the sample area; and
3. j—number of areas belonging to the same production level L_k.

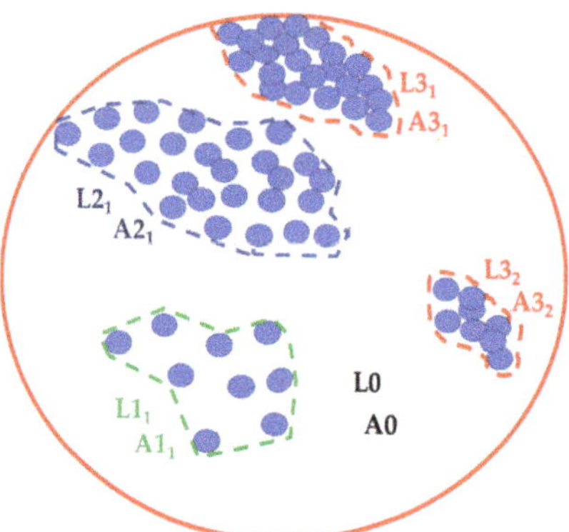

Figure 1. Graphic representation of the estimated areas (A_{kj}) corresponding to the biomass production levels (L_k) within a single sample area, calculated in relation to the vegetation density.

2.3. Transport Cost Evaluation Model

The analysis of the economic sustainability of the recovery of the biomass available on the territory of the small supply chain was based on the evaluation of the costs of the biomass transport operation (round trip of the truck) from the loading site to the biomass plant, including loading and unloading operations. For the acquisition of the biomass, the hypothesis adopted is that farmers provide free pruning biomass to avoid incurring the tariffs applied for the disposal of this material in landfills. Both the farmer and the manager of the plant benefit from the agreement—the farmer does not pay for the disposal of biomass residues; the latter does not pay for the recovered raw material.

In the case study, the pruning biomass should already be staked by the farmer and directly available for loading and transport (Figure 2a). An unloaded truck is assumed to leave the biomass plant and arrive at the biomass loading site, where a forest loader loads the stacked biomass (Figure 2b).

(a)   **(b)**

Figure 2. Pruning residues stacked (**a**) and biomass loading operation on the truck (**b**).

Once the loading operation is complete, the truck travels to its destination and unloads the biomass near the plant. The truck carries out subsequent travels until the schedule daily work time is completed, considering about 8 working hours per day. Given the small size of the processing plant, it was assumed that this operation was carried out using a truck with a total load capacity of 26 m^3, corresponding to an estimated residual biomass weight of about 8 Mg. A forest loader equipped with a grapple to carry out the loading/unloading of biomass must be daily transferred to the workplace and brought back using a dedicated truck. The hourly costs of the machines calculated with analytical methodology [27], and the principal economic and technical elements considered, are reported in Table 3.

Table 3. Main elements considered for the calculation of the hourly cost of machines and labor.

	Truck for Biomass Transport	Truck for Loader Transport	Forest Loader
Purchase price (€)	110,000	95,000	80,000
Salvage value (€)	7559	6528	8590
Life time (years)	12	12	10
Total time (h)	14,400	14,400	8000
Engine Power (kW)	309	280	88
Interest rate (%)	4.0	4.0	4.0
Fuel consumption (L h^{-1})	25.49	23.10	9.44
Fuel price (€ L^{-1})	1.50	1.50	1.10
Driver cost (€ h^{-1})	21.00	21.00	15.00
Machine cost (€ h^{-1})	71.00	64.00	35.00
Total operating cost (€ h^{-1})	92.00	85.00	50.00

The formula used to determine the unitary transport cost (Equation (2)), including the biomass loading, transport and unloading cost and the daily forest loader transfer cost, is the following:

$$BTC = \frac{[(Ttr \times Ctr] + (Tlu \times Clo) + (tc \times Ctl))}{bl} \tag{2}$$

where:

1. *BTC*—biomass transport cost per Mg (€ Mg^{-1});
2. *Ttr*—roundtrip travel time, obtained doubling the return travel time of the loaded truck (h);
3. *Tlu*—time required for loading and unloading operations (h);
4. *Ctr*—hourly cost of the truck (€ h^{-1});
5. *Clo*—hourly cost of the loader (€ h^{-1});
6. *tc*—loader transferring coefficient;

7. *Ctl*—hourly cost of the truck that transfer the loader to destination and return (€ h^{-1}); and
8. *bl*—average biomass load considered per travel (Mg).

The travel time (*Ttr*) was calculated by doubling the average time attributed to each of the five isochronous rings (h). An average time of 75 min was considered as the basis for calculating the *Tlu*, divided into about 56 min (75%) for loading and 19 min (25%) for unloading. To consider the influence of the different types and quantities of biomass on the loading phase, two multiplying coefficients were considered. *Tlu* was then calculated according to the following Equation (3).

$$Tlu = (Tl \times lc) \cdot (1 + yc) + Tu \tag{3}$$

where:

1. *Tl*—loading time (h);
2. *Tu*—unloading time (h)
3. *lc*—load coefficient depending on biomass type; and
4. *yc*—yield coefficient depending on the quantity of biomass per hectare.

The impact of the loader transfer on the total transport cost per Mg was estimated considering a loader transfer coefficient (*tc*) for each biomass class. In fact, the daily transfer time of the loader (one round trip) must be added to the total daily time used for transporting the biomass. The greater the number of biomass transport trips per day, the lower this coefficient, since the latter is calculated as the reciprocal of the number of daily trips. Table 4 shows the average values of the correction coefficients *lc*, *yc* and *tc*.

Table 4. Average coefficients used for the calculation of final travel time (*lc*, load coefficient; *yc*, yield coefficient; *tc*, loader transfer coefficient).

Typology	Coefficients		
	lc	*yc*	*Tc*
GUA	1.00	0.20	0.37
SLF	1.05	0.29	0.34
VIY	1.15	0.21	0.43
FTP	1.05	0.21	0.33
OGR	1.10	0.23	0.34
CCP	1.10	0.30	0.35
LOA	1.15	0.21	0.34
FOR	1.00	0.27	0.30

To assess the economic sustainability of the recovery and transport of biomass scattered over the territory of the small-scale energy chain, it was necessary to consider that the biomass unloaded at the plant had to be chipped before use. A positive judgment on economic sustainability was based on the positive difference between the average market value of the woodchip and the cost incurred for transport and chipping. For the chipping operation, a forest chipper available on the farm was considered. The average cost of this operation was estimated at 15 € Mg^{-1}. The market value of woodchip from biomass residues was instead estimated at around 45 € Mg^{-1}.

3. Results

The results are expressed in relation to the different typologies of biomass sources. Figure 3 shows the estimate of the quantity of biomass available in the territory examined in relation to the production levels (L) and the biomass classes. The land principally occupied by agriculture (LOA) class is the one with the highest average value (2.13 Mg ha^{-1} year^{-1}), with the highest incidence of L3 (3.07 Mg ha^{-1} year^{-1}). The class that contributes less is SLF with 0.25 Mg ha^{-1} year^{-1}. In Figure 4, we report the average time consumed (Figure 4a) and relative average costs (Figure 4b) for the load, transport, and unload operations of the residual biomass for each class. The highest time is request for the VIY class with 4.23 h trip^{-1}, while the shortest time is recorded for the FOR class, with 3.04 h trip^{-1}. The other biomass classes record intermediate times between 3.05 and 3.50 h trip^{-1}. The load/unload time is highest in the CCP class with 1.65 h, followed by LOA and VIY with 1.61 h, while GUA requires the lowest time of 1.44 h. The trend in average costs per trip reflects that the times with the highest value of €316.31 trip^{-1} was VIY, corresponding to €39.54 Mg^{-1}, and the lowest value was €213.84 trip^{-1} for FOR, that is €26.73 Mg^{-1}.

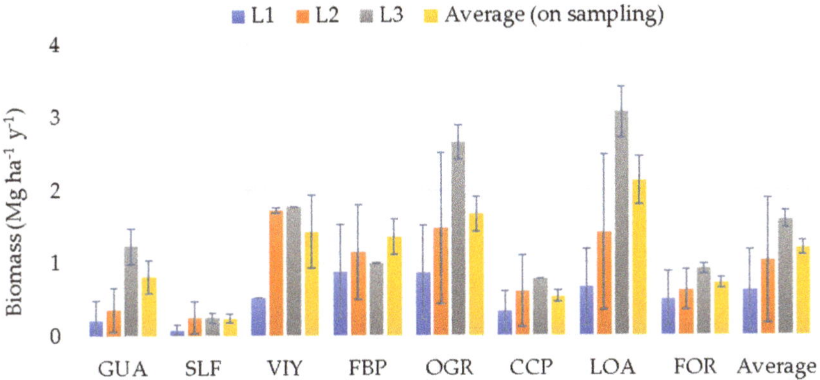

Figure 3. Average value (±standard error) of biomass available (Mg ha^{-1} year^{-1}) in reference to the production level (L) and biomass class.

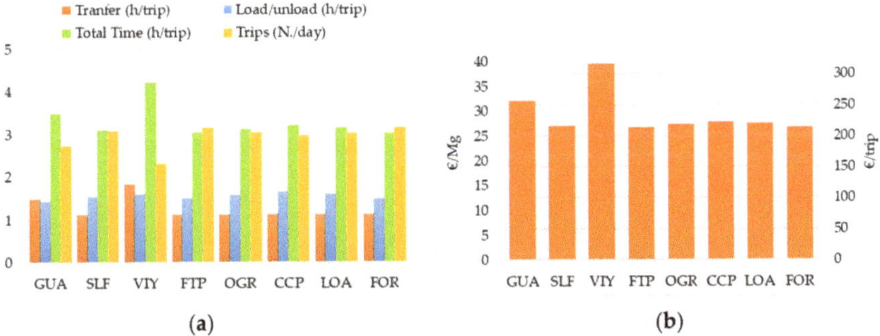

Figure 4. Average time consumption (**a**) and costs (**b**) for the recovery and transport of residual biomass per class.

It should be noted that, with regard to the VIY class, it is not widespread in the observed area, therefore it is not to be considered a representative figure for this class. For the other classes, on the other hand, there is a greater homogeneity of results for the classes considered in the agricultural field such as FTP, OGR, CCP and LOA. Figure 5 shows the matrix graphs of the average transport cost (Figure 5a) and economic sustainability (Figure 5b) per Mg, in relation to the biomass classes and isochronous rings. In this figure, it is clearly seen how the distance from the biomass plant is very important. The cost increases by proceeding from the 5th isochronous ring (travel time 0–20 min) to the 1st (50–60 min).

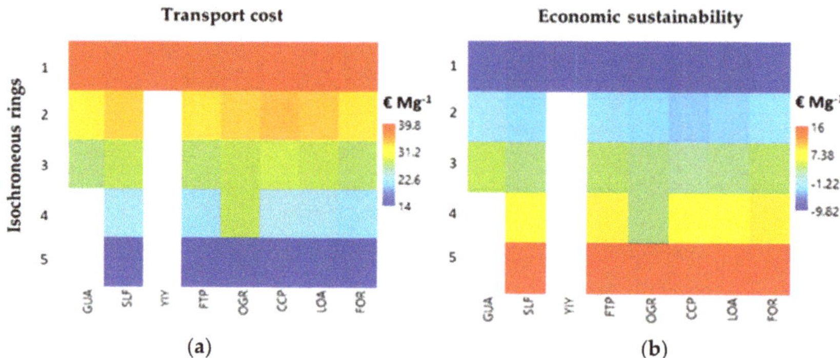

Figure 5. Matrix plots of the biomass transport cost (**a**) and economic sustainability of the transport operation (**b**), in relation to biomass classes and isochronous rings.

This is valid for all classes even if in a different way. The average costs vary from a minimum of about €14 Mg^{-1} in the area of the 5th ring, to a maximum of about €39.80 Mg^{-1} in the 1st ring. Economic sustainability is inversely proportional to the costs incurred. The red and yellow colors indicate the areas of greater sustainability, coincident with the 5th and 4th ring (positive values). The 3rd ring represents the boundary area between positive and negative values. The 2nd and 1st ring areas are always associated with negative values and indicate an uneconomic condition of the operation. Figure 6 shows the territorial map in which transport costs are associated with specific areas of the map. From this map it is possible to check the cost of transport (Figure 6a) in relation to the distance from the biomass loading point to the unloading point near the transformation plant. The conditions of economic sustainability (Figure 6b) show positive values in the areas ranging from yellow to blue (near the biomass plant).

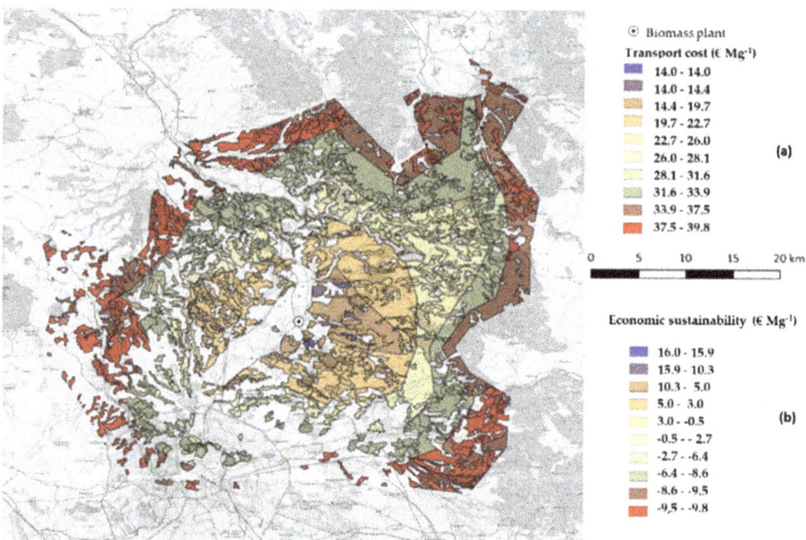

Figure 6. Territorial Map of the biomass transport cost (**a**) and economic sustainability of the operation (**b**) according to typological classes and isochronous rings.

In Figure 7, the total area (Figure 7a) and the related potential biomass available (Figure 7b) in the investigated area was compared with the area and quantity of biomass of greatest interest for the purposes of the study, where economic sustainability is verified, corresponding to the 4th and 5th isochronous rings.

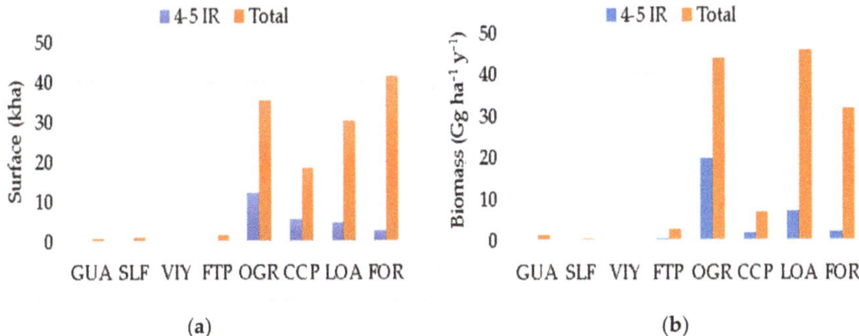

Figure 7. Total surface investigated and surface of the 4th and 5th isochronous ring (IR) (**a**) and total residual biomass available and biomass on 4th and 5th isochronous ring (IR) (**b**), where economic sustainability is verified, for each class considered.

The market price of the woodchip and the chipping cost can represent a critical aspect of the model. A sensitivity analysis on these parameters showed in Figure 8 reveals, for example, that a 20% reduction in the value of the woodchip (Figure 8a) would make most of the biomass classes uneconomic even in the ring closest to the biomass plant, while the same negative effect is obtained if the cost of the chipping operation increased by over 30% (Figure 8b). On the contrary, with the increase of the same percentage of the price, the recovery of biomass becomes convenient for almost all classes in all rings, while the same result is obtained if the cost of the chipping operation is reduced by more than 30%.

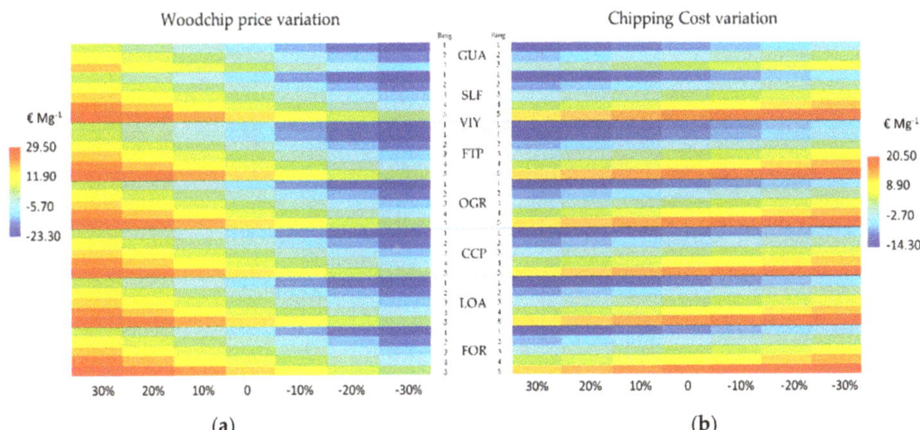

Figure 8. Sensitivity analysis related to economic sustainability (difference between woodchip price and recovery and chipping costs) in relation to the variation of woodchip price (**a**) and chipping cost (**b**), according to biomass classes and isochronous rings.

4. Discussion

The study has proposed a model to evaluate the cost and the relative economic sustainability of the recovery and transport of the potentially available residual biomass on a small-scale energy supply chain. The model was based on the cost evaluation of recovery of different potential biomass sources spread over the territory close to a small biomass plant. On a total observed area of 2276 km^2, about 57%, equal to 130 kha, was considered useful for the proposed model.

The annual residual biomass potentially available was about 134 Gg distributed on the territory observed. Much of the biomass available was classified within the agricultural area (LOA) for about 34.4%, followed by olive groves with 33.0% and forest area with 23.8%. The remaining 8.8% is mainly divided between associated crops and orchards. Of this total biomass available, only 24.0% falls within the area of the economic sustainability (5th and 4th isochronous ring) for a quantity of about 32 Gg, 62% of which is represented by olive grove pruning residues. Areas with agricultural crops and forests follow with an incidence of 23% and 7%, respectively. For the case study, only a small percentage of this available biomass can be used in the biomass plant, whose annual consumption in a cogeneration system does not exceed 1 Gg of woodchip. However, it is useful to have a tool to plan any biomass supplies under convenient conditions. The economic evaluation model is applied to the different types of biomass available, considering the various difficulties related to the quality of biomass residues and the influence this generates in the calculation of recovery and transport costs [28,29].

The cost of transporting biomass from the origin to its destination represents a significant share of the overall costs of supplying woody biomass in an energy chain. Some studies attribute to this cost a percentage between 25% and 50% of the total cost of the delivered biomass, depending on the transport distance, the bulk density of the load and the moisture content of the biomass transported [30,31]. Increasing the efficiency of transport operations should therefore lead to a significant reduction in related costs and with it also a probable reduction in environmental impacts.

The model, as expected, returns the highest cost values, for all biomass classes, correlating to the greatest distances to travel. For the examined small-scale energy chain, the economic sustainability for the supply of biomass to feed the power plant is verified when the travel distance does not exceed 20 km, with a travel time from the place where the biomass is loaded to the plant, no more than 35 min. At the same distance, the pruning residues of orchards and forest biomass are slightly more advantageous than the other classes, while the pruning of vineyards, being represented only within the most distant isochronous ring, is always uneconomical.

The economic sustainability in the model is calculated, for each class, considering the difference between the market price of the woodchip and the biomass transport and chipping costs. This last operation is very critical, and the average cost considered in this study could significantly increase in the case of agricultural pruning waste compared, for example, to forest residual biomass, reducing the economic convenience range [32].

The sensitivity analysis applied to these two parameters highlighted a greater sensitivity of the model to a change in the price of the woodchip compared to the same percentage change (of opposite sign) in the cost of the chipping operation. To maintain the same level of economic sustainability returned by the proposed model, a possible increase/decrease in the market value of the woodchip (decrease/increase in the cost of the woodchip) would correspond to an increase/decrease in the transport distance and, consequently, in the area affected by the recovery of the biomass.

5. Conclusions

The study carried out was aimed at implementing a geographic evaluation model capable of providing a mapping of the costs of transporting biomass (including loading and unloading) from production sites to processing sites. By mapping the cost of biomass transport, it is possible to orient the choices in relation to the size of the energy transformation plants to be considered also in a project to enhance the local resources available. The small-scale energy supply chain, in fact, currently represents a model to be encouraged and applied in farms that want to make a qualitative leap towards a bioenergy company. This model can be implemented on larger areas and on different territorial conditions and can be a useful knowledge base in the management of the bioenergy resources of a territory. In these cases, however, it is necessary to verify from time to time the boundaries of the potential biomass supply area, the road network, the type of biomass resources available and the level of productivity obtained in the investigated area to calibrate the coefficients of the economic model, on which the calculation of the final cost of transporting biomass depends.

Finally, a very interesting aspect of the proposed model concerns the possibility of starting a virtuous process of mutual benefit between the farmers of a territory and the bioenergy company, which translates into a recovery of residual biomass, otherwise destined for landfill, or burned in field. In this way, environmental impacts are also reduced thanks to a more controlled combustion process in small biomass plants. The model of the small-scale energy supply chain can represent the most suitable solution for the development of sustainable systems compatible with the availability of bioenergy that the territory is able to supply.

Author Contributions: Conceptualization, G.S.; methodology, G.S., M.B. and S.B.; formal analysis, G.S.; data curation, G.S.; writing—original draft preparation, G.S.; writing—review and editing, G.S., A.A., V.C., M.B. and S.B. All authors have read and agreed to the published version of the manuscript.

Funding: This research was funded by Italian Ministry of Agriculture, Food and Forestry Policies (Mi-PAAF), grant D.D. n. 26329, 1 April 2016, project AGROENER "Energia dall'agricoltura: innovazioni sostenibili per la bioeconomia".

Acknowledgments: We thank the anonymous reviewers for their valuable and constructive comments.

Conflicts of Interest: The authors declare no conflict of interest.

References

1. Padilla-Rivera, A.; Barrette, J.; Blanchet, P.; Thiffault, E. Environmental Performance of Eastern Canadian Wood Pellets as Measured Through Life Cycle Assessment. *Forests* **2017**, *8*, 352. [CrossRef]
2. Nishiguchi, S.; Tabata, T. Assessment of social, economic, and environmental aspects of woody biomass energy utilization: Direct burning and wood pellets. *Renew. Sustain. Energy Rev.* **2016**, *57*, 1279–1286. [CrossRef]
3. Tziolas, E.; Manos, B.; Bournaris, T. Planning of agro-energy districts for optimum farm income and biomass energy from crops residues. *Oper. Res.* **2017**, *17*, 535–546. [CrossRef]
4. Martín-Lara, M.A.; Ronda, A.; Zamora, M.C.; Calero, M. Torrefaction of olive tree pruning: Effect of operating conditions on solid product properties. *Fuel* **2017**, *202*, 109–117. [CrossRef]
5. United Nations. *Kyoto Protocol to the United Nations Framework Convention on Climate Change*; United Nations: Rome, Italy, 1998.
6. Wang, Y.; Wu, K.; Sun, Y. Effects of raw material particle size on the briquetting process of rice straw. *J. Energy Inst.* **2018**, *91*, 153–162. [CrossRef]
7. Civitarese, V.; Acampora, A.; Sperandio, G.; Assirelli, A.; Picchio, R. Production of Wood Pellets from Poplar Trees Managed as Coppices with Different Harvesting Cycles. *Energies* **2019**, *12*, 2973. [CrossRef]
8. Nurek, T.; Gendek, A.; Roman, K. Forest Residues as a Renewable Source of Energy: Elemental Composition and Physical Properties. *BioResources* **2018**, *14*, 6–20. [CrossRef]
9. Purohit, P.; Chaturvedi, V. Biomass pellets for power generation in India: A techno-economic evaluation. *Environ. Sci. Pollut. Res.* **2018**, *25*, 29614–29632. [CrossRef]
10. Han, H.; Chung, W.; Wells, L.; Anderson, N. Optimizing biomass feedstock logistics for forest residue processing and transportation on a tree-shaped road network. *Forests* **2018**, *9*, 121. [CrossRef]
11. Costa, C.; Sperandio, G.; Verani, S. Use of Multivariate approaches in biomass energy plantation harvesting: Logistic advantages. *Agric. Eng. Int. CIGR J.* **2014**, 70–79.
12. Cambero, C.; Sowlati, T.; Marinescu, M.; Röser, D. Strategic optimization of forest residues to bioenergy and biofuel supply chain. *Int. J. Energy Res.* **2015**, *39*, 439–452. [CrossRef]
13. Paulo, H.; Azcue, X.; Barbosa, A.P.; Relvas, S. Supply chain optimization of residual forestry biomass for bioenergy production: The case study of Portugal. *Biomass Bioenergy* **2015**, *83*, 245–256. [CrossRef]

14. Sierk de Jong, S.; Hoefnagels, R.; Wetterlund, E.; Pettersson, K.; Faaij, A.; Junginger, M. Cost optimization of biofuel production—The impact of scale, integration, transport and supply chain configurations. *Appl. Energy* **2017**, *195*, 1055–1070. [CrossRef]
15. Proto, A.R.; Sperandio, G.; Costa, C.; Maesano, M.; Antonucci, F.; Macrì, G.; Scarascia Mugnozza, G.; Zimbalatti, G. A three-step neural network artificial intelligence modelling approach for time, productivity and costs prediction: A case study in Italian forestry. *Croat. J. For. Eng.* **2020**, *41*, 35–47. [CrossRef]
16. Searcy, E.; Flynn, P.; Ghafoori, E. The relative cost of biomass energy transport. *Appl. Biochem. Biotechnol.* **2007**, *136–140*, 639–652.
17. Möller, B.; Nielsen, P.S. Analysing transport costs of Danish forest wood chip resources by means of continuous cost surfaces. *Biomass Bioenergy* **2007**, *31*, 291–298. [CrossRef]
18. Ruiz, J.A.; Juarez, M.C.; Morales, M.P.; Munoz, P.; Mendivil, M.A. Biomass logistics: Financial & environmental costs. Case study: 2 MW electrical power plants. *Biomass Bioenergy* **2013**, *56*, 260–267.
19. Zhang, Y.; Qinb, C.; Liuc, Y. Effects of population density of a village and town system on the transportation cost for a biomass combined heat and power plant. *J. Environ. Manag.* **2018**, *223*, 444–451. [CrossRef]
20. Sokhansanj, S.; Mani, S.; Tagore, S.; Turhollow, A.F. Techno-economic analysis of using corn stover to supply heat and power to a corn ethanol plant—Part 1: Cost of feedstock supply logistics. *Biomass Bioenergy* **2010**, *34*, 75–81. [CrossRef]
21. Cambero, C.; Sowlati, T. Assessment and optimization of forest biomass supply chains from economic, social and environmental perspectives—A review of literature. *Renew. Sustain. Energy Rev.* **2014**, *36*, 62–73. [CrossRef]
22. Yue, D.; You, F.; Snyder, S.W. Biomass-to-bioenergy and biofuel supply chain optimization: Overview, key issues and challenges. *Comput. Chem. Eng.* **2014**, *66*, 36–56. [CrossRef]
23. Verani, S.; Sperandio, G.; Civitarese, V.; Spinelli, R. Harvesting mechanization in plantations for wood production: Working productivity and costs. *For. Riv. Selvic. Ecol. For.* **2017**, *14*, 237–246. [CrossRef]
24. Verani, S.; Sperandio, G.; Picchio, R.; Marchi, E.; Costa, C. Sustainability Assessment of a Self-Consumption Wood-Energy Chain on Small Scale for Heat Generation in Central Italy. *Energies* **2015**, *8*, 5182–5197. [CrossRef]
25. Bascietto, M.; Sperandio, G.; Bajocco, S. Efficient Estimation of Biomass from Residual Agroforestry. *ISPRS Int. J. Geo Inf.* **2020**, *9*, 21. [CrossRef]
26. Luxen, D.; Vetter, C. Real-time routing with OpenStreetMap data. In Proceedings of the 19th ACM SIGSPATIAL International Conference on Advances in Geographic Information Systems, Chicago, IL, USA, 1–4 November 2011; ACM: New York, NY, USA, 2011; pp. 513–516.
27. Miyata, E.S. *Determining Fixed and Operating Costs of Logging Equipment*; General Technical Report NC-55; Department of Agriculture, Forest Service, North Central Forest Experiment Station: St. Paul, MN, USA, 1980.
28. Paolotti, L.; Martino, G.; Marchini, A.; Boggia, A. Economic and environmental assessment of agro-energy wood biomass supply chains. *Biomass Bioenergy* **2017**, *97*, 172–185. [CrossRef]
29. Zahraee, S.M.; Shiwakoti, N.; Stasinopoulos, P. Biomass supply chain environmental and socio-economic analysis: 40-Years comprehensive review of methods, decision issues, sustainability challenges, and the way forward. *Biomass Bioenergy* **2020**, *142*, 105777. [CrossRef]
30. Pan, F.; Han, H.S.; Johnson, L.; Elliot, W. Production and cost of harvesting and transporting small-diameter trees for energy. *For. Prod. J.* **2008**, *58*, 47–53.
31. Han, S.K.; Murphy, G.E. Solving a woody biomass truck scheduling problem for a transport company in Western Oregon, USA. *Biomass Bioenergy* **2012**, *44*, 47–55. [CrossRef]
32. Caputo, A.C.; Palumbo, M.; Pelagagge, P.M.; Scacchia, F. Economics of biomass energy utilization in combustion and gasification plants: Effects of logistic variables. *Biomass Bioenergy* **2005**, *28*, 35–51. [CrossRef]

 *forests*

Article

Environmental Sustainability of Heat Produced by Poplar Short-Rotation Coppice (SRC) Woody Biomass

Giulio Sperandio *, Alessandro Suardi, Andrea Acampora and Vincenzo Civitarese

Consiglio per la Ricerca in Agricoltura e l'Analisi dell'Economia Agraria (CREA), Centro di Ricerca Ingegneria e Trasformazioni Agroalimentari, Via della Pascolare 16, 00015 Roma, Italy; alessandro.suardi@crea.gov.it (A.S.); andrea.acampora@crea.gov.it (A.A.); vincenzo.civitarese@crea.gov.it (V.C.)

* Correspondence: giulio.sperandio@crea.gov.it

Abstract: As demonstrated for some time, the reduction of greenhouse gases in the atmosphere can also take place using agroforestry biomass. Short-rotation coppice (SRC) is one of the sources of woody biomass production. In our work, the supply of woody biomass was considered by examining four different cutting shifts (2, 3, 4 and 5 years) and, for each, the Global Warming Potential (GWP) was evaluated according to the IPCC 2007 method. Regarding the rotation cycle, four biomass collection systems characterized by different levels of mechanization were analyzed and compared. In this study, it was assumed that the biomass produced by the SRC plantations was burned in a 350 kWt biomass power plant to heat a public building. The environmental impact generated by the production of 1 GJ of thermal energy was assessed for each of the forest plants examined, considering the entire life cycle, from the field phase to the energy production. The results were compared with those obtained to produce the same amount of thermal energy from a diesel boiler. Comparing the two systems analyzed, it was shown that the production and use of wood biomass to obtain thermal energy can lead to a reduction in the Global Warming Potential of over 70% compared to the use of fossil fuel.

Keywords: biomass; poplar; SRC; thermal energy; life cycle assessment; GWP; wood energy supply chain

Citation: Sperandio, G.; Suardi, A.; Acampora, A.; Civitarese, V. Environmental Sustainability of Heat Produced by Poplar Short-Rotation Coppice (SRC) Woody Biomass. *Forests* **2021**, *12*, 878. https://doi.org/10.3390/f12070878

Academic Editor: František Kačík

Received: 13 May 2021
Accepted: 1 July 2021
Published: 5 July 2021

Publisher's Note: MDPI stays neutral with regard to jurisdictional claims in published maps and institutional affiliations.

1. Introduction

Given the current need for a progressive reduction in the use of fossil fuels, which are mainly responsible for CO_2 emissions and pollution in the atmosphere, support for the use of woody biomass as an alternative source to produce thermal and electrical energy represents an important aspect in the discussion on the supply of energy from renewable sources. Overall, compared to total primary energy, bioenergy accounts for around 9.5%, or 70% of the renewable energy consumed [1]. In the future, bioenergy consumption is expected to grow to up to 30% of renewables due to its significant use mainly in heat generation and the transport sector [2]. The European Union supports and promotes actions aimed at achieving a more sustainable economic-energy and environmental system aimed at the progressive reduction of the use of fossil fuels in favor of renewable energy sources such as bioenergy [3].

As reported by many studies, the use of bioenergy can contribute to a significant improvement in environmental impacts compared to that produced by fossil fuels [4,5]. Furthermore, the wider diffusion of biofuels will lead to a substantial reduction in greenhouse gas emissions, eutrophication, pollution, acidification and depletion of the ozone layer, with a consequent reduction in damage to human health [6]. There are several sources of biomass that can be used to generate different final forms of bioenergy (thermal, electrical, liquid fuels and biogas). Among these, an interesting source of biomass is represented by short-rotation coppice (SRC) plantations, characterized by a high planting density, and made with fast-growing species, such as poplar, willow and eucalyptus. These crops, although currently covering only a few tens of thousands of hectares in Europe, can still represent an interesting production option for the purposes of achieving the objectives set by the European Union in terms of improving future environmental conditions. In particular, SRC plantations can play an interesting role in energy chains developed on a small scale in local rural districts mainly to produce thermal energy. In these cases, when planning the activation of energy chains, it is very important and useful to carry out an assessment of sustainability not only in economical but also in terms of environmental impact generated using bioenergy. In recent years, the life cycle assessment (LCA) methodology has been mainly used to estimate the positive or negative environmental impacts of processes associated with the production and use of biofuels [7–11]. The LCA methodology, albeit with its limitations, if supported by an inventory of primary data that allows multi-criteria analysis on the different phases of the production chains, can represent a very useful tool to provide indications and allow comparisons based on the different externalities deriving from different scenarios.

The aim of this study is to assess the carbon footprint of small-scale self-consumption wood-energy chains for heat generation based on SRC poplar plantations. The analysis is developed following the LCA method, applied to different scenarios concerning harvesting logistics and plantation cutting cycles. The biomass obtained is used to produce thermal energy in a local 350 kWt biomass plant. The carbon footprint of the energy chain, which includes the production of biomass and its transformation in a boiler, is compared with that of a conventional diesel-based boiler to produce the same amount of thermal energy.

2. Materials and Methods

2.1. Study Area and Poplar SRC Plantations

The experimental field was located in the North-East of Rome, within the farm of the CREA Research Centre for Engineering and Agro-Food Processing of Monterotondo, Italy (42°6'2.63″ N; 12°37'37.36″ E). The SRC poplar plantation of reference for the evaluation of the environmental impact analysis model with the LCA method was planted in 2005 on a flat surface of a total of 4.5 ha on a clayey soil with low organic matter content and phosphorus [12]. Three poplar clones were used: AF2 (*Populus* × *canadensis* Moenech), AF6 (*Populus nigra* L. × *Populus* × *generosa* A. Henry) and Monviso (*Populus* × *generosa* A. Henry × *Populus nigra* L.) [13,14]. The plantation was in single rows, spaced 2.80 m, while the cuttings on the row were 0.5 m apart, obtaining a density of 7140 trees ha^{-1}. For the purposes of this study, two periods of the production cycle were applied, namely 16 and 15 years. With reference to the first period, the harvesting operations were considered every 2 and 4 years. For the 15-year period, however, the harvesting operations were carried out every 3 and 5 years. A total of four biomass harvesting systems were considered: two systems applied in the cutting shifts of 2 and 3 years and other two different systems in the cutting shifts of 4 and 5 years. The harvesting systems in the 2- and 3-year cutting cycle were a two-step tractor-based harvesting system (TBHS) and a forage-based harvesting system (FBHS). The TBHS uses different equipment to perform the tree felling, then the extraction of whole trees and then chipping at the landing site. FBHS is a one-stage harvesting system that produces fresh wood chips directly in the field, where the biomass is unloaded into a trailer alongside the harvester and transported to the landing site for storage (Figure 1a) [15]. The two options considered for the 4- and 5-year cutting shifts were represented on the one hand by a manual felling of the trees with a chainsaw, the extraction of whole trees with a tractor equipped with a winch and subsequent chipping with a forest chipper at the landing site (Chainsaw-Based Harvest System—CBHS), and, on the other hand, by a feller-buncher for felling (Figure 1b), a skidder with grapple for trees extraction and a chipper before using the wood chips in the boiler (Shear head-Based Harvesting System—SBHS). It was considered that the whole biomass produced was used in the biomass boiler within the farm. From the combination of the cutting cycles and the types of mechanization adopted for the collection of the biomass, eight case studies were considered. In Scheme 1, the field operations on the plantations and the harvesting options considered, together with the operations regarding the boilers (biomass and diesel boilers), are reported.

(a)   (b)

Figure 1. Mechanized harvesting systems on poplar rotation coppice plantations: (**a**) Forage-Based Harvesting System (FBHS); (**b**) Shear head-Based Harvesting System (SBHS).

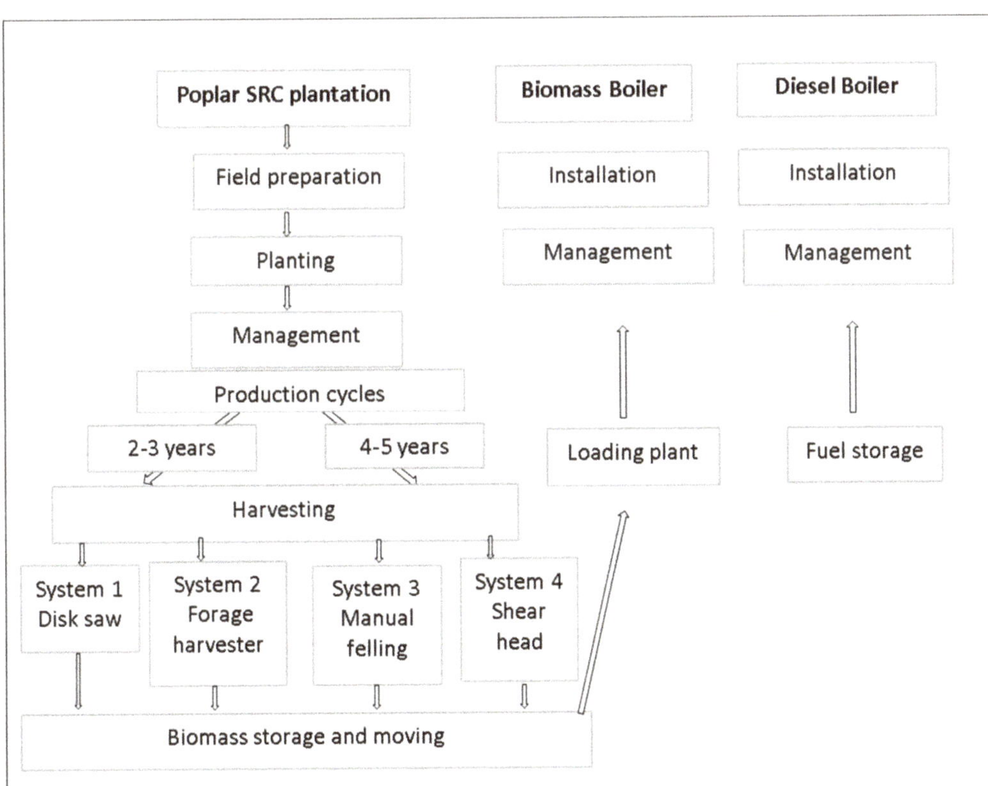

Scheme 1. Scheme of the poplar energy supply chain boundaries considered in the study.

2.2. Biomass and Diesel Boilers

The proposed environmental assessment model referred to the entire life cycle of the poplar plantation and the biomass power plant installed within the CREA farm. The biomass production energy system in the farm, and its use to produce thermal energy, was compared with a heating system of equivalent energy production but powered by fossil fuels (diesel).

This plant was used to heat the research center buildings, which were characterized by a potential volume to be heated of around 10,000 m^3. The biomass plant was equipped with a mobile grid, with a nominal power of 350 thermal kW. For the heating of the buildings, a period of 130 days per year was considered, calculating an average annual biomass supply of around 290 Mg, with a water content of 35%. For the purposes of this work, carbon footprint assessment generated by the thermal energy production of the biomass boiler was compared with that of the diesel boiler. The main parameters considered for the two boilers in comparison are shown in Table 1.

Table 1. Main parameters considered to evaluate the annual biomass or diesel consumption in the two boilers compared.

	Boilers	
	Biomass	**Diesel**
Building volume (m^3)	9450	9450
Operating period (days y^{-1})	130	130
Heating period (h y^{-1})	3120	1560
Rated thermal power (kWt)	350	315
Thermal efficiency of the boiler (%)	81%	90%
Lower heating value (LHV) (kWh kg^{-1})	3.11	11.86
Water content (%)	35.00%	$\leq$0.05%
Average biomass/diesel consumption (Mg y^{-1})	290.1	41.4

2.3. Environmental Analysis

The study evaluated the quantity of greenhouse gases emitted by poplar short- and medium-rotation coppices to produce thermal energy according to the LCA methodology. LCA is an in-depth "cradle-to-grave" analysis of the environmental impact of products or processes, and for this study, the impact category considered was the 100-year time horizon Global Warming Potential (GWP) based on the Intergovernmental Panel on Climate Change (IPCC) Fourth Assessment Report (AR4), published in 2007 [16]. In Table 2, for the eight different scenarios examined, the technical elements and the inputs used in the life cycle of the poplar plantations are reported. The CO$_2$ equivalent emissions per unit of thermal energy produced (1 GJ) downstream of each scenario were compared. The system evaluated the impact generated to produce 1 GJ of equivalent thermal energy from the agricultural, transport and transformation processes along the life cycle of the poplar groves, with reference to each cutting cycle and harvesting system considered. The functional unit was chosen to guarantee the comparison of the results obtained with other energy production systems, such as that from fossil sources. In the case of a small supply chain, the impact deriving from the production of 1 GJ of thermal energy produced by poplar wood chips in the biomass boiler was compared with 1 GJ produced by a diesel boiler. The system boundaries, i.e., the process units included in the LCA study, involved all the agricultural phases, the subsequent transport and transformation processes.

Table 2. Principal technical elements considered in the poplar plantations life cycle.

Operation	Period (Years)	Machine						Equipment		Technical Input		
		Operation (n./ha)	Power (kW)	Weight (kg)	Work Time (h/ha)	Fuel (L/ha)	Machine (N.)	Weight (kg)	Type	Type	Quantity (kg)	Rates (kg/ha)
Field preparation, planting and management												
- Deep scarification	1	1	199	8700	3.50	136	1	800	Ripper			
- Light ploughing	1	1	199	8700	1.60	67	1	1100	Plowshares			
- Fertilization (pre- and post-planting)	1	2	59	3100	0.60	5	2	200	Fertilizer spreader	N-P-K	800.00	500 KP; 300N
- Mechanized transplantation	1	1	73	3800	4.00	39	1	380	Transplanter	Cuttings	n. 7000	
- Chemical weeding post-planting	1	1	59	3100	0.80	6	1	250	Sprayer	Goal	2.00	
- Irrigation	1	1	59	3100	7.00	56	1	300	Pump and sprinkler	Water	400,000	
- Milling	1-2-3-4-5-6-7-8-9-10-11-12-13-14-15-16	1	26	2035	6.00	30	2	380	Milling machine			
- Harrowing	1-2-3-4-5-6-7-8-9-10-11-12-13-14-15-16	1	80	4100	1.20	14	1	500	Harrow			
- Stump grinding at the end cycle	15 or 16	1	199	8700	8.75	340	1	500	Stump grinder			
Harvesting												
-Option 1 (2y)—harvesting every 2 years		1										
- Felling (tractor with disksaw)		1	59	3100	2.13	17	1	180	Disk saw			
- Extraction (tractor with grapple)	2-4-6-8-10-12-14-16	1	80	5500	2.84	37	1	150	Log grapple			
- Chipping (farm chipper)		1	106	5500	10.22	195	1	1870	Chipper			
- Moving and load (chipwood)		1	74.50	7130	8.76	98	1					
-Option 2 (2y)—harvesting every 2 years		1										
- harvesting (forage harvester)		1	350	12,000	1.25	72	1					
- Extraction (tractor with trailer)	2-4-6-8-10-12-14-16	1	73	3800	1.25	14	2	600	Trailer			
- Moving stored chipwood			74.5	7130	2.84	35	1					
- Load chipwood (biomass plant)		1	125	8	74.50	35	1					
-Option 1 (3y)—harvesting every 3 years		1										
- Felling (tractor with disksaw)		1	59	3100	2.90	26	1	180	Disk saw			
- Extraction (tractor with grapple)	3-6-9-12-15	1	80	5500	3.76	50	1	150	Log grapple			
- Chipping (farm chipper)		1	106	5500	13.94	244	1	1870	Chipper			
- Moving and load (chipwood)			74.5	7130	13.15	147	1					
-Option 2 (3y) harvesting every 3 years		1										
- harvesting (forage harvester)		1	350	12,000	1.78	103	1					
- Extraction (tractor with trailer)	3-6-9-12-15	1	73	3800	1.78	19	2	600	Trailer			
- Moving stored chipwood		1	74.5	7130	4.26	52	1					
- Chipwood load		1	74.5	7130	13.15	147	1					
-Option 3 (4y)—harvesting every 4 years		1										
- Felling (manual with chainsaw)		1	1.7	4	85.20	43	1					
- Extraction (tractor winch)	4-8-12-16	1	70	3800	24.34	281	1	330	Winch			
- Chipping (farm chipper)		1	106	5500	17.53	307	1	1870	Chipper			
- Moving and load chipwood		1	74.5	7130	17.53	196	1					

Table 2. *Cont.*

Operation	Period (Years)	Machine						Equipment		Technical Input		
		Operation (n./ha)	Power (kW)	Weight (kg)	Work Time (h/ha)	Fuel (L/ha)	Machine (N.)	Weight (kg)	Type	Type	Quantity (kg)	Rates (kg/ha)
-Option 4 (4y)—harvesting every 4 years		1										
- Felling (shear head)		1	69	17,000	17.04	194	1	1350	Shear head			
- Extraction (skidder)	4-8-12-16	1	90	8000	5.68	84	1					
- Chipping (farm chipper)		1	106	5500	17.53	307	1	330	Chipper			
- Moving and load (chipwood)			74.5	7130	17.53	196	1					
-Option 3 (harvesting every 5 years)		1										
- Felling (manual with chainsaw)		1	1.7	4	88.75	45	1					
- Extraction (tractor winch)	5-10-15	1	70	3800	28.03	324	1		Winch			
- Chipping (farm chipper)		1	106	5500	20.72	362	1	330	Chipper			
- Moving and load (chipwood)		1	74.5	7130	21.91	245	1					
-Option 4 (harvesting every 5 years)												
- Felling (shear head)		1	90	8000	19.36	220	1	1350	Shear head			
- Extraction (skidder)	5-10-15	1	90	8000	6.66	99	1					
- Chipping (farm chipper)		1	106	5500	20.72	362	1		Chipper			
- Moving and load (chipwood)		1	74.5	7130	21.91	245	1					

For the construction of the inventory data, all inputs and outputs were collected and analyzed as primary and secondary data. The primary data were obtained directly from years of experimentation on the cultivation of SRC poplar. For some data not easily available, the database of the SimaPro 8.0.1 code, Ecoinvent 3 dataset (secondary data), was used. For each mechanical operation, the main technical characteristics were considered, such as the type of machine and equipment used, the engine power, the hours of work performed, the fuel and lubricant consumption, to evaluate the direct emissions of exhausted gases generated by the tractors, and the indirect emissions generated by the materials used for the construction of the agricultural machines used. The production processes were initially extrapolated from the SimaPro database and then modified, according to data collected directly in the field, only in the part of the tractors and equipment used and the consumption of diesel and lubricating oil, leaving unchanged the data relating to emissions into the air and onto the soil. All the operations necessary for the establishment of the plantations were considered and analyzed, as well as the post-planting management phases over the years, including the restoration of the field at the end of the cycle with grinding of the stumps [17].

Emissions related to the use of fertilizers and herbicides were determined based on the data available in the literature and of the outputs returned by the scientific software EFE-So (v 2.0.0.6; Fusi and Fusi) according to the model in [18]. CO_2 emissions from urea fertilization were calculated according to [19]. Herbicide emissions to air, surface water and groundwater were assessed using PestLCI 2.0 model [20]. A dry matter loss of 7% [21] was considered for wood harvesting systems that involved extracting the whole tree, drying at the landing site and chipping with a forest chipper when the moisture content of biomass reached 35%. Biomass storage was considered in the form of stacked and branchless trees. For these harvesting systems, 14.3 Mg per hectare per year of wood chips were produced. FBHS, on the other hand, considers the storage of fresh wood chips, with an average moisture content of 53%, in covered piles, which, during storage, are subjected to an average dry matter loss of 22% [22]. For the latter harvesting system, the quantity of wood chips obtainable from one hectare of poplar was $12 \, \text{Mg}^{-1} \, \text{ha}^{-1} \, \text{y}^{-1}$ (35% M.C.) after storage. Considering the data reported by various authors on biomass production from poplar groves subjected to different cutting cycles [16,23–25], in Figure 2, a prudential estimate of the dry biomass production for the four cycles considered in the study is shown. The average biomass production at farm gate was assumed for all the cases examined to be equal to 10 Mg of dry matter per hectare, per year.

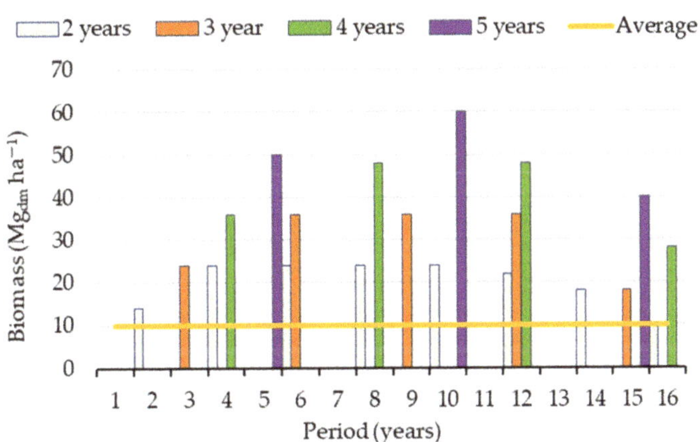

Figure 2. Estimate of dry biomass production in relation to the four cutting cycles considered.

Soil carbon (C) sequestration contributes to around 89% of the global mitigation potential of agriculture [26]. The amount of SOC stored in each soil is dependent on the balance between the amount of C entering the soil and the amount of C leaving the soil. The capacity of soils to store C is limited and the organic C content of soils tends towards equilibrium [27]. There is a high level of uncertainty regarding SOC estimation in soils, and there is no consensus or standard procedures on how to account for atmospheric carbon removals and releases [28–30]. According to the model reported by Whittaker et al. [31], an SRC plantation can store 8 Mg of stable organic carbon (SOC) at the end of the life cycle (corresponding to around 29 Mg CO_2 ha^{-1}). According to the model cited above, in the present research, the total C sequestered per hectare during the life of the SRC plantations studied, with an average yield of 10 Mg$_{dm}$ ha^{-1} y^{-1}, would correspond to 7.61 Mg C ha^{-1} (27.9 Mg CO_2 ha^{-1}). Although, in this study, SOC was not included in the calculation of the CO_2eq emitted, some considerations were subsequently reported to understand the positive impact provided by the SOC item in the final environmental performance results of the analyzed supply chains.

After considering all the agricultural phases, the impacts and resources of these phases (initially referred to a hectare of land) were compared to 1 GJ of biomass produced. This was possible by transforming the total production (Mg ha^{-1}) into energy (GJ ha^{-1}) since the low heating value (LHV) of poplar wood was calculated according to the Hartmann formula and considered equal to 11.2 MJ per kg of wood chips. The total inputs and emissions referring to one hectare were then divided by the production per hectare expressed in equivalent energy. In this way, it was possible to obtain the share of each agricultural phase to be attributed to 1 GJ of biomass produced. Average annual emissions and inputs were increased by the amount of inputs used and emissions generated over the years for planting, management, harvesting and removal, divided by the estimated life years of the crop (15 and 16 years). Reference was also made to an average annual production, calculated considering the yields obtained from poplar plantations during the years of their life cycle.

As the final phase, the results were evaluated, the weaknesses of the production phases were identified and the possibilities for improvement were defined.

3. Results and Discussion

The results in terms of emissions of kg CO_2eq per GJ of thermal energy produced are shown in Figure 3. The differences are highlighted above all between the two cases referring to the two-year cutting cycles compared to all the other cases. Despite the small number of observations without repetitions, which increases the margin of error, we still wanted to proceed with a statistical analysis by processing the data divided into four groups coinciding with the four cutting cycles. Welch's F test for unequal variance was performed and the results showed statistically significant differences ($p < 0.05$) between the first group (two-year cutting cycle) compared to the other three groups (cutting cycle of three, four and five years). From the first observations, it can be stated that more frequent harvesting operations contribute to increasing the number of agricultural practices adopted. This aspect is more evident especially in the case of the SRC plantations with a two-year cutting cycle (Figure 3), where the fertilization represented 49% of the overall emissions of the wood chip production.

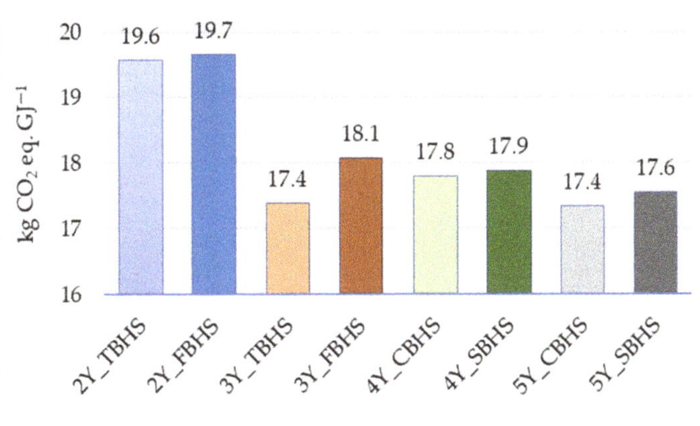

Figure 3. Emissions per GJ of thermal energy produced (in kg CO_2eq) with reference to the TBHS, FBHS, CBHS and SBHS harvesting systems (IPCC GWP 100a).

With specific reference to nitrogen fertilization (over 1 Mg of N in 16 years), it is responsible for 33% of the overall emissions referring to the production cycle of wood chips (34.6 kg CO_2eq per Mg of biomass at 35% of MC). A lower nitrogen intake was recorded in the five-year cutting cycle (around 0.4 Mg of N in 15 years), as it was administered only twice in correspondence with the first two harvests. This helped to reduce emissions attributable to fertilizers by nearly 60% compared to those of two-year cycles. The choice of the type of fertilizers to be used during the management of the plantations is therefore extremely important, as well as the application of good agricultural practices to ensure the maintenance of soil fertility, aimed at achieving a balance between the organic substance removed and the inputs of fertilizers, avoiding as much as possible the loss of nitrates by leaching. These results must be considered well-established when discussing emissions in agriculture [10,32–36]. In the analysis of the four different production cycles, slightly lower CO_2 emissions were recorded in reference to the use of a lower level of mechanization. More evident differences were recorded for the two-year production cycles (19.6–19.7 kg CO_2 GJ^{-1}), compared to all the other cases, which, on the other hand, had a lower impact in environmental terms (17.4-18.1 kg CO_2 GJ^{-1}).

According to the results of the study, the highest CO_2 emissions were therefore attributable to the two cases of the biennial poplar chain. In fact, for each GJ of thermal energy produced by the combustion of the biomass obtained from these cycles, a maximum of 19.7 kg CO_2eq was generated (Figure 3). However, within the two-year supply chain, the two cases examined were practically similar, showing differences contained within 1%.

This minimal difference is essentially attributable to two aspects, which, in the biennial poplar supply chains, are compensated: on the one hand, in the 2Y_FBHS case, there were greater storage losses for fresh wood chips, which led to greater emissions from the supply chain, while, on the other hand, in the 2Y_TBHS case, higher emissions were produced, mainly due to the increase in the number of operations to be carried out for the production of wood chips.

Figure 4 shows the CO_2 emissions per unit of dry biomass for the eight reference cases. Additionally, based on these values, within the limits already described above, a statistical analysis was performed with Welch's F test for unequal variance, returning a non-significant result ($p > 0.05$). The values referring to the four groups did not therefore show statistically significant differences between them. The CO_2 emissions related to the FBHS harvesting system (Figure 4) were due to the field harvesting phase of the forage harvester for 73% (18.8 kgCO_2eq Mg_{dm}^{-1}) and to the movement of wood chips in the piles for the remaining 27% (6.9 kgCO_2eq Mg_{dm}^{-1}). In the TBHS system, which involved the use of three different machines, most of the emissions, equal to 78.2%, were attributable to the chipping operation performed with a forest chipper at the landing site (33 kgCO$_2$eq Mg$_{dm}^{-1}$). Regarding the remaining part, 14.4% was due to the extraction of whole trees with a tractor with a winch from the plantation to the landing site (6.1 kgCO$_2$eq Mg$_{dm}^{-1}$), and 7.4% to the cutting of the trees with the TBHS system (3.1 kgCO$_2$eq Mg$_{dm}^{-1}$).

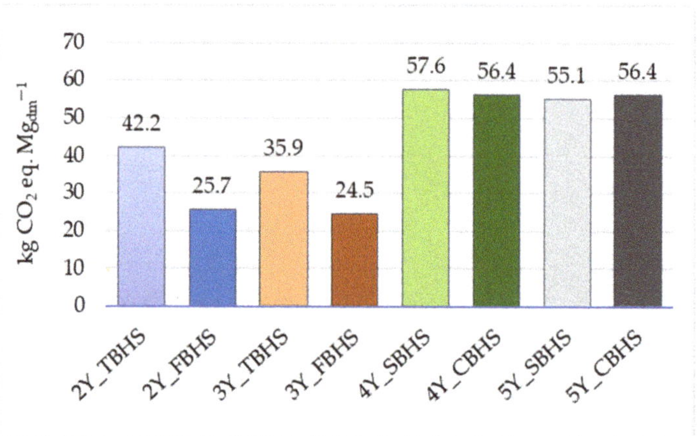

Figure 4. Emissions per Mg$_{dm}$ of biomass produced (in kg CO$_2$eq) with reference to the TBHS, FBHS, CBHS and SBHS harvesting systems, considering storage losses and excluding emissions from all other agricultural practices (IPCC GWP 100a).

In the FBHS scenario, much higher dry matter losses were found than in the TBHS scenario, due to the different storage system used. An optimization of this phase could lead the system to a reduction in emissions, making it more competitive with respect to the TBHS. This effect can also be applied in the case of three-year cutting cycle that applies the same harvesting logistics as the two-year one, even if, here, the differences to be bridged were slightly more marked (4%). In this case, therefore, it would be more difficult to obtain a possible reduction in emissions to the level of those recorded for the TBHS. In general, cutting shifts longer than two years showed lower emissions. Figure 3 shows that the two supply chains generated a very similar quantity of CO$_2$ in terms of GJ produced (an average of 17.6 kg CO$_2$eq GJ^{-1}) because the harvesting methods of the four-year and five-year cycles were the same. The harvesting systems used in four-year and five-year cycles produced, on average, 43% and 124% more CO$_2$ than the TBHS and FBHS systems, respectively, both used in the two-year and three-year cycles. From an environmental point of view, the results of the study show that the emissions of greenhouse gases produced by the analyzed wood-energy supply chains range from a maximum of 19.7 (biennial supply chain) to a minimum of 17.4 kg CO$_2$eq per GJ of thermal energy produced by the biomass boiler considered. This result, although higher than that reported in other studies [25,37–41], is still well below the CO$_2$ emissions emitted by a boiler of the same size fueled by fossil fuels.

In Figure 5, it can be seen how the transition from a diesel boiler to a biomass-fueled biennial poplar wood chip (which was the least efficient compared to the other scenarios analyzed) allows a 77% reduction in greenhouse gas emissions. In the study, CO$_2$ stored as stable soil organic carbon (SOC) was not considered in the calculation, although the literature reports various environmental studies in which the immobilized carbon in the soil is considered [42–47]. If, in the present study, the estimated 7.61 Mg C ha^{-1} (corresponding to 27.9 Mg CO$_2$ ha^{-1}) accumulated at the end of the cycle is included, then the greenhouse gas emissions produced to generate 1 GJ of thermal energy would be even lower. In fact, considering a life cycle of 16 years and a production amount of 130 GJ, 13.4 kg CO$_2$ would have been saved per GJ produced. In this case, the 2Y_TBHS scenario reported in Figure 4 would result in the emission of only 6.3 kg CO_2eq per thermal GJ generated.

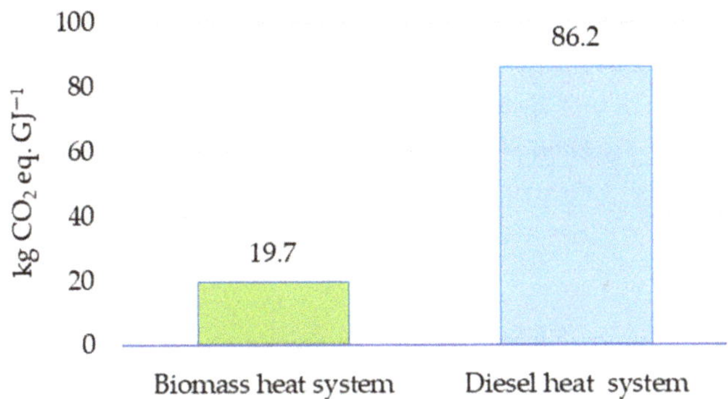

Figure 5. Comparison between the emissions (in kg CO_2eq) generated by a biomass boiler connected to the less efficient biennial poplar supply chain and a diesel boiler to produce 1 GJ of thermal energy.

4. Conclusions

The use of fossil fuels for energy production has led over time to high greenhouse gas emissions. This is why we are seeing increasing interest in renewable energy. This includes biomass from short-rotation agroforestry crops such as poplar due its energy potential, as an alternative to fossil fuels and the consequent production of GHG.

With the aim of evaluating the environmental sustainability of small-scale self-consumption wood-energy chains based on poplar SRC for heat generation, this study focused on the carbon footprint of various crop management scenarios represented by different harvesting systems and cutting cycles.

The energy production obtained from the poplar three-year and five-year cutting cycles (TBHS and CBHS harvesting systems) was found to be the most sustainable, but the most evident differences were highlighted only between the two-year supply chains and those of the other cutting cycles. Several studies have highlighted how the type and quantity of fertilizer applied is important in environmental performance and, in this context, the optimization of the inputs used becomes fundamental.

Clearly, the most sustainable harvesting method is characterized by fewer production steps. In our case, the FBHS harvesting system, characterized by chipping plants in the field, was the most sustainable compared to the two-stage harvesting systems (TBHS, CBHS and SBHS). However, it should be noted that the FBHS system results in higher dry matter losses during the storage phase and that the benefit in terms of emissions is not very evident when considering the entire energy chain.

It is therefore essential to optimize the storage of fresh wood chips to reduce dry matter losses and thus obtain a further reduction in total emissions from the energy supply chain.

Analyzing our results, we can affirm that producing thermal energy from biomass, compared to that obtained from fossil fuel, allows a reduction in greenhouse gas emissions equal to 77%. A further improvement (93% reduction of emissions) is possible if we consider the CO_2 stored in the soil in the form of SOC at the end of SRC life cycle. The stabilization of CO_2 in soil as soil organic carbon should be further investigated as this is the aspect that makes a bioenergy supply chain more sustainable.

Author Contributions: Conceptualization, G.S.; methodology, G.S., A.S.; software, A.S.; formal analysis, G.S., A.S.; data curation, G.S., A.S.; writing—original draft preparation, G.S., A.S., A.A.; writing—review and editing, G.S., A.S., A.A., V.C. All authors have read and agreed to the published version of the manuscript.

Funding: This research was funded by the Italian Ministry of Agriculture, Food and Forestry Policies (MiPAAF), grant D.D. n. 26329, 1 April 2016, project AGROENER "Energia dall'agricoltura: innovazioni sostenibili per la bioeconomia".

Institutional Review Board Statement: Not applicable.

Informed Consent Statement: Not applicable.

Data Availability Statement: Data is contained within the article.

Conflicts of Interest: The authors declare no conflict of interest.

References

1. IEA Renewables. *Analysis and Forecasts*; Executive Summary; IEA: Paris, France, 2017. Available online: https://www.iea.org/ (accessed on 13 May 2021).
2. IEA Renewables. *Analysis and Forecasts*; Executive Summary; IEA: Paris, France, 2018. Available online: https://www.iea.org/ (accessed on 13 May 2021).
3. Scarlat, N.; Dallemand, J.-F.; Monforti-Ferrario, F.; Banja, M.; Motola, V. Renewable energy policy framework and bioenergy contribution in the European Union—An overview from National Renewable Energy Action Plans and Progress Reports. *Renew. Sustain. Energy Rev.* **2015**, *51*, 969–985. [CrossRef]
4. Punter, G.; Rickeard, D.; Larivé, J.F.; Edwards, R.; Mortimer, N.; Horne, R.; Bauen, A.; Woods, J. *Well-to-Wheel Evaluation for Production of Ethanol from Wheat*; A Report by the LowCVP Fuels Working Group, WTW Sub-Group; Zemo Partnership: London, UK, 2004; Volume 40. Available online: http://www.rms.lv/bionett/Files/BioE-2004-001%20Ethanol_WTW_final_report.pdf (accessed on 13 May 2021).
5. Kim, S.; Dale, B.E. Life cycle assessment of various cropping systems utilized for producing biofuels: Bioethanol and biodiesel. *Biomass Bioenergy* **2005**, *29*, 426–439. [CrossRef]
6. Farrell, A.E.; Plevin, R.J.; Turner, B.T.; Jones, A.D.; O'hare, M.; Kammen, D.M. Ethanol can contribute to energy and environmental goals. *Science* **2006**, *311*, 506–508. [CrossRef]
7. Fritsche, U.R.; Hünecke, K.; Hermann, A.; Schulze, F.; Wiegmann, K.; Adolphe, M. *Sustainability Standards for Bioenergy*; Final Report; Öko-Institute e.V.: Darmstadt, Germany, 2006; p. 39. Available online: http://np-net.pbworks.com/f/OEKO+(2006)+Sustainability+Standards+for+Bioenergy.pdf (accessed on 13 May 2021).
8. Scharlemann, J.P.W.; Laurance, W.F. How Green Are Biofuels? Environmental science? *Science* **2008**, *319*, 43–44. [CrossRef]
9. Chiaramonti, D.; Recchia, L. Is life cycle assessment (LCA) a suitable method for quantitative CO_2 saving estimations? the impact of field input on the LCA results for a pure vegetable oil chain. *Biomass Bioenergy* **2010**, *34*, 787–797. [CrossRef]
10. Dias, G.M.; Ayer, N.W.; Kariyapperuma, K.; Thevathasan, N.; Gordon, A.; Sidders, D.; Johannesson, G.H. Life cycle assessment of thermal energy production from short-rotation willow biomass in Southern Ontario, Canada. *Appl. Energy* **2017**, *204*, 343–352. [CrossRef]
11. Sanz Requena, J.F.; Guimaraes, A.C.; Quirós Alpera, S.; Relea Gangas, E.; Hernandez-Navarro, S.; Navas Gracia, L.M.; Martin-Gil, J.; Fresneda Cuesta, H. Life Cycle Assessment (LCA) of the biofuel production process from sunflower oil, rapeseed oil and soybean oil. *Fuel Process. Technol.* **2011**, *92*, 190–199. [CrossRef]
12. Di Matteo, G.; Sperandio, G.; Verani, S. Field performance of poplar for bioenergy in southern Europe after two coppicing rotations: Effects of clone and planting density. *IForest* **2012**, *5*, 224–229. [CrossRef]
13. Picco, F.; Giorcelli, A.; Castro, G. *Dichotomous Key for the Nursey Recognition of the Main Poplar Clones Grown in the European Union*; Volume II Clonal cards CRA-PLF; Research Units for Intensive Wood Production: Casale Monferrato, Italy, 2007; Volume 352.
14. Di Matteo, G.; Nardi, P.; Verani, S.; Sperandio, G. Physiological adaptability of Poplar clones selected for bioenergy purposes under non-irrigated and suboptimal site conditions: A case study in Central Italy. *Biomass Bioenergy* **2015**, *81*, 183–189. [CrossRef]
15. Costa, C.; Sperandio, G.; Verani, S. Use of multivariate approaches in biomass energy plantation harvesting: Logistics advantages. *Agric. Eng. Int. CIGR J.* **2014**, 70–79.
16. Lemke, P.; Ren, J.; Alley, R.B.; Allison, I.; Carrasco, J.; Flato, G.; Fujii, Y.; Kaser, G.; Mote, P.; Thomas, R.H.; et al. *Climate Change 2007: The Physical Science Basis: Contribution of Working Group I to the Fourth Assessment Report of the Intergovernmental Panel on Climate Change*; Solomon, S., Qin, D., Manning, M., Chen, Z., Marquis, M., Averyt, K.B., Tignor, M., Miller, H.L., Eds.; Cambridge University Press: Cambridge, UK; New York, NY, USA, 2007.
17. Verani, S.; Sperandio, G.; Picchio, R.; Marchi, E.; Costa, C. Sustainability assessment of a self-consumption wood-energy chain on small scale for heat generation in central Italy. *Energies* **2015**, *8*, 5182–5197. [CrossRef]
18. Brentrup, F.; Kusters, J.; Lammel, J.; Kuhlmann, H. Methods to estimate on-field nitrogen emissions from crop production as an input to LCA studies in the agricultural sector. *Int. J. Life Cycle Assess.* **2000**, *5*, 349–357. [CrossRef]

19. De Klein, C.; Novoa, R.S.A.; Ogle, S.; Smith, K.A.; Rochette, P.; Wirth, T.C.; McConkey, B.G.; Mosier, A.; Rypdal, K.; Walsh, M. N_2O emissions from managed soils, and CO_2 emissions from lime and urea application. In *2006 IPCC Guidelines for National Greenhouse Gas Inventories*; IPCC: Geneva, Switzerland, 2006; Volume 4, pp. 1–54.

20. Dijkman, T.J.; Birkved, M.; Hauschild, M.Z. PestLCI 2.0: A second generation model for estimating emissions of pesticides from arable land in LCA. *Int. J. Life Cycle Assess.* **2012**, *17*, 973–986. [CrossRef]

21. Pecenka, R.; Lenz, H.; Hering, T. Options for Optimizing the Drying Process and Reducing Dry Matter Losses in Whole-Tree Storage of Poplar from Short-Rotation Coppices in Germany. *Forests* **2020**, *11*, 374. [CrossRef]

22. Lenz, H.; Idler, C.; Hartung, E.; Pecenka, R. Open-air storage of fine and coarse wood chips of poplar from short rotation coppice in covered piles. *Biomass Bioenergy* **2015**, *83*, 269–277. [CrossRef]

23. Facciotto, G.; Bergante, S.; Rosso, L.; Minotta, G. Comparison between two and five years rotation models in poplar, willow and black locust Short Rotation Coppices (SRC) in North West Italy. *Ann. Silvic. Res.* **2020**, *45*, 12–20.

24. Manzone, M.; Bergante, S.; Facciotto, G. Energy and economic evaluation of a poplar plantation for woodchips production in Italy. *Biomass Bioenergy* **2014**, *60*, 164–170. [CrossRef]

25. Njakou Djomo, S.; Ac, A.; Zenone, T.; De Groote, T.; Bergante, S.; Facciotto, G.; Sixto, H.; Ciria Ciria, P.; Weger, J.; Ceulemans, R. Energy performance of intensive and extensive short rotation cropping systems for woody biomass production in the EU. *Renew. Sustain. Energy Rev.* **2015**, *41*, 845–854. [CrossRef]

26. Palmieri, N.; Forleo, M.B.; Giannoccaro, G.; Suardi, A. Environmental impact of cereal straw management: An on-farm assessment. *J. Clean. Prod.* **2017**, *142*, 2950–2964. [CrossRef]

27. Sommer, R.; Bossio, D. Dynamics and climate change mitigation potential of soil organic carbon sequestration. *J. Environ. Manag.* **2014**, *144*, 83–87. [CrossRef]

28. Brandão, M.; Milà i Canals, L.; Clift, R. Soil organic carbon changes in the cultivation of energy crops: Implications for GHG balances and soil quality for use in LCA. *Biomass Bioenergy* **2011**, *35*, 2323–2336. [CrossRef]

29. Brandão, M.; Levasseur, A.; Kirschbaum, M.U.F.; Weidema, B.P.; Cowie, A.L.; Jørgensen, S.V.; Hauschild, M.Z.; Pennington, D.W.; Chomkhamsri, K. Key issues and options in accounting for carbon sequestration and temporary storage in life cycle assessment and carbon footprinting. *Int. J. Life Cycle Assess.* **2013**, *18*, 230–240. [CrossRef]

30. Petersen, B.M.; Knudsen, M.T.; Hermansen, J.E.; Halberg, N. An approach to include soil carbon changes in life cycle assessments. *J. Clean. Prod.* **2013**, *52*, 217–224. [CrossRef]

31. Whittaker, C.; Macalpine, W.; Yates, N.E.; Shield, I. Dry matter losses and methane emissions during wood chip storage: The impact on full life cycle greenhouse gas savings of short rotation coppice willow for heat. *BioEnergy Res.* **2016**, *9*, 820–835. [CrossRef] [PubMed]

32. Nordborg, M.; Berndes, G.; Dimitriou, I.; Henriksson, A.; Mola-Yudego, B.; Rosenqvist, H. Energy analysis of poplar production for bioenergy in Sweden. *Biomass Bioenergy* **2018**, *112*, 110–120. [CrossRef]

33. Fernando, A.L.; Rettenmaier, N.; Soldatos, P.; Panoutsou, C. Sustainability of Perennial Crops Production for Bioenergy and Bioproducts. In *Perennial Grasses for Bioenergy and Bioproducts*; Alexopoulou, E., Ed.; Elsevier Inc.: Amsterdam, The Netherlands, 2018; pp. 245–283.

34. Schweier, J.; Molina-Herrera, S.; Ghirardo, A.; Grote, R.; Díaz-Pinés, E.; Kreuzwieser, J.; Haas, E.; Butterbach-Bahl, K.; Rennenberg, H.; Schnitzler, J.P.; et al. Environmental impacts of bioenergy wood production from poplar short-rotation coppice grown at a marginal agricultural site in Germany. *GCB Bioenergy* **2017**, *9*, 1207–1221. [CrossRef]

35. Njakou Djomo, S.; Witters, N.; Van Dael, M.; Gabrielle, B.; Ceulemans, R. Impact of feedstock, land use change, and soil organic carbon on energy and greenhouse gas performance of biomass cogeneration technologies. *Appl. Energy* **2015**, *154*, 122–130. [CrossRef]

36. Gabrielle, B.; Nguyen The, N.; Maupu, P.; Vial, E. Life cycle assessment of eucalyptus short rotation coppices for bioenergy production in southern France. *GCB Bioenergy* **2013**, *5*, 30–42. [CrossRef]

37. Forleo, M.B.; Palmieri, N.; Suardi, A.; Coaloa, D.; Pari, L. The eco-efficiency of rapeseed and sunflower cultivation in Italy. Joining environmental and economic assessment. *J. Clean. Prod.* **2018**, *172*, 3138–3153. [CrossRef]

38. Palmieri, N.; Forleo, M.B.; Suardi, A.; Coaloa, D.; Pari, L. Rapeseed for energy production: Environmental impacts and cultivation methods. *Biomass Bioenergy* **2014**, *69*, 1–11. [CrossRef]

39. Fazio, S.; Monti, A. Life cycle assessment of different bioenergy production systems including perennial and annual crops. *Biomass Bioenergy* **2011**, *35*, 4868–4878. [CrossRef]

40. Roedl, A. Production and energetic utilization of wood from short rotation coppice-a life cycle assessment. *Int. J. Life Cycle Assess.* **2010**, *15*, 567–578. [CrossRef]

41. Krzyżaniak, M.; Stolarski, M.J.; Warmiński, K. Life cycle assessment of poplar production: Environmental impact of different soil enrichment methods. *J. Clean. Prod.* **2019**, *206*, 785–796. [CrossRef]

42. Searchinger, T.; Heimlich, R.; Houghton, R.A.; Dong, F.; Elobeid, A.; Fabiosa, J.; Tokgoz, S.; Hayes, D.; Yu, T.-H. Use of US croplands for biofuels increases greenhouse gases through emissions from land-use change. *Science* **2008**, *319*, 1238–1240. [CrossRef] [PubMed]

43. Fargione, J.; Hill, J.; Tilman, D.; Polasky, S.; Hawthorne, P.P. Land clearing and the biofuel carbon debt. *Science* **2008**, *319*, 1235–1238. [CrossRef] [PubMed]

44. Reinhardt, G.A.; Von Falkenstein, E. Environmental assessment of biofuels for transport and the aspects of land use competition. *Biomass Bioenergy* **2011**, *35*, 2315–2322. [CrossRef]
45. Milà i Canals, L.; Rigarlsford, G.; Sim, S. Land use impact assessment of margarine. *Int. J. Life Cycle Assess.* **2013**, *18*, 1265–1277. [CrossRef]
46. Arzoumanidis, I.; Fullana-i-Palmer, P.; Raggi, A.; Gazulla, C.; Raugei, M.; Benveniste, G.; Anglada, M. Unresolved issues in the accounting of biogenic carbon exchanges in the wine sector. *J. Clean. Prod.* **2014**, *82*, 16–22. [CrossRef]
47. Hamelin, L.; Jørgensen, U.; Petersen, B.M.; Olesen, J.E.; Wenzel, H. Modelling the carbon and nitrogen balances of direct land use changes from energy crops in Denmark: A consequential life cycle inventory. *GCB Bioenergy* **2012**, *4*, 889–907. [CrossRef]

*forests*

Article

Empowering Forest Owners with Simple Volume Equations for Poplar Plantations in the Órbigo River Basin (NW Spain)

Roberto Blanco and Juan A. Blanco *

Department Ciencias, Universidad Pública de Navarra, Campus de Arrosadía s/n, Pamplona, 31006 Navarra, Spain; Roberto_magister@yahoo.es
* Correspondence: juan.blanco@unavarra.es; Tel.: +34-948-16-9859

Abstract: Hybrid poplar plantations are becoming increasingly important as a source of income for farmers in northwestern Spain, as rural depopulation and farmers aging prevent landowners from planting other labor-intensive crops. However, plantation owners, usually elderly and without formal forestry background, lack of simple tools to estimate the size and volume of their plantations by themselves. Therefore, farmers are usually forced to rely on the estimates made by the timber companies that are buying their trees. With the objective of providing a simple, but empowering, tool for these forest owners, simple equations based only on diameter were developed to estimate individual tree volume for the Órbigo River basin. To do so, height and diameter growth were measured for 10 years (2009–2019) in 404 trees growing in three poplar plantations in Leon province. An average growth per tree of 1.66 cm year^{-1} in diameter, 1.52 m year^{-1} in height, and 0.03 m^3 year^{-1} in volume was estimated, which translated into annual volume increment of 13.02 m^3 ha^{-1} year^{-1}. However, annual volume increment was different among plots due to their fertility, with two plots reaching maximum volume growth around 11 years since planting and another at 13 years, encompassing the typical productivity range in plantations in this region. Such data allowed developing simple but representative linear, polynomial and power equations to estimate volume explaining 93%–98% of the observed variability. Such equations can be easily implemented in any cellphone with a calculator, allowing forest owners to accurately estimate their timber existences by using only a regular measuring tape to measure tree diameter. However, models for height were less successful, explaining only 75%–76% of observed variance. Our approach to generate simplified volume equations has shown to be viable for poplar, but it could be applied to any species for which several volume equations are available.

Keywords: tree plantations; growth equations; rotation length; growth rates; poplar productivity

Citation: Blanco, R.; Blanco, J.A. Empowering Forest Owners with Simple Volume Equations for Poplar Plantations in the Órbigo River Basin (NW Spain). *Forests* **2021**, *12*, 124. https://doi.org/10.3390/f12020124

Academic Editors: Angela Lo Monaco, Cate Macinnis-Ng and Om P. Rajora

Received: 4 January 2021
Accepted: 21 January 2021
Published: 23 January 2021

1. Introduction

Increasing wood and paper demand is fueling an expansion of fast-growing tree plantations around the world, and particularly in southwestern Europe [1]. Such short-term plantations cover 7% of planted areas provide 33% of wood volume used in the industry [2]. Among other species, poplar (*Populus* sp.) plantations have expanded due to the quality of their timber for pulp and fiber, and their new usages as raw material for high-end designer's furniture, as well as an incipient cellulose-based chemical industry. Poplar plantations, usually considered as short-term silviculture, are typically established in former agricultural lands or in fertile forestlands with little or none degradation [3].

In 2016, poplar stands in Spain accounted for 130,000 ha [4], being one-third natural stands and two-thirds plantations. Such plantations have been traditionally used for pulp, plywood or veneers. Poplars are also preferred in the rural landscape of southwestern Europe, as they are natural forest formations in riversides. Hence, they contribute to a more diverse landscape in cropland-dominated regions while supporting riparian biodiversity [5]. In addition, poplars are more resistant to frost than other fast-growing species such as radiata pine or eucalyptus, while reaching good profitability levels [6]. These facts make poplars ideal for continental-like climates with cold winters and hot summers, such as in interior Spain.

Poplar plantations are usually monoclonal, and although several hybrids are used in Spain, *Populus* × *euramericana* (Dode) Guinier clone I-214 (*P. deltoides* Marsh. ♀× *P. nigra* L. ♂) is the most common. This clone represents about 70% of the total area covered by poplar plantations in the Duero River basin [7]. Poplar plantations in this region are typically managed in 15-year rotations, demanding labor only for the first two years (plantation establishment and weed control) and once or twice more in the rotation when pruning. The rest of the time, management is reduced to sporadic irrigation during the driest part of the summer and opportunistic weed control (mechanical, chemical or by cattle or sheep grazing). Poplar plantations also provide additional income from low-productive croplands to farmers or urban owners not willing to work their farmlands, usually inherited [8,9]. In Spain, only about one third of owners consider themselves as professionally engaged with their land, being the rest either retired or amateurs who just take their forests as a hobby [10]. In addition, absent owners due to migration to urban centers and rural depopulation, combined with aging of forest owners [11], has resulted in a lack of land owners' interest in farming.

Consequently, many landowners (both public and private) are planting poplars as a way to keep their farmlands productive, but with much less involvement from their parts than regular crop-farming activities. However, this change has brought a different challenge for landowners: the estimation of their timber stocks. Due to lack of formal training in forestry, most owners do not know how much volume their plantations have, or how to estimate it. Hence, when selling the timber to loggers, plantation owners have to accept the prices offered by sawmills, which in this region are dominated by one large transnational company. Such lack of negotiation power reduces the profits for forest owners and creates an unbalanced market in which timber companies and sawmills impose their economic objectives over the plantation owners' [12].

A simple way to empower forest owners is by providing simple, easy-to-use tools to estimate standing timber volume. Although multitude of volume equations have been developed over the years for poplar (see for example [13–18]), they have always been created to reach objectives dictated by professional foresters of forest scientists. Hence, their objectives usually result in a preference for equations using several predictors combined (e.g., tree diameter, tree height, stocking density, climate, etc.), as they provide more accurate estimates. Such equations follow statistical procedures to estimate tree stem volume that sometimes can be quite complex (see, for example, [16,18]). However, such models are non-practical for forest owners, particularly in this region, where their average age is 60 years [19], and the average education level is primary school or less [20]. Hence, plantation owners could estimate standing timber volume in their properties, if they are provided with simple volume equations that use as predictor a single, easy to measure variable, such as tree diameter. In addition, knowing tree heights can also become part of the negotiations among owners and loggers, as height defines the number of pieces in which a stem is cut, and therefore the number of truck trips that must be done when hauling the harvested trees.

Previous research in poplar plantations has shown that simplifications in model structure to estimate plantation productivity [17], or alternative simple approaches to volume equations [21] are possible without losing predictive accuracy. Therefore, our initial hypothesis is that simple, diameter-based volume equations with a high predictive accuracy can be developed for poplar plantations. To test this hypothesis, we first selected several plots near the Órbigo River in its middle basin, and then tested for differences in soil fertility among plots to ensure we covered the typical range of productivity levels present at this region. Finally, we monitored planted trees for 10 years, measuring their diameter and heights annually. Hence, if such hypothesis were accepted, our main objective was to develop simple volume equation for any age during the rotation length. As a secondary objective, we also aimed to develop simple height equations following the same approach.

2. Materials and Methods

2.1. Study Sites

The Iberian Peninsula is surrounded by coastal mountain ranges, with an interior plateau giving most of Spain a continentalized Mediterranean climate. The landscape in the northern half of this region is dominated by the Duero River basin, from which the Órbigo River is a 2nd-order tributary. In this region, three plots were used for this research, located in the municipality of Villarejo de Órbigo (León province, 42°26′46″ N, 5°54′15″ W), and placed at 820 m.a.s.l. Climate is temperate with dry summers (Csb in the Köppen-Geiger classification [22]). Mean annual temperature is 11.2 °C, being July the warmest month (19.9 °C on average) and January the coldest (3.2 °C). Average annual precipitation is 562 mm, being November the rainiest month (74 mm) and July the driest (23 mm) [23]. Such temperature and precipitation patterns create a precipitation deficit in July and August [24], with 0.5–0.75 aridity index (P/ETP, [25]). Dominant winds are South to West (45% frequency), with moderate speed (3–4 m s^{-1}). The plots are flat and placed on former agricultural lands. Soils are calcic fluvisols (FAO classification), developed on the alluvial floodplain of the Órbigo River, at about 3–5 km from the riverbed, being loam to clay-loam in texture. pH is about neutral (7.2–7.3), with moderate organic matter content in the top layer (1.5%–1.8%) and high calcium concentration (0.20%–0.53%) [26].

Selected plots covered a typical range of previous land management history in the region. Plot 1 was planted after a previous poplar plantation in the same plot was harvested. Plot 2 was previously a pasture composed by local grasses, and Plot 3 was previously planted with a rotation of cereal and clover for two decades following reparcelling. These sites are representative of extensive areas in the floodplains of the middle Órbigo River basin.

In late autumn 2005, plots were plowed, and in March 2006, 2-year old saplings of *Populus × euroamericana*, clonal variety I-214 were planted in plots 1 and 2. Plot 3 was planted in December of the same year. A total of 430 trees were planted in holes about 1.2–1.4 m deep, in a 5 × 5 m spacing. Trees were tended following usual management practices in the region (localized fertilization with a commercial NPK fertilizer 12N-24P-12K, 0.3 kg tree^{-1} at plantation establishment and again in August 2008. Weeds were controlled with a commercial chemical herbicide at plantation time and by manual removal of weeds with scythe. Irrigation was applied twice each summer: in June and August. Trees were pruned in autumn 2008, and again in autumn 2011. Due to wind breakage, poor growth and other mechanical damage, 404 undamaged trees survived to 2019. A detailed description of the sites and additional analyses can be found in [27].

2.2. Trees and Soil Measurements, and Data Analysis

Diameters at breast height (1.30 m, DBH) were measured annually in August for each tree using a tree caliper, and estimated as the average of two perpendicular measurements. Tree heights were measured for each tree annually in August using an ultrasonic hypsometer (Vertex IV, Haglöf, Sweden), and estimated as the average of six measurements. Tree slenderness index was calculated as average height (m)/average DBH (m). Due to the impossibility of using destructive sampling methods, as we were monitoring the growth of the same trees over time, individual tree volume was estimated by averaging estimates from an ensemble of the five best volume equations for poplar plantations in the Órbigo River basin, published by [3,28,29].

To account for potential differences in tree growth due to site fertility, 12 soil samples of the first 10 cm in the soil profile were taken per plot in August 2019, being homogeneously distributed in each plot. Samples were air-dried, grounded, and sieved with a 2-mm sieve, prior to sending them for chemical analysis at the CEBAS-CSIC (Murcia, Spain). Soil conductivity and salinity were measured after diluting soil samples in water at the

1:2.5 proportion (10 g soil sample in 25 mL distilled water). A pH-meter Micro pH-2001 and a conductivimeter Micro-CM-2202 were used for measurements (both from Crison, Barcelona, Spain).

Diameter and height data were tested for normality with the Shapiro-Wilk and Kolmogorov-Smirnov tests. Homoscedasticity of diameter and height data was tested with the Levene and Bartlett tests. As data passed both tests, differences among plots for diameter, height, volume and soil chemical components were carried out with one-way ANOVA [30]. For trees with DBH $\geq$ 10 cm, least squares regressions between untransformed data for DBH and tree volume, and for DBH and height were carried out using linear, quadratic and power functions. Additional functions (such as cubic, logarithmic, and others) were also tested but as their complexity was higher and they did not provide better estimates than the three selected basic models, results are not reported here. As the number of trees in each plot was different, to keep a balanced design we created data subsets for each plot with equal number of trees by randomly selecting trees from the original database, and stratifying the selection by diameter to ensure that all DBH sizes were equally represented in the models. Then, we merged the three plot datasets, and the resulting database was randomly split into two sets: one set used for model building or parameterization (80% of the trees) and another set used for evaluating model performance (20% of trees).

Model predictions were evaluated against the validation dataset by comparing model estimations of volume and height generated with the different equations based on DBH with the observed values recorded at different ages for the trees in the validation dataset. Model performance was analyzed using goodness-of-fit indexes (coefficient of determination R^2, Theil's inequality coefficient U (Equation (1)) [31], modelling efficiency ME (Equation (2)) [32], bias estimators (average bias (Equation (3)), mean absolute difference (MAD, Equation (4)), root mean square error (RMSE, Equation (5)), and exploring the normality of residual distributions.

$$U = \sqrt{\frac{\sum(O_i - P_i)}{\sum O_i^2}} \tag{1}$$

$$ME = 1 - \frac{\sum(O_i - P_i)^2}{\sum(O_i - \hat{P})^2} \tag{2}$$

$$Average\ Bias = \frac{\sum(O_i - P_i)}{n} \tag{3}$$

$$Mean\ Absolute\ Difference = \frac{\sum|O_i - P_i|}{n} \tag{4}$$

$$Root\ Mean\ Square\ Error = \sqrt{\frac{\sum(O_i - P_i)^2}{n}} \tag{5}$$

where O_i stands for individual observation i, P_i is the corresponding model prediction for the i observation, $\hat{P}$ is the average of all predictions, and n is the number of observation.

The accuracy of model predictions was determined using critical errors, following the technique described by [33] and modified by [34]. The critical error e^* can be interpreted as the smallest error level, in absolute terms, which will lead to the acceptance of the null hypothesis (i.e., that the model is within e units of the true value) at the given level. Then, if a user specifies an e value (difference between real and modelled data) higher than e^*, the conclusion will be that the model is adequate for that user. Therefore, critical errors relate model accuracy to user's requirements. With this test, the model is judged accurate unless there is strong evidence to the contrary [35,36]. The critical error test was done at 5% and 20% error levels ($\alpha = 0.05$ and $\alpha = 0.20$), corresponding to an exigent and a less demanding model user, respectively.

3. Results

Tree diameter in 2019 showed no significant differences among plots ($F_{2,401} = 2.050$, $p = 0.130$). At plantation age 13 years, diameters ranged from 12.8 to 39.1 cm, but most trees were close to an average diameter of 25 cm (Table 1). There were significant differences among plots on tree height ($F_{2,401} = 53.684$, $p < 0.001$), with trees in Plot 1 growing significantly taller than those in Plot 2, and those taller that trees in Plot 3 (Table 1). Tree heights ranged from 11.93 to 37.17 m, with an average of approximately 22 m. The same trend was found for the slenderness index, with trees in Plot 1 being significantly more slender than those in Plot 2, and those more slender that trees in Plot 3 ($F_{2,401} = 65.235$, $p < 0.001$, Table 1). In all the plots, trees were quite slender, with indexes close to 100. Individual tree volume was very variable (ranging from 0.038 to 1.343 m^3) with an average of 0.458 m^3 (Table 1).

Table 1. Summary data for the measured trees in August 2019 (mean ± standard deviation). Different letters indicate differences among plots Tukey's HSD at $\alpha = 0.05$.

Variable	Plot 1	Plot 2	Plot 3	All Plots
Number of trees	51	79	274	404
DBH (cm)	24.64 ± 7.05a	25.70 ± 5.49a	24.55 ± 3.48a	24.78 ± 4.52
Height (m)	25.20 ± 5.48a	22.68 ± 3.30b	20.72 ± 2.13c	21.67 ± 3.36
Slenderness index (m/m)	105.2 ± 16.6a	89.9 ± 10.8b	85.4 ± 10.3c	88.8 ± 13.1
Tree volume (m^3)	0.492 ± 0.386a	0.457 ± 0.271a	0.356 ± 0.149b	0.393 ± 0.226
Tree basal area (m^2)	0.051 ± 0.029a	0.054 ± 0.023a	0.048 ± 0.014a	0.050 ± 0.018

When considering all the plots together, an average increment per tree of 1.66 cm year^{-1} in diameter, 1.52 m year^{-1} in height, and 0.03 m^3 year^{-1} in volume was estimated, which translated into mean annual increment of 13.02 m^3 ha^{-1} year^{-1}. Plot 1 and Plot 2 reached the maximum annual increment rates for diameter, height and volume in years 9 to 11, whereas Plot 3 reached maximum annual increment for diameter and height earlier (years 6 and 9, respectively), but maximum annual volume increment later (year 15) (Table 2). However, annual volume increment was different among plots likely due to their significantly different soil fertility, with the maximum being reached earlier in the most fertile plots (Table 2).

Table 2. Maximum values of annual increments in dimensional variables (mean ± standard deviation). The values between brackets indicate plantation age (tree age—2 years) at which the maximum value was reached.

Variable	Plot 1	Plot 2	Plot 3	All Plots
Annual DBH increment (cm year^{-1})	2.33 ± 0.86 (9)	2.16 ± 0.71 (7)	2.90 ± 0.83 (4)	2.39 ± 0.60 (8)
Annual height increment (m year^{-1})	2.22 ± 0.93 (9)	2.10 ± 0.84 (9)	1.80 ± 0.50 (7)	1.82 ± 0.60 (9)
Annual volume increment (m^3 year^{-1})	0.077 ± 0.064 (8)	0.067 ± 0.034 (9)	0.092 ± 0.050 (13)	0.081 ± 0.040 (13)

Soil in Plot 3 had consistent lower values of organic matter, soil carbon, soil nitrogen and all the other macronutrientes, and significantly higher C/N ratio. Plot 3 also had higher pH than Plot 2 and significantly higher conductivity than Plot 1 (Table 3).

Table 3. Soil chemical analysis in August 2019 (mean $\pm$ standard deviation). Different letters indicate differences among plots with Tukey's HSD at $\alpha = 0.05$.

Variable	Plot 1	Plot 2	Plot 3
Soil pH	7.346 ± 0.286ab	7.210 ± 0.164b	7.515 ± 0.212a
Conductivity (μs cm^{-1})	66.400 ± 10.511b	97.900 ± 31.3804a	94.775 ± 14.076a
Organic Matter (%)	6.880 ± 0.850a	7.180 ± 0.537a	3.130 ± 0.450b
Total C (%)	4.270 ± 0.425a	4.530 ± 0.453a	1.970 ± 0.242b
Organic C (%)	3.990 ± 0.492a	4.160 ± 0.311a	1.820 ± 0.277b
CaCO$_3$ (%)	2.290 ± 2.549a	3.050 ± 3.253a	1.270 ± 1.663a
N (%)	0.400 ± 0.045a	0.430 ± 0.057a	0.170 ± 0.035b
Ca (%)	0.514 ± 0.095a	0.415 ± 0.126ab	0.287 ± 0.109b
K (%)	0.640 ± 0.066a	0.561 ± 0.114ab	0.447 ± 0.088b
Mg (%)	0.214 ± 0.030a	0.181 ± 0.041a	0.131 ± 0.027b
Na (%)	0.038 ± 0.006a	0.036 ± 0.007a	0.028 ± 0.005b
P (%)	0.056 ± 0.009a	0.046 ± 0.012a	0.027 ± 0.007b
S (%)	0.052 ± 0.011a	0.044 ± 0.011a	0.025 ± 0.005b
C/N	10.656 ± 0.242a	10.561 ± 0.486a	11.367 ± 0.694b

Estimations of stem volume based solely on DBH showed a high model accuracy, as all equations reached adjusted coefficients of determination (R^2_{adj}) above 0.94, with all parameters being significant (Table 4). Not surprisingly, the linear model for volume underestimated values for the smallest and largest trees, whereas the power equation overestimated volume for the largest trees. However, the quadratic model showed balanced predictions along the whole data range (Figure 1). On the other hand, the higher data dispersion for tree height values caused that height models reached lower coefficient of determination than volume models (0.75–0.76, Table 4), although all the models' parameters were significant (except slope in the quadratic equation). There were not visible differences among equations in model behavior, with the three height models failing to estimate heights of the tallest trees in the DBH 15–25 cm range.

Table 4. Parameter values, their significance and standard errors for the equations tested. For volume, linear: $V(m^3) = Y_0 + A \cdot DBH(m)$; Quadratic: $V(m^3) = Y_0 + A + B \cdot DBH(m)$; Power: $V(m^3) = A \cdot DBH(m)^B$. For height, linear: $H(m) = Y_0 + A \cdot DBH(m)$; Quadratic: $H(m) = Y_0 + A + B \cdot DBH(m)$; Power: $H(m) = A \cdot DBH(m)^B$. Significancy levels *: significant with $p > 0.05$; **: significant with $p < 0.001$ at $\alpha = 0.05$.

Equation	Y_0	A	B	R^2_{adj}	SE$_{estimate}$
Volume					
Linear	-0.5038 ** ± 0.0071	0.0379 ** ± 0.0003	-	0.9418	0.0577
Quadratic	-0.0308 * ± 0.0013	-0.0088 ** ± 0.0013	0.0011 ** $\pm 2.9 \cdot 10^{-5}$	0.9782	0.0353
Power	-	0.000056 ** $\pm 3.7 \cdot 10^{-6}$	2.7473 ** ± 0.0196	0.9709	0.0408
Height					
Linear	5.0331 ** ± 0.2994	0.6935 ** ± 0.0143	-	0.7521	2.4357
Quadratic	0.4433 ± 0.9425	1.1472 ** ± 0.0896	-0.0102 ** ± 0.0020	0.7600	2.3966
Power	-	2.1061 ** ± 0.0982	0.7356** ± 0.0151	0.7575	2.4092

As described before, the validation dataset was created by randomly selecting 10–11 trees in each plot. The attributes for those trees were very similar to the whole data set, maintaining the same differences among plots observed in the full dataset (Table 5), which indicated such validation dataset was a good representation of the trees in these sites.

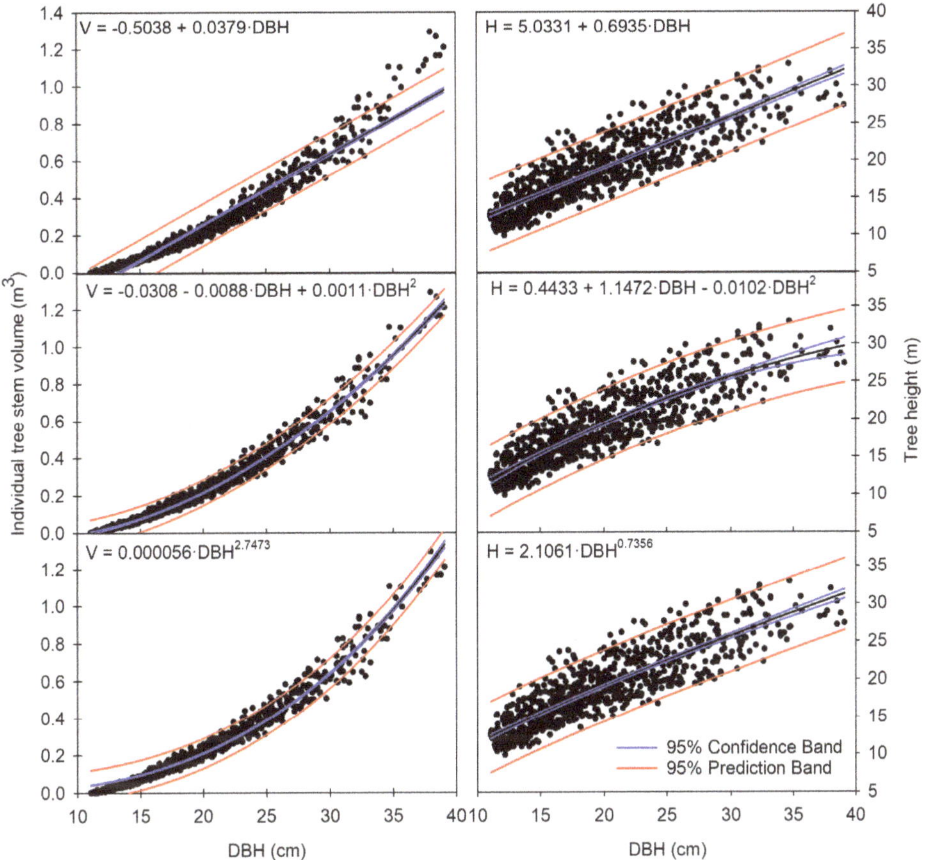

Figure 1. Estimation of tree volume (left panels) and tree height (right panels) under three different equations (showed in each panel) with tree DBH as the only predictor. Dots indicate observed values from the parameterization dataset.

Table 5. Summary data for the trees in the validation dataset in August 2019 (mean $\pm$ standard deviation). Different letters indicate differences among plots Tukey's HSD at $\alpha = 0.05$.

Variable	Plot 1	Plot 2	Plot 3	All Plots
Number of trees	10	10	11	31
DBH (cm)	24.93 ± 7.543a	24.30 ± 3.41a	24.37 ± 3.61a	24.83 ± 4.98
Height (m)	25.25 ± 5.54a	21.58 ± 4.02b	20.26 ± 2.16c	22.01 ± 4.29
Slenderness index (m/m)	104.1 ± 16.9a	92.3 ± 7.5b	87.6 ± 8.2c	93.4 ± 13.3
Tree volume (m³)	0.519 ± 0.405a	0.454 ± 0.145a	0.337 ± 0.139b	0.395 ± 0.257
Tree basal area (m²)	0.053 ± 0.031a	0.053 ± 0.011a	0.044 ± 0.013a	0.046 ± 0.020

Model performance for volume showed the smallest average bias for the power equation, but the smallest mean absolute difference and RMSE for the quadratic model. The quadratic equation also showed low U and high ME coefficients, being closely followed by the power model, and at some distance the linear equation. RMSE values were quite small, ranging from 0.034 m³ for the quadratic model to 0.056 m³ for the linear model. For the quadratic and power models, Freese's critical errors e^* were 0.062 and 0.065 m³ for the exigent user, and 0.042 y 0.045 m³ for the relaxed user, respectively. Hence, only model users demanding accuracy levels lower than those errors would reject those models (Table 6).

Table 6. Models' performance indicators when comparing recorded and simulated trees in the validation dataset. MAD: Mean Absolute Difference; RMSE: root mean square error; U: Theil's coefficient; ME: modelling efficiency; e^* = Freese's critical error.

Equation	Average Bias	MAD	RMSE	U	ME	$e^* \alpha = 0.05$	$e^* \alpha = 0.20$
Volume							
Linear	0.032	0.041	0.056	0.287	0.918	0.101	0.069
Quadratic	0.067	0.026	0.034	0.176	0.969	0.062	0.042
Power	0.011	0.029	0.037	0.188	0.965	0.066	0.045
Height							
Linear	−0.003	2.058	2.547	0.555	0.691	4.610	3.134
Quadratic	0.002	2.046	2.531	0.552	0.695	4.582	3.115
Power	0.000	2.053	2.534	0.553	0.694	4.588	3.119

The three height equations showed small average bias (less than one centimeter), indicating that the models could capture the average of the observed dispersion acceptably well. However, the mean absolute differences (about 2.05 m for all models), and RMSE values (about 2.53 m for all models) were notably higher and quite similar among models, indicating that the models were less able to capture the variability around the mean of the observed tree heights. Similarly, Theil's coefficients were moderate (0.552–0.555), as well as modelling efficiencies (0.691–0.694). Freese's critical values also indicated low predictive power of the height models, as even the less demanding user would reject the equivalence of the predicted and observed values if the expected accuracy were less than 3.12–3.13 m, a noticeable value (Table 6).

Differences between volume and height models were also evident when comparing residual distributions. Although in all cases distributions significantly departed from normality, the volume models showed more normal-like distributions, particularly the quadratic model. For height, all models showed residuals clearly different from normal distributions, although the quadratic model was also the one with a more balanced residual distribution (Figure 2).

Not surprisingly, the differences among model performances were reflected in the estimated values for some standard DBHs (Table 7). All models showed better accuracy for large than for small trees. Estimated volumes for the typical harvest sizes (DBH > 30 cm and higher) showed relatively narrow prediction intervals, particularly for the quadratic model. However, prediction intervals for tree height were quite large, reaching almost a 10 m difference between the lowest and the highest interval limits (Table 7).

Table 7. Tree volume and height for predicted intervals at 95% probability for selected DBH values.

DBH (cm)	Volume (m^3 stem^{-1})			Height (m)		
	Linear	Quadratic	Power	Linear	Quadratic	Power
10	0.000–0.009	0.000–0.062	0.000–0.113	7.10–16.84	6.09–15.68	6.64–16.27
15	0.000–0.180	0.014–0.155	0.014–0.177	10.56–20.31	10.55–20.13	10.62–20.26
20	0.139–0.370	0.163–0.304	0.129–0.293	14.03–23.77	14.49–24.09	14.26–23.90
25	0.328–0.559	0.366–0.507	0.308–0.471	17.50–27.24	17.93–27.52	17.66–27.30
30	0.518–0.749	0.625–0.766	0.561–0.725	20.97–30.71	20.86–30.44	20.89–30.53
35	0.707–0.938	0.938–1.079	0.900–1.064	24.43–34.18	23.27–32.86	23–98–33.61
40	0.897–1.128	1.307–1.448	1.335–1.499	27.90–37.64	25.18–34.76	26.95–36.58

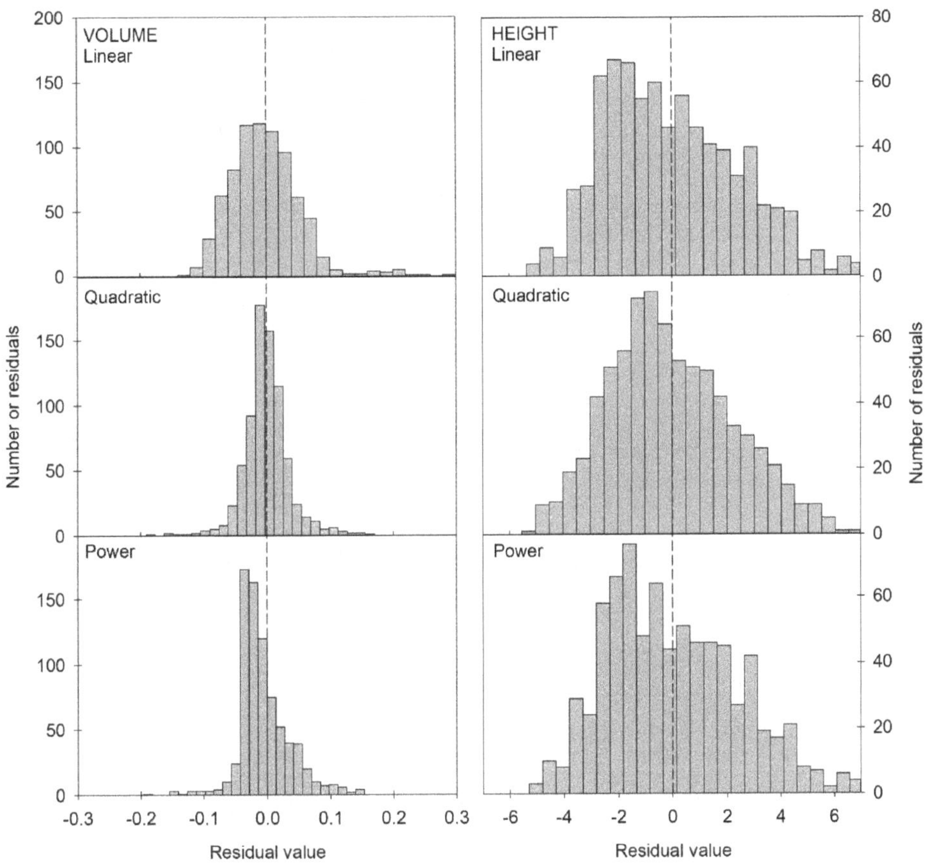

Figure 2. Distribution of residuals for the three models for each variable with the parameterization dataset (left panels: stem volume; right panels: tree height).

4. Discussion

4.1. Simplified Volume and Height Equations

Relationships among tree attributes are determined by tree architecture and growing conditions. As seen in this work, some poplar trees planted in the Órbigo River basin can reach commercial (harvesting) sizes (DBH > 30 cm) 13 years after planting. However, the stands as a whole had average DBH of 24.78 cm at plantation age 13 years, and assuming annual growth rates between the average and the maximum observed for the new following years (1.66 to 2.39 cm^{-1} year^{-1}), trees may need 15.2–16.1 years to reach commercial size (DBH > 30 cm). Hence, our results also show that the typical rotation length used by plantation owners in the region (which is about 15 years) is well fitted with the timing of maximum annual volume increment. Those relatively long rotation times for a fast-growing species such as hybrid poplar are typical for this region. Even so, the Duero River basin is considered as a region of good potential productivity for poplars [13]. However, our results also indicate soil fertility as the most likely factor determining differences in growth rates among plots. Although the climate, geology and parent soil material are virtually the same, Plots 1 and 2 showed higher productivity than Plot 3. Plot 2 was established on a farmland used for decades as pasture before plantation establishment. Plot 1was also used for long time as a pasture before planting a poplar stand 15 years prior to current plantation establishment. Consequently, these plots have kept organic matter levels much higher than Plot 3, which was used in intensive farming since reparcelling in the late 1980s (which involved soil movement and mixing), until plantation establishment in 2005. All these processes have resulted in the loss of soil organic matter and its associated fertility. In spite of this, the three plots reached an annual productivity of about 12 m^3 ha^{-1} year^{-1}, considered good for this region [13], and reached the typical productivity range recorded in poplar plantations in the Duero River basin [14].

Although soil fertility has been shown to be the main influential factor on growth rates at these sites, other plot-specific factors may have affected the shape and growth rates of the monitored trees. Trees in Plot 1 had the highest slenderness index, whereas trees in Plot 3 had the lowest. Such differences are likely due to the effect of neighboring farmlands. Plot 3 is surrounded by annual crops in all sides that produced no shading interference with poplar saplings as they grew. Plot 2 was bordered by poplar plantations in the southern side at planting time that provided shade during sapling development. Plot 1 (the most fertile of the three plots) was surrounded by other poplar plantations in its east, south and west sides at planting time. As poplar is a shade-intolerant pioneer genus, it is sensitive to light competition [37]. Hence, shade from neighboring stands had likely forced height growth in those saplings planted under partial shade [38–40]. However, surrounding plantations may not have been always a negative influence, as they could have also offered saplings shelter from wind, hence favoring height growth [41].

Another indication of light limitation could be the longer time needed in Plots 1 and 2 to reach maximum annual growths in diameter and height, compared to Plot 3 (about 2 to 5 years longer, Table 2). Such delay could also be related to more intense effects of pruning in reducing self-shading in those plots with other neighboring poplar plantations [42]. Finally, although the three plots were irrigated the same number of times at similar dates during summers, Plot 1 was the closest one to a small permanent stream, while Plot 3 was the furthest. This fact could have forced trees in Plot 3 to invest more in root development. However, trees in Plot 1 could reach water closer to the soil surface, therefore releasing biomass to invest in height growth, as it has been proven in other plantations [43–45]. All this plot-level variability in growing conditions has caused a high variability in tree height among trees of the same age. Such variability has been translated in height models with wide confidence intervals for estimated heights.

Because of the intense height growth, combined with pruning [46], some trees became quite slender [47]. This fact, combined with the lengthy plantation time and dense plantation spacing, can make trees quite susceptible to windthrow, as indicated by slender indexes above 100 [48,49]. Trees broken or fallen by wind are economic losses that may cancel the benefit of extending rotation times to reach larger diameters. Therefore, although some authors suggest 35 cm DBH as the optimum for harvesting [7], and suggest harvesting rotations up to 18 years depending on site quality [50], waiting the expected 17–18 years needed at these plots to reach such diameters may be risky. Hence, we suggest that after passing 14–15 years since plantation, owners in the Órbigo River basin must evaluate the potential benefit of letting trees grow another year (which could bring about 1.7–2.3 cm potential DBH increase) versus potential losses by wind. On the other hand, the high slenderness values reached in these plots could also be used as an argument to suggest forest owners to avoid planting poplar next to other plantations already established, because such neighboring plantations could force new saplings to grow very slender to avoid shading [51–53]. Later, when neighboring plantations are harvested, such slender saplings are suddenly exposed to wind, increasing windthrow risk.

The diameter and height ranges observed in the monitored plots are inside the reported range for poplar plantations in the Duero River Basin [1,28,29], demonstrating the representability of the selected plots as typical poplar plantations in the floodplains of the Órbigo River basin. However, in spite of the observed differences in soil fertility and tree height, we must highlight the lack of differences among plots in DBH. This fact corroborates the assumption that poplars have some capacity to maintain high productivity rates even when growing in former agricultural soils with low productivity [54].

Keeping in mind the initial hypothesis and the purpose of this research, such lack of differences in DBH among plots is vital for forest owners, as diameter is the only variable that they can easily monitor, and hence the only stand attribute used to make management decisions. Therefore, from the plantation owner's point of view, the three plots were equivalent, even though the significant differences in height and volume reported here. Following this argument, we developed equations to estimate tree volume from DBH, combining data from trees belonging to all the three plots. We are aware that, from an academic point of view, such combination of plots and ages can be criticized given the significant differences found among plots. However, here again we must remind readers that our objective was developing simple equations usable by non-professionals in a large range of site productivities, rather than creating the most accurate equations to estimate tree volume. In addition, we used data from the age range 5 to 14 years-old and not only from the final harvesting time, to better represent tree development in the models. Although trees change their allometry as they age [55], the obtained volume models have shown to be robust. Hence, we can accept our initial hypothesis, as the quadratic volume equation has shown a high level of accuracy and acceptable model performance. Second in performance was the power equation, while the linear equation displayed important biases for the smallest and largest trees in the observed range. However, we have to recognize that our secondary objective was not met, as the developed equations for tree height have shown only moderate predictive capacity. In spite of that, our results also show that trees have the capacity to maintain similar annual volume growths by changing their allometry, even when height growths are significantly different among plots, as shown by the much robust models for volume than for height.

For the forest owners to make management decisions, it is crucial to know timber market prices at the time of harvest. However, even more important is to know how much standing volume is present in their plantation. Although there are already several volume equations for poplar plantations in Spain [14–17], they have always been created to reach maximum accuracy by following statistical procedures available only to specialists, as they all use two or more tree attributes (e.g., diameter and height). In fact, our results for the height models also show the difficulty of estimating tree heights as it is a variable very dependent on plot-level environmental variables. This fact only reinforces the need for developing generalized volume equations that use DBH as the only predictor. In addition, most of the plantation owners are seniors already retired or that will reach retirement age during the current rotation, and with little capacity to follow advanced technical procedures by lack of equipment or training. As tree height is difficult to be precisely measured without specific (and usually expensive) equipment such hypsometers, two-variable equations are out of reach in practical terms for most forest owners. Other regular approaches that do not need calculations such as growth and yield tables for plantations densities and by given DBH and height values are also available for poplar [56]. However, tree height is not easy to measure (or even estimate, as our results show) with enough accuracy to be used in those tables with some reliability.

In the Órbigo River basin, most current plantation owners were born and raised in 20th Century post-Civil War Spain, with little opportunity to study beyond primary school, if they were even able to complete it [19,20]. Hence, applying any technical leaflet, report or guideline that displays traditional volume equations is likely out of their reach. Only in the last years, these owners are starting to have access to smartphones, but they still mistrust (or just do not understand) apps or other software that could implement such complex equations.

The accuracy of the equations developed here should be more than enough to provide volume estimations adequate for forest owners to make decisions on harvesting schedules, as indicated by the reduced values of Freese's critical errors. This is especially useful, as estimated volumes are a better information on timber stocks than just tree diameters. In addition, the equations provided here are simple enough (particularly the linear and quadratic models) that can be written in any regular calculator integrated in a smartphone. Such simplicity allows the forest owner to measure their trees with a regular measuring tape, introduce the circumference in the smartphone's calculator and just converting it into diameter and then into tree volume. Such simple calculations will provide non-trained forest owners with the capacity to initiate negotiations with timber companies on the time and cost of plantation harvesting from a much stronger position than if volume estimation is left into the hands of the same company that has to buy and harvest the trees.

4.2. Limitations of the Simplification Approach

The approach used here to provide simplified and generalized volume equations has been developed with practicality and usefulness for the intended final user in mind. This means that a sacrifice has been done in terms of statistical or biological rigor. However, it has been argued that cross-validation (which has generated acceptable values for our volume models) is a better approach for model evaluation than residual distributions [57]. In addition, it could been argued that more accurate volume estimations could be reached by using traditional destructive sampling in the target trees, rather than our approach of using an ensemble of already published volume equations. However, using destructive sampling to increase volume estimation accuracy would have limited the application of the resulting models only to the experimental plots. On the other hand, both methods for volume estimation (destructive sampling or ensemble of equations) are conceptually similar, as both assume than an irregular object such as a tree stem has the same volume of a regular solid defined by a mathematical equation. Hence, using an ensemble of models (volume equations in this case) has been shown to be a valid method to encompass uncertainty due to both sampling and model selection [58,59] when estimating growth poplar plantations. In addition, our approach could also be considered inside the current trend that proposes model simplification as a way to increase model usefulness [60], as in essence is a process in which one predictor (DBH) is used to substitute results obtained by several predictor (DBH, height, etc.). In the end, we should not forget that models are neither right nor wrong, but useful (or not) for the task they were designed, Hence, complex models should be used when simple ones cannot do the work [61,62].

As for the biological representativeness of the proposed equations, both the linear and the quadratic equations could be considered as not realistic enough. In the first case, a linear equation assumes constant growth rates independently of tree size, which is obviously not the case. Bigger trees have more access to light, nutrients and water and more photosynthetic capacity, and therefore grow faster (in the absence of other limiting factors) than smaller trees [63,64]. On the other hand, a quadratic model may be more realistic as volume or height growth would accelerate faster than diameter increases due to the squared term. However, as quadratic equations by definition depict parabolas, it may be the case that the estimated height or volume reaches a minimum for intermediate diameters while increasing for small DBHs. To avoid this issue, the quadratic equations were created and defined only for DBH $\geq$ 10 cm. In any case, we have to remind again that our objective was to define simple equations that would capture as much observed variability as possible, not to create biologically realistic models.

Regardless of our results, we must point out again that the equations shown here do not pretend to substitute those previously developed for poplar by other researchers for this or other regions, such as the dynamic model by [65]. As those equations incorporate more predictors related to tree shape, management schemes or environmental variables, they are indeed better suited for applications that require more accurate estimation of volume, biomass or carbon stocks in poplar forests, and whose expected user base is composed by technicians or forest specialists. Hence, we caution against using the volume equations showed here for objectives other than the stated: a rapid on-the-fly estimation of standing volume in poplar plantations in the Órbigo River basin by non-trained plantation owners.

5. Conclusions

Our work has demonstrated that simplified volume equations can be developed for hybrid poplar plantations in northwestern Spain. In an increasingly urbanized world with wood and paper products demand on the rise, aging rural owners are turning into hybrid poplar plantations that require minimal maintenance as a way to keep getting income from their farmlands. Our equations provide a simple but acceptably accurate tool to empower these rural owners when estimating standing volume by themselves. Our results also demonstrate the viability of an approach that could be used for any species for which several volume equations are available.

Author Contributions: Conceptualization, J.A.B.; methodology, R.B. and J.A.B.; formal analysis, J.A.B. and R.B.; writing—original draft preparation, R.B.; writing—review and editing, J.A.B.; supervision, J.A.B. All authors have read and agreed to the published version of the manuscript.

Funding: This research received no external funding.

Data Availability Statement: The data presented in this study are available on request from the corresponding author. The data are not publicly available due to the study plots' owner preferences.

Acknowledgments: The authors are thankful to the research plots owner (Enrique Blanco), for allowing the research work in his property.

Conflicts of Interest: The authors declare no conflict of interest.

References

1. Rueda Fernández, J.; García Caballero, J.L. *Parcela de Experimentación de Clones de Chopos LE-3 Gradefes*; Junta de Castilla y León: Valladolid, Spain, 2013.
2. Freer-Smith, P.; Muys, B.; Bozzano, M.; Drössler, L.; Farrelly, N.; Jactel, H.; Korhonen, J.; Minotta, G.; Nijnik, M.; Orazio, C. Plantation forests in Europe: Challenges and opportunities. *Sci. Policy* **2019**, *9*. [CrossRef]
3. Christersson, L. Silvicultura de rotación corta: Un complemento de la silvicultura "convencional". *Unasylva* **2006**, *57*, 223.
4. Vadell, E.; De-Miguel, S.; Pemán, J. Large-scale reforestation and afforestation policy in Spain: A historical review of its underlying ecological, socioeconomic and political dynamics. *Land Use Policy* **2016**, *55*, 37–48. [CrossRef]
5. Martín-García, J.; Barbaro, L.; Diez, J.J.; Jactel, H. Contribution of poplar plantations to bird conservation in riparian landscapes. *Silva Fennica.* **2013**, *47*, 4. [CrossRef]
6. Díaz, L.; Romero, C. *Caracterización económica de las choperas en Castilla y León: Rentabilidad y turnos óptimos*; Libro de actas del I Simposio del Chopo: Zamora, Spain, 2001; pp. 489–500.
7. Fernández Manso, A.; Hernanz Arroyo, G. *El Chopo (Populus sp.): Manual de Gestión Forestal Sostenible*; Junta de Castilla y León: Valladolid, Spain, 2004.
8. Ambrosio Torrijos, Y.; Picos Martín, J.; Valero Gutiérrez del Olmo, E. Small non industrial forest owners cooperation examples in Galicia (NW Spain). In Proceedings of the FAO Workshop on Forest Operations Improvements in Farm Forest, Logarska Dolina, Slovenia, 9–14 September 2003.
9. Dominguez, G.; Shannon, M. A wish, a fear and a complaint: Understanding the (dis)engagement of forest owners in forest management. *Eur. J. For. Res.* **2011**, *130*, 435–450. [CrossRef]
10. Wiersum, K.F.; Elands, B.H.M.; Hoogstra, M.A. Small-scale forest ownership across Europe: Characteristics and future potential. *Small Scale For. Econ. Manag. Policy* **2005**, *4*, 1–9. [CrossRef]
11. Rodriguez-Vicente, V.; Marey-Pérez, M.F. Characterization of nonindustrial private forest owners and teir influence on forest management aims and practices in Northern Spain. *Small Scale For.* **2009**, *8*, 479–513. [CrossRef]
12. Vainio, A.; Paloniemi, R. Forest owners and power: A Foucauldian study on Finnish forest policy. *Forest Policy Econ.* **2012**, *21*, 118–125. [CrossRef]
13. Emil Fraga, E.; Fidalgo, L.; Álvarez Rodríguez, E.; Rodriguez Soalleiro, R.; Oliveira, N.; Sixto, H. Crecimiento a medio turno de plantaciones madereras del clon RASPAJE en suelo ácido en Galicia. In *Proceedings of the II Simposio del Chopo*; Sociedad Pública de Infraestructuras y Medio Ambiente de Castilla y León: Valladolid, Spain, 2018; pp. 63–72.
14. Bravo, F.; Grau, J.M.; González Antoñanzas, F. Análisis de modelos de producción para *Populus x euroamericana* en la cuenca del Duero. *For. Syst.* **1996**, *5*, 77–95.
15. Rodríguez, F.; Molina, C. Análisis de modelos de perfil del fuste y estudio de la cilindricidad para tres clones de chopo (*Populus x euramericana*) en Navarra. *For. Syst.* **2003**, *12*, 73–85.
16. Barrio-Anta, M.; Sixto Blanco, H.; Cañellas Rey de Viñas, I.; González Antoñanzas, F. Sistema de cubicación con clasificación de productos para plantaciones de *Populus x euramericana* (Dode) Guinier cv. 'I-214' en la meseta norte y centro de España. *For. Syst.* **2007**, *16*, 65–75.

17. Rodríguez, F.; Pemán, J.; Aunços, A. A reduced growth model based on stand basal area. A case for hybrid poplar plantations in northeast Spain. *For. Ecol. Manag.* **2010**, *259*, 2093–2102. [CrossRef]
18. Hjelm, B. Stem taper equations for poplars growing on farmland in Sweden. *J. For. Res.* **2013**, *24*, 15–22. [CrossRef]
19. Instituto Nacional de Estadística–INE. Encuesta Sobre la Estructura de las Explotaciones Agrícolas. Available online: https://www.ine.es/dyngs/INEbase/es/operacion.htm?c=Estadistica_C&cid=1254736176854&menu=ultiDatos&idp=1254735727106 (accessed on 3 January 2021).
20. Carreras, A.; Tafunell, X. *Estadísticas Históricas de España: Siglos XIX-XX*, 2nd ed.; Fundación BBVA: Bilbao, Spain, 2005.
21. Marziliano, P.A.; Russo, D.; Altieri, V.; Macrì, G.; Lombardi, F. Optimizing the sample size to estimate growth in I-214 poplar plantations at definitive tree density for bioenergetic production. *Agron. Res.* **2018**, *16*, 821–837.
22. Köppen, W.; Geiger, R. *Das Geographische System der Klimate*; Borntraeger: Berlin, Germany, 1936.
23. Agencia Española de Meteorología–AEMET. 2020. Regionalización AR4-IPCC. Gráficos de evolución. Regionalización estadística análogos. Castilla y León. Accedido el 3 de febrero de 2020. Available online: http://www.aemet.es/es/serviciosclimaticos/cambio_climat/result_graficos?w=0&opc1=cle&opc2=P&opc3=Anual&opc4=0&opc6=1 (accessed on 10 December 2020).
24. Gaussen, H.; Bagnouls, F. *Saison Seche et Indice Xerotermique*; Université de Toulouse, Faculté des Sciences: Toulouse, Francia, 1953.
25. *World Atlas of Desertification*; Cherlet, M.; Hutchinson, C.; Reynolds, J.; Hill, J.; Sommer, S.; von Maltitz, G. (Eds.) Publication Office of the European Union: Luxembourg, 2018.
26. Instituto Tecnológico Agrario de Castilla y León–ITACYL. *Visor de Datos de Suelo*. Available online: suelos.itacyl.es/visor_datos (accessed on 3 January 2021).
27. Blanco, R. Estimación de Tasas de Crecimiento y Producción de Plantaciones de Chopo (Populus x Euroamericana) en la Ribera Media del Duero. Master's Thesis, Public University of Navarre, Pamplona, Spain, 2020.
28. Rueda, J.; García Caballero, J.R. *Parcelas de Experimentación de Clones de Chopos LE-4 La Milla del Río*; Consejería de Fomento y Medio Ambiente, Junta de Castilla y León: Valladolid, Spain, 2018.
29. Rueda, J.; García Caballero, J.L.; López Negredo, L.; Gómez Cáceres, C. *Parcela de Experimentación de Clones de Chopos LE-1 Valencia de Don Juan*; Consejería de Fomento y Medio Ambiente, Junta de Castilla y León: Valladolid, Spain, 2006.
30. Quinn, G.P.; Keough, M.J. *Experimental Design and Data Analysis for Biologists*; Cambridge University Press: Cambridge, UK, 2002.
31. Theil, H. *Applied Econometric Forecasting*; North-Holland: Amsterdam, The Netherlands, 1966.
32. Vanclay, J.; Skovsgaard, J.P. Evaluating forest growth models. *Ecol. Model.* **1997**, *98*, 1–12. [CrossRef]
33. Freese, F. Testing accuracy. *For. Sci.* **1960**, *6*, 139–145.
34. Reynolds, M.R. Estimating the error in model predictions. *For. Sci.* **1984**, *30*, 454–469.
35. Blanco, J.A.; González, E. Exploring the sustainability of current management prescriptions for *Pinus caribaea* plantations in Cuba: A modelling approach. *J. Trop. For. Sci.* **2010**, *22*, 139–154.
36. Blanco, J.A.; Seely, B.; Welham, C.; Kimmins, J.P.; Seebacher, T.M. Testing the performance of FORECAST, a forest ecosystem model, against 29 years of field data in a *Pseudotsuga menziesii* plantation. *Can. J. For. Res.* **2007**, *37*, 1808–1820. [CrossRef]
37. Welham, C.; Van Rees, K.C.J.; Seely, B.; Kimmins, J.P. Projected long-term productivity in Saskatchewan hybrid poplar plantations: Weed competition and fertilizer effects. *Can. J. For. Res.* **2007**, *37*, 356–370. [CrossRef]
38. Benomar, L.; DesRochers, A.; Larocque, G.R. The effects of spacing on growth, morphology and biomass production and allocation in two hybrid poplar clones growing in the boreal region of Canada. *Trees* **2012**, *26*, 939–949. [CrossRef]
39. Oliver, C.D.; Larson, B.C. *Forest Stand Dynamics: Update Edition*; John Wiley and Sons: New York, NY, USA, 1996.
40. Rodríguez Pleguezuelo, C.R.; Durán Zuazo, V.H.; Bielders, C.; Jiménez Bocanegra, J.A.; PereaTorres, F.; Francia Martínez, J.R. Bioenergy farming using woody crops: A review. *Agron. Sustain. Dev.* **2015**, *35*, 95–119. [CrossRef]
41. Gardiner, B.; Berry, P.; Moulia, B. Review: Wind impacts on plant growth, mechanics and damage. *Plant Sci.* **2016**, *245*, 94–118. [CrossRef] [PubMed]
42. Rauscher, H.M.; Isebrands, J.G.; Host, G.E.; Dickson, R.E.; Dickmann, D.I.; Crow, T.R.; Michael, D.A. ECOPHYS: An ecophysiological growth process model for juvenile poplar. *Tree Physiol.* **1990**, *7*, 255–281. [CrossRef]
43. Asadi, F.; Alimohamadi, A. Assessing the performance of *Populus caspica* and *Populus alba* cuttings under different irrigation intervals. *Agric. For.* **2019**, *65*, 39–51. [CrossRef]
44. Bagheri, R.; Ghasemi, R.; Calagari, M.; Merrikh, F. *Effect of Different Irrigation Interval on Superior Poplar Clones Yield*; Research Institute of Forest and Rangeland—RIFR; FAO: Tehran, Iran, 2005.
45. Lorenc-Plucinska, G.; Walentynowicz, M.; Lwandowski, A. Poplar growth and wood production on a grassland irrigated for decades with potato starch wastewater. *Agrofor. Syst.* **2017**, *91*, 307–324. [CrossRef]
46. Petty, J.A.; Swain, C. Factors influencing stem breakage in high winds. *Forestry* **1985**, *58*, 75–84. [CrossRef]
47. Orzel, S. A comparative analysis of slenderness of the main tree species of the Niepolomice forest. *Electron. J. Pol. Agric. Univ.* **2007**, *10*, 13.
48. Wang, Y.; Titus, S.J.; LeMay, V. Relationships between tree slenderness coefficients and tree or stand characteristics for major species in boreal mixedwood forests. *Can. J. For. Res.* **1998**, *28*, 1171–1183. [CrossRef]
49. Navratil, S. Silvicultural systems for managing deciduous and mixedwood stands with white spruce understory. In *Silvicultural of Temperate and Boreal Broadleaf-Conifer Mixture*; Comeau, P.G., Thomas, K.D., Eds.; B.C. Ministry of Forests: Victoria, BC, Canada, 1996; pp. 35–46.

50. Rueda, J.; García Caballero, J.L.; Cuevas, Y.; García-Jiménez, C.; Villar, C. *Cultivo de chopos en Castilla y León*; Consejería de Fomento y Medio Ambiente, Junta de Castilla y León: Valladolid, Spain, 2019.
51. DeBell, D.S.; Harrington, C.A. Productivity of Populus in monoclonal and polyclonal blocks at three spacings. *Can. J. For. Res.* **1997**, *27*, 978–985. [CrossRef]
52. Nguyen, T.H. Effects of Thinning on Growth and Development of Second Poplar Generations. Master's Thesis, Southern Swedish Forest Research Centre, Swedish University of Agricultural Sciences, Alnarp, Sweden, 2018. no. 303.
53. Feng, J.; Huang, P.; Wan, X. Interactive effects of wind and light on growth and architecture of poplar saplings. *Ecol. Res.* **2019**, *34*, 94–105. [CrossRef]
54. United States Department of Agriculture. USDA. *Hybrid Poplar: An Alternative Crop for the Intermountain West*; USDA Technical note on Plant Materials no. 37; USDA: Aberdeen, WA, USA, 2001.
55. Truax, B.; Gagnon, D.; Fortier, J.; Lambert, F. Biomass and volume yield in mature hybrid poplar plantations on temperate abandoned farmland. *Forests* **2014**, *5*, 3107–3130. [CrossRef]
56. Montoya Oliver, J.M. *Chopos y Choperas*; Mundi Prensa: Madrid, Spain, 1988.
57. Sileshi, G.W. A critical review of forest biomass estimation models, common mistakes and corrective measures. *For. Ecol. Manag.* **2014**, *329*, 237–254. [CrossRef]
58. Wang, F.; Mladenoff, D.J.; Forrester, J.A.; Blanco, J.A.; Schelle, R.M.; Peckham, S.D.; Keough, C.; Lucash, M.S.; Gower, S.T. Multimodel simulations of forest harvesting effects on long-term productivity and CN cycling in aspen forests. *Ecol. Appl.* **2014**, *24*, 1374–1389. [CrossRef]
59. Lo, Y.H.; Blanco, J.A.; González de Andrés, E.; Imbert, J.B.; Castillo, F.J. CO_2 fertilization plays a minor role in long-term carbon accumulation patterns in temperate pine forests in the Pyrenees. *Ecol. Model.* **2019**, *407*, 108737. [CrossRef]
60. Wu, S.; Harris, T.J.; McAuley, K.B. The use of simplified or misspecified models: The linear case. *Can. J. Chem. Eng.* **2007**, *85*, 386–398. [CrossRef]
61. Kimmins, J.P.; Blanco, J.A.; Seely, B.; Welham, C.; Scoullar, K. Complexity in Modeling Forest Ecosystems; How Much is Enough? *For. Ecol. Manag.* **2008**, *256*, 1646–1658. [CrossRef]
62. Blanco, J.A. Modelos ecológicos: Descripción, explicación y predicción. *Ecosistemas* **2013**, *22*, 1–5. [CrossRef]
63. Dietrich, R.; Anand, M. Trees do not always act their age: Size-deterministic tree ring standardization for long-term trend estimation of shade-tolerant trees. *Biogeosciences* **2019**, *16*, 4815–4827. [CrossRef]
64. Blanco, J.A.; Lo, Y.H.; Welham, C.; Larson, B. Productivity of forest ecosystems. In *Sustainable Forest Management: From Principles to Practice*; Innes, J.I., Tikina, A., Eds.; Earthscan: London, UK, 2017; pp. 72–100.
65. Barrio-Anta, M.; Sixto-Blanco, H.; Cañellas-Rey de Viñas, I.; Castedo-Dorado, F. Dynamic growth model for I-214 poplar plantations in the northern and central plateaus in Spain. *For. Ecol. Manag.* **2008**, *255*, 1167–1178. [CrossRef]

Article

Analysis of the Influence That Parameters Crookedness and Taper Have on Stack Volume by Using a 3D-Simulation Model of Wood Stacks

Felipe de Miguel-Díez [1,2,*], **Eduardo Tolosana-Esteban** [3], **Thomas Purfürst** [2,*] **and Tobias Cremer** [1]

[1] Department of Forest Utilization and Timber Markets, Eberswalde University for Sustainable Development, 16225 Eberswalde, Germany; Tobias.Cremer@hnee.de
[2] Chair of Forest Operations, University of Freiburg, Werthmannstr. 6, 79085 Freiburg, Germany
[3] E.T.S.I. Montes, Forestal y del Medio Natural, Universidad Politécnica de Madrid, 28040 Madrid, Spain; eduardo.tolosana@upm.es
* Correspondence: Felipe.Diez@hnee.de (F.d.M.-D.); thomas.purfuerst@foresteng.uni-freiburg.de (T.P.)

Abstract: The influence that parameters crookedness and taper have on the stack volume was analyzed by using a 3D-simulation model in this study. To do so, log length, diameters at the midpoint and both ends, crookedness, bark thickness, taper and ovality were measured in 1000 logs of Scots pine. From this database, several data sets with different proportions of crooked and tapered logs in stack as well as with different degrees of taper and crookedness were created and taken as basis to simulate the stacks and carry out the analysis. The results show how the variation of these parameters influences the stack volume and provide their volume variation grades. These rates of variation were compared with measurement guidelines of some countries and previous research works. In conclusion, the parameters crookedness and taper influence the stack volume to a considerable extent. Specifically, the stack volume is increased as the crookedness degree or the proportion of crooked logs increases. In contrast, the stack volume is reduced as the taper degree or the proportion of tapered logs increases. Furthermore, the results demonstrate the capability of this simulation model to provide accurate results which can serve as a basis for future studies.

Keywords: log properties; stack volume; solid wood content; roundwood measurement

Citation: de Miguel-Díez, F.; Tolosana-Esteban, E.; Purfürst, T.; Cremer, T. Analysis of the Influence That Parameters Crookedness and Taper Have on Stack Volume by Using a 3D-Simulation Model of Wood Stacks. *Forests* **2021**, *12*, 238. https://doi.org/10.3390/f12020238

Academic Editors: Angela Lo Monaco, Cate Macinnis-Ng and Om P. Rajora

Received: 21 January 2021
Accepted: 16 February 2021
Published: 20 February 2021

Publisher's Note: MDPI stays neutral with regard to jurisdictional claims in published maps and institutional affiliations.

1. Introduction

In many countries, round wood is sold in stacks. The most important factor for the purchaser when buying in this way is to know precisely the acquired amount. There are two relevant parameters here: stack volume and solid wood content. The stack volume is normally measured by multiplying the length, width and height of a stack to obtain the cubic area occupied by the stack of wood [1]. Although the sectional volumetric measurement method is frequently used, there are differences between countries in terms of some points such as the section length [2]. In Europe there is no standardized stack volume measurement method. European standards recognize only the national round timber measurement and volume calculation rules of several European countries [3].

Two common units to express the stacked wood measurements are the stere and the cord [1]. However, when expressing the volume of a stack in this way, not only the wood volume but also the bark portion and the air space are included [1].

The solid wood content corresponds to the real roundwood volume which is the basis of the sales process. To determine the latter, it is necessary to measure the volume of every log. However, due to the huge volumes of round wood purchases in many countries, this procedure is unfeasible. Therefore, this value is normally estimated. Calculation of the solid wood content in a stack has been an issue in forest research for the last two centuries [4]. The first publications related to this topic date back to the end of the 19th century, such as the research conducted by Bauer in Germany in 1879 and von Senkendorf in Austria in 1881 [5].

The basis for those studies was the increasing demand for wood and the diminishing supply of desirable wood species [6]. Recent publications have been dated from 2016 onwards. Some of these are related to the implementation and development of photo-optical systems for measuring stack volume and the accuracy and efficiency of these methods when compared with each other and with the traditional methods. The implementation of these methods led to a higher accuracy and they are faster than the traditional measurement methods [2,7]. Other studies included the determination of solid wood content as well as the influence of some parameters on the conversion factors [8,9]. Nowadays, the main reason for the recent research studies has been the need to manage financial resources more efficiently when purchasing raw material. A calculation error at this stage of the sales process can cause significant economic losses, which in the current context of a global market can mean a considerable loss of competitiveness. The most recent document was published by The Food and Agriculture organization, the International Tropical Timber Organization and the United Nations Economic Commission for Europe [10]. It takes as reference the Swedish model [11] to estimate the solid wood content in a stack by taking into consideration several influencing parameters on the conversion factors, and, in turn, on the stack volume.

The influencing parameters have been the basis of research on an equally important and related topic with the first publications dating back to the beginning of the 20th century [12]. Although these parameters are already considered in the roundwood measurement guidelines in some countries such as Sweden [11], their influence has not yet been analyzed individually with reference to a broad statistical basis.

The collection of data for such research requires the measurement of the stack volume and the solid wood content of a considerable number of log piles in order to obtain a sufficiently broad statistical basis. Additionally, the quantification of the solid wood content must be accurate. The most precise method for measuring the real volume of a log is the water displacement method [6]. Considering that the application for such an enormous number of logs would require a long period of time and significant financial resources, it can be stated that the whole process is very costly in practice.

It can hence be assumed that another means for making these calculations is necessary given that the collection of data is, at that scale, practically unfeasible. Using the latest IT technologies and simulation approaches could be a more time and cost-effective solution. Consequently, the approach based on the implementation of a 3D simulation model to carry out these analyses was tested. The simulation model was programmed by "Dr. Philippe Guigue Software Artisan" as an important part of the project "Optimization of the wood supply chain-through analysis, evaluation and further development of log measuring methods and logistics processes in the round wood trade". The programming of the simulation model as software was performed in the ".Net Framework" in C# programming language. It is based on a cross-platform game engine known as Unity software (version 2019.4.9f1), developed by Unity Technologies (San Francisco, CA, US). It makes it possible to reproduce virtually innumerable stacked piles (see Figure 1) and consequently it is not necessary to measure so many stacks in the field.

Figure 1. Simulated stack.

The objective of this study was to test the simulation model for analyzing the effect of two influencing parameters on the stack volume and implicitly, on the conversion factors as well. In this exemplary case, the effect on the stack volume of the parameters taper and crookedness is analyzed, as well as the proportions of crooked and tapered logs in the stack by using the simulation model.

2. Materials and Methods

The model can simulate an enormous number of stacks in 3D and measure them in accordance with one of the manual measurement methods which is described in the German framework agreement for the roundwood trade (RVR) [13]. It means that the simulation model reproduces virtually the execution of this measurement method to calculate the stacked cubic volume. The virtual reproduction of the measurement according to the manual method included in the RVR is represented in Figure 2.

Figure 2. Virtual reproduction of the measurement method of the RVR.

In addition, the model calculates the solid wood cubic volume of stack by adding up the volume of each one of the logs which constitutes the stack and provides the conversion factors according to those methods.

The program consists of several interfaces: the first ones are designed to introduce the data related to the log and the pile to be simulated such as the minimum length, the angles of the stack sides and the proportion of logs stacked in a certain direction. The data may be actual or selected arbitrarily by the user. Another interface corresponds to the simulation parameters, specifically, the number of stacks to be simulated. The last interface shows the results which can be exported as, e.g., a.csv table.

A preliminary model validation was conducted using measurements of real stacks. To do so, 405 logs of Norway spruce (*Picea abies* L.) were measured according to the methodology described below. After that, the logs were randomly stacked 10 times by a forwarder and the stacks were measured according to one of the manual methods outlined in the RVR [13]. These data were introduced into the simulation model. After that, 10 simulations with 10 simulated stacks respectively were carried out. The simulation results were compared with the measurements of the real stacks by assessing the error through calculating RMSE [14] and Mean Bias Error (MBE) [15].

The results demonstrate a small deviation. The average RMSE is 1.2 m^3 (st) and ranges from 0.9 m^3 (st) to 1.6 m^3 (st). It implies an average deviation of 2.6% which ranges from 2.0% to 3.4%. The average MBE is -0.02 m^3 (st) and ranges from -0.4 m^3 (st) to 0.5 m^3 (st). It implies an average deviation of -0.06% which ranges from -0.9% to 1.1%. The average MBE indicates that the simulation results are underestimated. In the near future, it is intended to carry out more analysis in order to establish a more reliable validation of the model.

To analyze the influence of crookedness and taper, a database of real logs of Scots pine (*Pinus sylvestris* L.) was created with the following parameters, which were measured on 1000 logs:

- The log length, measured by means of a forest tape. These data are expressed in meters, rounded to two decimal places.
- The diameters at both ends and at the midpoint, measured by means of a caliper. For each point two measurements were taken perpendicularly. These data are expressed in millimeters.
- The logs' crookedness, measured by setting a levelling rule on the curved log side and measuring the distance between the levelling rule and the log at the deepest point in millimeters. After that, this distance was divided by the log length. The parameter is then expressed in millimeters per meter.
- The bark thickness, measured by means of a Swedish bark gauge. For some logs, where the bark was too thin, a ruler was used for the measurement at both ends of the log and the mean value of both measurements was computed. It is expressed in millimeters.
- The taper of the logs, calculated by subtracting the average small end diameter from the average large end diameter and dividing it by the log length. This parameter is expressed in millimeters per meter.
- Finally, log ovality was computed by dividing the smaller diameter by the bigger diameter of each measurement point and calculating the average of the three resulting values (large end, small end and midpoint of the log). The measurement of this parameter results in a factor between 0 and 1, where 1 represents a round log.

The logs measured belonged to an assortment of "industry wood". Their average midpoint diameter was 23.1 cm, ranging from 10.6 to 43.3 cm, while their length was 3.01 m.

The stack parameters remained unmodified for all simulation sets. The sides' angle for each stack was set to 45°. The minimum stack length was 10 m for the taper analysis and 12 m for the crookedness analysis. This difference in length is due to the different number of logs in the data sets. The proportion of the stacking direction of the logs on each side was set to 50%. This means 50% of the logs were stacked forwards and the others were stacked backwards.

From the original database, 750 logs were selected at random by taking into consideration their crookedness. This means those logs whose values, taking into consideration this parameter, were more than zero (i.e., the logs were crooked) were chosen. The selection of this particular number was chosen in order to form three different data sets with the same number of logs, 250, and a similar crookedness degree. As a result, 250 logs from the original database were discarded. In doing so, the first group corresponded to logs with a crookedness degree less than 10 mm/m. The second group corresponded to logs with a crookedness degree from 10 to 20 mm/m. The third group corresponded to logs with a crookedness degree of more than 20 mm/m. This classification is based on real measured data and was used to analyze the influence of those three crookedness degrees on the stack volume. The average values of the main parameters of those three data sets are represented in Table 1. In combination with these three groups of different crookedness levels, five data sets were created in order to analyze the influence of the proportion of crooked logs in a stack: the first data set contains the original data from real logs, where 100% of the logs are crooked. In the second data set, the crookedness for 20% of the logs was modified to zero and 80% remained unaltered (i.e., crooked). The selection of these logs was performed randomly. According to this method, three additional data sets were created with (i) 50% of the logs being crooked, (ii) 20% of the logs being crooked and the last one with (iii) only straight logs. This was performed for each of the three crookedness degrees selected (see Table 2).

Table 1. Average values of the parameters of each data set to analyze the influence of different crookedness and taper degrees on the stack volume.

	Data Set	Average Length (m)	Average Crookedness of Logs (mm/m)	Average Midpoint Diameters o.b. (mm)	Average Taper (mm/m)	Average Bark Thickness (mm)	Average Ovality Factor
Crookedness	1st	3.01	5	228	11	1	0.95
	2nd	3.01	15	228	11	1	0.95
	3rd	3.01	37	228	11	1	0.95
Taper	1st	3.01	17	225	6	1	0.95
	2nd	3.01	17	225	17	1	0.95

Table 2. Average values of the parameters for each data set to analyze the influence of the proportion of tapered and crooked wood in a stack on the stack volume.

		Data Set	Proportion of Crooked Logs (%)	Average Length (m)	Average Crookedness (mm/m)	Average Midpoint Diameters o.b. (mm)	Average Taper (mm/m)	Average Bark Thickness (mm)	Average Ovality Factor
Crookedness Degree	<10 mm/m	1st	100	3.01	5	228	11	1	0.95
		2nd	80	3.01	4	228	11	1	0.95
		3rd	50	3.01	3	228	11	1	0.95
		4th	20	3.01	1	228	11	1	0.95
		5th	0	3.01	0	228	11	1	0.95
	10–20 mm/m	1st	100	3.02	15	216	11	2	0.94
		2nd	80	3.02	12	216	11	2	0.94
		3rd	50	3.02	8	216	11	2	0.94
		4th	20	3.02	3	216	11	2	0.94
		5th	0	3.02	0	216	11	2	0.94
	>20 mm/m	1st	100	3.01	37	234	12	2	0.94
		2nd	80	3.01	30	234	12	2	0.94
		3rd	50	3.01	19	234	12	2	0.94
		4th	20	3.01	8	234	12	2	0.94
		5th	0	3.01	0	234	12	2	0.94
Taper Amount	<10 mm/m	1st	100	3.01	17	234	5	1	0.95
		2nd	80	3.01	17	234	5	1	0.95
		3rd	50	3.01	17	234	3	1	0.95
		4th	20	3.01	17	234	1	1	0.95
		5th	0	3.01	17	234	0	1	0.95
	> 10 mm/m	1st	100	3.01	21	230	17	2	0.94
		2nd	80	3.01	21	230	14	2	0.94
		3rd	50	3.01	21	230	8	2	0.94
		4th	20	3.01	21	230	3	2	0.94
		5th	0	3.01	21	230	0	2	0.94

To analyze the influence of taper, 700 logs were selected at random from the original data base by taking into consideration their taper values. The selection of that number aimed to form two different data sets with the same number of logs, 350, and similar taper degree. Consequently, 300 logs from the original database were discarded. The first class corresponded to those logs with a taper degree less than 10 mm/m. The logs from the second class presented a taper degree of more than 10 mm/m. This classification was done based on the measured logs. Still, the differentiation of taper in two classes corresponds to the division of the Swedish model [11] or to the classification done by Richter [16]. The analysis was performed in the same way as described before for the crookedness parameter. The average values of the main parameters of each data set to analyze the influence of those two taper degrees on stack volume are represented in Table 1 and the influence of the proportion of tapered wood in pile on the stack volume in Table 2.

In total, 150 stacks were simulated for each data set. After that, the average stack volumes, which were measured according to the RVR method, were analyzed using the programme RStudio (version 1.4.1103) developed by RStudio, Inc. (Boston, MA, US) [17]. For the visualization of the simulation results the package *ggplot 2* was used [18]. The calculation of the RMSE and MBE was conducted using the packages *hydroGOF* [19] and *tdr* [20].

3. Results

The graph represented in Figure 3 provides information on the variation of the stacked cubic volume (m^3 (st)), represented on the y-axis, when the proportion of the crooked and tapered logs in a stack varies. The proportion of the crooked and tapered logs in percentage is represented in the graph on the x-axis. The three different degrees represented in the graph correspond to the crookedness degrees represented in Table 2: high crookedness degree corresponds to crookedness values over 20 mm/m, medium crookedness degree corresponds to crookedness values between 10 and 20 mm/m and low crookedness degree corresponds to crookedness values under 10 mm/m. The obvious difference concerning the initial volume between the data sets to analyze the taper and crookedness is due more to the greater number of logs in the data set on which the taper analysis is based, than the number of logs in the data set to analyze the crookedness. The most obvious trends in the graph illustrate a reduction of the stack volume when the proportion of tapered logs increased and an increment of the stack volume when the proportion of crooked logs increased. Different variation grades can be deduced from that plot according to the different degrees of taper and crookedness.

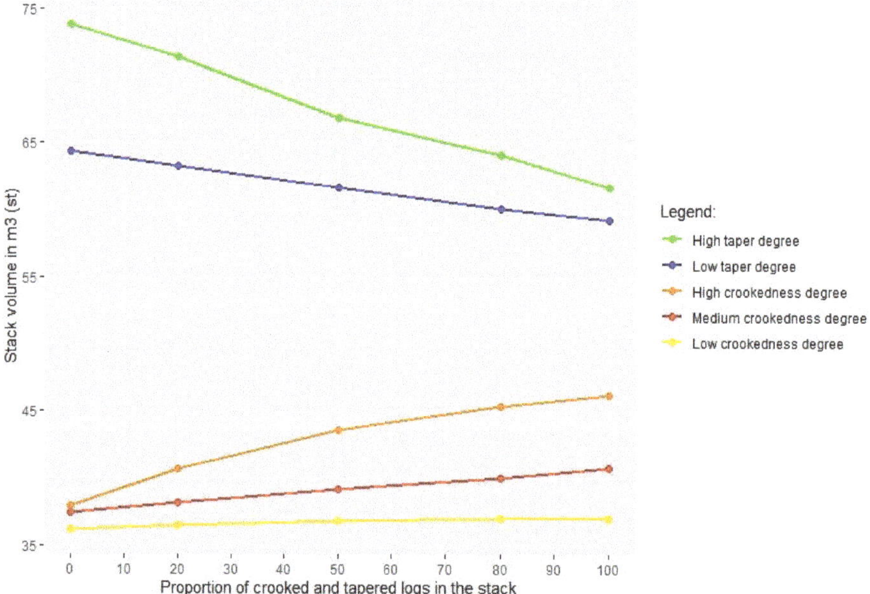

Figure 3. Stack volume variation according to increasing proportion of crooked and tapered wood.

The graph represented in Figure 4 shows the variation of the stacked cubic volume (m^3 (st)), represented on the y-axis, when the degree of crookedness and taper vary and the proportion of the crooked and tapered logs in stack remains unaltered according to Table 1. The different degrees of crookedness and taper are represented on the x-axis and correspond to the ranges defined above, more specifically the first degree of crookedness and taper corresponds to values under 10 mm/m, the second degree of crookedness and taper includes values between 10 and 20 mm/m and the third degree of crookedness corresponds to values over 20 mm/m.

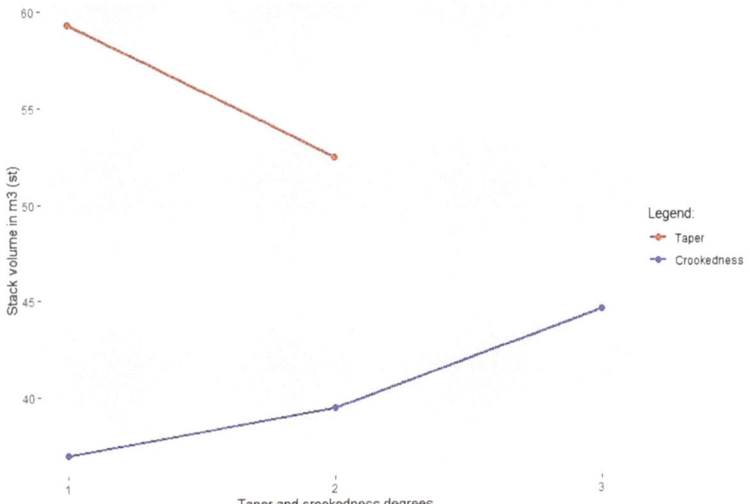

Figure 4. Stack volume variation according to increasing tapering and crookedness degrees.

4. Discussion

The simulation results show that crookedness exerts a direct influence on the stack volume: The higher the proportion of crooked logs in the stack, the larger the stack volume. The results confirm the hypothesis that curved logs occupy a larger space in the stack than do the straight ones, producing voids. This same hypothesis can be found in the conclusions of several previous research results [6,9,12,21,22] as well as in the roundwood measurement guidelines for some countries such as Ireland [23], The United Kingdom [24,25] or Sweden [11]. Moreover, from the simulation outcomes, the grade of variation from the influence exerted by the different crookedness degrees and the proportion of crooked logs on the stack volume was accurately predicted.

In regard to precise stack volume variation ratios according to the different crookedness degrees, only a few references could be found. Heinzmann concluded that a moderately negative correlation coefficient exists for this parameter based on data from 33 stacks [9]. He did not define variation ratios nor crookedness degrees. In addition, his conclusion was also affected by other parameters, e.g., the mean diameter, the ovality and the taper degree. Due to this fact, the results are not comparable, even though they could be interpreted as having a moderate influence on the stack volume, which would correspond to the simulation results.

The Swedish model determines different crookedness degrees based on visual assessment and provides variation ratios for each crookedness degree [11]. Those ratios indicate a linear relation between the variations of the conversion factors and crookedness degrees. This linear relation can also be deduced from the English method. It determines a reduction in the conversion factors by 2% for every 5% of average curvature [25]. The variations from the conversion factors can be interpreted as variation of the stack volume assuming there is a constant solid wood content. Thus, those relations which appear in the references can be compared with the simulation results which consider the stack volume variation. The simulation results support the Swedish and English statements only concerning the linear relation for a crookedness degree between 10 and 20 mm/m. Concerning the other two crookedness degrees, the results showed that the lines do not demonstrate a linear relation but rather a decreasing slope as the proportion of crooked logs increases which can be seen in Figure 3. Furthermore, the increment of the stack volume when increasing the proportion of crooked logs for a low crookedness degree was insignificant.

In addition, the results indicated that whereas the stack volume variation is barely noticeable for the smallest crookedness degree, the variation increases considerably for larger degrees. No reference to previous analysis of the stack volume variation according to different proportions of crooked logs in the stack was found.

According to the simulation results, taper exerts an indirect influence on the stack volume. The simulation results revealed a reduction in the stack volume as the proportion of tapered logs was increased. This fact is due to a reduction of the solid wood content in the stack to a similar degree as well. However, the outcomes indicate different relations for each variation. The increasing variation of proportion of tapered logs in the stack with a taper degree less than 10 mm/m showed that such a decrease in the stack volume follows a linear relation. In contrast, in the case of a higher taper degree, the stack volume decreases at a more rapid rate than a low taper degree when the proportion of tapered logs in the stack is increased as depicted in Figure 3.

A similar decreasing trend was obtained as the taper degree was increased, and the proportion of tapered logs and other parameters remained unaltered as depicted in Figure 4. Considering that the stack volume and the solid wood content are reduced to a similar degree, this parameter has a modest effect on the conversion factors according to the simulation outcomes. This slight influence from the taper on the conversion factors is reflected in the English and Swedish method as well [11,25].

In regard to the stack volume variation for different tapering degrees, the results from Heinzmann's study showed a negative correlation [9]. That statement could be understood as a small decreasing variation. Thus, it is in line with the simulation results for a high taper degree. However, the estimation of that correlation was also affected by other parameters. Thus, a comparison between both results is only partially possible.

According to the simulation results the parameter which most influences stack volume for values less than 20 mm/m is the taper. A comparison which considers taper and crookedness values over 20 mm/m cannot be achieved based on the obtained results.

Finally, it must be pointed out that the simulation model cannot reproduce burls or infrequent crookedness forms which can be found in the natural environment. Neither can it take into consideration the singularities of the terrain, such as an uneven or rough ground under the stack. In such specific cases, the model is unable to accurately provide the stack volume.

Based on the simulation results, purchasers of round wood can estimate the degree of influence a parameter will have on the stack volume and, in turn, on the solid wood content. In theory, this will allow them to estimate the solid wood content more efficiently by means of a quick visual assessment of the logs that constitute the stack. In addition, these simulation results might serve to further develop new guidelines for measuring stack volumes.

5. Conclusions

As shown, the crookedness exerts a considerable influence on the stack volume. Therefore, provenance regions, where roundwood is more crooked because of genetics patterns or site conditions, should be considered when purchasing roundwood, since the solid wood content of a stack will be much lower in comparison to regions where the trees present lower crookedness degrees. In a similar fashion, the tapering exerts a considerable influence on the stack volume as well. However, its effect is irrelevant in the conversion factors.

The results of this work demonstrate furthermore the capacity of the simulation model to obtain large databases for statistical analyses, further reinforcing the solidity of the results. The simulation model provides credible and accurate results which can serve as a basis for future investigations and the creation of a measurement method which unites the different stack volume measurement methods, e.g., in the European countries. However, it is necessary to perform more simulations with further data sets to determine the influence other parameters might have on the stack volume such as length of the logs or midpoint diameters and the resulting influence a combination of the aforementioned parameters could provoke.

Author Contributions: Conceptualization, F.d.M.-D.; T.C. and E.T.-E.; methodology, F.d.M.-D. and T.C.; validation, F.d.M.-D., T.C. and E.T.-E.; formal analysis, F.d.M.-D.; investigation, F.d.M.-D. and T.C.; data curation, F.d.M.-D.; writing—original draft preparation, F.d.M.-D.; writing—review and editing, F.d.M.-D., T.C., E.T.-E. and T.P.; supervision, T.C., E.T.-E. and T.P.; project administration, T.C.; funding acquisition, T.C. All authors have read and agreed to the published version of the manuscript.

Funding: This study was undertaken in the framework of the project "HoBeOpt", which was financially supported via the Fachagentur Nachwachsende Rohstoffe (FNR), Germany, by the Federal Ministry of Food and Agriculture (BMEL) (Grant number: 22007918). The article processing charge was funded by the Baden-Wuerttemberg Ministry of Science, Research and Art and the University of Freiburg in the funding programme Open Access Publishing.

Data Availability Statement: The data that support the findings of this study are available from the corresponding author, F.d.M.-D., upon reasonable request.

Acknowledgments: We thank Tim Pettenkofer (Arbeitsgemeinschaft Rohholz e.V.) for his support during the development the simulation model, Reiner Mohrlock (ForstBW AöR) for his support by organizing the data collection and Lubomir Blasko and his team (Landesbetrieb Forst Brandenburg) for their technical support.

References

1. Fonseca, M. *The Measurement of Roundwood: Methodologies and Conversion Ratios*; CABI Publishing: Oxfordshire, UK, 2005; pp. 79–80.
2. Pasztory, Z.; Heinzmann, B.; Barbu, M.C. Comparison of different stack measuring methods. *Sib. J. For. Sci.* **2019**, *3*, 5–13.
3. European Committee for Standardization. *EN 1309-2:2006 Round and Sawn Timber—Method of Measurement of Dimensions—Part 2: Round Timber—Requirements for Measurement and Volume Calculation Rules*; CEN; European Committee for standardization: Brussels, Belgium, 2006.
4. Schnur, G.L. Converting factors for some stacked cords. *J. For.* **1932**, *30*, 814–820.
5. Graves, H.S. *Forest Mensuration*; John Wiley and Sons, Inc.: New York, NY, USA, 1906; p. 103.
6. Keepers, C. A new method of measuring the actual volume of wood in stacks. *J. For.* **1945**, *43*, 16–22.
7. Jodłowski, K.; Moskalik, T.; Tomusiak, R.; Sarzyński, W. The use of photo-optical systems for measurement of stacked wood. In Proceedings of the Conference: From Theory to Practice: Challenges for Forest Engineering 49th Symposium on Forest Mechanization (FORMEC), Warsaw, Poland, 4–7 September 2016.
8. Pásztory, Z.; Polgár, R. Photo Analytical Method for Solid Wood Content Determination of Wood Stacks. *J. Adv. Agric. Technol.* **2016**, *3*, 54–57.
9. Heinzmann, B.; Barbu, M.C. Effect of mid-diameter and log-parameters on the conversion factor of cubic measure to solid measure concerning industrial timber. *Pro Ligno* **2017**, *13*, 39–44.
10. FAO; ITTO; United Nations. *Forest Product Conversion Factors*; Food and Agriculture Organization of the United Nations; International Tropical Timber Organization; United Nations Economic Commission for Europe: Rome, Italy, 2020; p. 10.
11. SDC ek för. SDC's instructions for timber measurement. In *Measurement of Roundwood Stacks*; SDC: Sundsvall, Sweden, 2014; pp. 1–15.
12. Zon, R. Factors influencing the volume of solid wood in the cord. *J. For.* **1903**, *1*, 126–133.
13. RVR. *Rahmenvereinbarung für den Rohholzhandel in Deutschland (RVR)*, 3rd ed.; Fachagentur für Nachwachsende Rohstoffe e.V. (FNR): Gülzow-Prüzen, Germany, 2020; pp. 34–35.

14. Fox, D.G. Judging air quality model performance. *Bull. Am. Meteorol. Soc.* **1981**, *62*, 599–609. [CrossRef]
15. Addiscott, T.M.; Whitemore, A.P. Computer simulation of changes in soil mineral nitrogen and crop nitrogen during autumn, winter, and spring. *J. Agric. Sci.* **1987**, *109*, 141–157. [CrossRef]
16. Richter, C. Holzmerkmale. In *Beschreibung der Merkmale, Ursachen, Vermeidung, Auswirkungen auf die Verwendung des Holzes, Technologische Anpassung*; DRW: Leinfelden-Echterdingen, Germany, 2007; p. 38.
17. RStudio Team. *RStudio: Integrated Development Environment for R. RStudio*; PBC: Boston, MA, USA, 2021; Available online: http://www.rstudio.com/. (accessed on 8 February 2021).
18. Wickham, H. *ggplot2: Elegant Graphics for Data Analysis*; Springer: New York, NY, USA, 2009.
19. Zambrano-Bigiarini, M. hydroGOF: Goodness-of-Fit Functions for Comparison of Simulated and Observed Hydrological Time Series. R Package Version 0.4-0. Available online: https://cran.r-project.org/web/packages/hydroGOF/hydroGOF.pdf (accessed on 24 October 2020).
20. Perpinan Lamigueiro, O. tdr: Target Diagram. R Package Version 0.13. Available online: https://cran.r-project.org/web/packages/tdr/tdr.pdf (accessed on 24 October 2020).
21. Eucalyptus Newsletter n°48. Available online: http://www.eucalyptus.com.br/news/pt_out15.pdf (accessed on 24 October 2020).
22. Răzvan Câmpu, V.; Dumitrache, R.; Borz, S.A.; Timofte, A.I. The impact of log length on the conversion factor of stacked wood to solid content. *Wood Res. Slovakia* **2015**, *60*, 503–518.
23. Purser, P. *Timber Measurement Manual. Standard Procedures for the Measurement of Round Timber for Sale Purposes in Ireland*; COFORD, The Council for Forest Research and Development: Dublin, Ireland, 1999; pp. 32–36.
24. Edwards, P.N. *Timber Measurement. A Field Guide*, 4th ed.; Forestry Commission Booklet 49: Edinburgh, UK, 1998; pp. 58–59.
25. Hamilton, G.J. *Forest Mensuration Handbook*; Forestry Commission Booklet No. 39: London, UK, 1975; pp. 27–35.

Article

Growth and Potential of *Lomatia hirsuta* Forests from Stump Shoots in the Valley of El Manso/Patagonia/Argentina

Hendrik Kühn [1,*], Gabriel A. Loguercio [2], Marina Caselli [3] and Martin Thren [1]

[1] Faculty of Natural Resource Management, University of Applied Sciences and Arts (HAWK), 37077 Göttingen, Germany; martin.thren@hawk.de
[2] Centro de Investigación y Extensión Forestal Andino Patagónico (CIEFAP) and Ordenación Forestal, Facultad de Ingeniería, National University of Patagonia San Juan Bosco, Esque l9200, Argentina; gloguercio@ciefap.org.ar
[3] CIEFAP and Consejo Nacional de Investigaciones Científicas y Técnicas (CONICET), Esque l9200, Argentina; mcaselli@ciefap.org.ar
* Correspondence: Hendrik.kuehn@t-online.de

Abstract: *Lomatia hirsuta* (Lam.) Diels is a pioneer tree species that develops after wildfires, and in advanced successional stages, it is often found as a secondary species in Patagonian forests. However, in El Manso Valley, Province of Río Negro in Western Argentina, *L. hirsuta* forms mature pure stands, originated from stump shoots. The wood is very attractive for its colourful appearance and beautiful grain. Nevertheless, these forests are not managed for timber production, they are mostly strong thinned for grazing, and the wood is mainly used as firewood. The objective of this study was to evaluate the possibility to improve quality wood production in stands through silvicultural interventions in a sustainable way. Samples have been carried out in stands of different developmental stages. We evaluated the state and quality of the trees, and their growth has been studied by means of trunk analysis. The results indicate that there is significant potential to improve the production of quality wood in dense stands by thinning to release crop trees. Thinning should start in young stands. It also became apparent that forest management is first necessary to stabilise these nearly unattended forests.

Keywords: *Lomatia hirsuta*; Patagonia; pioneer tree species; stump shoots; quality wood; trunk analysis; stability of stands

Citation: Kühn, H.; Loguercio, G.A.; Caselli, M.; Thren, M. Growth and Potential of *Lomatia hirsuta* Forests from Stump Shoots in the Valley of El Manso/Patagonia/Argentina. *Forests* **2021**, *12*, 923. https://doi.org/10.3390/f12070923

Academic Editor: Angela Lo Monaco

Received: 28 May 2021
Accepted: 12 July 2021
Published: 15 July 2021

Publisher's Note: MDPI stays neutral with regard to jurisdictional claims in published maps and institutional affiliations.

1. Introduction

Lomatia hirsuta is an evergreen pioneer tree species and belongs to the family Proteaceae [1]. It has a disjunct distribution area, occurring in the north-western Andes in Ecuador and Peru [2] and also in the eastern side of the Andes in Patagonia (Argentina). There, *L. hirsuta* is normally associated with species from Gavileo-Austrocedretum [3]. For example, it often appears in forests dominated by *Austrocedrus chilensis* (D. Don) Pic. Serm. & Bizzarri or *Nothofagus dombeyi* (Mirb.) Oerst [4,5]. *Lomatia hirsuta* has a wide ecological range and is able to cope with warm and cold climate conditions [2]. In the abovementioned forests with high precipitation, *L. hirsuta* grows in the form of a tree. However, it is also able to exist in dry areas, and in the forest-steppe ecotone, it appears as a shrub of 5–6 m in height [4].

Lomatia hirsuta regenerates from stump shoots after severe forest fires (Figure 1, left). It can also colonise areas after disturbances like road cuttings or abandoned fields, growing from stump shoots or seeds [6], and rarely grows in pure stands. *Lomatia hirsuta* is a light-demanding species but can also germinate under mid-shady conditions [7].

Figure 1. *Lomatia hirsuta* pure forest (**left**) and its wood (**right**).

Little is known about *L. hirsuta's* natural dynamic and growth. Dasometric descriptions of mixed shrublands, where *L. hirsuta* is present, have been made [8], volume and biomass equations have been developed [9], the diameter increment has been measured [10], and first recommendations for the management of these mixed shrubs have been made [11]. In general, in those places with less precipitation, the species grows to lower heights than the area of the present study [8,9]. Furthermore, there is no silvicultural knowledge of pure forests of *L. hirsuta* and no large-scale use of its wood.

Nevertheless, the wood of *L. hirsuta* has very attractive characteristics because of its colourful look, chestnut-pink to dark brown with violet tints (Figure 1, right). Furthermore, it is light to moderately heavy (0.53–0.57 g/cm^3) and flexible [1]. The contraction of *L. hirsuta* has medium values, so it is somewhat unstable, but not extremely. It has clearly visible rays, which are especially apparent in radial sectioning [1]. The wood also has good workability because of its straight grain. For these reasons, it is often used for works of art, handicraft, and sometimes also for fine furniture. In short, it is a wood that could be considered of very high quality. However, the traditional uses of this species do not take full advantage of these characteristics. Normally the stands are thinned intensely, and the brushwood is eliminated to convert the forest into wood pasture: the traditional "parquizado", and the main current destination of *L. hirsuta* is fuelwood.

The objective of this study was to determine the possibility to improve quality wood production through silvicultural management, evaluating the actual vigour and quality of the trees and studying their individual growth.

2. Materials and Methods

2.1. Study Area

The study area is located in El Manso Valley in the southwest of the Province of Río Negro at the international border of Chile (Figure 2). The studied tree stands are particularly in the south of the river and mostly have a northern or north-western exposition. The El Manso river is the southern limit of the Nahuel Huapi National Park.

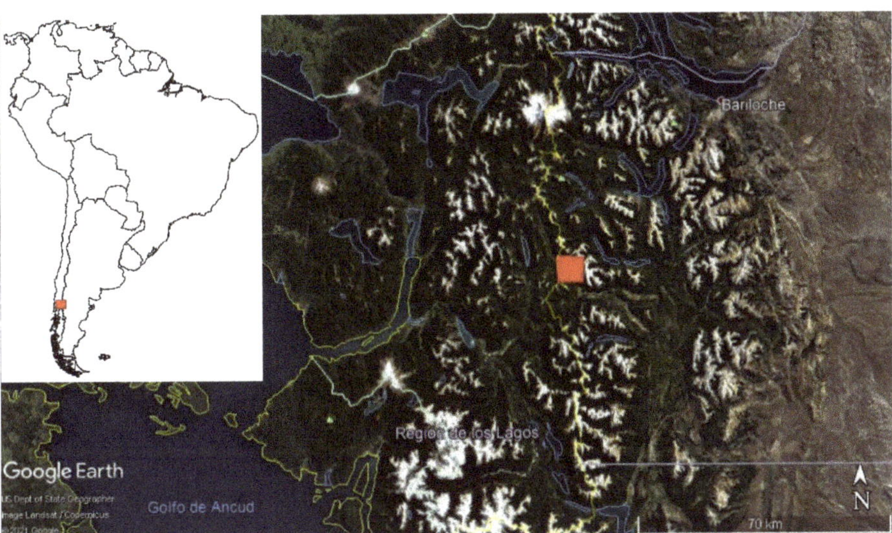

Figure 2. Location of the study area (red) (41°31′ S and 71°49′ W) in El Manso Valley, Province of Río Negro. South America's map source: [12].

The climate in this region is cold temperate under a maritime buffer effect because of the proximity to the Pacific Ocean. The annual temperature of the region averages 9.3 °C. There is precipitation all year, and around 1600 mm/year are reached [13].

The soils in the study area are young Andosols (FAO legend) [14] of volcanic origin [15]. The soils have a high cation exchange capacity and a great water storage capacity; both properties are increased by allophane contents in the soil. Furthermore, the soils were profound, free from stones and had a moderate content of nitrogen and a high content of bases (K, Ca and Mg). All these characteristics were confirmed for the study area by four soil profiles.

In summary, the trees in the studied stands have very good climate and soil conditions for their growth.

2.2. Data Acquisition

2.2.1. Structures of the *L. hirsuta* Stands

The pure stands of *L. hirsuta* in El Manso Valley represent a secondary forest after past wildfires. The stands were differentiated into three stages by the dimensions of the trees: young, intermediate, and old. In every stage, three sample plots were carried out. In addition, three plots were installed in a less dense stand under the traditional use as pastures ("parquizado"). The plots were rectangular and had a size of 1000 m^2, except in the young stands, which had 300 m^2, because of the very high stocking density. Inside the plots, every tree and every trunk was accounted for, and the diameter at breast height (DBH) of all trunks from 7 cm and upwards was collected. In the young stands, all trunks from 4 cm upwards were measured. Furthermore, all trunks were described concerning their social position, their health condition and their form (Table 1). Furthermore, some tree heights per plot were measured along with the range of DBHs.

Table 1. Quality criteria for the trunks inside the plots.

Criterion	Categories		
Social position	Dominant	Codominant	Suppressed
Health condition	Fit	Injured	Dying
Form of the trunk	Straight	Curved	Crooked

We also registered the volume of quality logs; for this, the mid-diameters of good logs were measured with a Finn calliper, and their lengths were estimated in fixed sections. Logs had to have a length of 1.2 m, or its multiples, and a top diameter of at least 15 cm. In the young stands, the potential to produce quality wood in the future was estimated by only taking into account the length of good straight logs, although the top diameter did not reach 15 cm yet. The quality wood of some fallen trees, which were found in the intermediate and old stands, were also measured.

2.2.2. Thinning Simulation

After collecting the data of the individual trunks, a crop tree orientated thinning was simulated in every plot. The criteria to choose the crop trees were vigour, quality, and distribution (in order of importance). Concerning the quality, a good log (sound and fit) of at least 2.4 m long was required; distribution had secondary importance. It was only ensured that two crop trees had no contact with their crowns. For each crop tree, the main competitor, or in some cases two competitors, were identified and scheduled for the cutting.

In addition to the crop tree orientated thinning, a catch-up low thinning was also simulated. It had the objective to obtain the wood of dying and damaged trees (salvage and improvement cutting).

2.2.3. Growth of the Crop Trees

In every plot of the young, intermediate, and old structures, two crop trees were selected and cut for the trunk analysis. Altogether 18 trees were considered (6 trees per structure). Before cutting the selected tree, its DBH, height, and crown radii were measured. The number of measured radii depended on the crown's form. Furthermore, the competitors were identified (the ones that pressured the crop tree's crown), and their DBHs and heights were measured, too. Furthermore, the distances from the crop tree's trunk to the crown border and to the trunk of the competitors were collected.

After cutting the selected tree, the length was measured, and trunk discs were extracted at the height of 0.3 m and every two meters after (2.3 m, 4.3 m, etc.) until the beginning of the crown. The beginning of the crown was considered to be the point in which the trunk forked, and both branches were alive. Additionally, the disc in the height of the DBH (1.3 m) was extracted. In the laboratory, all discs were ground and then analysed by identifying, counting, and measuring the annual rings. Moreover, the height increment of the last year was collected by measuring the length between shoot stigmas from the top of the tree downwards. Between two and seven shoot lengths per tree could be identified and measured. Furthermore, the whole crown material was weighed together, and then the thick branches were weighed separately. Moreover, samples of thick branches, thin branches, and leaves were weighed in fresh conditions and then taken to the laboratory. There, the samples were dried in an oven until a constant weight was reached, so conversion factors from fresh to dry biomass could be determined.

2.3. Processing and Data Analysis

2.3.1. Structures of the *L. hirsuta* Stands and Thinning Simulation

For every age class and for the parquizado, a hypsometric height function through linear regression analysis, using the coefficient of determination (r^2) and the distribution of residuals as a fit indicator, was generated. Since the available volume equations of

L. hirsuta [9] correspond to sites of lower quality (= less height), to quantify the plots' volume, functions for the total volume and the trunk volume of individual trees was adjusted. For this, the destructive sampling trees for the study of growth through stem analysis were used. The trunk volume was determined by means of Smalian's formula [16]. This formula allows us to calculate the volume by using the given diameters and the extraction heights of the discs for the trunk analysis. To get the tree's total volume, the trunk volume and the crown volume (branches of 2 cm upwards) were summed. The crown volume was determined by conversion of the fresh weight of biomass to dry weight and by dividing this with the dry density of *L. hirsuta*, which is 0.55 g/cm^3 [1].

Furthermore, the volumes of the logs of good quality were calculated by means of the mid-diameter and the lengths by using the formula of Huber [17]. For the young stands, the potential of future quality wood was estimated. Therefore, the measured lengths of the straight trunks and the average mid-diameter of the quality logs of the intermediate and old stands, which averaged around 20 cm, were used.

In this way, it was also possible to quantify the volume of the different products which would be extracted in the thinning simulation. The volume of badly formed stems was determined by subtracting the volume of quality wood from the volume of stems with a DBH $\geq$ 20 cm. The rest of the total volume, which would be extracted, was declared as firewood from the crowns and thin stems (<20 cm). To convert the volume in m^3 of standing wood over bark (m^3 o.b.) into timber harvested under bark (m^3 u.b.), the factor of 0.8 was used [18].

Differences among dasometrics parameters from each structure and the results of the cutting simulations (number of trees/ha, number of stems/ha, number of stems/tree, quadratic mean diameter (QMD), dominant height, basal area, total volume, volume of mature stems firewood from the crowns and thin stems, badly formed stems and quality wood) were analysed using ANOVA techniques and the Tukey test to determine significant differences at $p \leq 0.05$. We verified that the models comply with the assumptions of normality through the Shapiro-Wilks test and of homoscedasticity through residual analysis. We made corrections to the models when these assumptions were not met. Contingency tables and Pearson's Chi-square statistic were used to determine if there is a relationship between the forest structure (young, intermediate, old, and parquizado) and the quality parameters.

2.3.2. Growth of the Crop Trees

To evaluate the competition of the selected trees, the A-value of Johann was used [19]. This is a proportionality factor to determine the minimum distance between trees without competition (Equation (1)). Normally it is used as an objective parameter to regulate the intensity of a thinning. If the distance between two trees is measured, the A-value can be used to quantify the competition. The smaller A, the lower the competition is. As an example, in even-aged pure stands of *Picea abies* (L.) H.Karst, for which the A-value was developed, values from 4, 5, and 6 are recommended, which equates to heavy, moderate, and light releases, respectively [19]. The A-value was determined for every competitor of the selected crop trees.

$$A_{ij} = \frac{h_j}{dist_{ij}} * \frac{d_i}{d_j},\tag{1}$$

where A_{ij} is the A-value between crop tree j and competitor i, h_j is the height of the crop tree j, $dist_{ij}$ is the distance from the crop tree j to the competitor i, d_i is the DBH of the competitor i and d_j is the DBH of the crop tree j.

Analysing the trunk discs, some corrections were necessary because some annual rings were missed due to years of extreme conditions or very high stand density during the last years. By means of the discs, the diameter growth was analysed, and also, the height growth could be estimated.

Moreover, the degree of slenderness (h/d ratio), which is an indicator of the tree's stability, was determined for all analysed crop trees. If the h/d ratio is above the critical value of 80, the tree is presumed to be vulnerable to damages caused by wind, wet snow, and ice accumulation [18–20]. The development over time of the h/d ratios of all the selected trees was analysed, and different types of growth were identified. Forest management promotes trees to have h/d values lower than 80 for most of their life to ensure stability, which is achieved by releasing the diameter growth. For this reason, to analyse the evolution of the growth of the crop trees studied, they were grouped into three classes:

1. Dominant type: Trees whose h/d ratios, in general, were below 80, already in early years, and remained below the limit value throughout their lives (5 trees).
2. Variable type: Trees whose h/d ratios varied throughout their lives, generally reducing the h/d value with age, reaching less than 80; but the opposite could also have happened (5 trees).
3. Suppressed type: Trees whose h/d ratios, in general, were always over 80 (8 trees).

For every type of growth, the characteristic development of height, diameter, and volume was examined.

3. Results

3.1. Structures of the L. hirsuta Stands

The determined functions for total volume and trunk volume, which were used to calculate the stands volumes, are shown in Table 2. The DBH was the only independent variable used since the height coefficient was not statistically significant.

Table 2. Functions of total volume and individual trunk volume.

Volume	Function	n	REE	r^2adj
total [m³cc]	0.0007737 (DBH (cm))²	18	0.07747	0.963
trunk [m³cc]	0.0004338 (DBH (cm))²	18	0.0421	0.965

REE: residual standard error; r^2adj: adjusted determination coefficient.

Lomatia hirsuta occupies the whole site in a high density. Most of the dasometric parameters differed significantly between the young and old stands, while in the intermediate structure, they had values between them. Only the basal area and the total volume did not have significant differences between the structures without intervention (Table 3). If the density of the old stand is assumed as 100%, in the parquizado, the number of stems/ha was reduced at 28% and the basal area at 32%. The total volume reaches its maximum in the intermediate structure, with 560 m³/ha and the mature stem volume participation in the total volume increases from 4 to 46% between the young and the old structure (Table 3).

Table 3. Parameters of the different structures.

Variables	Young	Intermediate	Old	Parquizado
Number of trees/ha	2244 a	1167 ab	600 b	223 c
Number of stems/ha	6278 a	2180 b	1157 b	300 c
Number of stems/tree	2.8 a	1.9 b	1.9 b	1.3 b
Quadratic mean diameter (QMD) [cm]	10.7 c	18.2 bc	24.5 ab	26.8 a
Dominant QMD [cm]	24.3 b	34.9 ab	43.7 a	36 ab
Height of the tree of QMD [m]	9.5 b	12.1 ab	13.9 a	14.5 a
Mean dominant height [m]	13.9 b	16.4 ab	18.2 a	16.7 ab
Basal area [m²/ha]	56.1 a	56.9 a	54.4 a	16.9 b
Total volume [m³ o.b./ha]	553 a	560 a	536 a	167 b
Volume of mature stems [m³ o.b./ha] *	24 c	173 ab	249 a	82 bc

* DBH ≥ 20 cm. Letters indicate differences between groups based on a significance level ($p \leq 0.05$).

The diameter distributions of all stands are similar to a bell-shaped curve and only show little skewness to the small diameters (Figure 3). The fact that the diameter distribution of the parquizado is very similar to the others shows that by cutting, trees of all diameters were removed.

Looking at the development of the heights over diameter, it can be seen that the gradient is bigger in the smaller diameter classes, reaching its maximum at a DBH between 20 and 30 cm. After that, the curve continues approximately like an asymptote (Figure 4). Furthermore, the trees in the parquizado have minor heights at the same diameters.

The frequency of trees of each health condition, social position, and form of the trunk was significantly related with the forest structure Pearson's Chi-square statistics ($p < 0.0001$ for the three quality parameters). This means that social position, health condition, and form of the trunk are significantly related to the forest structure. The structure with relatively more fit and straight trees was parquizado, and that with relatively more dying trees was the young (Table 4). Intermediate trees were more frequently in the intermediate structure, and suppressed trees were more frequently in the young and old structures.

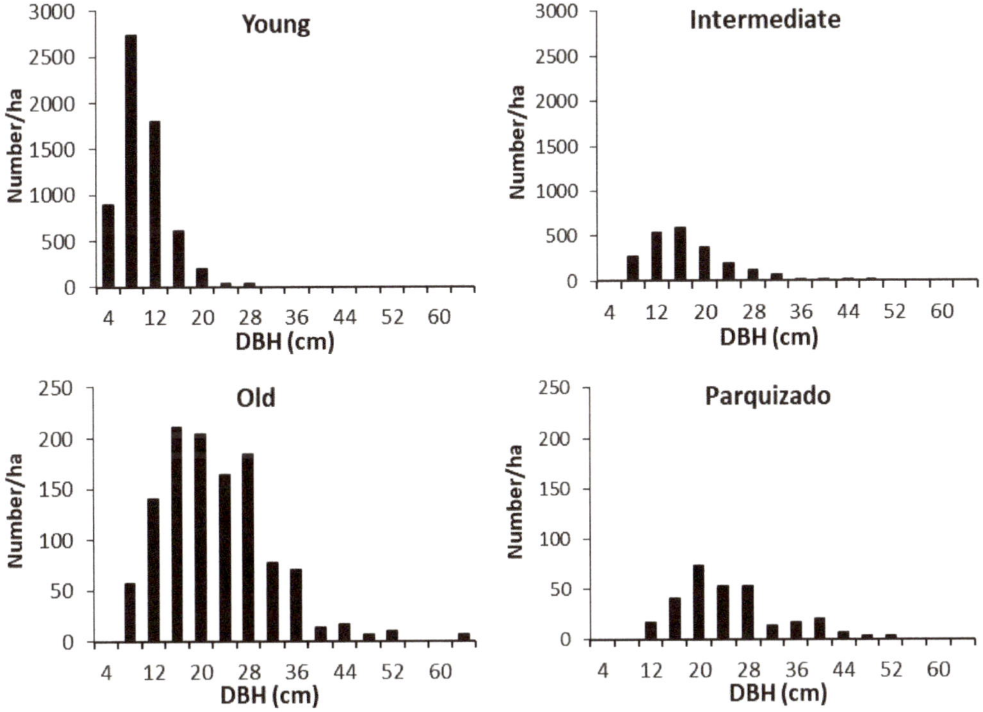

Figure 3. Diameter distributions of the different structures; the ordinates are differently scaled because otherwise, the charts with low number of stems/ha (old stands and parquizado) would not be readable.

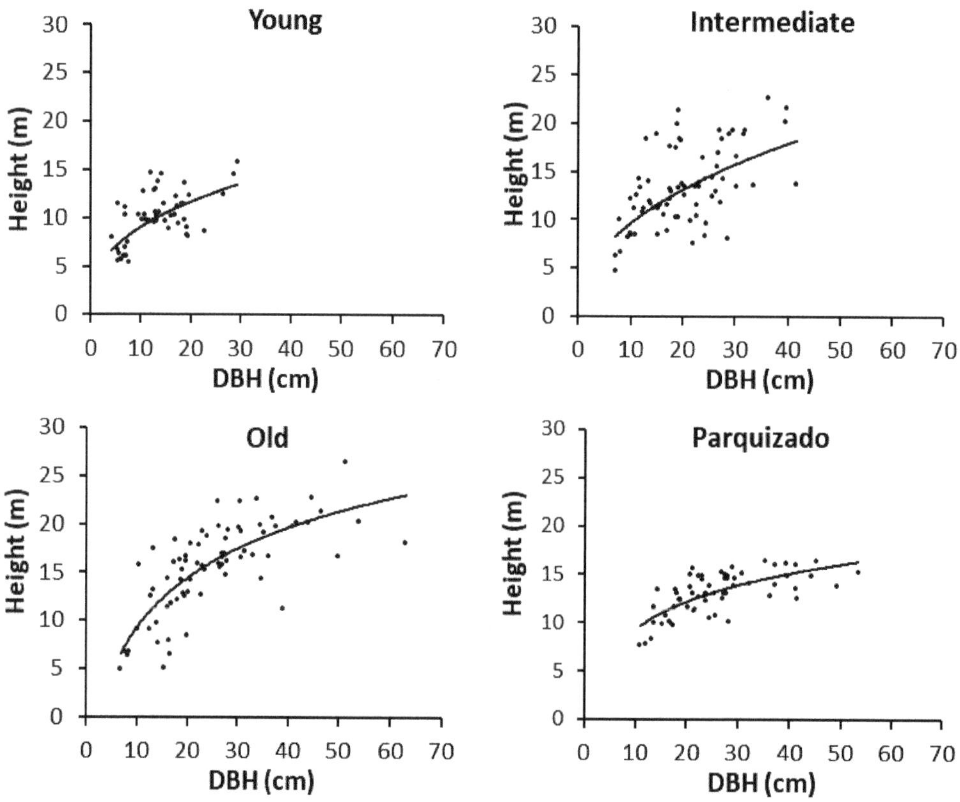

Figure 4. Height curve of the different structures.

Table 4. Quality parameters of the different structures: Absolute and relative values are shown.

Variables		Young Nha	%	Intermediate Nha	%	Old Nha	%	Parquizado Nha	%
Health condition	Fit	1622	25.8	663	30.4	150	11.5	157	52.2
	Injured	1844	29.4	883	40.5	627	50.0	133	44.5
	Dying	2811	44.8	633	29.1	380	38.5	10	3.3
Social position	Dominant	367	5.8	183	8.4	147	9.1	-	
	Codominant	2022	32.2	1080	49.5	440	36.5	-	
	Suppressed	3889	62.0	917	42.1	570	54.4	-	
Form of the trunk	Straight	1089	17.3	317	14.5	207	15.9	120	40.0
	Curved	1544	24.6	527	24.2	353	28.7	93	31.1
	Crooked	3644	58.1	1337	61.3	597	55.4	87	28.9
Volume [m³ o.b./ha]	crowns and thin stems *	529 a		386 b		288 b		85 c	
	badly formed stems	18 c		139 ab		194 a		78 bc	
	quality wood	6 ab (169 **)		35 ab		55 a		4 b	

Nha: Number of trees per hectare; * DBH < 20 cm; ** estimated volume of quality wood (future) when the mean dimensions, which were observed in the intermediate and old stands, will be reached. Letters indicate differences between groups based on a significance level ($p \leq 0.05$).

3.2. Thinning Simulation

The trees, which were marked for cutting in the simulation were dominated by small trees (Figure 5). While it was possible to select an average of 300 crop trees/ha to release in the young stands, in the intermediate and old ones, only 60 and 47 trees/ha were found. The proportion between the number of crop trees/ha and the total tree number/ha in the young (1:6.4) and in the old stands (1:5.4) was very similar.

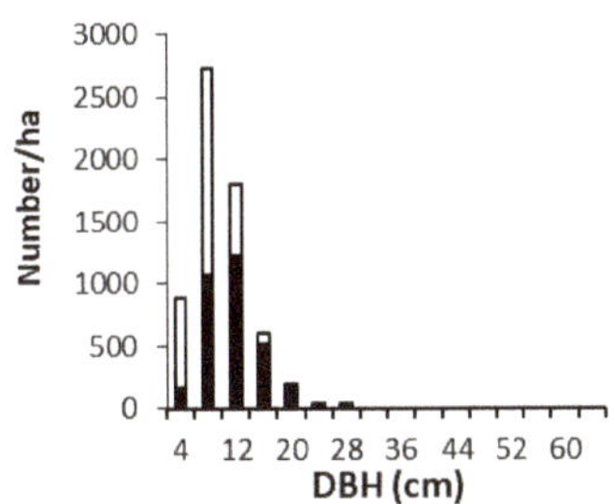
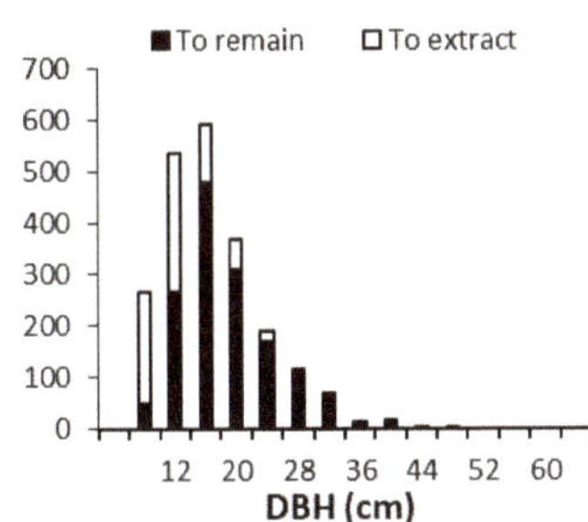
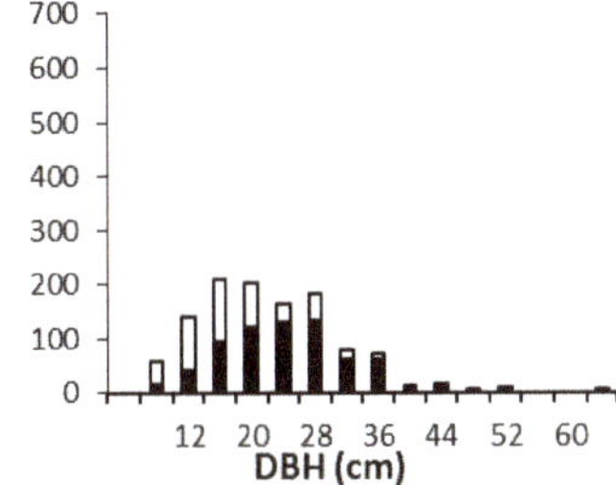

Figure 5. Number of stems to extract and to remain in the thinning simulation per diameter class. **Left**: young, **centre**: intermediate, and **right**: old stands; the ordinates are differently scaled because otherwise, the charts with low number of stems/ha (intermediate and old stands) would not be readable.

The number of trees to extract would be significantly higher in the young structure compared to the intermediate and old structures, while the basal area and the total volume to extract are only significantly different between the young and intermediate structures (Table 5). The volume of mature stems to extract increases significantly between the young and old structures. On the other hand, the QMD and mean height increased significantly between extract and remained in all structures (Table 5).

Table 5. Parameters of the trees to be extracted and to be alive in the simulation in every structure. The density variables (1–4) were compared between structures, while the size variables (5–6) within each structure (extract vs. remain).

	Variables	Young		Intermediate		Old	
		Extract	Remain	Extract	Remain	Extract	Remain
1	Number of stems/ha	3078 a	3200 a	697 b	1483 ab	450 b	707 b
2	Basal area [m²/ha]	18.3 a	37.8 a	11.0 b	45.9 a	13.6 ab	40.8 a
3	Total volumen [m³ o.b./ha]	180 a	373 a	109 b	452 a	134 ab	402 a
4	Volume of mature stems [m³ o.b./ha] (*)	5 c	18 b	20 ab	154 a	48 a	201 a
5	QMD [cm]	8.7 b	12.3 a	14.2 b	19.8 a	19.6 b	27.1 a
6	Height of QMD trees [m]	8.5 b	9.7 a	10.3 b	13.1 a	11.6 b	16.7 a

* DBH $\geq$ 20 cm. Letters indicate differences between groups based on a significance level ($p \leq 0.05$).

Most of the extracted wood in the thinning simulation would be firewood (Table 6). Only little quality wood could be harvested. In the old stands, an appreciable share of quality wood is from fallen trees.

Table 6. Product classes and volumes of the extractable wood of the thinning simulation (m³ u.b./ha).

Range of Products	Young	Intermediate	Old
Firewood from the crowns and thin stems *	140 a	71 b	69 b
Badly formed stems	4 b	13 ab	32 a
Quality wood	-	3 b	11 (5 **) a

* DAP < 20 cm; ** correspond to quality logs of fallen trees. Letters indicate differences between groups based on a significance level ($p \leq 0.05$).

3.3. Growth of the Crop Trees

The average A-value of the analysed trees reached 8.9 in the young and old stands and 7.2 in the intermediate. These are quite high values compared to the recommended ones (cf. Section 2.3.2), showing a high level of individual competition in all structures.

Considering the types of growth described in Section 2.3.2, the crop trees were grouped, and every ten years, the mean value of the variables were determined (Figure 6). Comparing the evolutions of the variables in each growth type, there was practically no difference in the mean height development, while the diameter increment and the stem volume development shows clear differences (Figure 6). The trees of the dominant type had a greater diameter growth since their early years, which caused their h/d value to remain below 80 for most of their life (Figure 6, top left).

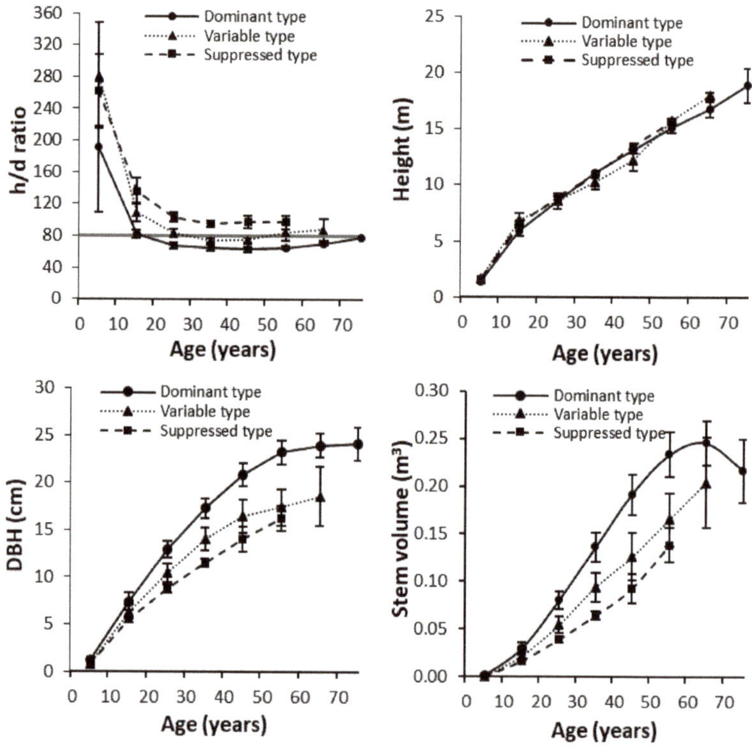

Figure 6. Development of each variable (mean value and standard error every ten years) from studied crop trees of each growth type: dominant type, variable type, and suppressed type. **Top left**: h/d ratio (80 indicates the limit value of stability), **top right**: height development; **bottom left**: diameter development and **bottom right**: volume development. The stem volume at 75 years of trees of dominant type has a lower number of trees than in previous ages (in the age of 75 years there was data from fewer trees to be considered in the mean value).

The increment of the dominant trees with dominant type growth is a good approximation of the growth that could be reached in managed stands. The periodic annual diameter increment increased in the juvenile stage and then decreased quickly when the stand was closing and equalled the mean annual increment at 25 years of age (Figure 7).

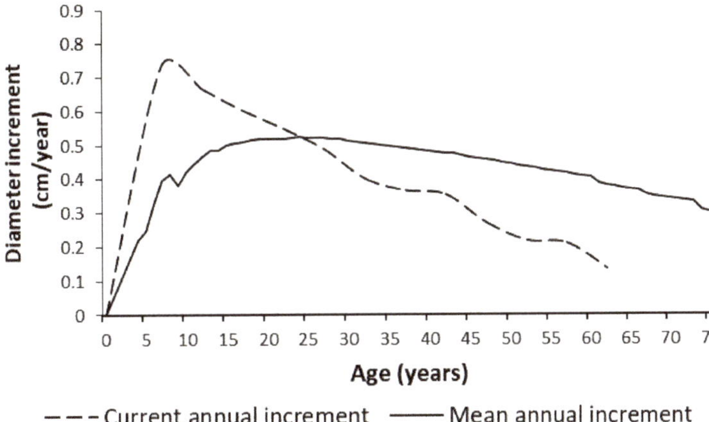

Figure 7. Development of the current annual increment (CAI) and the mean annual increment (MAI) of tree diameter of the dominant type of growth.

4. Discussion

4.1. Credibility of Collected Data and What We Do and Don 't Know

The selection and classification of the different age stages of stands were made in consideration of the trees' dimensions but without knowing the real age. Because of that, the distinction between the intermediate and the old stands is not always clear. Another uncertainty is caused by the fact that it was often very difficult to identify all the annual rings at the discs. Therefore, slight differences in the real age in the analysed trunks cannot be ruled out.

The reduction in the number of trees in the unmanaged stands between the young and old structure (Table 3) is explained by the mortality due to competition, and this is also expressed in the high percentage of injured and dying trees in each structure (Table 4). The damages in the stems and crowns were probably caused by snow and wind in trees with high h/d ratios due to the high stand density, expressed in high A-values, and the lack of light that controls the diameter increment, especially for the short trees. This is also the reason why the number of selected crop trees is only optimal in the juvenile phase, but without applying silviculture, it is reduced considerably in older stages.

Another aspect that may have an effect on vulnerability to damage, known from other species in temperate climates, is that the vigour, sprouting capacity, and growth of stump shoots decrease after three or more generations [21–23]. *Lomatia hirsuta* is a pioneer species, which forms post-fire mixed shrublands together with others such as *Schinus patagonicus* (Phil.) I. M. Johnst. ex Cabrera, *Maitenus boaria* (Mol.), and *Nothofagus antarctica* (Forst.) Oerst. [5,11,24]. In the process of succession, they are gradually invaded and replaced by species from the high forest, such as *Austrocedrus chilensis* and/or *Nothofagus dombeyi* [5,24]. It was not possible to know the type of forest that existed before the last wildfire. At present, the almost absolute dominance of *L. hirsuta* of stump regrowth, practically without other tree species or its regeneration, suggests that the previous forest was also dominated by the same species. And if so, it would then be possible that the studied *L. hirsuta* forests also started after two or more previous fire events. This might be also the reason for the loss of vitality and stability of the *L. hirsuta* with old rootstocks in the intermediate and old stands in El Manso Valley.

4.2. Possibility of Growth for High-Quality Trees

In relation to the simulation of cutting applying the concept of crop trees, the fact that the QMD increases due to the thinning, especially in the intermediate and old stands

(Table 5), shows that the intervention resembled a low thinning more than the liberation of the crop trees. This can be traced back to the low number of crop trees found. Although the optimal phase to start the application of silviculture for quality improvement is already over, the implementation of a crop tree-orientated management in the young stands, with three hundred registered aspirants to crop trees per hectare, should be still possible and successful.

The trunk analysis of the dominant type of growth in dominant trees (Figure 7) shows that it is possible to reach an MAI in DBH of 0.35 cm/year at the age of 70. Furthermore, trunks with a DBH of 45 cm were registered, which had to have nearly the same age. This means that an MAI of 0.6 cm/year seems to be achievable, too. This could be taken as an indicator for the magnitude of the target diameter. Nevertheless, during the first pre-commercial thinning in the young stands, almost no wood of sawn timber quality will be harvested. On the other hand, the high volume of badly formed stems (139–194 m^3/ha) in intermediate and old stands (Table 4) is an indicator of the potential volume to improve their quality through the application of forest management from an early age.

4.3. Parquizado Management and Fire Risk Reduction

The named problems of falling trees do not appear that intensely in the parquizado, probably because in the intense thinning, the trees with less vitality were eliminated. The fact that the diameter distribution of the parquizado is very similar to the others (Figure 3) shows that by cutting, trees of all diameters were removed. The trees in the parquizado presented minor heights at the same diameters (Figure 4). This could be a sign that the remaining trees could have reacted to the intervention by bigger diameter increments. On the other hand, in the parquizado, the total volume and the continuity of biomass fuel is much lower than in the other structures (visually appreciated). This leads to lower flammability and lower fire intensity in case of a fire event and results in a lower risk of wildfires [25]. Moreover, for the parquizado, a forest regeneration concept must be developed to ensure sustainability.

4.4. Recommendations for the Intermediate and Old Stands

Most of the wood harvested during the conversion to managed forests in the intermediate stands will be firewood and timber for handicraft or glued-wood products. For the conversion, the following steps are recommended:

1. Salvage and improvement cutting: This includes the extraction of fallen and unstable trees or great damages to be obtained for their wood in favour of the better trees.
2. Remove patchy clearances to create space for natural regeneration or plantings.
3. Planting *L. hirsuta* in irregular holes in order to renew the stand and with the expectation of improving the vigour and quality of trees in the old phase. The question of whether the vitality and productive quality of the stand can be improved through the planting of seed trees is subject to research. *L. hirsuta* and seedlings of other native species that are later successional stage of this forests, like *A. chilensis* or *N. dombeyi*, could also be integrated. To succeed, it would be necessary to stop the grazing in the stands with plantings; further, the stumps of the cut trunks have to be killed to prevent new shoot growth. In this context, knowledge about the biological-technical maximum rotation period of *L. hirsuta* should be researched in further studies.

4.5. Recommendations for Young Stands

Two hundred crop trees should be selected and favoured by extracting 1–3 competitors per tree. The objective is to have around 150 crop trees/ha at the age of harvesting. It should be possible to reach a target diameter of 40 cm in 70 years by favouring the growth of crop trees. Additionally, at least in the first interventions, the worst trees concerning health conditions and quality should be extracted to reduce the general competition.

4.6. Additional Considerations

The proposed management is aimed to maintain the dominance of *L. hirsuta* in the stand to promote the production of its quality wood. However, the entry of natural regeneration of other shrub species also with the potential to produce quality wood, such as *M. boaria*, or others of a later stage of the succession, such as *A. chilensis* or *N. dombeyi*, as they occur, could be incorporated into this forest management. In this case, it should be taken into account that *A. chilensis* and especially *N. dombeyi*, due to their greater growth and development in height, will tend to dominate in the stand [5,24]. In this sense, new silvicultural research will be required for those conversion and management processes.

It is also suggested to apply the proposed thinning method in experimental units on an operational scale. There, the responses to the crop trees could be monitored and the results validated. Furthermore, in the experimental units, the predictions and their relationships with other variables could be adjusted, for example, the effect of wet and dry years. In that sense, the applicability in other site conditions, especially those with less precipitation, should be evaluated.

5. Conclusions

It can be concluded that the pure *L. hirsuta* stands, as in El Manso Valley, have worthwhile potential and that it would be possible to improve their quality and vigour. Giving consideration to the actual situation, especially in the intermediate and old stands, it is clear that it will be a slow and long process. The main problems are the very high density that contributes to individual growth reduction and high h/d ratio, which cause vitality loss and instability. Because of these aspects, it is necessary to implement forest management to improve first their stability and then their quality. For the production of quality wood through thinning oriented to crop trees, it is essential to start at the early age of the stand when there is a sufficient number of selectable trees. On the other hand, the use as parquizado should mainly take place in areas with high fire risk, for example, along roadsides.

Author Contributions: Conceptualisation, G.A.L. and M.T.; methodology, G.A.L., H.K. and M.T.; software, H.K. and G.A.L.; validation, M.C. and G.A.L.; formal analysis, G.A.L., M.C. and H.K.; investigation, G.A.L.; resources, G.A.L.; data curation, H.K.; writing—original draft preparation, H.K.; writing—review and editing, G.A.L., M.C. and M.T.; visualisation, H.K. and G.A.L.; supervision, M.T.; project administration, G.A.L.; funding acquisition, H.K. and G.A.L. and M.T. All authors have read and agreed to the published version of the manuscript.

Funding: This research was funded by the PROMOS Scholarship program of the German Academic Exchange Service (DAAD) for travel costs and by UVT-CIEFAP for the fieldwork and by HAWK for the publication.

Data Availability Statement: The data presented in this study are available on request from the corresponding author.

Acknowledgments: This study was realised as part of a cooperation agreement between CIEFAP in Argentina and the University of Applied Sciences and Arts (HAWK) of Göttingen in Germany. We would like to express our thanks to the owners of forests for the permission to realise the fieldwork in their properties and also to the Servicio Forestal Andino de Río Negro for granting the permits for the tree cutting. Further, our thanks go to the forest engineer Pedro Pantaenius for his help cutting the trees, to the student's Candela Rodríguez and Diego Apablaza for their active support during the fieldwork.

Conflicts of Interest: The authors declare no conflict of interest. The funders had no role in the design of the study, in the collection, analyses, or interpretation of data, in the writing of the manuscript, or in the decision to publish the results.

References

1. Tortorelli, L. *Maderas y Bosques Argentinos*, 2nd ed.; Orientación Gráfica Editora: Buenos Aires, Argentina, 2009; pp. 329–332.
2. Weinberger, P. Verbreitung und Wasserhaushalt araukano-patagonischer Proteaceen in Beziehung zu mikroklimatischen Faktoren. *Flora* **1974**, *163*, 251–264. [CrossRef]
3. Eskuche, U. Estudios fitosociológicos en el norte de la Patagonia. I. Investigación de algunos factores de ambiente en comunidades de bosque y de chaparral. *Phytocoenologia* **1973**, *1*, 64–113. [CrossRef]
4. Seibert, P. Die Vegetationskarte des Gebietes von El Bolsón, Prov. Río Negro, und ihre Anwendung in der Landnutzungsplanung. In *Bonner Geographische Abhandlungen*; Hahn, H., Kuls, W., Lauer, W., Höllermann, P., Boesler, K.A., Ruckert, H.-J., Eds.; Ferd. Dümmlers Verlag: Bonn, Germany, 1979; Volume 62, pp. 14–16.
5. Kitzberger, T. *Ecotones Between Forest and Grassland*; Springer: New York, NY, USA, 2012; p. 63. [CrossRef]
6. Lusk, C.H.; Corcuera, L.J. Effects of light availability and growth rate on leaf lifespan of four temperate rainforest Proteaceae. *Rev. Chil. Hist. Nat.* **2011**, *84*, 269–277. [CrossRef]
7. Figueroa, J.A.; Lusk, C.H. Germination requirements and seedling shade tolerance are not correlated in a Chilean temperate rain forest. *New Phytol.* **2001**, *152*, 483–489. [CrossRef] [PubMed]
8. Reque, J.; Sarasola, M.; Gyenge, J.; Fernández, M.E. Caracterización silvícola de los ñirantales de la cuenca central del Río Foyel (Río Negro, Patagonia Argentina). *Bosque* **2006**, *28*, 33–45.
9. Gyenge, J.; Fernández, M.E.; Sarasola, M.; de Urquiza, M.; Schlichter, T. Ecuaciones para la estimación de biomasa aérea y volumen de fuste de algunas especies leñosas nativas en el valle del río Foyel, NO de la Patagonia argentina. *Bosque* **2009**, *30*, 95–101. [CrossRef]
10. Gyenge, J.; Fernández, M.E.; Sarasola, M.; Schlichter, T. Testing a hypothesis of the relationship between productivity and water use efficiency in Patagonian forests with native and exotic species. *For. Ecol. Manag.* **2008**, *255*, 3281–3287. [CrossRef]
11. Goldenberg, M.G.; Oddi, F.J.; Gowda, J.H.; Garibaldi, L.A. Shrubland Management in Northwestern Patagonia: An Evaluation of Its Short-Term Effects on Multiple Ecosystem Services. In *Ecosystem Services in Patagonia: A Multi-Criteria Approach for an Integrated Assessment*; Peri., P.L., Martínez-Pastur, G., Nahuelhual, L., Eds.; Springer Nature: Cham, Switzerland, 2021; Chapter 5; ISBN 978-3-030-69165-3.
12. Porto Tapiquén, C.E. "South America" Layer (Shape File). Orogenesis Geographic Solutions. Porlamar, Venezuela 2015. Based on layers from Enviromental Systems Research Institute (ESRI). Free Distribution. Available online: http://tapiquen-sig.jimdo.com (accessed on 8 July 2021).
13. Rogel, M. (Animal Breeder, El Manso, Río Negro, Argentina). Personal communication, 2020.
14. IUSS Working Group WRB. *World Reference Base for Soil Resources. International Soil Classification System for Naming Soils and Creating Legends for Soil Maps—Update 2015*; World Soil Resources Reports No. 106; FAO: Rome, Italy, 2014; updated 2015; ISBN 978-92-5-108369-7.
15. Buduba, C.; La Manna, L.; Irisarri, J. El suelo y el bosque en la Región Andino Patagónica. In *Suelos y Vulcanismo: Argentina, 1st ed*; Imbellone, P., Barbosa, O., Eds.; Asociación Argentina de la Ciencia del Suelo: Buenos Aires, Argentina, 2020; pp. 361–390. [CrossRef]
16. Smalian, H.L. *Beitrag zur Holzmeßkunst*; C. Löffler Verlag: Stralsund, Germany, 1837.
17. Huber, F.X. *Hilfstafeln für Bedienstete des Forst- und Baufaches: Zunächst zur Leichten und Schnellen Berechnung des Massengehaltes Roher Holzstämme und der Theile Derselben, und Auch zu Anderm Gebrauche für Jedes Landesübliche Maaß Anwendbar*; Fleischmann E.A.: München, Germany, 1828.
18. Pretzsch, H. *Grundlagen der Waldwachstumsforschung*, 2nd ed.; Springer-Verlag: Berlin, Germany, 2019. [CrossRef]
19. Pretzsch, H. *Forest Dynamics, Growth and Yield. From Measurement to Model*; Springer: Berlin, Germany, 2009. [CrossRef]
20. Šēnhofa, S.; Katrevičs, J.; Adamovičs, A.; Bičkovskis, K.; Bāders, E.; Donis, J.; Jansons, A. Tree Damage by Ice Accumulation in Norway Spruce (*Picea abies* (L.) Karst.) Stands Regarding Stand Characteristics. *Forests* **2020**, *11*, 679. [CrossRef]
21. Nicolescu, V.-N.; Carvalho, J.; Hochbichler, E.; Bruckman, V.; Piqué-Nicolau, M.; Hernea, C.; Viana, H.; Štochlová, P.; Ertekin, M.; Tijardovic, M.; et al. *Silvicultural Guidelines for European Coppice Forests*; Albert-Ludwigs-Universität Freiburg: Freiburg, Germany, 2017; pp. 4–5.
22. Burschel, P.; Huss, J. *Grundriss des Waldbaus. Ein Leitfaden für Studium und Praxis*; Parey Buchverlag: Berlin, Germany, 1997; p. 487.
23. Nyland, R.D. *Silviculture: Concepts and Applications*, 2nd ed.; Waveland Press, Inc.: Long Grove, USA, 2002; p. 682. ISBN 978-1-57766-527-4.
24. Veblen, T.T.; Kitzberger, T.; Lara, A. Disturbance and forest dynamics along a transect from Andean rain forest to Patagonian shrublands. *J. Veg. Sci.* **1992**, *3*, 507–520. [CrossRef]
25. Corona, P.; Ascoli, D.; Barbati, A.; Bovio, G.; Colangelo, G.; Elia, M.; Garfi, V.; Iovino, F.; Lafortezza, R.; Leone, V.; et al. Integrated forest management to prevent wildfires under Mediterranean environments. *Ann. Silvic. Res.* **2015**, *39*, 1–22. [CrossRef]

Article

Age-Based Survival Analysis of Coniferous and Broad-Leaved Trees: A Case Study of Preserved Forests in Northern Japan

Pavithra Rangani Wijenayake [1,*] and Takuya Hiroshima [2]

[1] Department of Forest Science, Graduate School of Agricultural and Life Sciences, The University of Tokyo, Bunkyo-ku, Tokyo 113-8657, Japan

[2] Department of Global Agricultural Sciences, Graduate School of Agricultural and Life Sciences, The University of Tokyo, Bunkyo-ku, Tokyo 113-8657, Japan; hiroshim@g.ecc.u-tokyo.ac.jp

* Correspondence: rangani@g.ecc.u-tokyo.ac.jp

Abstract: Scientifically sound methods are essential to estimate the survival of trees, as they can substantially support sustainable management of natural forest resources. Tree mortality assessments have mainly been based on forest inventories and are mostly limited to planted forests; few studies have conducted age-based survival analyses in natural forests. We performed survival analyses of individual tree populations in natural forest stands to evaluate differences in the survival of two coniferous species (*Abies sachalinensis* (F. Schmidt) Mast. and *Picea jezoensis* var. microsperma) and all broad-leaved species. We used tree rings and census data from four preserved permanent plots in pan-mixed and sub-boreal natural forests obtained over 30 years (1989–2019). All living trees (diameter at breast height $\geq$ 5 cm in 1989) were targeted to identify tree ages using a Resistograph. Periodical tree age data, for a 10-year age class, were obtained during three consecutive observation periods. Mortality and recruitment changes were recorded to analyze multi-temporal age distributions and mean lifetimes. Non-parametric survival analyses revealed a multi-modal age distribution and exponential shapes. There were no significant differences among survival probabilities of species in different periods, except for broad-leaved species, which had longer mean lifetimes in each period than coniferous species. The estimated practical mean lifetime and diameter at breast height values of each coniferous and broad-leaved tree can be applied as an early identification system for trees likely to die to facilitate the Stand-based Silvicultural Management System of the University of Tokyo Hokkaido Forest. However, the survival probabilities estimated in this study should be used carefully in long-term forest dynamic predictions because the analysis did not include the effects of catastrophic disturbances, which might significantly influence forests. The mortality patterns and survival probabilities reported in this study are valuable for understanding the stand dynamics of natural forests associated with the mortality of individual tree populations.

Keywords: mean lifetime; natural forest; survival analysis; tree age distribution

Citation: Wijenayake, P.R.; Hiroshima, T. Age-Based Survival Analysis of Coniferous and Broad-Leaved Trees: A Case Study of Preserved Forests in Northern Japan. *Forests* **2021**, *12*, 1014. https://doi.org/10.3390/f12081014

Academic Editor: Daniele Castagneri

Received: 30 June 2021
Accepted: 26 July 2021
Published: 30 July 2021

1. Introduction

Changes in the survival probability of forests have severe environmental [1] and economic consequences [2] and can influence forest management decisions [3,4]. Hence, long-term tree mortality studies are needed [5–8] considering the threat of global climate change, which is forecasted to increase tree mortality in forests worldwide [9–13]. Because of the difficulty in identifying the exact ages of trees, researchers mainly use diameter at breast height (DBH), dominant height, basal area, growth rate, or competition index, avoiding age-based measures to determine mortality probabilities [14] Popular methods for identifying tree mortality are logistic models [15], Weibull [16], the Gamma [17], Richard's function [18], and exponential or probit models [19]. Tree age can be used for the accurate prediction of tree mortality [20–22]. Generally, the lifespan of a tree is unknown; unlike logistic regressions, the right censoring and left truncation approaches of survival analysis can handle tree age data [14,23]. Several survival analysis techniques have been suggested to determine forest mortality [24]; however, these are limited to even-aged forest stands [25–27]. Woodall et al. [14] carried out a survival analysis based on DBH. More recently, survival analysis of natural forests has been successfully applied based on inventory measurements (i.e., [28–30]). Nothdurft [28] provided spatiotemporal predictions of tree mortality based on parametric frailty modeling. Moreover, survival models have been fitted based on inventory data by omitting the first 20 years of a tree's life [31]. However, these studies emphasized the importance of determining exact tree ages for the survival analysis of forests [14,29,30].

The principles underlying the approach presented in this paper are related to the idea presented by Hiroshima [32], which involved more flexible techniques to find tree age for survival analysis based on non-parametric and parametric approaches. The presented method is based on annual tree ring measurements obtained using a semi-nondestructive device, as well as periodic inventory data of secondary natural forests in Japan. However, previous studies focusing on the age-based survival analysis based on uneven-aged natural forests are comparatively rare, because they require unique methods to detect tree age and conduct long-term, large-scale observations [33,34].

In this study, we used a device to determine tree age at the long-term research site in the preserved area in the University of Tokyo Hokkaido forest (UTHF), which is considered ideal for performing investigations similar to the present one, because plots in this area do not undergo any management following natural disturbance events [35]. Overall, the purpose of this study was to perform a survival analysis of individual tree populations in natural forest stands of the UTHF, Japan, and estimate and compare the age distribution of two major coniferous species (*Abies sachalinensis* (F. Schmidt) Mast. and *Picea jezoensis* var. microsperma) and broad-leaved species. We applied survival analysis techniques to each of three observational periods in a 10-year age class to determine the non-parametric survival probabilities of the species. Specifically, we revealed the multi-temporal age distribution changes and mean lifetime changes in two groups of trees (coniferous and broad-leaves) during the three observational periods. Survival probability changes over the periods can be applied to determine the tree harvesting strategy of the UTHF.

2. Materials and Methods

2.1. Inventory and Site Data

Retrospective inventory and site data were derived from UTHF, located in Furano (central Hokkaido, Japan). This site experiences a mean annual temperature and precipitation of 6.4 °C and 1297 mm, respectively [35]. Dominant stratified soils include brown forest soil, dark brown forest soil, black soil, and podzol [36]. Mixed conifer–hardwood forest, typical of the cool temperate zone, covers most of the site area. Tree species commonly found in this forest area are *Fraxinus mandshurica*, *Ulmus davidiana* var. *japonica*, *Alnus hirsuta*, and *Salix* spp. in deciduous swamp forests at lower elevations of less than 300 m; a coniferous and broad-leaved mixed forest dominated by *A. sachalinensis* at middle elevations (300–600 m), scattered forests mixed with *P. jezoensis*, *Picea glehnii*, and *Betula ermanii* at upper elevations (800–1200 m), and alpine vegetation (e.g., *Pinus pumila*) in the upper forest limit (>1200 m).

The Stand-based Silvicultural Management System (SSMS), in other words, a natural forest management system based on selected cutting and natural regeneration, is currently being employed extensively in UTHF except for research plots [37,38]. In SSMS, 10–17% of the stand volume is harvested by single-tree selection, with a cutting cycle of 15–20 years, by removing defective (e.g., diseased, senescent, non-vigorous, and twisted) and over-matured trees. This system helps maintain tree health and productivity in the stand and controls the stand composition [37]. To determine cutting rates, permanent plots and long-term ecological research plots have been established and periodically assessed. Among these permanent plots, 25 are located in the preserved area, ranging in size from 0.04 to 2.25 ha, and have an elevation between 380 and 1290 m. Within these plots, DBH measurements of all trees with a DBH $\geq$ 5 cm are regularly performed by UTHF staff; in most cases, 5-year interval and census data are available for the last five decades. Other than these periodical measurements, no human intervention has occurred in this preserved area for several decades.

2.2. Survival Data

Survival data were derived from periodic surveys on preserved permanent sample plots. We selected four plots, ranging in size from 0.04 to 2.25 ha, located at an elevation range of 570 to 690 m with similar slope aspects and slope angles (Figure 1; Table 1). The main soil types found in the plots are brown forest and podzolic soils [35]. In addition, the stands in the four plots are all classified as "coniferous selective cutting with poor regeneration" in UTHF, where no continuous sufficient and new in-growth trees are expected. All four plots were within close proximity and had similar species composition. The typical vegetation of the plots was coniferous and broad-leaved mixed forests dominated by *A. sachalinensis*, *P. jezoensis*, *Acer* spp., and *Tilia* spp. Therefore, the four plots were aggregated in further analyses on tree survival.

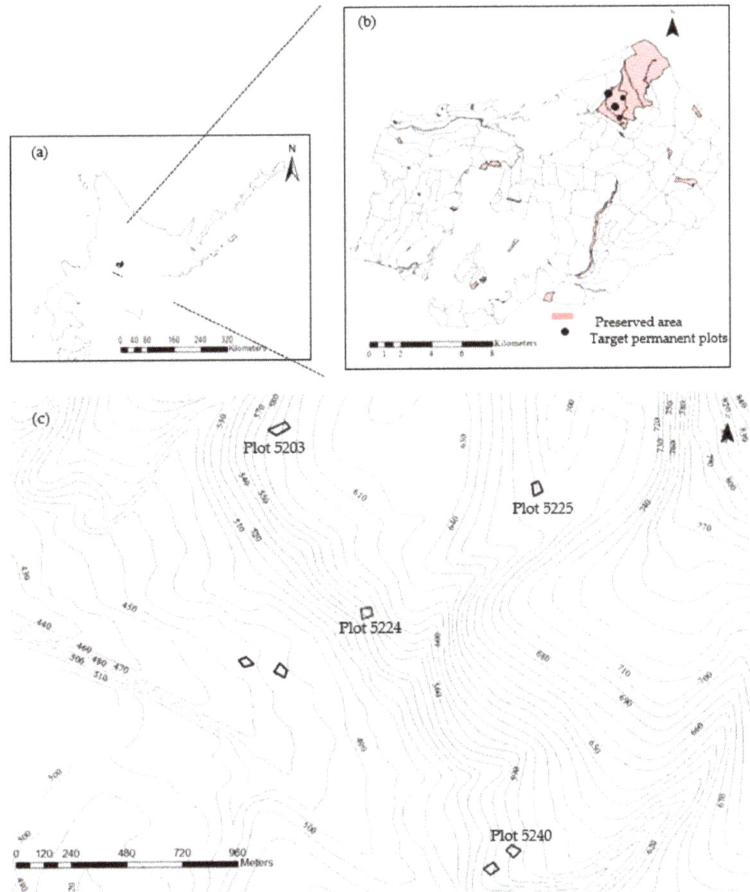

Figure 1. Locations of the study area and plots. (**a**) University of Tokyo Hokkaido Forest (UTHF) of Hokkaido island of Japan; (**b**) preserved area and permanent plots of the UTHF; (**c**) contour map of the four preserved permanent plots of UTHF.

Table 1. Stand characteristics of selected plots.

Plot	Plot Size (ha)	Elevation (m)	No. of Target Trees in Period 3	Slope Aspect	Mean Slope Angle (°)
5203	0.40	580	327	Southwest	15
5224	0.25	570	150	Southwest	18
5225	0.25	690	237	Southwest	20
5240	0.25	600	214	Southwest	25

We used tree census data (species, DBH, living or dead status, cause of death, etc.) of the plots, obtained between 1989 and 2019. Within this period, we set three observation periods of 1989–1999 (period 1), 1999–2009 (period 2), and 2009–2019 (period 3). We detected the number of tree rings at breast height (1.3 m) in 2019 using a semi-nondestructive device, RESISTOGRAPH® [39]. All target trees were living and had a DBH $\geq$ 5 cm in 1989; some of these trees died by 2019. Live/dead trees and new in-growth trees were counted at the end of each observation period. The connected RESISTOGRAPH (Heidelberg, Germany) data logger recorded measurements, which were transferred to a computer for further analyses. The field measurement tree ring data (radius at breast height -2.5 cm) in 2019 were extracted via the DECOM^{TM} software (Philadelphia, PA, USA), which is used for annual tree ring detection. The "radius at breast height -2.5 cm" reveals the tree age after in-growth if we set DBH = 5 cm as an in-growth border.

Figure 2 shows the annual rings in one living *A. sachalinensis* tree detected using the DECOM^{TM} software. RESISTOGRAPH measurements were ineffective for severely rotten and center-decayed trees, for which we established the regression equation between the age after in-growth (y) and the radius -2.5 cm (w) [40]. Next, the age after in-growth was estimated by inputting the radius data into simple three-dimensional equations fitted to a scatter diagram (Table 2). This method was followed for 20% of the sample trees. Table 3 further illustrates the method applied to determine the in-growth years of trees.

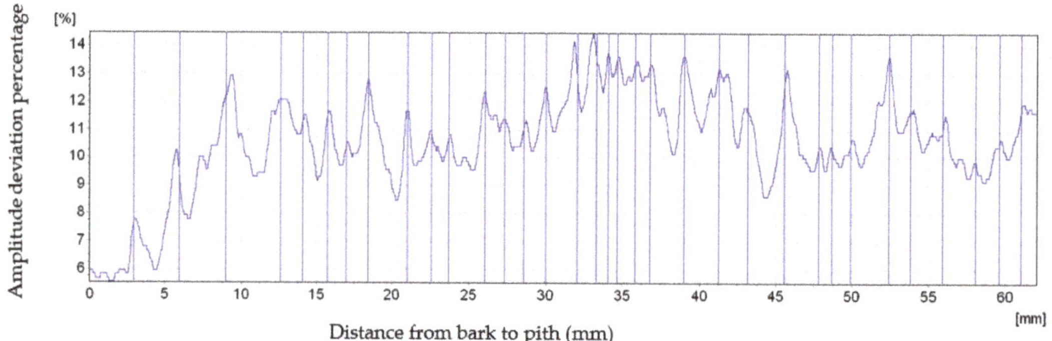

Figure 2. Manually detected annual rings of *A. sachalinensis* tree in plot 5240 detecting using DECOM^{TM} software.

Table 2. Regression models developed based on RESISTOGRAPH measurements.

Tree Species	No. of Tree Samples	Equation	Coefficient of Determination
A. sachalinensis	51	$y = -0.0000008w^3 + 0.0005w^2 + 0.1995w + 32.294$	0.8379
P. jezoensis	59	$y = -0.000002w^3 + 0.0008w^2 + 0.2545w + 30.32$	0.8506
A. ukurunduense	41	$y = -0.00001w^3 + 0.0043w^2 + 0.1416w + 41.991$	0.7668
T. japonica	32	$y = 0.000002w^3 - 0.0012w^2 + 0.5815w + 13.189$	0.7803
Other broad-leaved species	45	$y = -0.000005w^3 + 0.0023 w^2 + 0.0088 w + 42.386$	0.8592

Table 3. Methodologies used to determine the age of each tree species.

Species	Census Data		Direct AResistograph Measurements		Developed ARegression Models		Total No. of Trees	
	Living Trees	Dead Trees	Living Trees	Dead Trees	Living Trees	Dead Trees	Living Trees	Dead Trees
A. sachalinensis	113	9	58	16	48	16	219	41
P. jezoensis	18	4	55	11	49	6	117	21
Broad-leaved trees	289	67	76	21	55	10	423	98
Percentage	45.70	8.71	20.57	5.22	16.54	3.48		

Survival models were built for two major coniferous species present in UTHF, *A. sachalinensis* and *P. jezoensis*. All broad-leaved trees were considered collectively to develop survival models. The model data of period 3 were for 260 *A. sachalinensis*, 138 *P. jezoensis*, and 521 broad-leaved trees (see Table 3 for the number of dead and living trees in period 3 included in this study). The ages of target trees in period 3 were first identified using the above-mentioned methods. The ages in periods 1 and 2 were calculated by deducting 10 and 20 years from those in period 3. Finally, tree ages were classified, with 10 years per age class.

2.3. Fundamentals of Survival Analysis of Major Species

Survival probability functions were the essential elements of this analysis, as they reflect the overall performance of the major species; they were developed based on previous studies [20,22,32,41–43]. The survival data demonstrated that both dead and living trees were present in the study plots at the end of the observation period. When we carry out survival analysis, the issue of missing data appears most of the time. Therefore, the censoring and truncation techniques must be incorporated to overcome this issue. Right censoring occurs when a target leaves the study before an event occurs or the follow-up ends before mortality happens. In contrast, left truncation occurs when an object is not observed from the beginning of the study, but rather enters the study at a later time in the observation period [20,44]. Furthermore, if target trees survived at the t-th age class in the observation period, then observed mortalities were truncated and censored at both beginning and end of the observation period.

Mortality rates (p_t) were calculated using age class after in-growth (T), according to Equation (1):

$$\Pr(t-1 < T \leq t | T > t-1) = p_t \tag{1}$$

where p_t is the mortality rate in the t-th age class.

If a tree survives the t-th age class during the observation period, the conditional probability is defined as follows:

$$\Pr(T > t | T > t-1) = 1 - p_t. \tag{2}$$

Following previous methods [45–47] and considering Equations (1) and (2), we described the likelihood function (L) of the observation as follows:

$$L = \prod_t \Pr(t-1 < T \leq t | T > t-1)^{d_t} \Pr(T > t | T > t-1)^{a_t} = \prod_t p_t^{d_t}(1 - p_t)^{a_t}, \tag{3}$$

where d_t is the number of dead trees and a_t is the number of surviving trees in the t-th age class during the observation period.

The mortality probability (q_t) of a new in-growth tree in the t-th age class was defined as follows:

$$\Pr(t-1 < T \leq t) = q_t. \tag{4}$$

The survival probability (r_t) in the t-th age class was defined as follows:

$$\Pr(T > t-1) = r_t. \tag{5}$$

Therefore, based on Equation (1), the mortality rate (p_t) can be expressed as follows:

$$p_t = \frac{q_t}{r_t}.$$ (6)

The maximum likelihood estimators of p_t can be calculated by the first-order derivation of Equation (3), as shown in Equation (7):

$$\hat{p}_t = \frac{d_t}{a_t + d_t}.$$ (7)

Considering that Equations (5) and (6) can be combined as follows:

$$r_t - r_{t+1} = q_t,$$ (8)

the survival function can be converted into:

$$r_{t+1} = r_t - q_t = r_t\left(1 - \frac{q_t}{r_t}\right) = r_t(1 - p_t).$$ (9)

Equation (9) can also be expressed as follows:

$$r_t = \prod_{k<t}(1 - p_k).$$ (10)

The survival function is described by the following formula:

$$\hat{r}_t = \prod_{k<t}\left(1 - \hat{p}_k\right) = \prod_{k<t}\left(1 - \frac{d_k}{a_k + d_k}\right),$$ (11)

where $\hat{r}_t$ represents the survival function and $d_k/(a_k + d_k)$ represents the hazard function in the t-th age class.

This $\hat{r}_t$ is the Kaplan–Meier estimate, which is an important tool for analyzing censored data [23]. Survival analysis is performed to describe the distribution of tree mortality using Kaplan–Meier estimates in the form of step-wise curves [23].

It is meaningful to assess whether there are differences in survival (or cumulative incidence of the event) among different groups. For this purpose, there are several tests available to compare survival among independent groups.

Survival probabilities can be compared using the log-rank test [48], which tests a null hypothesis (i.e., no significant difference in survival between consecutive periods in this study) and the expectation of an equal number of deaths (E) in each of the two groups of each species. The observed (i.e., real) number of deaths is indicated by O in the following equation:

$$\text{Logrankstatistic} = (O - E)^2 / \text{Var}(O - E).$$ (12)

Wilcoxon test [49] can be used as an alternative to the log-rank test. It emphasizes the information at the beginning of the survival curve where the number at risk is significant, allowing early failures to receive more weight than later failures:

$$\text{Wilcoxon test statistics} = \frac{\left(\sum_f w(t_{(f)})(O - E)\right)^2}{\left(\sum_j w(t_{(f)})(O - E)\right)}.$$ (13)

It weights the observed minus expected score at the time t_f by the number at risk, $w(t_f)$ overall groups at time t_f.

Mean lifetime, which is also expressed as mean longevity or mean lifespan, can be considered as the area under survival curve. It is mathematically calculated as the weighted sum of the age and estimated mortality probability described previously herein and can be expressed as follows:

$$\text{Mean lifetime} = \sum tq_t / \sum q_t. \tag{14}$$

It is meaningful to calculate not only mean lifetime but also "practical" mean lifetime from the forest management perspective based on the survival estimates for the two groups of trees (coniferous and broad-leaved). Common mean lifetime was derived by considering all the age classes whereas practical mean lifetime used only specific age class following the Kaplan–Meier estimates. The particular point of the practical mean lifetime is that it has avoided the drastic decrease of survival probabilities in younger age classes, which commonly happen in natural forest stands due to suppression. We intended to explore the possibility of using the practical mean lifetime for kinds of harvesting standards in tree marking process, so that too short mean lifetime was inconvenient. Finally, these practical mean lifetime values were converted into DBH values by using the equations of Table 2 to facilitate practical application in the tree marking process of SSMS.

3. Results

3.1. Temporal Forest Structure Changes

The dominant species with relatively large sample sizes are listed in Table 4. The demographic parameters of all the species were analyzed; however, only the major species in each period are presented in Table 4. In 1989 (period 1), there were 28 species, and the majority of living trees were broad-leaved species (49.78%). The percentage of coniferous species was highest in period 1, with the highest number of trees being that of *A. sachalinensis*. Following the same tendency of period 1, periods 2 and 3 also had higher percentages of broad-leaved species than of coniferous species. In 1999 (period 2), 27 species were sampled in the four permanent sample plots, and the dominant living species were *A. sachalinensis* (22.16%) and *Tilia japonica* (17.63%). In period 3, the dominant species among living trees was *A. sachalinensis* (23.60%). The number of dead coniferous trees increased over the three periods, and most of them were *A. sachalinensis*. *Tilia japonica* was the most common species among dead broad-leaved trees, and it showed an increasing trend, except in period 3.

Table 4. Summary of the dominant tree species in the four plots in the three observation periods.

Species	No. of Trees (%)					
	Period 1		Period 2		Period 3	
	Living Trees	Dead Trees	Living Trees	Dead Trees	Living Trees	Dead Trees
Conifer						
A. sachalinensis	201 (21.90)	22 (2.40)	215 (22.16)	22 (2.27)	219 (23.60)	41 (4.42)
P. jezoensis	148 (16.12)	14 (1.53)	135 (13.92)	16 (1.65)	117 (12.61)	22 (2.37)
Other conifers	4 (0.44)	2 (0.22)	1 (0.10)	3 (0.31)	1 (0.11)	0 (0.00)
Total conifer	353 (38.45)	38 (4.14)	351 (36.19)	41 (4.23)	337 (36.31)	63 (6.79)
Broad-leaved						
Ulmus laciniata	56 (6.10)	8 (0.87)	52 (5.36)	10 (1.03)	46 (4.96)	8 (0.86)
Sorbus commixta	35 (3.81)	3 (0.33)	27 (2.78)	11 (1.13)	25 (2.69)	8 (0.86)
Acer ukurunduense	21 (2.29)	9 (0.98)	48 (4.95)	11 (1.13)	50 (5.39)	19 (2.05)
Acer mono var. *myrii*	56 (6.10)	2 (0.22)	57 (5.88)	6 (0.62)	54 (5.82)	7 (0.75)
Tilia japonica	174 (18.95)	29 (3.16)	171 (17.63)	52 (5.36)	153 (16.49)	32 (3.45)
Other broad-leaved	115 (12.53)	19 (2.07)	113 (11.65)	20 (2.06)	101 (10.88)	25 (2.69)
Total broad-leaved	457 (49.78)	70 (7.63)	468 (48.25)	110 (11.34)	429 (46.23)	99 (10.67)
Total	810 (88.24)	108 (11.76)	819 (84.43)	151 (15.57)	766 (82.54)	162 (17.46)

No. of trees (%) = Number of trees in each period of major species and percentage of it within brackets.

A. *sachalinensis* and *P. jezoensis* accounted for approximately 99% of the coniferous species in each period, and the results for these species, with relatively large sample sizes, were considered for further analysis. Due to the insufficient number of dead trees of each broad-leaved tree species, all broad-leaved trees were combined in further analyses.

3.2. Changes in Age Distributions by Species Groups

In the three periods (periods 1–3), tree age-class distributions, including dead and living trees by three species groups, are presented in Figure 3. New in-growths found in the census data were aggregated in the 1st age class of each period.

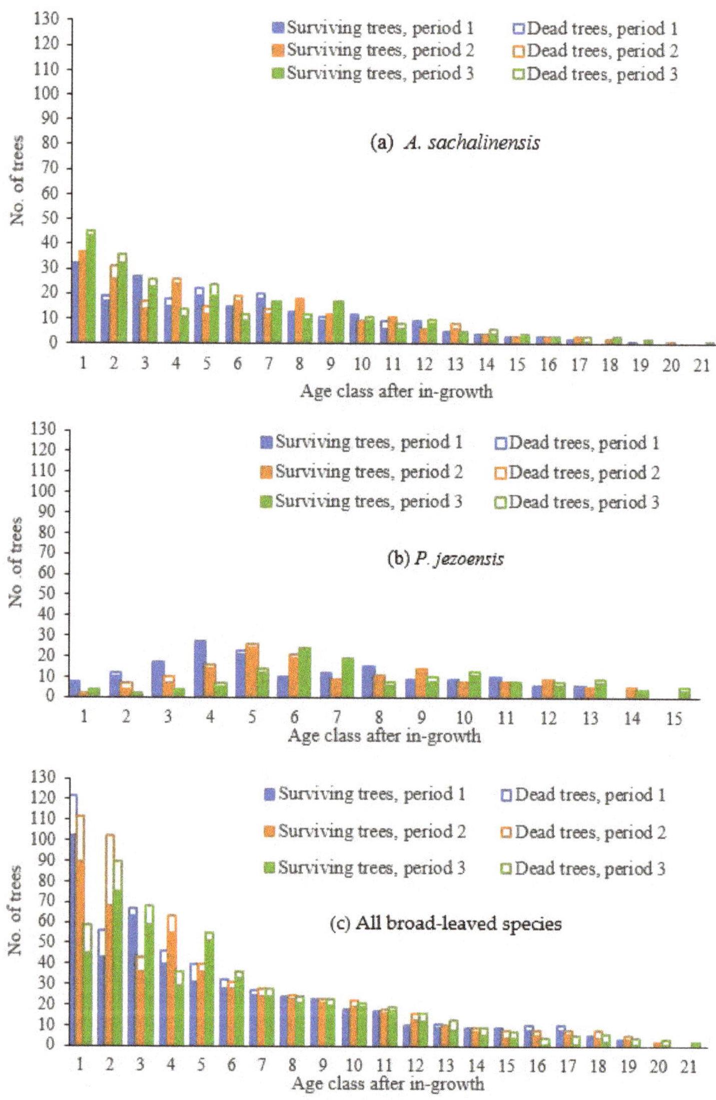

Figure 3. Age class distributions of living and dead trees of major species: (**a**) A. *sachalinensis*, (**b**) P. *jezoensis*, and (**c**) all broad-leaved species.

In Figure 3a, the highest number of new in-growths can be seen in the 1st age class of period 3, whereas the lowest number of in-growths can be seen in period 1. The dead *A. sachalinensis* trees were distributed almost equally across all age classes. The distribution of *A. sachalinensis* trees showed an exponential shape, with many young trees and few old trees, in each period, with a few exceptions in period 1; for example, the 3rd age class of period 1 had a higher number of trees than the 2nd age class.

For *P. jezoensis* in Figure 3b, the largest number of new in-growth trees was observed in the 1st age class of period 1 among the three periods. Dead trees were distributed equally across those present in most age classes. *P. jezoensis* had a relatively low number of new in-growths in each period compared with *A. sachalinensis*, which might lead to a multi-modal age distribution, different from that of *A. sachalinensis*.

Figure 3c shows the distribution of all broad-leaved trees found in each period, as mentioned above. The largest number of new in-growths can be seen in the 1st age class of period 1. The 1st age class of period 3 had a lower number of new in-growths than those in other periods. Dead trees were distributed among almost all the age classes across the periods. Periods 1 and 2 exhibited an exponential age class distribution; however, the 1st age class of period 3 had a relatively low number of new in-growth trees. Thus, an exponential age class distribution was observed in broad-leaved species, except in the 1st age class of period 3.

3.3. Kaplan–Meier Curve Differences among the Periods

Figure 4 shows the estimated Kaplan–Meier curves for the three major species over the three observation periods. Consecutive periods (periods 1–3) are compared here. For *A. sachalinensis*, Figure 4a shows similar curves over the age classes in periods 1 and 2, with constant survival probabilities after age classes 15 and 14. In contrast, period 3 had a decreasing survival probability throughout age classes 1 to 20. The Kaplan–Meier curves for *P. jezoensis* show three different patterns for the three periods in Figure 4b. Remarkably, period 1 had higher survival probabilities throughout all age classes. The curve of period 3 shows a lower survival probability than that of period 2 after the 10th age class. The curves of broad-leaved species in Figure 4c show similar distributions, though the values remained slightly higher for period 1 than for the others after the 10th age class. In the results of the Log-rank and Wilcoxon statistical tests (Table 5), the differences in mortality of broad-leaved species between periods 1 and 2 were only statistically significant based on the log-rank test (p-value = 0.0194) Coniferous species did not show statistically significant differences, in spite of the differences in their appearance, such as in the case of *P. jezoensis*.

Table 5. Log-rank test results comparing the three observational periods.

	Tree Species											
	A. sachalinensis				*P. jezoensis*				All Brzoad-Leaved Species			
Period	Log-Rank Test		Wilcoxon Test		Log-Rank Test		Wilcoxon Test		Log-Rank Test		Wilcoxon Test	
	p-Value	Chi-Square Value	p-Value	Chi-Square Value	p-Value	Chi-Square Value	p-Value	Chi-Square Value	p-Value	Chi-Square Value	p-Value	Chi-Square Value
Periods 1 and 2	0.9161	0.0111	0.6542	0.2007	0.5023	0.4502	0.3674	0.8124	0.0194 *	5.4637	0.0772	3.1236
Periods 2 and 3	0.1034	2.6524	0.5685	0.3252	0.1538	2.0337	0.2549	1.2961	0.2691	1.2215	0.0610	3.5101

* Indicates a significant difference ($p < 0.05$).

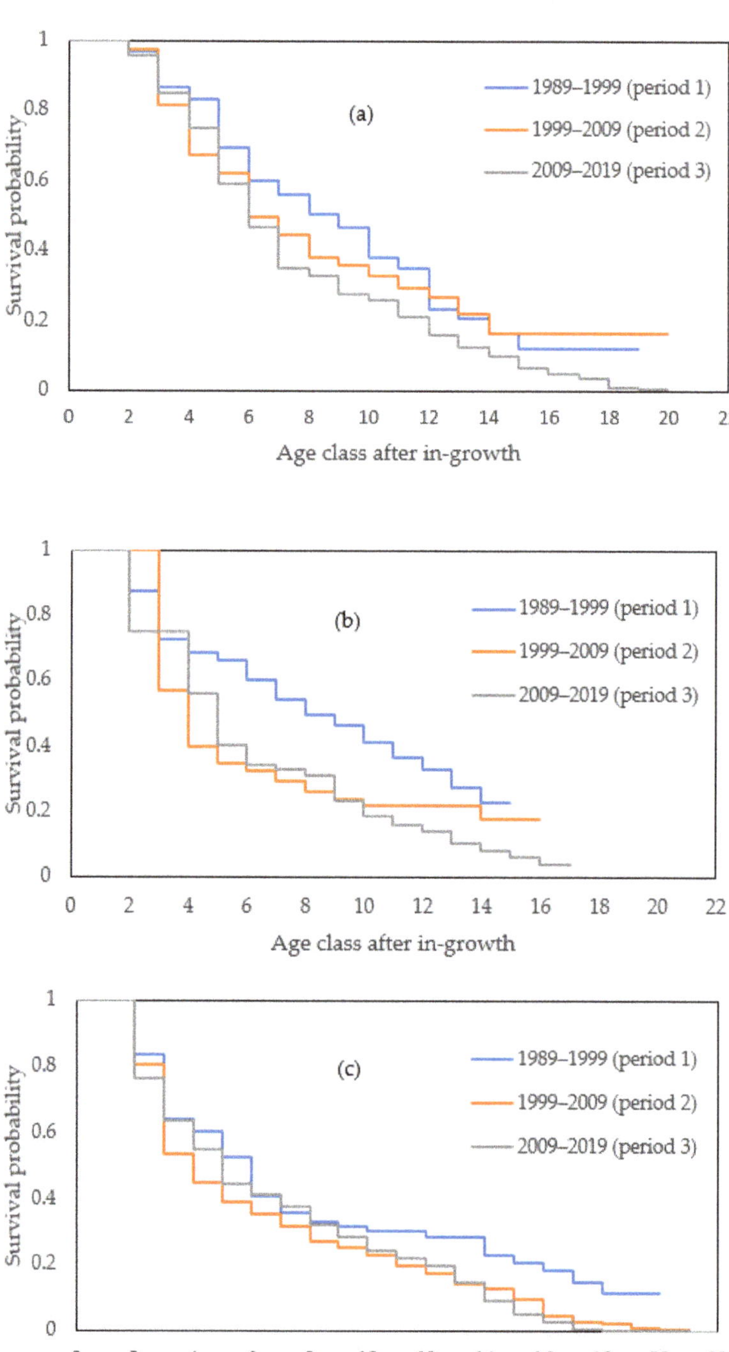

Figure 4. Kaplan–Meier curves showing tree survival probabilities during the three observational periods: (**a**) *A. sachalinensis*, (**b**) *P. jezoensis*, and (**c**) all broad-leaved species.

3.4. Kaplan–Meier Curve Differences by Species

Figure 5 shows the Kaplan–Meier curves of the two major conifer species, *A. sachalinensis* and *P. jezoensis*, for the three periods. In addition, Table 6 shows the results of statistical tests for each period. There was no statistically significant difference for any combination. Therefore, we combined the two conifer species to determine species group differences in the next step.

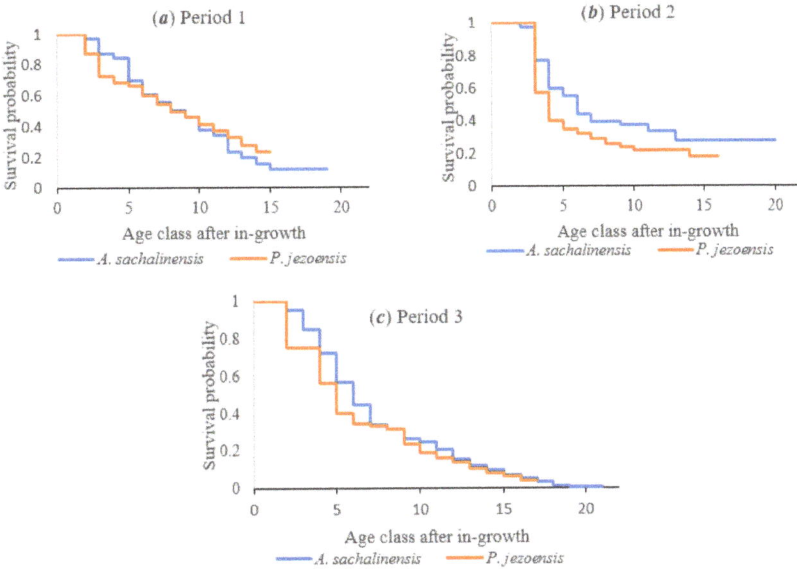

Figure 5. Kaplan–Meier curves showing tree survival probabilities of *A. sachalinensis* and *P. jezoensis* during the three observational periods: (**a**) period 1, (**b**) period 2, and (**c**) period 3.

Table 6. Log-rank test results of *Abies sachalinensis* and *Picea jezoensis* in each period.

	A. sachalinensis **and** *P. jezoensis*					
	Period 1		Period 2		Period 3	
Statistical Test	*p*-Value	**Chi-Square Value**	*p*-Value	**Chi-Square Value**	*p*-Value	**Chi-Square Value**
Log-rank test	0.4985	0.4581	0.5751	0.3142	0.9959	0.0000
Wilcoxon test	0.4827	0.4928	0.5720	0.3193	0.8692	0.0271

Figure 6 shows the estimated Kaplan–Meier curves for the new two groups of all conifers and all broad-leaved species over the three observation periods. For period 1, Figure 6a shows that the curve of broad-leaved trees declined considerably more than that of the conifers due to the high mortalities of young and old trees. No decline was observed in the curve of coniferous trees for age class 15 or higher because no dead trees were observed after these age classes. For period 2, Figure 6b shows different Kaplan–Meier curves from those of period 1; broad-leaved trees reached nearly zero at older age classes, whereas conifers remained constant after age class 14, as there were no dead trees after that point. For period 3, Figure 6c shows that both of the curves declined constantly because of young and older tree deaths across age classes. Notably, both curves reached almost zero for the oldest existing age class. Table 7 shows the results of statistical tests. Overall, differences in mortality between conifer and broad-leaved trees in each period were statistically significant in either log-rank or Wilcoxon tests.

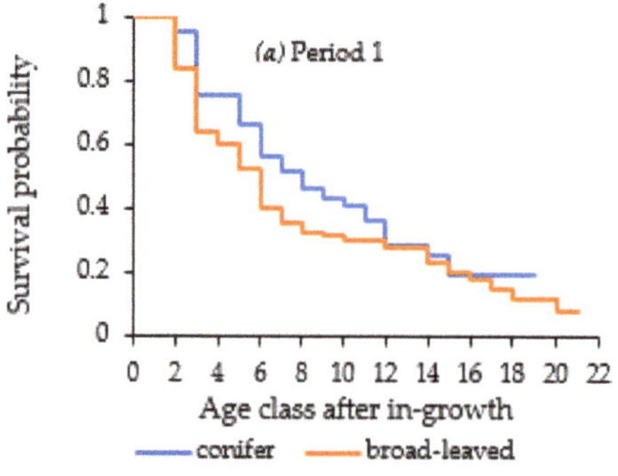

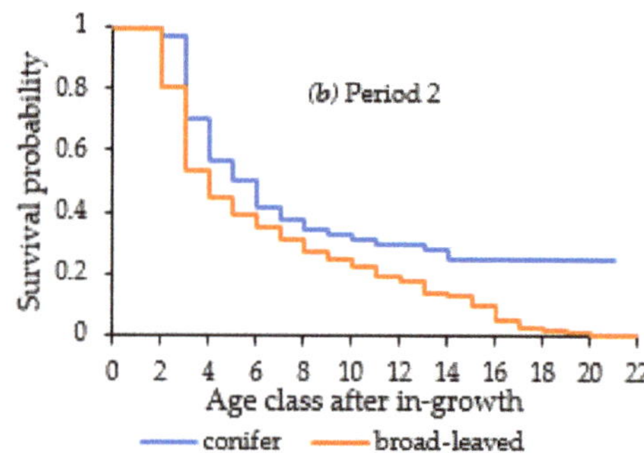

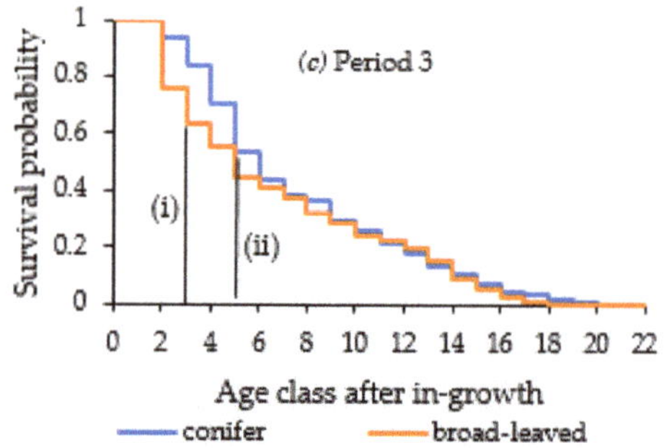

Figure 6. Kaplan–Meier curves for each period for conifers and broad-leaved trees: (**a**) period 1, (**b**) period 2, and (**c**) period 3 (i and ii denotes the starting age classes of "practical" mean lifetime of each broad-leaved and coniferous species, respectively).

Table 7. Log-rank test and Wilcoxon test comparing the conifer and broad-leaved trees in each period.

	Conifer and broad-Leaved Species					
	Period 1		Period 2		Period 3	
Statistical Tzst	***p*-Value**	**Chi-Square Value**	***p*-Value**	**Chi-Square Value**	***p*-Value**	**Chi-Square Value**
Log-rank test	0.1614	1.9607	**0.0097 ***	6.6870	**0.0381 ***	4.3018
Wilcoxon test	**0.0386 ***	4.2797	**0.0463 ***	3.9719	0.3035	1.0586

* indicates a significant difference ($p < 0.05$).

3.5. Mean Lifetime over Three Periods

The (common) mean lifetimes after in-growth were calculated based on non-parametric estimates to investigate the temporal changes between coniferous and broad-leaved species (Table 8). The mean lifetimes of all conifer species were 39, 36, and 43 years for periods 1–3, respectively, and those of all broad-leaved species were 44, 36, and 42 years, respectively. Table 7 shows the results of statistical tests. Overall, differences in mortality between conifers and broad-leaved trees in each period were statistically significant in either log-rank or Wilcoxon tests.

Table 8. Mean lifetime and standard deviation (in parenthesis) of two groups of trees (coniferous and broad-leaves) in each period.

Period	Conifer Species	Broad-Leaved Species
Period 1	39 (4)	44 (5)
Period 2	36 (4)	36 (5)
Period 3	43 (5)	42 (5)

Next, the "practical" mean lifetimes were calculated for period 3 only. We calculated this for period 3 only, because the Kaplan–Meier curves for that period appeared to almost converge to time-stable states, with probabilities reaching almost zero (Figure 6) and no significant difference between periods 2 and 3 in either tree category (Table 5). In addition, because there was no significant difference between the two major coniferous species (Table 6), it was suitable to derive one practical mean lifetime value for all coniferous species. Thus, the practical mean lifetime of all conifer groups was calculated as 72 years when avoiding the first five age classes, illustrated as line (ii) in Figure 6c (Table 9). For all broad-leaved species, it was 80 years when avoiding trees of the first three age classes illustrated as line (i) in Figure 6c (Table 9). These two lines were identified as the age classes up to which tree mortality due to suppression occurred, and survival probability curves drastically declined in a single step-wise manner.

Table 9. The predicted practical mean lifetime and corresponding diameter at breast height value of period 3.

Species	Practical Mean Lifetime in Years (SD)	Practical Mean DBH in cm (SD)
Conifer species	72 (7)	31 (7)
Broad-leaved species	80 (5)	36 (8)

Diameter at breast = DBH; standard deviation = SD.

Finally, the practical mean lifetimes were converted to the corresponding DBH of 31 cm and 36 cm using the DBH-age equations developed (Table 9).

4. Discussion

4.1. Temporal Changes in Age Distribution and Species Composition

In Figure 3, the age distributions of *A. sachalinensis* and broad-leaved species show exponential shapes, with few outliers in some periods, whereas *P. jezoensis* shows an irregular multi-modal shape in each period. Remarkably, fewer new in-growths of coniferous species were found compared with those of broad-leaved species. Considering these matters, it appears that *A. sachalinensis* might not keep the current exponential shape and change to the multi-modal shape in the upcoming periods. In addition, larger numbers of dead trees and new in-growth trees were observed in broad-leaved species compared with conifer species over these periods. These results could not fully support the suggestion that the dominance of coniferous species would decrease further with climate change in northern Hokkaido, as found by Hiura et al. [50].

4.2. Survival Analyses over the Observational Periods

Regarding period differences, no significant differences were found between the Kaplan–Meier curves of two consecutive periods for the coniferous species, although the Kaplan–Meier curves declined along with the periods. Broad-leaved species in period 1 had a significantly greater survival probability than those in period 2. The only significant difference in the Kaplan–Meier curves was found between periods 1 and 2 for broad-leaved species, which might have been caused by the decrease in mortality in period 1. These survival probability trends correspond to the findings of Sato [51] and Hiura et al. (1996) [52]. Moreover, Kaplan–Meier curves between periods 2 and 3 were not significantly different for either tree category. The findings of Wijenayake and Hiroshima [40] suggested that the stand could be considered as an almost matured state when there was no statistical difference among Kaplan–Meier curves over time and the survival probabilities of old age classes reached nearly zero. Along with this, the survival probability values of period 3 can be applied to predict future age distributions.

Regarding species differences, the survival of coniferous and broad-leaved trees was statistically different over the three periods. Nakashizuka [53] also reported that the growth and survival of coniferous and broad-leaved species reflected large variabilities along with time. Therefore, it is reasonable to differentiate tree groups as coniferous and broad-leaved trees in further analyses.

4.3. Mean Lifetime Values and Harvesting Decisions

Selvin [54] discussed the importance of analyzing age at the time of death to provide more accurate lifetime estimations on survival probability. Yamamoto [55] indicated that natural mortalities can be minimized based on SSMS, although SSMS could fail if it does not adequately identify over-matured trees prior to decay. The early identification of over-matured trees may improve the SSMS practices, as well as sustainable natural forest management. Based on this belief, selection cutting can be planned and implemented to improve the quality and quantity of crop trees. The early identification of over-matured trees facilitates the maintenance of stands in continuously healthy and productive states [37].

The mean lifetimes did not differ greatly between conifers and broad-leaves; both were approximately only 40 years (Table 8). Instead of these mean lifetimes, the developed practical mean lifetimes might be applied for harvesting decisions in SSMS because they exclude juvenile tree mortalities due to suppression. These values were estimated based on the step-down width of each Kaplan–Meier curve and were inherent to species. In natural forest management, it is much more sensible in terms of DBH values rather than age values to facilitate harvesting decisions. The estimated DBH values of coniferous and broad-leaved species based on practical mean lifetime were 31 cm and 36 cm, respectively. These DBH values could be supplementary tools for the prediction of over-mature trees in SSMS. For example, with broad-leaved trees, it is best to pay attention when their DBH reaches 36 cm. This attention allows for selecting the harvestable trees in each cutting period by maintaining vigorous trees in the stand. Therefore, these developed practical mean lifetime and DBH values would be a supplementary tool in the practice of SSMS for early identification of the trees likely to die in the near future.

Based on the principles of SSMS, forest stands must be maintained in the pre-climax stage, which enables a high growth rate at the stand level [56]. Forest managers need a comprehensive understanding of natural stand development processes when designing silvicultural systems that integrate ecological and economic objectives [57]. Foresters usually manage stands that are developing as part of secondary succession. The significant tree species in terms of economic importance are commonly part of numerous seral stages towards the climax. Therefore, foresters manage their forests by controlling the tendency of that population to move toward a climax species forest [58]. By following the DBH, which is based on practical mean lifetime, stands can be maintained in the pre-climax successional stage.

The present analysis might not precisely reflect the mean lifetime in the future because the analysis primarily relied on non-parametric analyses. Parametric analysis approaches such as Weibull and Gamma could facilitate more precise future predictions, considering species differences, to identify likely dead trees for harvesting in SSMS. Therefore, in future research models, parametric analyses could be applied to predict age distributions.

5. Conclusions

This study showed the age class distribution of two coniferous tree species and all broad-leaved species at the long-term research site in the preserved area in UTHF over three observation periods and estimated survival probabilities using non-parametric methods. The resulted mean lifetime values were included suppressed young tree mortalities as well.

Therefore, "practical" mean lifetimes were introduced by avoiding the drastic decrease in survival probabilities of Kaplan–Meier estimates. Furthermore, the developed values were converted to DBH values to enhance the practical use in the field. For example, when preparing the management plan, foresters should pay special care for the broad-leaved trees where DBH is 36 cm or more, and 31 cm for coniferous trees. Therefore, these developed practical mean lifetime and DBH values would be a supplementary tool in the practice of SSMS for early identification of the trees likely to die in the near future. Moreover, the mortality patterns and survival probabilities reported in this study would constitute a valuable reference for future studies to understand the stand dynamics of natural forests associated with the mortality of individual tree populations.

Author Contributions: Data analysis, writing—original draft preparation, P.R.W.; Supervision, review and editing, T.H. Both authors have read and agreed to the published version of the manuscript.

Funding: This study was funded by the JSPS KAKENHI, Grant Number JP19K06142.

Data Availability Statement: The datasets relevant to the current study are available from the corresponding author on reasonable request.

Acknowledgments: We thank the technical staff of the UTHF for their technical support throughout the fieldwork and their assistance with the tree census data.

Conflicts of Interest: The authors declare no conflict of interest.

References

1. Seidl, R.; Schelhaas, M.J.; Rammer, W.; Verkerk, P.J. Increasing forest disturbances in Europe and their impact on carbon storage. *Nat. Clim. Chang.* **2014**, *4*, 806–810. [CrossRef]
2. Neuner, S.; Knoke, T. Economic consequences of altered survival of mixed or pure Norway spruce under a dryer and warmer climate. *Clim. Chang.* **2017**, *140*, 519–531. [CrossRef]
3. Griess, V.C.; Knoke, T. Bioeconomic modeling of mixed Norway spruce-European beech stands: Economic consequences of considering ecological effects. *Eur. J. For. Res.* **2013**, *132*, 511–552. [CrossRef]
4. Paul, C.; Brandl, S.; Friedrich, S.; Falk, W.; Härtl, F.; Knoke, T. Climate change and mixed forests: How do altered survival probabilities impact economically desirable species proportions of Norway spruce and European beech? *Ann. For. Sci.* **2019**, *76*, 14. [CrossRef]
5. Adams, H.D.; MacAlady, A.K.; Breshears, D.D.; Allen, C.D.; Stephenson, N.L.; Saleska, S.R.; Huxman, T.E.; McDowell, N.G. Climate-induced tree mortality: Earth system consequences. *EOS Trans. Am. Geophys. Union* **2010**, *91*, 153–154. [CrossRef]
6. Dietze, M.C.; Moorcroft, P.R. Tree mortality in the eastern and central United States: Patterns and drivers. *Glob. Chang. Biol.* **2011**, *17*, 3312–3326. [CrossRef]
7. Pfeifer, E.M.; Hicke, J.A.; Meddens, A.J.H. Observations and modeling of aboveground tree carbon stocks and fluxes following a bark beetle outbreak in the western United States. *Glob. Chang. Biol.* **2011**, *17*, 339–350. [CrossRef]
8. Trumbore, S.; Brando, P.; Hartmann, H. Forest health and global change. *Science* **2015**, *349*, 814–818. [CrossRef]
9. Runkle, J.R. Canopy tree turnover in old-growth mesic forests of eastern north America. *Ecology* **2000**, *81*, 554–567. [CrossRef]
10. Harcombe, P.A.; Bill, C.J.; Fulton, M.; Glitzenstein, J.S.; Marks, P.L.; Elsik, I.S. Stand dynamics over 18 years in a southern mixed hardwood forest, Texas, USA. *J. Ecol.* **2002**, *90*, 947–957. [CrossRef]
11. Van Mantgem, P.J.; Stephenson, N.L.; Byrne, J.C.; Daniels, L.D.; Franklin, J.F.; Fulé, P.Z.; Harmon, M.E.; Larson, A.J.; Smith, J.M.; Taylor, A.H.; et al. Widespread increase of tree mortality rates in the Western United States. *Science* **2009**, *323*, 521–524. [CrossRef] [PubMed]
12. Allen, C.D.; Macalady, A.K.; Chenchouni, H.; Bachelet, D.; McDowell, N.; Vennetier, M.; Kitzberger, T.; Rigling, A.; Breshears, D.D.; Hogg, E.H.; et al. A global overview of drought and heat-induced tree mortality reveals emerging climate change risks for forests. *For. Ecol. Manag.* **2010**, *259*, 660–684. [CrossRef]
13. McDowell, N.G.; Allen, C.D. Darcy's law predicts widespread forest mortality under climate warming. *Nat. Clim. Chang.* **2015**, *5*, 669–672. [CrossRef]
14. Woodall, C.W.; Grambsch, P.L.; Thomas, W. Applying survival analysis to a large-scale forest inventory for assessment of tree mortality in Minnesota. *Ecol. Model.* **2005**, *189*, 199–208. [CrossRef]
15. Monserud, R. Simulation of Forest Tree Mortality. *For. Sci.* **1976**, *22*, 438–444. [CrossRef]
16. Hamilton, D.A. A logistic model of mortality in thinned and unthinned mixed conifer stands of northern Idaho. *For. Sci.* **1986**, *32*, 989–1000. [CrossRef]
17. Kobe, R.K.; Coates, K.D. Models of sapling mortality as a function of growth to characterize interspecific variation in shade tolerance of eight tree species of northwestern British Columbia. *Can. J. For. Res.* **1997**, *27*, 227–236. [CrossRef]
18. Buford, M.A.; Hafley, W.L. Probability distributions as models for mortality. *For. Sci.* **1985**, *31*, 331–341.

19. Finney, D.J. *Probit Analysis*, 3rd ed.; Cambridge University Press: London, UK, 1971.
20. Kalhfleisch, J.G.; Prentice, R.L. *The Statistical Analysis of Failure Time Data*; John Wiley: New York, NY, USA, 1980.
21. Fujikake, I. An Analyses of Harvesting Activities in a Forest Management Entity Using Harvesting Age Distribution (Tentative Translation by Author). Ph.D. Thesis, Kyoto University, Kyoto, Japan, 2000.
22. Kleinbaum, D.G.D.; Klein, M. *Survival Analysis: A Self-Learning Text*, 3rd ed.; Statistics for Biology and Health; Springer: Berlin/Heidelberg, Germany, 2011.
23. Kaplan, E.L.; Meier, P. Nonparametric Estimation from Incomplete Observations. *J. Am. Stat. Assoc.* **1958**, *53*, 457–481. [CrossRef]
24. Waters, W.E. Life-table approach to analysis of insect impact. *J. For.* **1969**, *67*, 300–304.
25. Morse, B.W.; Kulman, H.M. Plantation white spruce mortality: Estimates based on aerial photography and analysis using a life-table format. *Can. J. For. Res.* **1984**, *14*, 195–200. [CrossRef]
26. Zhang, S.; Amateis, R.L.; Burkhart, H.E. Constraining individual tree diameter increment and survival models for loblolly pine plantations. *For. Sci.* **1997**, *43*, 413–423. [CrossRef]
27. Wyckoff, P.H.; Clark, J.S. Predicting tree mortality from diameter growth: A comparison of maximum likelihood and Bayesian approaches. *Can. J. For. Res.* **2000**, *30*, 156–167. [CrossRef]
28. Nothdurft, A. Spatio-temporal prediction of tree mortality based on long-term sample plots, climate change scenarios and parametric frailty modeling. *For. Ecol. Manag.* **2013**, *291*, 43–54. [CrossRef]
29. Neuner, S.; Albrecht, A.; Cullmann, D.; Engels, F.; Griess, V.C.; Hahn, W.A.; Hanewinkel, M.; Härtl, F.; Kölling, C.; Staupendahl, K.; et al. Survival of Norway spruce remains higher in mixed stands under a dryer and warmer climate. *Glob. Chang. Biol.* **2015**, *21*, 935–946. [CrossRef] [PubMed]
30. Neumann, M.; Mues, V.; Moreno, A.; Hasenauer, H.; Seidl, R. Climate variability drives recent tree mortality in Europe. *Glob. Chang. Biol.* **2017**, *23*, 4788–4797. [CrossRef]
31. Brandl, S.; Paul, C.; Knoke, T.; Falk, W. The influence of climate and management on survival probability for Germany's most important tree species. *For. Ecol. Manag.* **2020**, *458*, 117652. [CrossRef]
32. Hiroshima, T. Applying age-based mortality analysis to a natural forest stand in Japan. *J. For. Res.* **2014**, *19*, 379–387. [CrossRef]
33. Manion, P.D.; Griffin, D.H. Large landscape scale analysis of tree death in the Adirondack Park, New York. *For. Sci.* **2001**, *47*, 542–549. [CrossRef]
34. Zens, M.S.; Peart, D.R. Dealing with death data: Individual hazards, mortality and bias. *Trends Ecol. Evol.* **2003**, *18*, 366–373. [CrossRef]
35. The University of Tokyo Hokkaido Forest. 2019. Available online: http://www.uf.a.u-tokyo.ac.jp/files/gaiyo_hokkaido.pdf (accessed on 26 July 2020).
36. Asahi, M. Studies on the classification of forest soils in the Tokyo University Forest, Hokkaido. *Bull. Tokyo Univ. For.* **1963**, *58*, 1–132.
37. Watanabe, S.; Sasaki, S. The silvicultural management system in temperate and boreal forests—A case history of the Hokkaido Tokyo University Forest. *Can. J. For. Res.* **1994**, *24*, 1176–1185. [CrossRef]
38. Owari, T.; Matsui, M.; Inukai, H.; Kaji, M. Stand structure and geographic conditions of natural selection orests in central Hokkaido, Northern Japan. *J. For. Plan.* **2011**, *16*, 207–214.
39. Rinn, F. Eine neue Bohrmethode zur Holzuntersuchung. *Holz-Zentralblatt* **1989**, *34*, 529–530.
40. Wijenayake, P.R.; Hiroshima, T. Survival analyses of individual tree populations in natural forest stands to evaluate the maturity of forest stands: A case study of preserved forests in Northern Japan). *J. For. Plan.* **2021**. [CrossRef]
41. Cox, D.R.; Oakes, D. *Analysis of Survival Data*; Chapman and Hall: London, UK, 1984.
42. Crowder, M.J.; Kimber, A.C.; Smith, R.L. *Statistical Analysis of Reliability Data.*; Chapman and Hall: London, UK, 1994; Volume 27.
43. Klein, J.P.; Moeschberger, M.L. *Survival Analysis: Techniques for Censored and Truncated Data*; Springer: New York, NY, USA, 1997.
44. Cox, D.R. Regression Models and Life-Tables. *J. R. Stat. Soc.* **1972**, *34*, 187–220. [CrossRef]
45. Fujikake, I. Estimation of Gentan probability based on the forest resource table. *Proc. Inst. J. Stat. Math.* **2003**, *51*, 95–109. [CrossRef]
46. Hiroshima, T. Study on the methodology of calculating mean and variance of felling age in forest planning. *Jpn. J. For. Plan.* **2006**, *40*, 139–149. [CrossRef]
47. Tiryana, T.; Tatsuhara, S.; Shiraishi, N. Empirical models for estimating the stand biomass of teak plantations in Java, Indonesia (Multipurpose Forest Management). *J. For. Plan.* **2011**, *16*, 35–44. [CrossRef]
48. Mantel, N. Evaluation of survival data and two new rank order statistics arising in its consideration. *Cancer Chemother. Rep.* **1966**, *50*, 163–170.
49. Gehan, E. A generalised Wilcoxon test for comparing arbitrarily singly-censored samples. *Biometrika* **1965**, *52*, 203–223. [CrossRef]
50. Hiura, T.; Go, S.; Iijima, H. Long-term forest dynamics in response to climate change in northern mixed forests in Japan: A 38-year individual-based approach. *For. Ecol. Manag.* **2019**, *449*, 117469. [CrossRef]
51. Sato, T. Time trends for genetic parameters in progeny test of *Abies sachalinensis* (Fr. Schm.) Mast. *Silvae Genet.* **1994**, *43*, 304–307.
52. Hiura, T.; Sano, J.; Konno, Y. Age structure and response to fine-scale disturbances of *Abies sachalinensis*, *Picea jezoensis*, *Picea glehnii*, and *Betula ermanii* growing under the influence of a dwarf bamboo understory in northern Japan. *Can. J. For. Res.* **1996**, *26*, 289–297. [CrossRef]

53. Nakashizuka, T. Population dynamics of coniferous and broad-leaved trees in a Japanese temperate mixed forest. *J. Veg. Sci.* **1991**, *2*, 413–418. [CrossRef]
54. Selvin, S. *Statistical Analysis of Epidemiologic Data*; Oxford University Press: Oxford, UK, 2004.
55. Yamamoto, H. Natural forest management based on selection cutting and natural regeneration. In Proceedings of the IUFRO International Workshop on sustainable Forest Managements, Furano, Japan, 17–21 October 1994; pp. 10–22.
56. Takahashi, N. *The Stand-Based Silvicultural Management System: Its Theory and Practices*, 1st ed.; Log Bee: Sapporo, Japan, 2001.
57. Franklin, J.F.; Spies, T.A.; Van Pelt, R.; Carey, A.B.; Thornburgh, D.A.; Berg, D.R.; Lindenmayer, D.B.; Harmon, M.E.; Keeton, W.S.; Shaw, D.C.; et al. Disturbances and structural development of natural forest ecosystems with silvicultural implications, using Douglas-fir forests as an example. *For. Ecol. Manag.* **2002**, *155*, 399–423. [CrossRef]
58. Daniel, T.W.; Helms, J.A.; Baker, S.F. *Principles of Silviculture*, 2nd ed.; McGraw-Hill: New York, NY, USA, 1979.

 *forests*

Article

The Impact of Air Pollution on the Growth of Scots Pine Stands in Poland on the Basis of Dendrochronological Analyses [†]

Longina Chojnacka-Ożga * and Wojciech Ożga

Department of Silviculture, Institute of Forest Sciences, Warsaw University of Life Sciences, 02-776 Warsaw, Poland; les_khl@sggw.edu.pl
* Correspondence: longina_chojnacka_ozga@sggw.edu.pl; Tel.: +48-698-561-411
† The present study is an extended version of a paper presented at the 1st International Electronic Conference on Forests (IECF2020), 15–30 November 2020.

Abstract: The aim of this study was to evaluate Scots pine stand degradation caused by the pollutants emitted from Zakłądy Azotowe Puławy, one of the biggest polluters of the environment in Poland for over 25 years (1966–1990). To assess the pollution stress in trees, we chose the dendrochronological analysis We outlined three directions for our research: (i) the spatio-temporal distribution of the growth response of trees to the stress associated with air pollution; (ii) the direct and indirect effects of air pollution which may have influenced the growth response of trees; and (iii) the role of local factors, both environmental and technological, in shaping the growth response of trees. Eight Scots pine stands were selected for study, seven plots located in different damage zones and a reference plot in an undamaged stand. We found that pollutant emission caused disturbances of incremental dynamics and long-term strong reduction of growth. A significant decrease in growth was observed for the majority of investigated trees (75%) from 1966 (start of factory) to the end of the 1990s. The zone of destruction extended primarily in easterly and southern directions, from the pollution source, associated with the prevailing winds of the region. At the end of the 1990s, the decreasing trend stopped and the wider tree-rings could be observed. This situation was related to a radical reduction in ammonia emissions and an improvement in environmental conditions. However, the growth of damaged trees due to the weakened health condition is lower than the growth of Scots pine on the reference plot and trees are more sensitive to stressful climatic conditions, especially to drought.

Keywords: Scots pine; tree-ring; air pollution; growth reduction; climate change; Poland

Citation: Chojnacka-Ożga, L.; Ożga, W. The Impact of Air Pollution on the Growth of Scots Pine Stands in Poland on the Basis of Dendrochronological Analyses. *Forests* **2021**, *12*, 1421. https://doi.org/10.3390/f12101421

Academic Editor: Riccardo Marzuoli

Received: 11 September 2021
Accepted: 12 October 2021
Published: 18 October 2021

1. Introduction

Air pollution and climate change are regarded as key stressors that entail a global threat to forest health and sustainability [1–3]. Interactions between these factors—synergistic on the one hand and antagonistic on the other—that result in direct and indirect changes to forest ecosystems have been described in many studies over the last few decades [4–11]. The synergistic, mutually reinforcing interaction of these factors results in a cumulative negative impact on the metabolism and physiological processes of trees [12]. High concentrations of air pollutants (mainly SO_2 and NO_2) can damage trees directly through the foliage and indirectly through the soil [1,3]. Typical symptoms include disturbances in photosynthesis and stomatal conductance, shifts in carbon dioxide allocation and water use efficiency, and leaf loss. The mechanism and effects of the impact of toxic chemicals on the course of physiological processes in plants have been described in numerous studies [12–18]. Additionally, the deposition of pollutants influences the way in which trees respond to other abiotic and biotic stressors, an example being increased sensitivity to drought and pathogen attacks [19].

Global warming is causing an increase in extreme weather events, and in particular in severe droughts—further compounded by unusually high temperatures [20]. Increasingly, trees are exposed to water stress which can cause physiological damage [21]. Trees weakened by the deposition of contaminants and, at the same time, by unfavourable weather conditions become predisposed to secondary stress (from insects, diseases, or fires). Interactions between these factors cause a gradual reduction in tree vigour and growth, which—in extreme cases—may lead to dieback of trees or certain species of trees and to changes in the ecosystem [22]. According to Manion's concept [22], air pollution can be a factor that both predisposes forests to dieback and initiates this process. A schematic view of the integrated impact of air pollution, climatic conditions, and pathogens is presented in Figure 1.

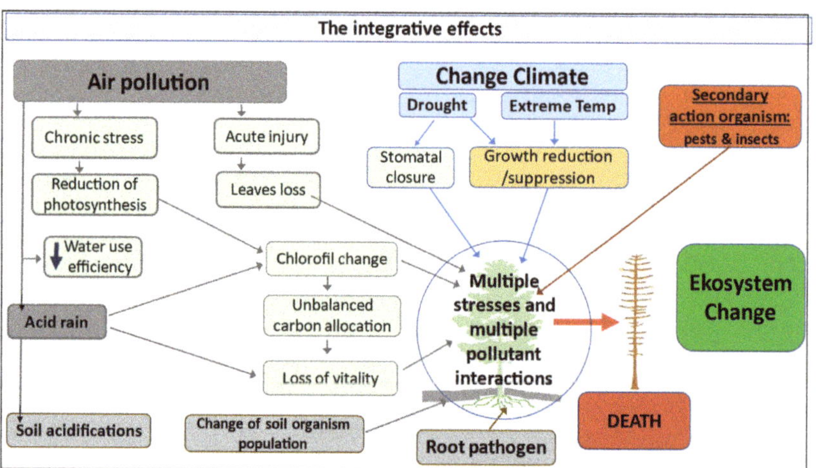

Figure 1. A schematic, integrated influence of air pollution, climatic conditions, and pathogens on trees. Explanation of the colours used in the figure: green—changes taking place in plants; grey—changes in the soil environment; brown—a growth reaction recorded in tree rings.

Despite the significant changes that have occurred over the recent decades in the chemistry, concentration, and geographical distribution of air pollutants (reduction in SOX emissions by 80% and NOX emissions by 30%), the influence of these pollutants on the structure and functioning of forest ecosystems is still visible [3,6,16,18,19,23,24]. A significant proportion of forest ecosystems remain at risk due to the excessive deposition of nitrogen compounds (ammonia NH3) and nitrogen oxides (NOX) [3,17,23–25].

To assess the pollution stress in trees, many biochemical, physiological, and morphological methods are used [14–19,26]. Among the morphological methods, the most common approach is tree-ring analysis [27–35]. A retrospective analysis of annual ring widths and other tree-ring parameters (e.g., chemical composition of the wood, stable isotopes) makes it possible to obtain information on the state of the environment at a high time resolution. Hence, tree rings are often used to assess the long-term effects of air pollution in the environment [29–38]. These studies have indicated that trees growing in a polluted environment have narrower increments or, in extreme cases, do not produce any increments in a given year [36–38].

It is commonly believed that on a global scale, the second half of the 20th century was the period with the greatest pressure of industrial pollution on forest ecosystems e.g., [1–3,23]. Since the 1960s, Poland has been classified as one of the countries with the worst air pollution indicators in Central Europe. For almost forty years, significant amounts of toxic pollutants have been emitted into the atmosphere, causing damage to forests and, in extreme cases, resulting in the dieback of entire stands [39]. The degree of forest stand degradation has varied spatially depending on the amount, type, and concentration of the pollutants as well as the duration of emission, distance from the emitter, and local orographic and climatic conditions. Most of the forest areas exposed to long-term stress related to environmental pollution suffered chronic damage and progressive deterioration of tree health, sometimes resulting in death [34–40]. The most severe damage affected forests growing in areas having a high concentration of industrial plants and subjected to the long-term negative influence of various toxic substances. This problem has been widely described in the literature [30,35–41].

Against this background, the rapid degradation of forest ecosystems caused by pollutant emissions from one plant—the nitrogen fertiliser factory in Puławy (Zakłady Azotowe Puławy)—was unique both in Poland and in Central Europe. The sudden interference with the environment in the form of high concentrations of toxic substances emitted by the plant resulted in rapid degradation of many hectares of forest within a period of several years and in the creation of a "biological death zone." Initially very abrupt, the changes in the forest environment later became chronic. A more detailed description of the damage is provided in the further sections of this paper. The degradation of the environment caused by the operation of the nitrogen fertiliser factory in Puławy has been described in many publications, e.g., [42–49]. However, the ecological interaction between air pollution and the resistance of forests to the abiotic stress that comes with climate change, especially drought, has not been examined. The role of other environmental and technological factors that could result in such extensive degradation of the forest has not been analysed to date, either.

Hence, in this study, we decided to expand the existing knowledge on the impact of pollutants emitted by the fertiliser plant in Puławy on the degradation of the pine stands growing in the surrounding area. We outlined three directions for our research: (i) the spatio-temporal distribution of the growth response of trees to the stress associated with air pollution; (ii) the direct and indirect effects of air pollution which may have influenced the growth response of trees; and (iii) the role of local factors, both environmental and technological, in shaping the growth response of trees. We set ourselves the goal of testing the following hypotheses: (i) that the emission of pollutants had caused a long-term reduction in the annual growth of pine, with the distribution and intensity of this reduction varying spatially and temporally; (ii) that the extent and spatial coverage of the reduction in the growth of pine stands can be attributed to the amount and type of pollutants, but also to local factors, especially anemometric and habitat conditions; and (iii) that the negative impact of pollutants on the growth response persists for a very long time and that even after a radical reduction in emissions, the trees continue to show reduced resistance to abiotic stress related to climate change (especially drought). To test the above hypotheses, we chose the dendrochronological analysis methods used in previous spatial-temporal studies on changes in the growth responses of trees growing in other industrial areas of Poland see [34–41]. The following reasons determined the choice of pine as the species to be studied: (i) pine is the dominant species in Poland and in the study area (accounting for 58% and 71% of forests in Poland and in the Puławy Forest District, respectively); (ii) pine as a species is very sensitive to air pollution (exposure of needles to pollution may lead to the decline of trees, e.g., [14,15,29–32,35,36]); and (iii) the spatial and temporal growth responses of pine to air pollution are being analysed for other regions of Poland. The last premise and the choice of the dendrochronological method also make it possible to use the results of this study to obtain a geographical and historical view of the impact of industrial pollution on the pine stands growing in Poland.

2. Study Area, Materials and Methods

2.1. Study Area

2.1.1. The Nitrogen Fertiliser Factory in Puławy

The nitrogen fertiliser factory in Puławy (Zakłady Azotowe Puławy) is located in central-eastern Poland (51°27′ N & 21°58′ E), in a region with prevailing westerly winds (Figure 2). The area had no significant air pollution prior to the launch of the factory. The industrial plant was built on the western edge of a large forest complex, in an area covered by oligotrophic habitats types, dominated by Scots pine, formed on sandy soils low in nutrients and suffering from periodical acute water deficits [45,50]. The facility was commissioned in the autumn of 1966, and the first signs of damage to pine stands were already observed in the early spring of 1967. In the same year, many trees died within an area of 70 hectares [41]. The zone of forest destruction kept rapidly increasing in size: within the next three years, an area of 500 hectares of the forest on the eastern side of the plant was completely degraded (with all trees being dead), creating a "biological death zone". The extent of damage to tree stands in the remaining areas was varied, with more than 75% of dry or severely damaged trees in the most affected zone. From 1970 onwards, as the devastation of the environment progressed, the zones of destruction continued to expand outwards from the fertiliser plant. Currently, severely damaged forests (with more than 75% of trees affected) cover an area of 1200 hectares, and forests with moderate damage (31% to 75% of trees) and minor damage (5% to 30% of trees) cover approximately 500 hectares and 7000 hectares, respectively [50]. The amount of air pollutants emitted from the factory in the different years is shown in Figure 2 (bottom).

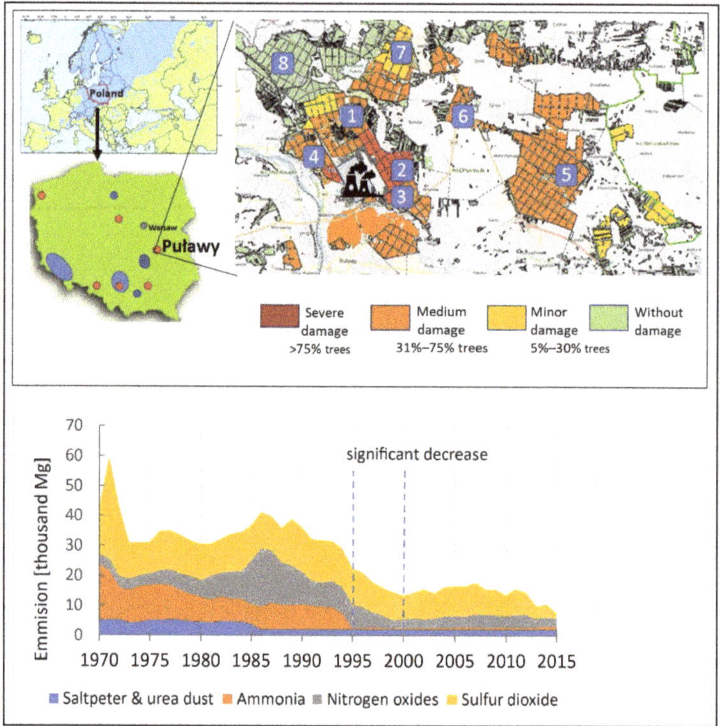

Figure 2. (Top): Location of the study plots. Also marked on the map of Poland are the locations of Polish industrial plants which emit toxic pollutants that cause damage to forests in Poland; Explanation: red circles—nitrogen plants, blue sign- other plants or the industrial region. Information on the impact of pollution on forests in the regions shown on the map can be found in the following studies: 30; 35–38; 40–49. **(Bottom)** Emission of gaseous pollutants (SO_2, NH_3, NO_x) from the emitters at Zakłady Azotowe Puławy in 1970–2015.

However, in addition to emissions resulting from the normal manufacturing process, there were also uncontrolled emissions associated with equipment failures as a result of which toxic substances were released into the atmosphere at concentrations well in excess of the annual average values. In the early years of the plant's operation, the deposition of pollutants reached 1000–1200 kg ha^{-1} yr^{-1} (42,46,48,51). Beginning in the early 1990s, pollutant emissions began to decrease gradually as a result of modernisation, with a radical decrease—especially in the amount of ammonia—taking place in 1995. In 2014, there was a further reduction in pollutant emissions. The current level of emissions is 15% of the value recorded at the beginning of the plant's operation [51].

2.1.2. Climatic Conditions

The meteorological data used in this study—originating from the Puławy Meteorological Station—was acquired courtesy of the Polish Institute of Meteorology and Water Management (IMGW-PIB). The monthly average air temperatures, monthly precipitation totals and monthly average wind speed and direction values for 1951–2015 were used to determine the climate characteristics.

The 1951–2015 period was characterised by an annual average temperature of approximately 8.2 °C (with data points ranging from 6.1 to 10.2 °C) and a mean annual precipitation total of around 572 mm (with data points ranging from 403 to 797 mm per year). The growing season (GS) began in April and lasted until the end of October. The average GS air temperature was 14.0 °C (with data points ranging from 12.4 to 15.3 °C), and the mean precipitation total in this period was 368 mm (with data points ranging from 207 to 570 mm). The lowest level of precipitation was observed between the mid-1980s and mid-1990s and in the first decade of the 21st century. The wind conditions in the region were characterised by a relatively low average wind speed (2.1 m/s) and a high frequency of calm periods, especially between June and September. The dominant wind directions were north-west and south (Figure 3).

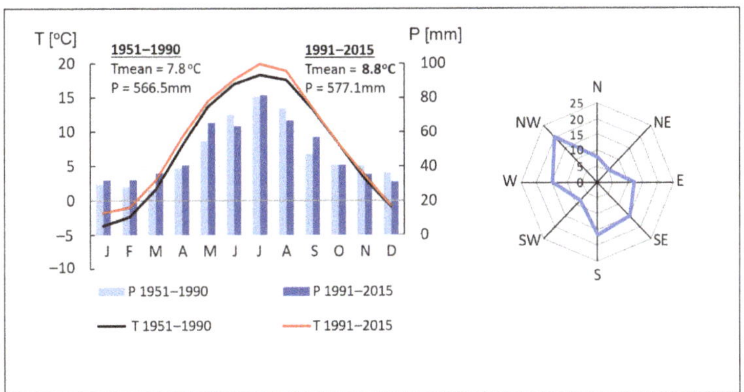

Figure 3. Climate diagram and wind rose for study area, constructed on the basis of data from meteorological station in Puławy for the years 1951–2015.

Within the research area, climate change mainly manifested itself in an increase in air temperature (Figure 3). In the years 1951–2015, a statistically significant increase in the annual average air temperature (0.26 °C per 10 years) and the average temperatures in the winter months (December to February), in spring (March) and in summer (June to August) was recorded in the study area. The greatest increase in air temperature was recorded for March, February and January (0.512, 0.49 and 0.33 °C over 10 years, respectively). In the case of precipitation, the changes were not statistically significant; there was a slight upward trend in the total annual rainfall with a negative trend in summer rainfall, especially in June and August.

2.2. Research Materials and Methods

2.2.1. Study Plots and Sample Acquisition

Eight Scots pine stands were selected for the study: seven plots located in different damage zones and a reference plot in an undamaged stand (Figure 2). When selecting the plots, we applied the following criteria: (i) all stands under study must be growing in the same habitat; (ii) the stands must be of similar age; (iii) the stands must be growing in different damage zones; and (iv) the stands must be located in different compass directions in relation to the emitter. The first two criteria made it possible to largely eliminate the impact of habitat and age, whereas the third and fourth criteria allowed us to determine the temporal and spatial distribution of the growth reduction and the influence of wind direction on the degree of that reduction. All the stands under study represented an oligotrophic mixed coniferous forest habitat and were about 120 to 130 years old (Supplementary Material, Table S1). At each site, 20 trees were sampled with a Pressler increment borer at breast height, one core per tree. The sampling was performed at the end of the 2015 growing season (in November). In total, we collected cores from 160 pines growing at different distances and in different compass directions from the factory. The cores were prepared for measurement using standard dendrochronological procedures [52,53]. The samples were dried, placed in wooden mounts and sanded with progressively finer abrasive paper (80, 120, 180, 240 and 400 grit).

2.2.2. Data Analyses

The sanded core samples were scanned at a resolution of 2400 dpi using a standard scanner (EPSON Expression 10,000XL). Tree-ring widths were measured to the nearest 0.01 mm using Coo Recorder software [54], and individual growth sequences were created for each tree. Cross-matching checks were performed using the COFECHA program [55]. Then, TRW chronologies (raw and residual) were produced for each test site using the ARSTAN program [56]. The residual chronologies were obtained from double detrending (using a negative exponential curve followed by a cubic spline function with a rigidity of 64 years and 50% frequency cut-off); autoregressive modelling was also applied [53]. The homogeneity of the growth reactions and the strength of the environmental signal in the chronologies were estimated using the expressed population signal (EPS), correlation coefficient, GLK (Gleichläufigkeit) coefficient and t values between all pairs of series included in the chronology [52,53,57]. The following statistics were used to characterise the site chronologies (raw and residual): mean value, measures of variability, mean sensitivity, and autocorrelation [52,53]. These statistics have been calculated for three periods: a period before starting factory (1931–1966); a period of extremely high air pollution (1967–1995) and a period with a decrease in the emission of ammonia and a gradual decrease in air pollution (1996–2015). The Kruskal–Wallis test was used to determine the significance of differences between the respective research and control plots [58].

The Schweingruber method [59] was used to evaluate the impact of air pollution on the pine stands under study. This method relies on the analysis of characteristic years and abrupt changes in tree-ring width. The abrupt changes reflect major shifts in eco-physiological conditions that lead to the stimulation or inhibition of cambial activity over several successive years [59]. By using this method, it is possible to determine the exact onset and duration of abrupt changes in tree-ring width. With this method, the duration and degree of TRW reductions are calculated from the ratio of the sum of the reduced ring widths to the sum of the ring widths from the period preceding the reduction. The size of the reduction is classified as follows: RI (30–50%)—low reduction; RII (50–70%)—strong reduction; and RIII (above 70%)—very strong reduction. This research methodology has been successfully used in numerous studies on the impact of industrial emissions on forest stands in Poland [30,35,36,40,41]. The reduction in annual ring width was determined for each sample using the Quercus program [60]. This was done by comparing the dendrogram of each individual sample with the chronology developed for the reference site (plot no. 8).

The climate–growth relationships were investigated by calculating bootstrapped multivariate response functions between residual chronology and climate variables: monthly average air temperatures and monthly precipitation totals [52,53]. Response function analysis is a correlation and multiple regression model that links growth indices (as dependent variables) with climate parameters (as explanatory variables). The analyses were performed with reference to each study plot for the period spanning from June of the year preceding ring formation to September of the year of the current growth (16 months in total) using DendroClim2002 software [61]. The significance of the correlations was determined at p = 0.05. The climate data originated from the meteorological station in Puławy (see 2.1.2). Taking note of reports in the literature that the deposition of pollutants may potentially be a factor in the varying sensitivity of pine to climatic conditions [31,38], we performed dendroclimatic analyses for all plots separately for three periods: 1931–1966 (a period of 35 years before the commissioning of the factory); 1967–1995 (a period of 30 years of extremely high air pollution) and 1996–2015 (with a decrease in the emission of ammonia and a gradual decrease in air pollution from 1995 onwards) (Figure 4).

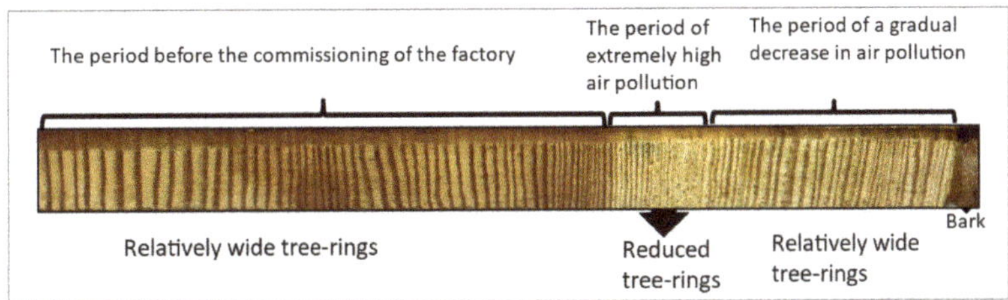

Figure 4. Illustration of a zone with reduced radial growth (sample taken from a pine growing on the 1st plot).

3. Results and Discussion

Our dendrochronological analyses have indicated the existence of significant and prolonged reductions in the annual ring widths of the Scots pine trees growing in all the forest stands under study (Figure 5). This points to a persistent, chronic decline in the vitality of the trees being examined. The decline began in the late 1960s, following the commissioning of the nitrogen fertiliser plant, which is when the pollutant emissions began. Reductions occurred in a majority of the trees under study (Table 1), although they were differentiated spatially and temporally.

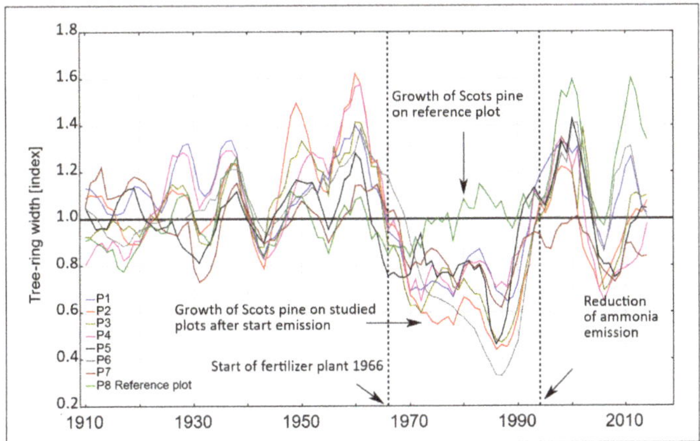

Figure 5. Chronologies of the Scots pines growing in the vicinity of Zakłady Azotowe Puławy. commissioning of the nitrogen fertiliser plant, which is when the pollutant emissions began. Reductions occurred in a majority of the trees under study (Table 1), although they were differentiated spatially and temporally.

Table 1. Characteristics of tree-ring chronologies, and results of calculation of growth changes under influence of air pollution from the nitrogen fertilizer factory in Puławy.

Plots	% of Absent Rings	% Trees with Reduction > 30%	Average Tree Ring Width a Period [index]		
			1931–1965	1966–1995	1996–2015
P1	5.7	66.7	1.14	0.72	0.89
P2	9.1	89.7	1.22	0.61	0.71
P3	6.7	76.9	1.2	0.66	0.82
P4	10.1	88.1	1.18	0.72	0.85
P5	3.9	62.5	1.04	0.8	0.96
P6	4.1	66.5	1.15	0.71	0.96
P7	1.7	41.2	1.00	0.85	0.97
P8-Reference	0.9	0	1.03	1.01	1.01

The most rapid and severe response to the pollution and the most pronounced reduction in growth (and thus decline in vitality) occurred within a radius of 3 km from the plant (plots 1 through 4) (Figure 6). In this zone, the share of trees with reduced growth was as high as 88%–89%. The ring width reductions occurred abruptly in the first two years after the commissioning of the plant and persisted over an extended period of time (more than 30 years); they were not directly linked to the prevailing wind directions. In plots 1 through 3, the reductions in growth persisted until around 1995 and were followed by an increase in growth. Then, after 2003, there occurred another growth reduction, although less pronounced. In plot 4, the reduction in growth began in 1968 and was still visible in 2015 (Figure 6).

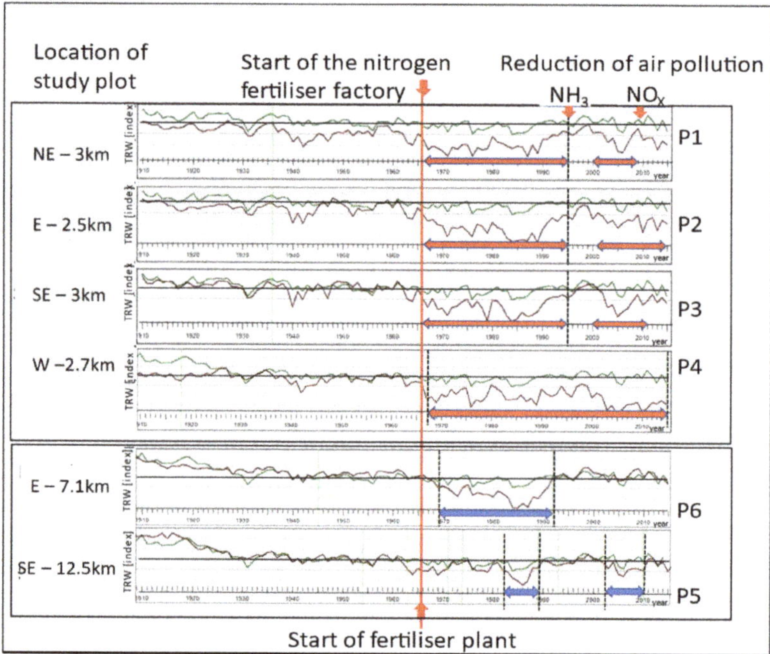

Figure 6. Reductions in the radial growth of the Scots pines growing in the vicinity of Zakłady Azotowe Puławy (selected plots). Explanation of markings: Brown lines—pine chronologies on

research plots, green line pine growth on the reference plot. Red horizontal arrows indicate the period of word reduction on surfaces located in the zone up to 3 km from the factory, blue arrows indicate the period of reduction in rings on surfaces located further from the factory. Vertical green lines indicate which the period to be taken into account when calculating the rate of growth reduction, reduction, following the Schweingruber method [59], e.g., diagram P1—the reduction period lasts from 1967 to 1995 (28 years), hence the preceding period 1938–1966 was used for the calculation of the reduction rate. See explanation in the text, methods section.

The spatial and temporal distribution of the growth reduction between the different study plots is shown in Figures 7 and 8. This distribution points to a clear relationship between the size of the reduction on the one part and the distance from the emitter and the prevailing wind direction on the other. Within a radius of 3 km from the source of the emissions, reductions in growth occurred almost simultaneously in a majority of the trees being studied (66.7% to 88.1%); the reductions lasted for many years and were strong or very strong (above 50%). The largest share of trees showing a very strong reduction in growth (above 70%) could be found in plot 2, which is located approximately 2.5 km east of the emitter, whereas the pines growing in the northern part of the zone (plot 1) showed a lesser degree of reduction. In the stands located further away from the emitter, the extent of damage to the trees generally decreased with distance from the emitter. However, trees growing in stands located in the prevailing downwind direction were affected more severely than those growing at a similar distance from the factory but located to the north of it (Figure 7).

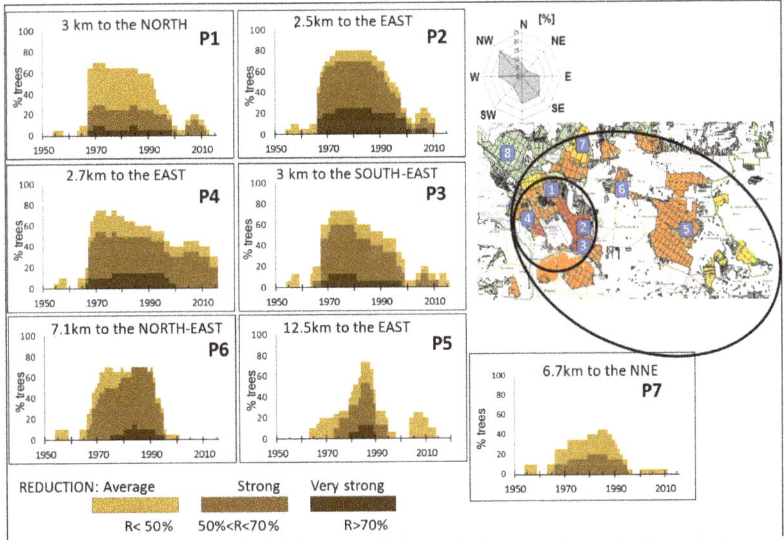

Figure 7. Graph showing growth reductions at the Scots pines in the different study plots.

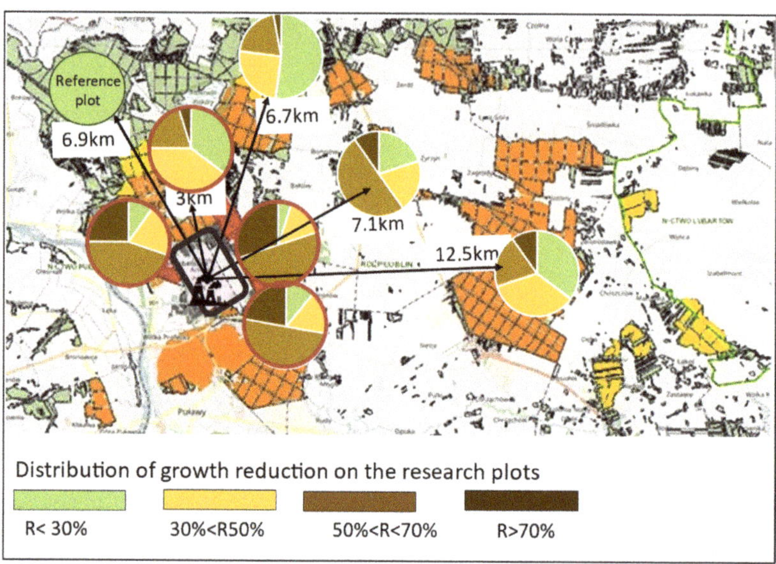

Figure 8. Spatial distribution of the reductions in the tree-ring width of the Scots pines.

In comparison with forest complexes growing near other nitrogen factories in Poland, the reductions in tree-ring width and the damage to pine stands in Puławy were much greater [30,62]. A similar degree of damage can be observed in forests growing in Silesia, which is the most polluted region of Poland. There, however, the negative effect of air pollution was more prolonged and the amount of pollution was significantly greater [36,63,64]. For that reason, we decided to search for an answer to the question of what other factors—apart from the excessive emissions of toxic substances—could have caused such reductions in growth and determined their spatial distribution.

We believe that the severe damage to the stands growing around the nitrogen factory in Puławy might have resulted from the synergistic influence of three factors: the emission of toxic substances, the relatively low height of the exhaust stacks that emit nitrogen compounds and the anemometric conditions that prevail in the region (Figure 9). The concentrations of gaseous pollutants are inversely proportional to the height of the emitter. Toxic nitrogen compounds were emitted from six stacks with a height of 47 m (ammonium nitrate) and five stacks with a height of 30 m (gaseous ammonia) [47]. Some of the nitrogen compounds also permeated into the atmosphere from the surface of industrial effluent tanks. The excessive emissions of nitrogen-based pollutants from the relatively low stacks under low wind speed conditions (with an average wind speed of 2.1 m/s) and the frequent periods of lull, especially in summer, resulted in the formation of particularly high concentrations of various pollutants near the factory itself (Figure 10). Consequently, the stands located in the immediate vicinity of the factory suffered the greatest amount of damage, regardless of their location in relation to the prevailing wind direction. Sulphur dioxide was emitted from a 160 metre-tall exhaust stack, which resulted in lower SO_2 levels in the vicinity of the nitrogen fertiliser factory itself, but the pollutants were transported over distances of up to 120 km (Figure 10). Therefore, one could conclude that the toxic nitrogen compounds (especially ammonia) were the main factors behind the degradation of the pine stands around Zakłady Azotowe Puławy.

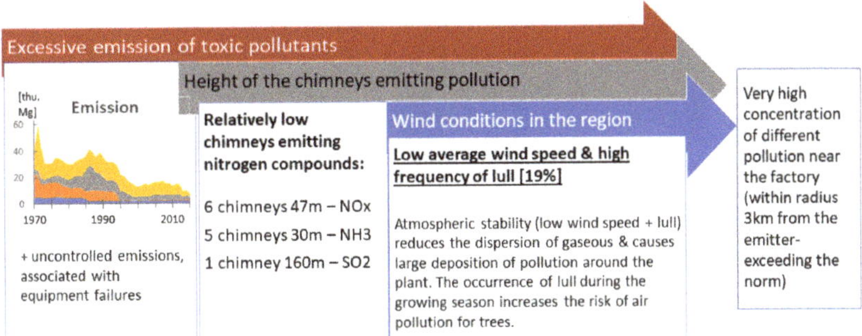

Figure 9. Possible factors determining the reductions in the annual ring width of the Scots pines.

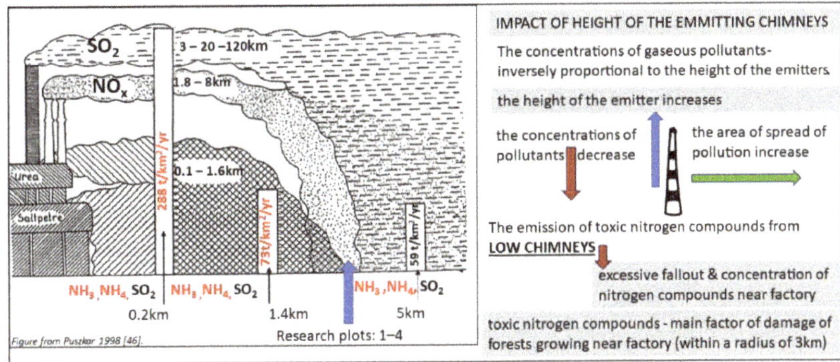

Figure 10. Extent and amount of air pollution emitted by Zakłady Azotowe Puławy.

In our opinion, this degradation—as recorded in the annual rings in the form of an abrupt and prolonged reduction in radial growth—is the result of both direct and indirect effects of toxic nitrogen compounds on the entire forest ecosystem. Our concept of this impact is shown in Figure 11. The sudden introduction of very high concentrations of nitrogen compounds into the forest environment caused a rapid, excessive toxic accumulation of nitrogen in pine needles, thus disturbing metabolic processes, causing a failure of the assimilation apparatus and death of apices, and ultimately leading to the dieback of some of the trees [43,44]. This, in turn, resulted in decreased stand density. In some of the pines, the dieback of needles and shoots caused a thinning of the crown. This permitted greater penetration of toxic pollutants and increased deposition of toxins on the needles, which was an ongoing process. At the same time, the decrease in stand density and the thinning of the crown allowed more sunlight to penetrate to the forest floor, causing changes in the microclimate—in particular by increasing the difference between the maximum and minimum air temperature. This entailed an increase in potential evaporation, decrease in mean air moisture level, increase in soil temperature, drying up of the soil, extension of the drought period and increase in water deficit [65].

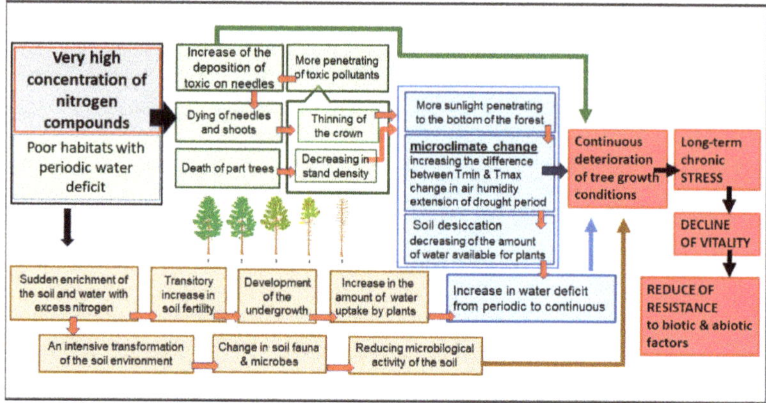

Figure 11. Concept of the direct and indirect impact of air pollution on the growth of the Scots pine stands under study. Explanation of the colours used in the figure: green—changes taking place in pine trees; blue—changes in the microclimate and water available; brown—changes in the soil environment; red—changes in vitality of Scots pine.

Very high concentrations of nitrogen compounds resulted in a sudden enrichment of the soil with excess nitrogen, which—as a consequence—caused an intensive transformation of the soil environment [46–48]. Prior to the construction of the factory, the soils in the area had generally been poor and characterised by periodic deficits of water and certain nutrients, especially nitrogen. The dieback of pine needles and shoots resulted in more sunlight penetrating to the forest floor, which—combined with a transitory increase in soil fertility—enabled extensive development of the undergrowth as well as bushes and birch and oak trees [42,46,47]. Changes to the composition of plant communities contributed to a rapid depletion of nutrients and an increase in water deficit. This was accompanied by changes to soil fauna and microbes and a reduction in the microbial activity of the soil [46].

Changes in the pine forest stands, microclimate and soil environment resulted in continuous deterioration of pine tree growth conditions, long-term chronic stress, a decline in vitality and decrease in resistance to biotic and abiotic factors. Interestingly, unlike the Scots pine, the emission from the nitrogen fertiliser factory in Puławy had a beneficial effect on oak and larches trees growing in the experimental Forest Range Ruda in Puławy (ca 3 km from factory), [66]. In the case of larch, an increase in the width of the annual rings was observed during the first decades of exposure to pollution, while oak increased its growth throughout the pollution period. According to Karolewski et al. [66], it is related both to the lower sensitivity of these trees to pollution than Scots pine, as well as to the fertilizing effect of nitrogen compounds.

As noted in the literature, pine trees weakened by air pollution are very often colonised by pests [22]. However, as the pest threat to the pine stands under study had been relatively low from the 1970s onwards [4,5], the impact of pests could not be investigated.

In regards to the climate, the question is whether the damaged trees are in fact more sensitive to climatic conditions.

A response function analysis performed for the 1951–2015 period showed that the main determinants of growth in the pines under study were temperature conditions in winter and early spring and precipitation in summer (Figure 12A). The above-average air temperatures of the January–March period contributed to the production of wide annual rings in the next growing season. Another factor that determined the trees' good health and the production of wide annual rings was the supply of water in June–August, that is during the period of greatest cambial activity in pine [31]. Droughts in that period were a strong inhibitor of growth. The dominant role of late-winter and early-spring temperatures in the development of annual rings in pine has also been noted by other authors [34,49,62–64,67,68]. The existing body of research confirms the existence of a common climate signal for Central Europe that differentiates the growth rhythm of this species [67]. As winter ends and daylight lengthens, pine loses its frost resistance and becomes sensitive to low temperatures. The freezing of needles, branches and trunks (resulting in frost damage), the drying-off caused by cold winds and the mechanical damage from snow all contribute to the deterioration of the trees' health and reduction in growth dynamics during the next growing season.

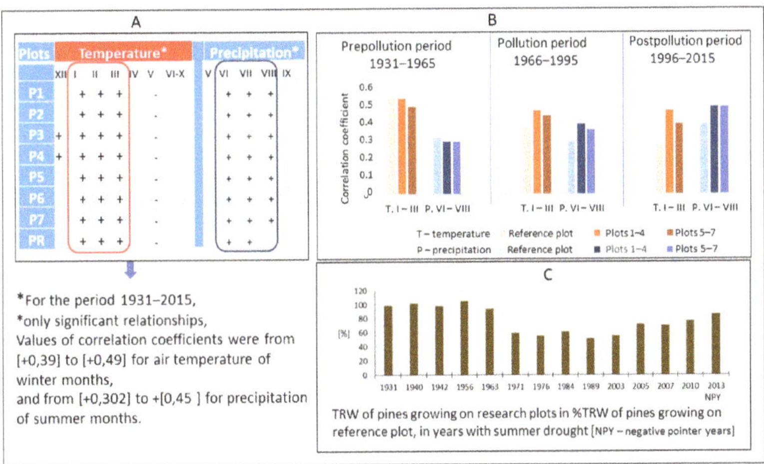

Figure 12. Climate—growth relationships for Scots pine growing on study plots. (**A**)—Results of correlation and response analyses for index pine chronologies from study plots (the values of correlation coefficients in Supplementary Material Figure S2); (**B**)—correlation coefficients between index chronologies, and winter temperature and summer precipitations, for different period; (**C**)—Tree-ring width of Scots pines growing on research plots (plots 1–4) in %TRW of Scots pine growing on reference plot, only years with extreme drought.

Analyses performed for shorter time periods—covering the years prior to the commissioning of the plant (1931–1965), the years of increased emission of airborne pollutants (1966–1995) and the years of successive reduction in toxic emissions (1996–2015)—demonstrated that the impact of climatic conditions on pine growth was similar in trend for all plots, but differences between the plots in terms of the strength of linkages began appearing in the late 1960s. Interestingly the correlation coefficient between TRW and mean air temperature of winter months decreases over time in the reference plot, while that with summer precipitation increases, witnessing the ongoing climate change with milder winters and drier summers (12B). Pines exposed to toxic pollutants were more vulnerable to cold winters and prolonged summer droughts. The reduction in industrial emissions and improvement of environmental conditions in the last decade of the 20th century resulted in the formation of wider annual rings.

Nevertheless, the trees are still weakened: they show reduced immunity to climatic stress and are more sensitive to adverse weather conditions, especially drought (Figure 12C). Similar observations concerning pine stands have been made by Oleksyn [69], Augustiastis [70,71], Vacek [26] and Putalova [34]. Furthermore, Vacek [26] demonstrated that the most severe damage can be attributed to the synergistic interaction between chemical stress and climatic stress, in particular in connection with a severe drought.

4. Conclusions

The extent and spatial coverage of forest ecosystem degradation in the Puławy area can be attributed to the amount and type of pollutants and to a number of local factors, especially the anemometric and habitat conditions and the height of the exhaust stacks (Figure 13). The high frequency of periods of lull combined with the low height of the stacks emitting the toxic pollutants multiplied the negative effects of the emissions. A radical reduction in pollutant emissions improved the environmental conditions, enabling the tress to grow once again, but the prolonged period of strong anthropopressure caused a long-term reduction in the trees' resistance to abiotic factors. Our research indicates that in areas with prolonged exposure to a high concentration of pollutants, the adverse impact of pollution on forests persists for a very long time and may be observed even 20 years after a radical reduction in emissions. These forests have reduced resistance to abiotic stress related to climate change, especially drought. Therefore, a greater impact of climate change—and in particular of extreme events—on the dieback of the trees growing in areas with strong anthropopressure can be expected.

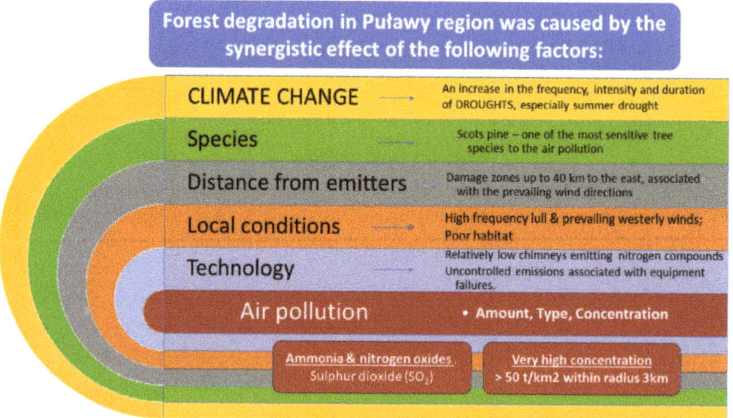

Figure 13. Factors determining forest degradation in Puławy region.

Supplementary Materials: The following are available online at https://www.mdpi.com/article/10.3390/f12101421/s1, Figure S1: The Tree-ring reduction caused by air pollution from Zakłądy Azotowe Puławy, Figure S2: Climate -growth relationships between residual chronologies and air temperature and precipitation, Table S1: Description of study plots in Puławy Forest District. GPS, Coordinates of the Study Plot.

Author Contributions: Conceptualization, L.C.-O. and W.O.; methodology, L.C.-O.; software, L.C.-O.; validation, W.O.; formal analysis, L.C.-O.; investigation, L.C.-O. and W.O.; writing—original draft preparation, L.C.-O. and W.O.; writing—review and editing, L.C.-O.; visualization, L.C.-O.; W.O.; supervision, L.C.-O. All authors have read and agreed to the published version of the manuscript.

Funding: This research received no external funding.

Institutional Review Board Statement: Not applicable.

Informed Consent Statement: Not applicable.

Data Availability Statement: Not applicable.

Conflicts of Interest: The authors declare no conflict of interest.

References

1. Paoletti, E.; Bytnerowicz, A.; Andersen, C.; Augustaitis, A.; Ferretti, M.; Grulke, N.; Günthardt-Goerg, M.S.; Innes, J.; Johnson, D.; Karnosky, D.; et al. Impacts of Air Pollution and Climate Change on Forest Ecosystems—Emerging Research Needs. *Sci. World J.* **2007**, *7*, 1–8. [CrossRef] [PubMed]

2. Matyssek, R.; Clarke, N.; Cudlin, P.; Mikkelsen, T.N.; Tuovinen, J.P.; Wieser, G.; Paoletti, E. *Climate Change, Air Pollution and Global Challenges. Understanding and Perspectives from Forest Research*; Developments in Environmental Science Series; Elsevier Ltd.: Oxford, UK, 2013; Volume 13, 648p.

3. *EEA Report. Effects of Air Pollution on European Ecosystems. Past and Future Exposure of European Freshwater and Terrestrial Habitats to Acidifying and Eutrophying Air Pollutants*; EEA: Copenhagen, Denmark, 2014. [CrossRef]

4. Bytnerowicz, A.; Omasa, K.; Paoletti, E. Integrated effects of air pollution and climate change on forests: A northern hemisphere perspective. *Environ. Pollut.* **2007**, *147*, 438–445. [CrossRef] [PubMed]

5. Lorenz, M.; Clarke, N.; Paoletti, E.; Bytnerowicz, A.; Grulke, N.; Lukina, N.; Sase, H.; Staelens, J. Air pollution impacts on forests in changing climate. In *Forest and Society—Responding to Global Drivers of Change*; IUFRO World Series; Mery, G.E.A., Ed.; International Union of Forest Research Organizations: Vienna, Austria, 2010; Volume 25, pp. 55–74.

6. Bytnerowicz, A.; Fenn, M.; McNulty, S.; Yuan, F.; Pourmokhtarian, A.; Driscoll, C.; Meixner, T. Interactive Effects of Air Pollution and Climate Change on Forest Ecosystems in the United States: Current Understanding and Future Scenarios. *Dev. Environ. Sci.* **2013**, *13*, 333–369. [CrossRef]

7. Paoletti, E.; Schaub, M.; Matyssek, R.; Wieser, G.; Augustaitis, A.; Bastrup-Birk, A.; Bytnerowicz, A.; Günthardt-Goerg, M.; Müller-Starck, G.; Serengil, Y. Advances of air pollution science: From forest decline to multiple-stress effects on forest ecosystem services. *Environ. Pollut.* **2010**, *158*, 1986–1989. [CrossRef] [PubMed]

8. Matyssek, R.; Wieser, G.; Calfapietra, C.; de Vries, W.; Dizengremel, P.; Ernst, D.; Jolivet, Y.; Mikkelsen, T.; Mohren, G.; Le Thiec, D.; et al. Forests under climate change and air pollution: Gaps in understanding and future directions for research. *Environ. Pollut.* **2012**, *160*, 57–65. [CrossRef]

9. Serengil, Y.; Augustaitis, A.; Bytnerowicz, A.; Grulke, N.; Kozovitz, A.; Matyssek, R.; Müller-Starck, G.; Schaub, M.; Wieser, G.; Coskun, A.A.; et al. Adaptation of forest ecosystems to air pollution and climate change: A global assessment on research priorities. *iForest—Biogeosciences For.* **2011**, *4*, 44–48. [CrossRef]

10. Sicard, P.; Dalstein-Richier, L. Health and vitality assessment of two common pine species in the context of climate change in southern Europe. *Environ. Res.* **2015**, *137*, 235–245. [CrossRef]

11. Sicard, P.; Augustaitis, A.; Belyazid, S.; Calfapietra, C.; de Marco, A.; Fenn, M.; Bytnerowicz, A.; Grulke, N.; He, S.; Matyssek, R.; et al. Global topics and novel approaches in the study of air pollution, climate change and forest ecosystems. *Environ. Pollut.* **2016**, *213*, 977–987. [CrossRef]

12. Woo, S.Y. Forest decline of the world: A linkage with air pollution and global warming. *Afr. J. Biotechnol.* **2009**, *8*, 7409–7414.

13. Bytnerowicz, A. Physiological aspects of air pollution stress in forests. *PHYTON-HORN* **1996**, *36*, 15–22.

14. Kurczyńska, E.; Bełtowski, M.; Włoch, W. Morphological and anatomical changes of Scots pine dwarf shoots induced by air pollutants. *Environ. Exp. Bot.* **1996**, *36*, 185–197. [CrossRef]

15. Kurczyńska, E.U.; Dmuchowski, W.; Włoch, W.; Bytnerowicz, A. The Influence of Air Pollutants on Needles and Stems of Scots Pine (*Pinus Sylvestris* L.) Trees. *Environ. Pollut.* **1997**, *98*, 325–334. [CrossRef]

16. de Vries, W.; Dobbertin, M.H.; Solberg, S.; van Dobben, H.F.; Schaub, M. Impacts of acid deposition, ozone exposure and weather conditions on forest ecosystems in Europe: An overview. *Plant Soil* **2014**, *380*, 1–45. [CrossRef]

17. Lilleskov, E.A.; Kuyper, T.W.; Bidartondo, M.I.; Hobbie, E.A. Atmospheric nitrogen deposition impacts on the structure and function of forest mycorrhizal communities: A review. *Environ. Pollut.* **2019**, *246*, 148–162. [CrossRef] [PubMed]

18. Davidson, C.J.; Foster, K.R.; Tann, R.N. Forest health effects due to atmospheric deposition: Findings from long-term forest health monitoring in the Athabasca Oil Sands Region. *Sci. Total Environ.* **2020**, *699*, 134277. [CrossRef] [PubMed]

19. Carter, T.S.; Clark, C.M.; Fenn, E.; Jovan, S.; Perakis, S.S.; Riddell, P.; Schaberg, P.G.; Greaver, T.L.; Hastings, M.G. Mechanisms of nitrogen deposition effects on temperate forest lichens and trees. *Ecosphere* **2017**, *8*, e01717. [CrossRef]

20. Masson-Delmotte, V.; Zhai, P.; Pörtner, H.-O.; Roberts, D.; Skea, J.; Shukla, P.R.; Pirani, A.; Moufouma-Okia, W.; Pean, C.; Pidcock, R.; et al. *Global Warming of 1.5 °C. An IPCC Special Report on the Impacts of Global Warming of 1.5 °C above Pre-Industrial Levels and Related Global Greenhouse Gas Emission Pathways, in the Context of Strengthening the Global Response to the Threat of Climate Change, Sustainable Development, and Efforts to Eradicate Poverty*; Report IPCC: Geneva, Switzerland, 2019.

21. Allen, C.D.; Breshears, D.D.; McDowell, N.G. On underestimation of global vulnerability to tree mortality and forest die-off from hotter drought in the Anthropocene. *Ecosphere* **2015**, *6*, 129. [CrossRef]

22. Manion, P.D.; Lachance, D. Forest decline concepts: An overview. In *Forest Decline Concepts*; Manion, P.D., Lachance, D., Eds.; APS Press: St. Paul, MN, USA, 1992; pp. 181–190.

23. IPC Forests. *Forest Condition in Europe. The 2020 Assessment. IPC Forests Technical Report under the UNECE Convention on Long-Range Transboundary Air Pollution (Air Convention)*; Thünen Institute: Braunschweig, Germany, 2020.

24. Dirnböck, T.; Djukic, I.; Kitzler, B.; Kobler, J.; Mol-Dijkstra, J.P.; Posch, M.; Reinds, G.J.; Schlutow, A.; Starlinger, F.; Wamelink, W.G.W. Climate and air pollution impacts on habitat suitability of Austrian forest ecosystems. *PLoS ONE* **2017**, *12*, e0184194. [CrossRef]

25. De Marco, M.; Proietti Ch Cionni, I.; Fischer, R.; Screpanti, A.; Vitale, M. Future impacts of nitrogen deposition and climate change scenarios on forest crown defoliation. *Environ. Pollut.* **2014**, *194*, 171–180. [CrossRef]

26. Vacek, S.; Vacek, Z.; Remes, J.; Bileki, L.; Hunova, I.; Bulušek, D.; Putalová, T.; Král, J.; Simo, J. Sensitivity of unmanaged relict pine forest in the Czech Republic to climate change and air pollution. *Trees* **2017**, *31*, 1599–1617. [CrossRef]

27. McLaughlin, S.B.; Shortle, W.C.; Smith, K.T. Dendroecological applications in air pollution and environmental chemistry: Research needs. *Dendrochronologia* **2002**, *20*, 133–157. [CrossRef]
28. Ferretti, M.; Innes, J.; Jalkane, R.; Saurer, M.; Schaeffer, J.; Spiecker, H.; von Wilpert, K. Air pollution &environmental chemistry-what role for tree-ring studies? *Dendrochronologia* **2002**, *20*, 159–174.
29. Sensuła, B.M.; Wilczyński, S.; Piotrowska, N. Application of dendrochronology and mass spectrometry in bio-monitoring of Scots pine stands in industrial areas. *Sylwan* **2016**, *160*, 730–740.
30. Szychowska-Krąpiec, E.; Wiśniowski, E. Tree-ring analysis of *Pinus sylvestris* as a method to assess the industrial pollution impact on the example of "Police" chemical factory (NW Poland). *Kwart. AGH Geol.* **1996**, *22*, 281–299.
31. Wilczyński, S. The variation of tree-ring widths of Scots pine (*Pinus sylvestris* L.) affected by air pollution. *Eur. J. For. Res.* **2006**, *125*, 213–219. [CrossRef]
32. Stravinskiene, V.; Bartkevicius, E.; Plausinyte, E. Dendrochronological research of Scots pine (*Pinus sylvestris* L.) radial growth in vicinity of industrial pollution. *Dendrochronologia* **2013**, *3*, 179–186. [CrossRef]
33. Juknys, R.; Augustaitis, A.; Vencloviene, J.; Kliučius, A.; Vitas, A.; Bartkevičius, E.; Jurkonis, N. Dynamic response of tree growth to changing environmental pollution. *Eur. J. For. Res.* **2014**, *133*, 713–724. [CrossRef]
34. Putalová, T.; Vacek, Z.; Vacek, S.; Štefančík, I.; Bulušek, D.; Král, J. Tree-ring widths as an indicator of air pollution stress and climate conditions in different Norway spruce forest stands in the Krkonoše Mts. *Cent. Eur. For. J.* **2019**, *65*, 21–33. [CrossRef]
35. Barniak, J.; Krąpiec, M. The tree-ring method of estimation of the effect of industrial pollution on pine (*Pinus sylvestris* L.) tree stands in the northern part of the Sandomierz Basin (SE Poland). *Water Air Soil Pollut.* **2016**, *227*, 166. [CrossRef]
36. Danek, M. The influence of industry on Scots pine stands in the south-eastern part of the Silesia-Krakow Upland (Poland) on the basis of dendrochronological analysis. *Water Air Soil Pollut.* **2007**, *185*, 265–277. [CrossRef]
37. Łuszczyńska, K.; Wistuba, M.; Malik, I. Reductions in tree-ring widths of silver fir (*Abies alba* Mill.) as an indicator of air pollution in southern Poland. *Environ. Socioecon. Stud.* **2018**, *6*, 44–51. [CrossRef]
38. Godek, M.; Sobik, M.; Błaś, M.; Polkowska, Ż.; Owczarek, P.; Bokwa, A. Tree rings as an indicator of atmospheric pollutant deposition to subalpine spruce forests in the Sudetes (Southern Poland). *Atmos. Res.* **2015**, *151*, 259–268. [CrossRef]
39. Greszta, J.; Gruszka, A.; Kowalkowska, M. *Wpływ Imisji na Ekosystem*; Wyd. Naukowe: Katowice, Poland, 2002; ISBN 8371643691. (In Polish)
40. Krąpiec, M.; Szychowska-Krąąiec, E. Tree-ring estimation of the effect of industrial pollution on pine (*Pinus sylvestris*) and fir (*Abies alba*) in the Ojców National Park (Southern Poland). *Nat. Conserv.* **2001**, *58*, 33–42.
41. Barniak, J.; Jureczko, A. Impact of air pollution on forest stands in the vicinity of Wodzisław Śląski and Rybnik, Poland. *Geol. Geophys. Environ.* **2019**, *45*, 283–290. [CrossRef]
42. Sokołowski, A.W.; Kawecka, A. *Effects of Industrial Air Pollution on Plants of the Puławy Forest Department*; Prace IBL: Sękocin Stary, Poland, 1973. (In Polish)
43. Kawecka, A. Changes of leaves of the Scots pine (*Pinus sylvestris* L.) due to pollution of the air with nitrogen compounds. *Ekol. Pol.* **1973**, *21*, 105–120.
44. Ostrowska, A. Nitrogen accumulation in pine needles, bark and wood within the impact area of the Nitrogen Works at Puławy. *Rocz. Glebozn.* **1980**, *31*, 117–131. (In Polish)
45. Sierpiński, Z. *Secondary Pine Pests in Stands within Area of Industrial Emissions Containing* Nitrogen. *Sylwan* **1971**, *10*, 11–18. (In Polish)
46. Puszkar, L.; Puszkar, T. Environmental impact of emissions from the nitrogen fertilizer plant in Puławy. *Int. Agrophys.* **1998**, *12*, 361–371.
47. Oleksyn, J.; Fritts, H.C.; Hughes, M.K. Tree−ring analysis of different *Pinus sylvestris* provenances, *Quercus robur*, *Larix decidua* and *Larix decidua* × *Larix kaempferi* affected by air pollution. *Arbor. Kórnickie* **1993**, *38*, 87–111.
48. Bielińska, E.J.; Domżał, H. Changes in the ecochemical condition of forest soils in the influence area of Zakłady Azotowe "Puławy" S.A. *Rocz. Glebozn.* **2008**, *59*, 29–36. (In Polish)
49. Kalbarczyk, R.; Ziemiańska, M.; Nieróbca, A.; Dobrzańska, J. The impact of climate change and strong anthropopressure on the annual growth of Scots Pine (*Pinus sylvestris* L.) wood growing in Eastern Poland. *Forests* **2018**, *9*, 661. [CrossRef]
50. Habitat Elaborat for Puławy Forest District. BULiGL Lublin, Poland. Unpublished work. 2017. (In Polish)
51. Grupa Azoty Zakłady Azotowe "Puławy" S.A. Emissions of Pollutants into the Air in the Years 1970—2015. Unpublished work. (In Polish)
52. Zielski, A.; Krąpiec, M. *Dendrochronologia*; Wydawnictwo Naukowe PWN: Warsaw, Poland, 2004.
53. Speer, J.H. *Fundamentals of Tree-Ring Research*; University of Arizona Press: Tucson, AZ, USA, 2010; ISBN B00GA42F4O.
54. Cybis Elektronik. CDendro and CooRecorder. 2010. Available online: http://www.cybis.se/forfun/dendro/index.htm (accessed on 25 August 2020).

55. Holmes, R.L. Quality control of crossdating and measuring. User's manual for computer program COFECHA. In *Tree-Ring Chronologies of Western North America: California, Eastern Oregon and Northern Great Basin with Procedures Used in the Chronology Development Work Including Users Manuals for Computer Programs COFECHA and ARSTAN*; Holmes, R.L., Adams, R.K., Fritts, H.C., Eds.; Chronology Series; Laboratory of Tree-Ring Research, University of Arizona: Tuscon, AZ, USA; pp. 41–49.
56. Cook, E.R.; Krusic, P.J. ARSTAN4.1b_XP. Available online: http://www.ldeo.columbia.edu (accessed on 25 August 2020).
57. Wigley, T.M.L.; Briffa, K.R.; Jones, P.D. On the average value of correlated time series, with applications indendroclimatology and hydrometeorology. *J. Clim. Appl. Meteorol.* **1984**, *23*, 201–213. [CrossRef]
58. Statgraphics Centurion 19 Software. Statgraphics Technologies, Inc.: The Plains, VA, USA, 2019.
59. Schweingruber, F.H. Abrupt growth changes in conifers. *IAWA Bull.* **1986**, *7*, 277–283. [CrossRef]
60. Walanus, A. *Quercus Software. Instrukcja Obsługi*; AGH-UST: Kraków, Poland, 2005.
61. Biondi, F.; Waikul, K. DENDROCLIM2002: A C++ program for statistical calibration of climate signals in tree-ring chronologies. *Comput. Geosci.* **2004**, *30*, 303–311. [CrossRef]
62. Sensuła, B.; Wilczyński, S. Climatic signals in tree-ring width and stable isotopes composition of *Pinus sylvestris* L. growing in the industrialized area nearby Kędzierzyn-Koźle. *Geochronometria* **2017**, *44*, 240–255. [CrossRef]
63. Sensuła, B.; Opała, M.; Wilczyński, S.; Pawełczyk, S. Long- and short- term incremental response of *Pinus silvestris* L. from industrial area nearby steelworks in Silesian Upland, Poland. *Dendrochronologia* **2015**, *36*, 1–12. [CrossRef]
64. Sensuła, B.; Wilczyński, S.; Opała, M. Tree growth and climate relationship: Dynamic of Scots pine (*Pinus sylvestris* L.) growing in the near-source region of the combined heat and power plant during the development of the pro-ecological strategy in Poland. *Water Air Soil Pollut.* **2015**, *226*, 220. [CrossRef]
65. Wołk, A. Climatic changes in relation to forest degradation degree in area of Nitrogen Works at Puławy. *Sylwan* **1977**, *7*, 33–46. (In Polish)
66. Karolewski, P.; Giertych, M.J.; Oleksyn, J.; Żytkowiak, R. Differential reaction of *Pinus sylvestris*, *Quercus robur*. and *Q. petraea* trees to nitrogen and sulfur pollution. *Water Air Soil Pollut.* **2005**, *160*, 95–108. [CrossRef]
67. Wilczyński, S. Uwarunkowania przyrostu radialnego wybranych gatunków drzew z Wyżyny Kieleckiej w świetle analiz dendroklimatologicznych. *Zesz. Nauk. Uniw. Rol. Im. Hugona Kołłątaja W Krakowie Rozpr.* **2010**, *464*, 1–221. (In Polish)
68. Cedro, A.; Cedro, B. Influence of climatic conditions and air pollution on radial growth of Scots pine (*Pinus sylvestris* L.) in Szczecin's city Forests. *For. Res. Pap.* **2018**, *79*, 105–112. [CrossRef]
69. Oleksyn, J. Zróżnicowanie wrażliwości na działanie szkodliwych czynników abiotycznych. In *Biologia Sosny Zwyczajnej*; Białobok, S., Boratyński, A., Bugała, W., Eds.; PAN ID Sorus Press: Poznań-Kórnik, Poland, 1993; pp. 395–404.
70. Augustaitis, A.; Juknys, R.; Kliučius, A.; Augustaitiene, I. The changes of Scots pine (*Pinus sylvestris* L.) tree stem and crown increment under decreased environmental pollution load. *Ekologia* **2003**, *22* (Suppl. 1), 35–41.
71. Augustaitis, A. Impact of meteorological parameters on responses of pine crown condition to acid deposition at Aukštaitija National Park. *Balt. For.* **2011**, *17*, 205–214.

 *forests*

Article

Assessing Black Locust Biomass Accumulation in Restoration Plantations

Gavriil Spyroglou [1,*], **Mariangela Fotelli** [1], **Nikos Nanos** [2] **and Kalliopi Radoglou** [3]

[1] Forest Research Institute, Hellenic Agricultural Organization DEMETER, Vasilika, 57006 Thessaloniki, Greece; fotelli@fri.gr
[2] School of Forestry and Natural Environment, Aristotle University of Thessaloniki, University Campus, 54124 Thessaloniki, Greece; nikosnanos@for.auth.gr
[3] Department of Forestry and Management of the Environment and Natural Resources, Democritus University of Thrace, Ath. Pandazidou 193 Str., P.O. Box 129, 68200 Orestiada, Greece; kradoglo@fmenr.duth.gr
* Correspondence: spyroglou@fri.gr; Tel.: +30-23-1046-1171 (ext. 233)

Abstract: Forests (either natural or planted) play a key role in climate change mitigation due to their huge carbon-storing potential. In the 1980s, the Hellenic Public Power Corporation (HPPC) started the rehabilitation of lignite post-mining areas in Northwest Greece by planting mainly black locust (*Robinia pseudoacacia* L.). Today, these plantations occupy about 2570 ha, but the accumulation of Above Ground Biomass (AGB) and deadwood has not been assessed to date. Therefore, we aimed at estimating these biomass pools by calibrating an allometric model for AGB, performing an inventory for both pools and predicting the spatial distribution of AGB. 214 sample plots of 100 m^2 each were set up through systematic sampling in a grid dimension of 500 × 500 m and tree dbh and height were recorded. AGB was estimated using an exponential allometric model and performing inventory measurements and was on average 57.6 t ha^{-1}. Kriging analysis reliably estimated mean AGB, but produced errors in the prediction of high and low biomass values, related to the high fragmentation and heterogeneity of the studied area. Mean estimated AGB was low compared with European biomass yield tables for black locust. Similarly, standing deadwood was low (6–10%) and decay degrees were mostly 1 and 2, indicating recent deadwood formation. The overall low biomass accumulation in the studied black locust restoration plantations may be partially attributed to their young age (5–30 years old), but is comparable to that reported in black locust restoration plantation in extremely degraded sites. Thus, black locust successfully adapted to the studied depositions of former mines and its accumulated biomass has the potential to improve the carbon footprint of the region. However, the invasiveness of the species should be considered for future management planning of these restoration plantations.

Keywords: climate change mitigation; forest restoration; forest biomass estimation; standing and lying dead wood; variogram model; kriging regression

Citation: Spyroglou, G.; Fotelli, M.; Nanos, N.; Radoglou, K. Assessing Black Locust Biomass Accumulation in Restoration Plantations. *Forests* **2021**, *12*, 1477. https://doi.org/10.3390/f12111477

Academic Editors: Angela Lo Monaco, Cate Macinnis-Ng and Om P. Rajora

Received: 20 September 2021
Accepted: 21 October 2021
Published: 28 October 2021

1. Introduction

Forests are considered important components in climate change mitigation strategies for being key drivers of greenhouse gas removals and for facilitating the global community target of net-zero emission by 2050 [1]. Indeed, the Land Use, Land Use Change and Forestry sector (LULUCF) may provide approximately 25% of emission reductions according to nationally determined contributions of the Paris Climate Agreement [2]. A large part of emission reductions is expected from the surplus accumulation of Above Ground Biomass (AGB), a carbon pool that represents 30% of the total carbon of terrestrial ecosystems [1]. In this sense, countries and companies worldwide need to adopt strategies towards enhancing forest carbon sequestration and reducing greenhouse gas emissions [3].

Among forest management actions aiming at maximizing the forest carbon sequestration potential, increasing forest cover in degraded non-forested land is one of the most promising approaches [4]. Major challenges to this respect include evaluating the growth potential and the associated AGB accumulation of species with high adaptation potential to adverse growth conditions and to changing climate [5]. To this direction, data-based information on growth potential, biomass accumulation and associated carbon assimilation capacity of forest plantations is essential to estimate their climate change mitigation potential and plan their future management.

Black locust is the second most often planted tree species worldwide, after eucalyptus [6]. It is native to North America and has been introduced in Europe at the beginning of the 17th century [7]. In Greece and Europe, black locust is characterized as a naturalized alien plant species [8–10]. According to Sitzia et al. [10], the core area of its distribution is in sub-Mediterranean to warm continental climates, characterized by a high heat-sum. Warmer and drier climate may cause a shift in the distribution limits of black locust (*Robinia pseudoacacia* L.), as it has been described for several native Mediterranean and other European forest species [11]. Among the alien species, black locust was foreseen to increase its distribution area under different climate change scenarios [12,13]. On the contrary, a recent study suggested that, by 2050, a decline of potential niches of this species may take place in Southern Europe [14]. However, black locust exhibits great plasticity of its growth under diverse climatic conditions, indicating a high acclimation potential to future climate changes [15].

There is great concern about the future management of black locust in Europe due to its ecological disadvantages related to its invasiveness, imposed threats for biodiversity and induced changes in light microclimate and soil traits, as discussed in Refs. [16–19]. At the same time, black locust has certain advantages, due to its drought-tolerance [20], tolerance of alkalic soils with pH up to 8 [7], N-fixation ability [21–23], fast growth and high root sucking ability resulting in dense root biomass [24]. Based on these beneficial traits, black locust is often used for the production of timber, firewood, animal forage and for apiculture, as well as for energy plantations [25]. In Greece, black locust has been planted for torrent stabilization on mountains and for soil erosion control on rivers, roads and railway banks. It has been also used as a fodder in silvopastures [26] and as an alternative crop for privately owned plantations, in line with the 2080/92 and 1257/99 European Union (EU) Regulations, due to its great adaptation to marginal agricultural lands [9]. Furthermore, black locust is a suitable non-native species for the restoration of sites extremely degraded by anthropogenic activities such as mining, as it can survive on nutrient-poor depositions like the ones of former open-cast mining areas [20].

The Hellenic Public Power Corporation (HPPC) installed four large power plants in Northwest Greece, close to the largest coal deposit of the country (2.3 billion t), while the open-cast mining activities began in the 1950s to cover the country's increasing electricity demands. In the 1980s, the HPPC started the rehabilitation of the open-cast mining fields in the Lignite Center of Northwest Greece by planting black locust in mixture with other species. To date, the establishment of plantations is still ongoing, while their current area is 2570 ha, making HPPC the largest private forest plantations owner in Greece, owning 26% of the country's black locust plantations area [27]. However, there is neither an inventory for those plantations, nor any allometric equations available for an accurate estimation of their AGB.

The estimation of AGB per forest unit area is a major challenge in the context of carbon stock estimation and dynamics. AGB estimation over a domain is based on a combination of ground reference measurements such as classical field-data (forest inventory data) and the harvest of selected trees that are subsequently oven-dried and weighed for deriving accurate estimation of above ground biomass [28,29]. In some cases, however, auxiliary variables derived from satellite-based products (including LIght Detection And Ranging (LIDAR) technology) are used to render more precise spatial estimation of AGB over the desired domain [30–32]. Ground reference measurements can provide accurate estimates for the calibration and validation of global satellite-based AGB models. In both cases, geostatistics are applied to estimate the spatial distribution of AGB over a forested area using both data-sources [33–35].

The aims of our study were to (a) calibrate an allometric above ground biomass model for the studied black locust plantations, (b) to provide a reliable estimation of the live and dead above ground biomass distribution across the restored lignite mining areas of Northwest Greece by conducting an inventory of black locust plantations, and (c) to map the AGB pools over the study area by using a geostatistical approach. Our hypotheses were: (1) that the adopted geostatistical analysis would provide a reliable estimation of AGB, in relation to the determined inventory measurements and (2) that the estimated AGB of the studied plantations would indicate the satisfactory development of the restoration at the Lignite Center of Northwest Greece.

2. Materials and Methods

2.1. Study Area

In the present study, we analyzed the above ground biomass data collected from the black locust restoration plantations of the former open-cast mining areas of the lignite center of Western Greece (Figure 1). The plantations are located near Amyntaio (40.56° to 40.61° N and 21.62° to 21.69° E) and Ptolemaida (40.39° to 40.51° N and 21.7° to 21.89° E), in NW Greece (Figure 1). Black locust covers more than 95% of the planted area, followed by weaver's broom (*Spartium junceum* L) and Arizona cypress (*Cupresus arizonica* Greene), covering 2.45% and 1.44%, respectively. Other planted species comprise oaks, maples, pines and various deciduous broadleaves in very small percentages. The total area of the black locust plantations is approximately 2570 ha, and the elevation ranges from 530 to 950 m. The plantations are established on open-cast mining depositions with varying topography (moderate to steep slopes, plains and terraces). The landscape is fragmented by different land uses, which, apart from the forest plantations, also includes grass lands, agricultural lands and bare lands for photovoltaic parks and recycling facilities. The temperature of the region ranges from 6.1 to 17.4 °C with a mean annual temperature of 12.2 °C, and the total annual precipitation is 664 mm (50 years average values).

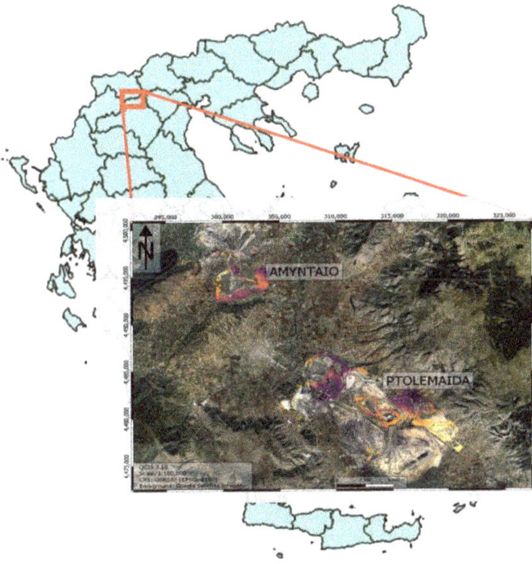

Figure 1. Location of the study area in Greece. Colored polygons indicate the studied restoration plantations of black locust in Amyntaio and Ptolemaida mine fields.

2.2. Inventory Data

For the estimation of the AGB distribution, in total 214 circular sample plots of 100 m^2 each were set up through systematic sampling in a grid dimension of 500×500 m in the open-cast mining fields of Amyntaio and Ptolemaida. The field campaign was carried out in the summer of 2019. In each sample plot, the coordinates in the Hellenic Geodetic Reference System (HGSA 87), the tree species, the tree status (dead or alive), the diameter at breast height (*dbh*) in cm using diameter measuring tape, the tree height (*Ht*), and the height to the base of live crown (*Hlc*) in m using Haglöf hypsometer were recorded for all trees. In addition, the lying dead wood (coarse woody debris) defined as dead, woody material of trees with a diameter of >1 cm was recorded for all fallen trunks, fragmented woody branches either lying on the ground or stuck above ground level. The diameter at two ends of the log (*d1* for the small and *d2* for the large, in cm) and the length of the log *L* in m were determined. Diameter measuring tape was used to measure diameters of pieces larger than 5 cm in diameter. A five-decay class system was used to classify the recorded woody debris based on morphology and hardness observed in the field following the methodology of Paletto and Tosi [36]. For samples smaller than 5 cm or rotten ones in decay class 5, only one horizontal measurement was taken by a caliper. The woody debris volume was estimated using the formula of the truncated cone and for the conversion of volume into dry biomass, the wood density reduction according to decay classes [30] was used.

2.3. Above Ground Biomass Estimation

The above ground biomass was estimated using the simple exponential allometric model of the form M = a*dbh^b. For the calibration of the allometric equation, 30 black locust trees covering all diameter range appeared in the study site were destructively sampled during the summers of 2019 and 2020. The diameter at stump height (D0.3 in cm) and at breast height (dbh in cm) was measured before the tree felling. The sampled trees were cut at the stump height and, after felling, total tree height (H in m), diameter at 50% of bole length (D0.5 in cm) and diameter at the base of live crown were recorded. Each one of the 30 stems was divided into six sections (including the stump). After felling, the fresh weight of each stem section was measured in the field. From each section, a 7 cm wide stem disk was taken, weighed, and oven-dried at the lab at 80 °C until a constant weight was reached to determine the fresh/dry biomass ratio. The fresh biomass of the whole crown was also weighed in the field and oven-dried at the lab at 80 °C until a constant weight was reached. For further details on the biomass estimation methodology, see Zianis et al. [37]. Linear regression was used for the parameterization of the log transformed allometric model.

2.4. Statistical Analysis

For the estimation of the spatial distribution of the aboveground biomass, a geostatistical approach was applied. For variogram modeling, the experimental variograms were assessed using the spherical and exponential approach. For the spatial interpolation of the above ground biomass at the points where there was no information, the ordinary kriging method was performed based on a grid size of 50×50 m for the predictions of above ground biomass in t ha^{-1} The R programming software (sp library) [38] was used in spatial analysis. For variogram analysis and variogram modeling the gstat and geor libraries were also used [39]. Finally, ordinary kriging regression [40] was applied using nugget, range and partial sill as parameters of the best variogram models. To test the quality of kriging regression, we performed a cross-validation using the leaving-one-out strategy.

3. Results

The allometric model for the estimation of tree above ground biomass, after the correction for log transformation bias (e$^{(mse/2)}$), where mse stands for mean square error, was

$$M = 0.1916 \cdot dbh^{2.2882} \cdot 1.00912$$

While the coefficient of determination (R^2) was 0.982, the residual standard error (RSE) was 0.1395 and the mean residual standard error (MSE) was 0.0182. The parameter estimates were statistically significant $F\,(1,28) = 1571$, $P < 2.2 \times 10^{-16}$. The residuals were normally distributed (Sapiro–Wilk test: $W = 0.97$, $P = 0.55$). There was no evidence of heterosctasticity, (Breusch–Pagan test: $BP = 0.52$, $df = 1$, $P = 0.469$) and autocorrelation (Durbin–Watson test: $D\text{-}W = 2.51$, $P = 0.921$).

The descriptive statistics, before kriging interpolation, of black locust stands data are presented in Table 1. Generally, black locust stands showed high heterogeneity with dbh ranging from 1.4 to 22.3 cm (mean 7.9 ± 0.25 cm) and AGB from 0.74 to 274.7 t ha^{-1} (mean 59.9 ± 2.77 t ha^{-1}). At the Amyntaio mine field, the inventoried trees had mean dbh 9.3 ± 0.31 cm, Ht 11.5 ± 0.31 m and Hlc 6.2 ± 0.24 m. The AGB ranged from 29.5 to 206 t ha^{-1} (mean 71.9 ± 3.73 t ha^{-1}). At Ptolemaida, the mean dbh was 7.2 ± 0.32 cm, Ht 9.0 ± 0.28 m and Hlc 4.7 ± 0.22 m. The AGB ranged from 0.74 to 274.7 t ha^{-1} (mean 54.4 ± 3.56 t ha^{-1}).

Table 1. Descriptive statistics of data measured in the forest inventory plots.

Parameter	N (tress ha^{-1})	dbh (cm)	Ht (m)	Hlc (m)	BA [1] (m^2 ha^{-1})	AGB$_{inv}$ [2] (t ha^{-1})	AGB$_{krig}$ [3] (t ha^{-1})
Amyntaio (788,08 ha, n_{inv} = 65, n_{krig} = 2631)							
Mean	1975	9.3	11.5	6.2	14.4	71.9	67.1
Range	800–3500	6.2–20.9	6.1–17.2	0.55–9.5	6.1–31.9	29.5–206.0	7.3–120.5
SD	590.2	2.3	2.3	1.8	4.6	28.1	19.3
SE (±)	78.2	0.31	0.31	0.24	0.61	3.73	0.38
RSE (%)	3.95	3.29	2.68	3.85	4.21	5.19	0.56
Ptolemaida (1782.00 ha, n_{inv} = 149, n_{krig} = 7088)							
Mean	2746	7.2	9.0	4.7	11.1	54.4	48.1
Range	300–30,078	1.4–22.3	2.5–16.9	0.14–9.9	0.25–43.6	0.74–274.7	17.5–81.3
SD	2896	3.5	3.1	2.5	6.8	39.8	16.7
SE (±)	259.0	0.32	0.28	0.22	0.61	3.56	0.20
RSE (%)	9.43	4.4	3.08	4.7	5.5	6.55	0.41
Whole study site (2570.08 ha, n = 214)							
Mean	2505	7.9	9.8	5.2	12.1	59.9	
Range	300–30,078	1.4–22.3	2.5–17.2	0.14–9.9	0.25–43.6	0.74–274.7	
SD	2445.9	3.4	3.1	2.4	6.4	37.4	
SE (±)	181.3	0.25	0.23	0.18	0.47	2.77	
RSE (%)	7.23	3.16	2.35	3.42	3.90	4.63	

[1] BA stands for stands for Basal Area, [2] AGB$_{inv}$ stands for above ground biomass from inventory plots, [3] AGBk$_{rig}$ stands for above ground biomass from kriging regression, n_{inv}: inventory sample size, n_{krig}: kriging sample size, SD: Standard Deviation, SE: Standard Error, RSE: Relative Standard Error.

Due to high biomass heterogeneity in the inventoried plots, the logarithmic transformation of biomass values was used for normality correction. The log transformed values follow a normal distribution with mean m = 4.13 and standard deviation $\sigma = 0.38$ for Amyntaio (Figure 2 left) and mean m = 3.53 and standard deviation $\sigma = 1.1$ for Ptolemaida (Figure 2 right).

3.1. Amyntaio Mine Field

For the estimation of the spatial distribution of the above ground biomass in the Amyntaio mine field using geostatistics, the first step was to analyze and estimate the experimental semivariogram. The exponential, spherical and Gaussian models were fitted to semivariagram data. The spherical model had the better fit with a nugget effect of 423.3, partial sill of 769.2, leveling off at the range of 1500 m (Figure 3).

The data showed a medium spatial autocorrelation with the nugget to total sill ratio being 0.35. The relatively high nugget effect suggests that there is a high measurement error in the data, possibly due to the short scale variation.

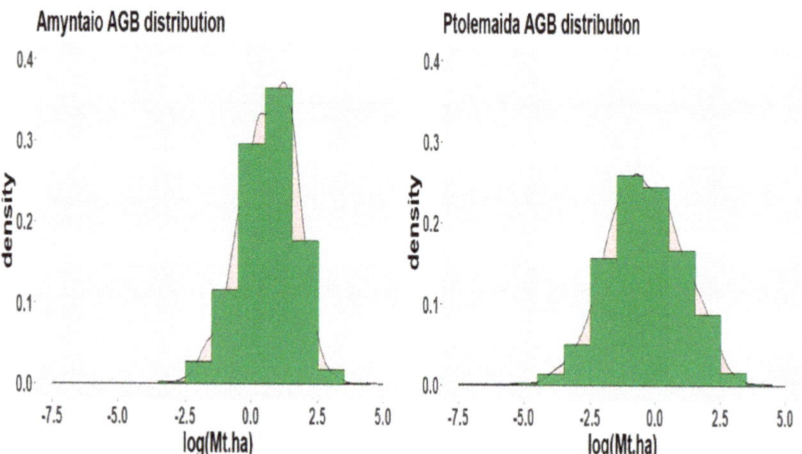

Figure 2. Histogram (frequency and density) of the logarithm of AGB (t ha^{-1}) at the study sites of Amyntaio (**left**) and Ptolemaida (**right**).

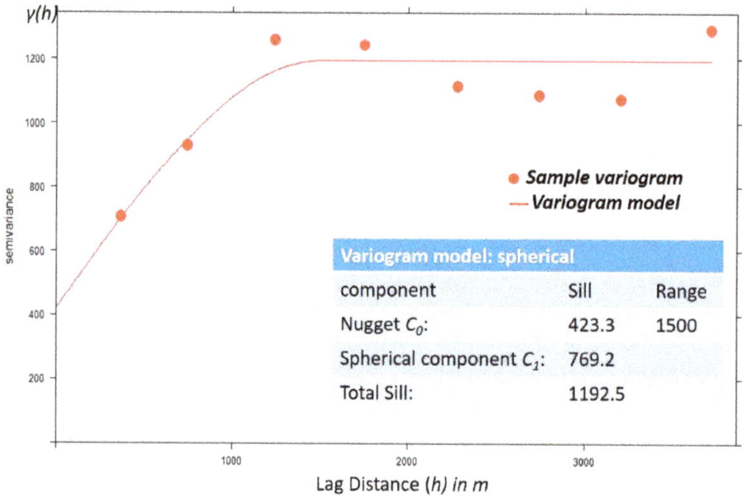

Figure 3. Experimental semivariogram with the spherical model fitted at Amyntaio mine area.

The total AGB estimated by ordinary kriging, resulted from a sample of 2631 points based on a raster consisting of 50 × 50 m squares, was 52,869.1 t with a mean of 67.1 ± 0.38 t ha^{-1}. The ABG distribution map derived by the ordinary kriging for Amyntaio mine field is presented in Figure 4.

Figure 4. Aboveground biomass distribution map at Amyntaio mining area. Coordinates in the Hellenic Geodetic Reference System (HGSA 87).

3.2. Ptolemaida Mine Field

For the estimation of the spatial distribution of the above ground biomass in the Ptolemaida mine field using geostatistics, the experimental variogram was also analyzed and estimated from the inventory data. The exponential, spherical and Gaussian models were fitted to semivariagram data. The spherical model was selected because it exhibited a better fit with a nugget effect of 920.4, partial sill of 753.9, leveling off at the range of 5000 m (Figure 5).

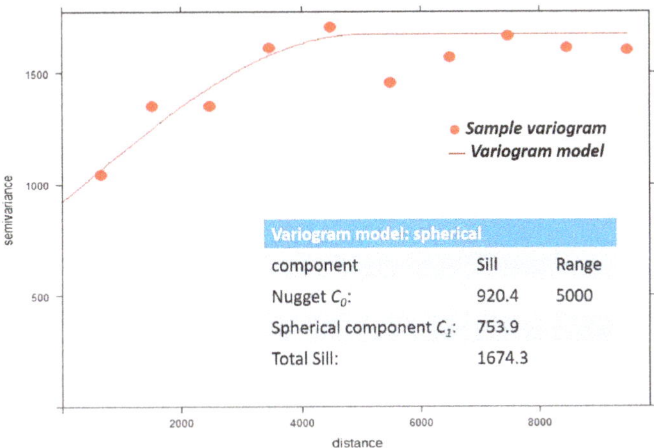

Figure 5. Experimental semivariogram with the spherical model fitted at Ptolemaida mine area.

The data showed a similar medium spatial autocorrelation, as found at Amyntaio, given that the nugget to total sill ratio was 0.55. The relatively high nugget effect, higher that the respective nugget effect in Amyntaio, suggests that there was a high measurement error present in the data, possibly due to the short scale variation and to higher degree of fragmentation that is observed in Ptolemaida compared to Amyntaio.

The total AGB estimated by ordinary kriging, resulted from a sample of 7088 points based on a raster consisting of 50×50 m squares, was 85,772 t with a mean of 48.1 ± 0.2 t ha^{-1}. The AGB distribution map for Ptolemaida is shown in Figure 6. The kriging estimates of AGB ranged from 17.5 to 81.3 t ha^{-1}.

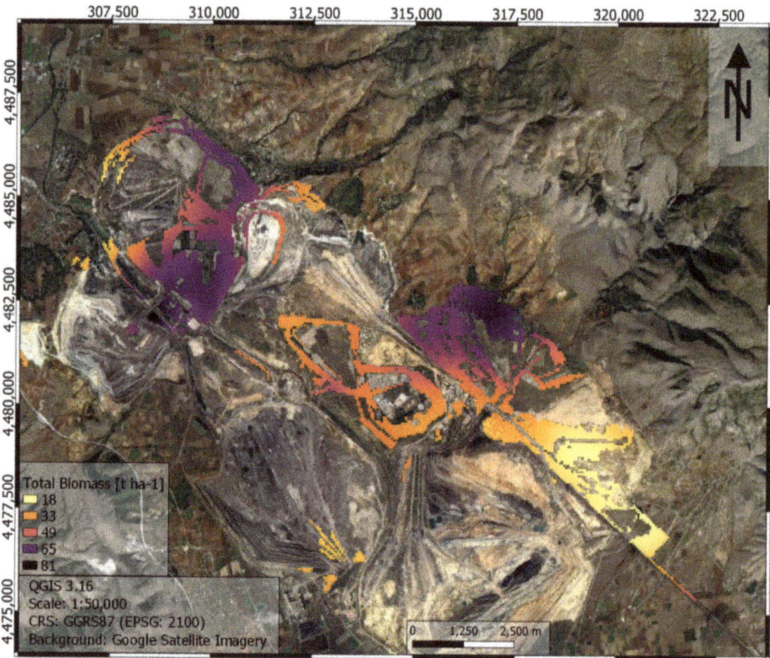

Figure 6. Aboveground biomass distribution map at Ptolemaida mining area. Coordinates in the Hellenic Geodetic Reference System (HGSA 87).

The cross-validation of the kriging analysis, which uses the leaving-one-out procedure, is shown in Table 2. The min and max values of kriging predictions were 49.0 and 109.3 t ha^{-1} for Amyntaio and 14.9 and 100.4 t ha^{-1} for Ptolemaida, while the mean observed inventory values were higher and lower, respectively, for both sites. However, the mean AGB estimated by kriging was 75.22 ± 12.5 for Amyntaio and 57.34 ± 20.7 t ha^{-1} for Ptolemaida and was similar to the observed AGB values, which were 74.84 ± 29.2 for Amyntaio and 57.8 ± 41.3 t ha^{-1} for Ptolemaida. On average, an error of 28.8 t ha^{-1} of mean AGB can be expected at a given location for Amyntaio and 38.58 t ha^{-1} for Ptolemaida with the application of the krigging analysis.

Table 2. Residual statistics of cross-validation results using the leaving-one-out procedure for AGB.

	Min	Max	Mean	SD	ME	RMSE
Amyntaio						
Observed data	29.5	206.0	74.84	29.2		
Kriging predictions	49.0	109.3	75.22	12.5	−0.383	28.77
Ptolemaida						
Observed data	0.74	275.4	57.08	41.3		
Kriging predictions	14.9	100.4	57.34	20.7	−0.267	38.58

SD: Standard Deviation, ME: Mean residual error, RMSE: Root Mean Square Error.

3.3. Standing and Lying Dead Wood

The standing dead wood ranged from 0.4 to 17.2 t ha^{-1} for Amyntaio, and 0.08 to 26.8 t ha^{-1} for Ptolemaida mine field respectively (Table 3). The lying dead wood ranged from 0.5 to 19.4 and from 0.5 to 66 m^3 ha^{-1}.

Table 3. Standing and lying black locust dead wood data at the study site.

Parameter	Lying Dead Wood [1] (m^3 ha^{-1})	Lying Dead Wood [2] (t ha^{-1})	Standing Dead Wood (t ha^{-1})	Live Wood (t ha^{-1})
Amyntaio (n = 65)				
Mean	5.2	1.6 (2.2%)	4.4 (6.2%)	71.9
Range	0.5–19.4	0.14 (0.5%)–6.9 (3.3%)	0.40 (1.4%)–17.2 (8.3%)	29.5–206.0
SD	5.1	1.7	4.0	28.1
SE (±)	1.0	0.33	0.65	3.7
RSE (%)	19.4	20.4	14.7	5.2
Ptolemaida (n = 149)				
Mean	9.3	2.9 (5.3%)	4.5 (8.3%)	54.4
Range	0.5–88.2	0.16 (22.8%)–28.2 (10.3%)	0.1 (14.3%)–26.8 (9.8%)	0.7–274.7
SD	15.2	5.1	4.7	39.9
SE (±)	1.9	0.65	0.54	3.6
RSE (%)	20.9	22.1	12.2	6.6
Whole study site (n = 214)				
Mean	8.1	2.6 (4.3%)	4.4 (7.3%)	59.9
Range	0.5–88.2	0.14 (20.0%)–28.2 (10.3%)	0.1 (14.3%)–26.8 (9.8%)	0.7–274.7
SD	13.1	4.4	4.5	37.4
SE (±)	1.4	0.47	0.42	2.8
RSE (%)	17.4	18.4	9.5	4.6

[1] wood volume m^3 ha^{-1}, [2] wood dry weight t ha^{-1}; n: sample size, SD: Standard Deviation, SE: Standard Error, RSE: Relative Standard Error.

The overall decay degree in the quality scale from 1 to 5 was ranged as: 10% for 1, 27% for 2, 45% for 3, 17% for 4 and 1% for 5 (Figure 7). Eighty five percent of the dead wood concentrates in decay classes 1 and 2 meaning an increase in mortality rate in the last 10 years.

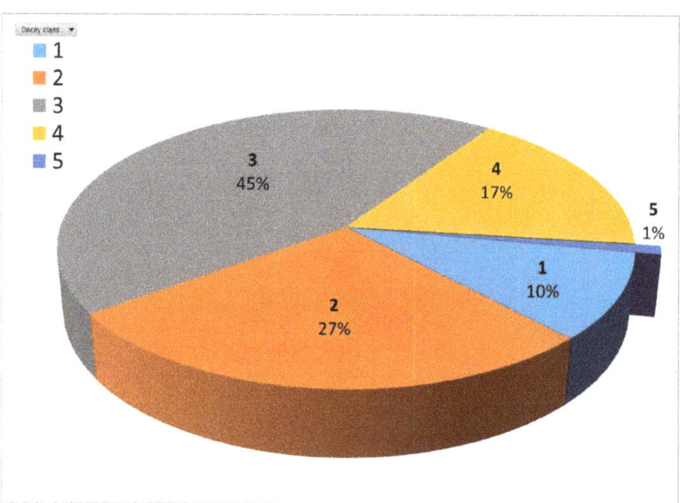

Figure 7. Overall distribution of decay classes of dead standing and lying wood in the restored open-cast lignite mining areas.

4. Discussion

Black locust is an alien forest species in Europe with controversial traits; it is an invasive species, threatening biodiversity and causing alterations in microclimate and soil conditions in forest plantations e.g., [19,25]. However, its fast growth potential, dense root system due to extensive root suckers [24] and N-fixation ability [35] allow it to survive, adapt and form dense stands under unfavorable conditions. Owing to these features, black locust has been extensively used by the HPPC for the restoration of open-cast mines at the Lignite Centers in Greece, aiming at improving the carbon footprint of these areas by increasing carbon sequestration, while also providing goods and ecosystem services such as fuel, timber, non-wood forest products and recreational opportunities to local populations. The AGB of black locust restoration plantations and its spatial distribution at the Lignite Center of Northwest Greece, established on post-mining depositions, was estimated by the development of an allometric model and the performance of inventory measurements and kriging analysis.

The spatial biomass distribution showed a tendency to increase from Southeast to Northwest in Ptolemaida and from West to North in the Amyntaio mine field, verifying the spatial direction of the planting process in the past years. The semi-variograms of above ground biomass in both Amyntaio and Ptolemaida mine fields were effectively described by the spherical model, although the exponential model has also been selected for broadleaf forests [35] and other ecosystems [41]. Still, the spherical model is one of the most widely applied in ecological studies [40]. The partial sill to total sill ratio was higher than that of nugget to total sill, verifying that the spatial variation of AGB was mainly affected by structural factors, such as the age of the plantations and competition among trees. However, the effects of random factors, such as the sampling resolution and inventory sample plot size cannot be excluded. In Ptolemaida, both structural and random factors equally affected the spatial distribution of AGB. The magnitude of the spatial correlation of AGB was medium at both field sites, with nugget to total sill ratio varying between 25 and 75%. However, the spatial autocorrelation range differed between the two sites. It was 1500 m in Ptolemaida and 5000 m in Amyntaio, indicating that the AGB at a given point affected the AGB of another point in a much lower radius in Amyntaio than in Ptolemaida. The observed high nugget effects could be partially due to the fact that the allometric equation was calibrated from a relatively small number of sampled trees, 30 in our case. Moreover, an increased sampling resolution of the systematic grid at the inventory field campaign, thus a higher number of sampling plots, may decrease the nugget effects, e.g., as seen in Ref. [42]. However, similarly high nugget effect and spatial autocorrelation range observed at the spatial prediction of forest biomass by Lamsal et al. [43] were attributed to the heterogenous studied landscape, which also applies in our case due to the fragmented and heavily impacted post-mining terrain.

The predicted mean AGB by kriging regression was similar to that measured by the inventory (Table 2). However, low values were overestimated, and high ones were underestimated, leading to a smoother spatial distribution of AGB in the study area. The heterogeneity and fragmentation of the post-mining landscape, together with the above-mentioned potential limitations, associated with the nugget effects and the resolution of the systematic grid, probably explain this difference in AGB estimation between the kriging and the inventory analysis. Jaja et al. [44] and Akhawan et al. [45] similarly concluded that kriging did not produce accurate estimations of AGB because of the high spatial heterogeneity and uneven age of the forests they studied, contrary to other reports [41]. Thus, our first hypothesis regarding the reliable estimation of AGB at the black locust plantations of the Lignite Center of Northwest Greece by means of kriging analysis is only partially confirmed.

The inventory revealed that the growth traits of the plantations varied greatly, due their different age (c. 5–35 years old). Tree density, dbh and height over the entire study area ranged from 300 to 30,078 trees ha^{-1}, 1.4 to 22.3 cm and 2.5 to 17.2 m, respectively. Moreover, total biomass fluctuated from 7.3 to 120.5 t ha^{-1}. Although the restoration was initiated at Ptolemaida, planting is still active there resulting in many young plantations and, thus, a lower mean dbh and height and a higher tree density (6.9 cm, 9.1 m, and 2746 trees ha^{-1} respectively), than at Amyntaio (9.3 cm, 11.5 m and 1975 trees ha^{-1}, respectively). Nicolescu et al. [6] reported that black locust can reach 14 m height in 10 years and dbh up to 20 cm in 25 to 30 years, under optimal conditions. Thus, the rather stressful conditions that the studied black locust plantations cope with explains their lower growth. In fact, our results correspond to the Vth growth class, the first of the two poorest classes, of the Hungarian black locust yield and biomass tables [46]. Similarly low was the above ground biomass estimated by Wang et al. [47] for 5 to 25 years old black locust restoration plantations on degraded agricultural lands at the Loess Plateau in China, which ranged from 4.1 to 27.5 t ha^{-1}

Deadwood represented 6 to 10% of the standing biomass at the studied plantations. Under unfavorable conditions, black locust cannot withstand shade, resulting in high self-thinning [6]; thus, the amount of dead wood is expected to increase in the future, considering that the studied plantations have not been managed and thinned to date. Consistently, deadwood accounted for 16% of the total carbon stock in 60 years old black locust restoration plantations in Serbia [48]. Moreover, 85% of the estimated dead wood in our study belonged to the decay degrees 1 and 2, indicating that the decay and thus tree mortality was initiated recently, as a result of self-thinning induced by competition for light and other resources.

Thus, the inventory analysis revealed that both the AGB and the deadwood biomass of the studied black locust restoration plantations were relatively low, compared to reported values under optimal conditions. This was both due to their young age and the harsh conditions of the post-mining depositions where the plantations grow. Still, as above-mentioned, the AGB of the studied plantations was comparable to other black locust restored sites developed at rather degraded sites.

Exotic species have been often used for the rehabilitation of bare coal mines due to their successful establishment and fast growth [49,50]. In this context, black locust has been planted for the rehabilitation of heavily degraded sites [23], such as the depositions of former coal mines [16,51] and combustion waste disposal sites [52]. Kraszkiewicz (2021) [53] concludes that planting black locust on various sites of wastelands establishes tree stands for medium size timber production, which perform better that other species such as poplar and willows. Filcheva et al. [50] confirmed the positive impact of afforestation on the initial soil-forming processes in coalmine spoils classifying black locust as being more beneficial than the black pine because it promotes less acidification, fixes nitrogen and incorporates more organic material into the soil. Thus, the morphological characteristics and the eco-physiological adaptations that the black locust exhibits allow it to grow successfully at extremely degraded post-mining sites [20,50]. However, the dangers that this invasive species imposes for biodiversity indicate that it should be used only for such restoration cases and point to the need for its proper future management.

5. Conclusions

Kriging analysis accurately estimated the mean AGB at the black locust restoration plantations of the Lignite Center of Northwest Greece, but presented large errors in the prediction of high and low biomass values. The inventory measurements estimated relatively low, but comparable tree biometric traits and total AGB, as in other black locust restoration plantations in heavily degraded regions. Deadwood biomass was also low, but is expected to increase if the plantations remain out of management in the future. Our results indicate that black locust plantations are successfully established at the studied coal mine spoils, and the produced biomass may contribute to the improvement of carbon footprint, to carbon storage and, thus, to climate change mitigation in such a degraded post-mining area. However, management measures should be taken to limit potential threats related to the invasiveness of the species.

Author Contributions: Conceptualization, G.S. and N.N.; methodology, G.S. and N.N.; software, G.S. and N.N.; validation, M.F., N.N.; formal analysis, G.S. and N.N.; investigation, G.S.; data curation, G.S. and N.N.; writing—original draft preparation, G.S. and M.F.; writing—review and editing, G.S.; M.F. and N.N.; supervision, K.R.; project administration, K.R.; funding acquisition, K.R. All authors have read and agreed to the published version of the manuscript.

Funding: This research was funded by Single RTDI state Aid Action Research–Create–Innovation with the co-financial of Greece and the European Union (European Regional Development Fund) in context with Operational Program Competitiveness, Entrepreneurship and Innovation (ΕΠΑΝΕΚ) of the NSRF 2014–2020 (project contribution of the tree planted land of West Macedonia lignite center to protection of environment and to mitigation of climate change T1EDK-02521).

Institutional Review Board Statement: Not applicable.

Informed Consent Statement: Not applicable.

Data Availability Statement: Not applicable.

Acknowledgments: The authors would like to acknowledge the Hellenic Public Power Corporation (HPPC S.A.) for its substantial contribution with the necessary staff and machinery in order to conduct the field campaigns and data collection. Special thanks are due to Marina Tentsoglidou, Aris Azas, Christos Papadopoulos and their technical team. Special thanks are also due to Stamatis Tziaferidis, ENA Development Consultants, for his valuable help in inventory and tree logging.

Conflicts of Interest: The authors declare no conflict of interest.

References

1. Kumar, L.; Mutanga, O. Remote Sensing of Above-Ground Biomass. *Remote. Sens.* **2017**, *9*, 935. [CrossRef]
2. Grassi, G.; House, J.; Dentener, G.G.F.; Federici, S.; Elzen, M.D.; Penman, J. The key role of forests in meeting climate targets requires science for credible mitigation. *Nat. Clim. Chang.* **2017**, *7*, 220–226. [CrossRef]
3. TSVCM. Taskforce on Scaling Voluntary Carbon Markets; Final Report. 2021. Available online: https://www.iif.com/Portals/1/Files/TSVCM_Report.pdf (accessed on 27 October 2021).
4. Tong, X.; Brandt, M.; Yue, Y.; Ciais, P.; Jepsen, M.R.; Penuelas, J.; Wigneron, J.-P.; Xiao, X.; Song, X.-P.; Horion, S.; et al. Forest management in southern China generates short term extensive carbon sequestration. *Nat. Commun.* **2020**, *11*, 1–10. [CrossRef] [PubMed]
5. Pecchi, M.; Marchi, M.; Burton, V.; Giannetti, F.; Moriondo, M.; Bernetti, I.; Bindi, M.; Chirici, G. Species distribution modelling to support forest management. A literature review. *Ecol. Model.* **2019**, *411*, 108817. [CrossRef]

6. Nicolescu, V.-N.; Rédei, K.; Mason, W.L.; Vor, T.; Pöetzelsberger, E.; Bastien, J.-C.; Brus, R.; Benčat, T.; Đodan, M.; Cvjetkovic, B.; et al. Ecology, growth and management of black locust (*Robinia pseudoacacia* L.), a non-native species integrated into European forests. *J. For. Res.* **2020**, *31*, 1081–1101. [CrossRef]

7. Peabody, F.J. A 350-Year-Old American Legume in Paris. *Castanea* **1982**, *47*, 99–104. Available online: http://www.jstor.org/stable/4033219 (accessed on 6 November 2020).

8. Arianoutsou, M.; Bazos, I.; Delipetrou, P.; Kokkoris, Y. The alien flora of Greece: Taxonomy, life traits and habitat preferences. *Biol. Invasions* **2010**, *12*, 3525–3549. [CrossRef]

9. Dini-Papanastasi, O.; Arianoutsou, M.; Papanastasis, V.P. *Robinia pseudoacacia* L.: A dangerous invasive alien or a useful multi-purpose tree species in the Mediterranean environment? In *Ecology, Conservation & Management of Mediterranean Climate Ecosystems, Proceedings of the 10th MEDECOS Conference, Rhodes, Greece, 25 April–1 May 2004*; Arianoutsou, M., Papanastasis, V.P., Eds.; Millpress: Rotterdam, The Netherlands, 2004; pp. 42–59.

10. Sitzia, T.; Cierjacks, A.; de Rigo, D.; Caudullo, G. Robinia pseudoacacia in Europe: Distribution, habitat, usage and threats. In *European Atlas of Forest Tree Species*; San-Miguel-Ayanz, J., de Rigo, D., Caudullo, G., Houston Durrant, T., Mauri, A., Eds.; Publications Office of the EU: Luxembourg, 2016; p. e014e79+.

11. Dorado-Liñán, I.; Piovesan, G.; Martínez-Sancho, E.; Gea-Izquierdo, G.; Zang, C.; Cañellas, I.; Castagneri, D.; Di Filippo, A.; Gutiérrez, E.; Ewald, J.; et al. Geographical adaptation prevails over species-specific determinism in trees' vulnerability to climate change at Mediterranean rear-edge forests. *Glob. Chang Biol.* **2019**, *25*, 1296–1314. [CrossRef] [PubMed]

12. Thurm, E.A.; Hernández, L.; Baltensweiler, A.; Ayan, S.; Rasztovits, E.; Bielak, K.; Zlatanov, T.M.; Hladnik, D.; Balic, B.; Freudenschuss, A.; et al. Alternative tree species under climate warming in managed European forests. *For. Ecol. Manag.* **2018**, *430*, 485–497. [CrossRef]

13. Dyderski, M.; Paź-Dyderska, S.; Frelich, L.; Jagodziński, A.M. How much does climate change threaten European forest tree species distributions? *Glob. Chang. Biol.* **2017**, *24*, 1150–1163. [CrossRef]

14. Puchałka, R.; Dyderski, M.K.; Vítková, M.; Sádlo, J.; Klisz, M.; Netsvetov, M.; Prokopuk, Y.; Matisons, R.; Mionskowski, M.; Wojda, T.; et al. Black locust (*Robinia pseudoacacia* L.) range contraction and expansion in Europe under changing climate. *Glob. Chang. Biol.* **2020**, *27*, 1587–1600. [CrossRef]

15. Klisz, M.; Puchałka, R.; Netsvetov, M.; Prokopuk, Y.; Vítková, M.; Sádlo, J.; Matisons, R.; Mionskowski, M.; Chakraborty, D.; Olszewski, P.; et al. Variability in climate-growth reaction of Robinia pseudoacacia in Eastern Europe indicates potential for acclimatisation to future climate. *For. Ecol. Manag.* **2021**, *492*, 119194. [CrossRef]

16. Keskin, T.; Makineci, E. Some soil properties on coal mine spoils reclaimed with black locust (*Robinia pceudoacacia* L.) and umbrella pine (*Pinus pinea* L.) in Agacli-Istanbul. *Environ. Monit. Assess.* **2008**, *159*, 407–414. [CrossRef] [PubMed]

17. Usuga, J.C.L.; Rodríguez-Toro, J.A.; Alzate, M.V.R.; Tapias, D.J.L. Estimation of biomass and carbon stocks in plants, soil and forest floor in different tropical forests. *For. Ecol. Manag.* **2010**, *260*, 1906–1913. [CrossRef]

18. Brundu, G.; Pauchard, A.; Pyšek, P.; Pergl, J.; Brunori, A.; Canavan, S.; Campagnaro, T.; Celesti-Grapow, L.; de Sá Dechoum, M.; Dufour-Dror, J.-M.; et al. Global guidelines for the use of non-native trees: Awareness, sustainable use, invasion risk prevention and mitigation. *NeoBiota* **2020**, *61*, 65–116. [CrossRef]

19. Vítková, M.; Sádlo, J.; Roleček, J.; Petřík, P.; Sitzia, T.; Müllerová, J.; Pyšek, P. *Robinia pseudoacacia*-dominated vegetation types of Southern Europe: Species composition, history, distribution and management. *Sci. Total. Environ.* **2019**, *707*, 134857. [CrossRef]

20. Mantovani, D.; Veste, M.; Bohm, C.; Vignudelli, M.; Freese, D. Spatial and temporal variation of drought impact on black locust (*Robinia pseudoacacia* L.) water status and growth. *iFor. Biogeosci. For.* **2015**, *8*, 743–747. [CrossRef]

21. De Gomez, T.; Wagner, M.R. Culture and Use of Black Locust. *HortTechnology* **2001**, *11*, 279–288. [CrossRef]

22. Moshki, A.; Lamersdorf, N.P. Symbiotic nitrogen fixation in black locust (*Robinia pseudoacacia* L.) seedlings from four seed sources. *J. For. Res.* **2011**, *22*, 689–692. [CrossRef]

23. Liu, Z.; Hu, B.; Bell, T.L.; Flemetakis, E.; Rennenberg, H. Significance of mycorrhizal associations for the performance of N2-fixing Black Locust (*Robinia pseudoacacia* L.). *Soil Biol. Biochem.* **2020**, *145*, 107776. [CrossRef]

24. Cierjacks, A.; Kowarik, I.; Joshi, J.; Hempel, S.; Ristow, M.; von der Lippe, M.; Weber, E. Biological Flora of the British Isles: *Robinia pseudoacacia*. *J. Ecol.* **2013**, *101*, 1623–1640. [CrossRef]

25. Vítková, M.; Müllerová, J.; Sádlo, J.; Pergl, J.; Pyšek, P. Black locust (*Robinia pseudoacacia*) beloved and despised: A story of an invasive tree in Central Europe. *For. Ecol. Manag.* **2016**, *384*, 287–302. [CrossRef]

26. Papachristou, T.G.; Platis, P.D.; Papachristou, I.; Samara, T.; Spanos, I.; Chavales, E.; Bataka, A. How the structure and form of vegetation in a black locust (*Robinia pseudoacacia* L.) silvopastoral system influences tree growth, forage mass and its nutrient content. *Agrofor. Syst.* **2020**, *94*, 2317–2330. [CrossRef]

27. Ministry of Rural Development and Food. *Forestry Activities from Third Framework Program Funding, Regional Operational Programmers and National Funds for the Year 2004*; Ministry of Rural Development and Food: Athens, Greece, 2004; p. 78.

28. Brown, S. *Estimating Biomass and Biomass Change of Tropical Forests: A Primer*; FAO Forestry Paper 134; Food and Agriculture Organization of the United Nations: Rome, Italy, 1997; p. 55.

29. Kim, T.J.; Bronson, P.B.; Wijaya, A. Spatial interpolation of above-ground biomass in labanan concession forest in east kalimantan, Indonesia. *Math. Comput. For. Nat. Res. Sci.* **2016**, *8*, 27–39.

30. Theofanous, N.; Chrysafis, I.; Mallinis, G.; Domakinis, C.; Verde, N.; Siahalou, S. Aboveground Biomass Estimation in Short Rotation Forest Plantations in Northern Greece Using ESA's Sentinel Medium-High Resolution Multispectral and Radar Imaging Missions. *Forests* **2021**, *12*, 902. [CrossRef]

31. Chen, Q.; McRoberts, R.E.; Wang, C.; Radtke, P.J. Forest aboveground biomass mapping and estimation across multiple spatial scales using model-based inference. *Remote. Sens. Environ.* **2016**, *184*, 350–360. [CrossRef]

32. Mitchard, E.T.A.; Feldpausch, T.R.; Brienen, R.J.W.; Lopez-Gonzalez, G.; Monteagudo, A.; Baker, T.R.; Lewis, S.L.; Lloyd, J.; Quesada, C.A.; Gloor, M.; et al. Markedly divergent estimates of Amazon Forest carbon density from ground plots and satellites. *Glob. Ecol. Biogeogr.* **2014**, *23*, 935–946. [CrossRef] [PubMed]

33. World Bank. *Assessment of Innovative Technologies and Their Readiness for Remote Sensing-Based Estimation of Forest Carbon Stocks and Dynamics*; World Bank: Washington, DC, USA, 2021. [CrossRef]

34. Sales, M.H.; Souza, C.M.; Kyriakidis, P.; Roberts, D.A.; Vidal, E. Improving spatial distribution estimation of forest biomass with geostatistics: A case study for Rondônia, Brazil. *Ecol. Model.* **2007**, *205*, 221–230. [CrossRef]

35. Li, Y.; Li, M.; Liu, Z.; Li, C. Combining Kriging Interpolation to Improve the Accuracy of Forest Aboveground Biomass Estimation Using Remote Sensing Data. *IEEE Access* **2020**, *8*, 128124–128139. [CrossRef]

36. Paletto, A.; Tosi, V. Deadwood density variation with decay class in seven tree species of the Italian Alps. *Scand. J. For. Res.* **2010**, *25*, 164–173. [CrossRef]

37. Zianis, D.; Spyroglou, G.; Tiakas, E.; Radoglou, K. Bayesian and classical models to predict aboveground tree biomass allometry. *For. Sci.* **2016**, *62*, 247–259. [CrossRef]

38. R Development Core Team. *A Language and Environment for Statistical Computing*; R Foundation for Statistical Computing: Vienna, Austria, 2020; Available online: http://www.rproject.org/ (accessed on 6 September 2020).

39. Ribeiro, P.J., Jr.; Diggle, P.J. geoR: A package for geostatistical analysis. *R-NEWS* **2001**, *1*, 15–18.

40. Dale, M.R.T.; Fortin, M.J. *Spatial Analysis: A Guide for Ecologists*; Cambridge University Press: Cambridge, UK, 2005; p. 365.

41. Du, H.; Zhou, G.; Fan, W.; Ge, H.; Xu, X.; Shi, Y.; Fan, W. Spatial heterogeneity and carbon contribution of aboveground biomass of moso bamboo by using geostatistical theory. *Plant Ecol.* **2009**, *207*, 131–139. [CrossRef]

42. Mallarino, A.P.; Wittry, D.J. Efficacy of Grid and Zone Soil Sampling Approaches for Site-Specific Assessment of Phosphorus, Potassium, pH, and Organic Matter. *Precis. Agric.* **2004**, *5*, 131–144. [CrossRef]

43. Lamsal, S.; Rizzo, D.M.; Meentemeyer, R. Spatial variation and prediction of forest biomass in a heterogeneous landscape. *J. For. Res.* **2012**, *23*, 13–22. [CrossRef]

44. Jaya, I.N.S. The Interpolation Method for Estimating the Above-Ground Biomass Using Terrestrial-Based Inventory. *J. Manaj. Hutan Trop.* **2014**, *20*, 121–130. Available online: https://journal.ipb.ac.id/index.php/jmht/article/view/8445 (accessed on 27 October 2021). [CrossRef]

45. Akhavan, R.; Amiri, G.Z.; Zobeiri, M. Spatial variability of forest growing stock using geostatistics in the Caspian region of Iran. *Caspian J. Environ. Sci.* **2010**, *8*, 43–53.

46. Rédei, K.; Gal, J.; Keserű, Z.; Antal, B. Above-Ground Biomass of Black Locust (*Robinia pseudoacacia* L.) Trees and Stands. *Acta Silv. Lign. Hung.* **2017**, *13*, 113–124. [CrossRef]

47. Wang, J.J.; Hu, C.X.; Bai, J.; Gong, C.M. Carbon sequestration of mature black locust stands on the Loess Plateau, China. *Plant Soil Environ.* **2016**, *61*, 116–121. [CrossRef]

48. Lukić, S.; Pantić, D.; Simić, S.; Borota, D.; Tubić, B.; Djukic, M.; Djunisijević-Bojović, D. Effects of black locust and black pine on extremely degraded sites 60 years after afforestation—A case study of the Grdelica Gorge (Southeastern Serbia). *iForest Biogeosci. For.* **2016**, *9*, 235–243. [CrossRef]

49. Dutta, R.K.; Agrawal, M. Restoration of opencast coal mine spoil by planting exotic tree species: A case study in dry tropical region. *Ecol. Eng.* **2003**, *21*, 143–151. [CrossRef]

50. Filcheva, E.; Noustorova, M.; Gentcheva-Kostadinova, S.; Haigh, M. Organic accumulation and microbial action in surface coal-mine spoils, Pernik, Bulgaria. *Ecol. Eng.* **2000**, *15*, 1–15. [CrossRef]

51. Carl, C.; Biber, P.; Landgraf, D.; Buras, A.; Pretzsch, H. Allometric Models to Predict Aboveground Woody Biomass of Black Locust (*Robinia pseudoacacia* L.) in Short Rotation Coppice in Previous Mining and Agricultural Areas in Germany. *Forests* **2017**, *8*, 328. [CrossRef]

52. Woś, B.; Pająk, M.; Krzaklewski, W.; Pietrzykowski, M. Verifying the Utility of Black Locust (*Robinia pseudoacacia* L.) in the Reclamation of a Lignite Combustion Waste Disposal Site in Central European Conditions. *Forests* **2020**, *11*, 877. [CrossRef]

53. Kraszkiewicz, A. Productivity of Black Locust (*Robinia pseudoacacia* L.) Grown on a Varying Habitats in Southeastern Poland. *Forests* **2021**, *12*, 470. [CrossRef]

 *forests*

Communication

Evaluation of Early Bark Beetle Infestation Localization by Drone-Based Monoterpene Detection

Sebastian Paczkowski [1,*], Pawan Datta [2], Heidrun Irion [2], Marta Paczkowska [1], Thilo Habert [1], Stefan Pelz [3] and Dirk Jaeger [1]

[1] Department of Forest Work Science and Engineering, Georg August University Göttingen, Büsgenweg 4, 37077 Göttingen, Germany; martapaczkowska@interia.pl (M.P.); t.habert@stud.uni-goettingen.de (T.H.); dirk.jaeger@uni-goettingen.de (D.J.)
[2] Chair of Remote Sensing and Landscape Information Systems, Albert-Ludwig-University Freiburg, Tennenbacherstr. 4, 79106 Freiburg, Germany; pawan.datta@felis.uni-freiburg.de (P.D.); heidrun.irion@students.uni-freiburg.de (H.I.)
[3] Chair of Forest Utilization, University of Applied Forest Science Rottenburg, Schadenweilerhof, 72108 Rottenburg, Germany; Pelz@hs-rottenburg.de
* Correspondence: Sebastian.Paczkowski@uni-goettingen.de; Tel.: +49-551-392-3574

Abstract: The project PROTECT^FOREST deals with improvements in early bark beetle (e.g., *Ips typographus* and *Pityogenes chalcographus*) detection to allow for fast and effective response to initial infestation. The removal of trees in the early infestation stage can prohibit bark beetle population gradation and successive timber price decrease. A semiconductor gas sensor array was tested in the lab and attached to a drone under artificial and real-life field conditions. The sensor array was able to differentiate between α-pinene amounts and between different temperatures under lab conditions. In the field, the sensor responded to a strong artificial α-pinene source. The real-life field trial above a spruce forest showed preliminary results, as technical and environmental conditions compromised a proof of principle. Further research will evaluate the detection rate of infested trees for the new proposed sensor concept.

Keywords: UAV; VOC; drone sensor; semiconductor metal oxide gas sensors; alpha pinene

Citation: Paczkowski, S.; Datta, P.; Irion, H.; Paczkowska, M.; Habert, T.; Pelz, S.; Jaeger, D. Evaluation of Early Bark Beetle Infestation Localization by Drone-Based Monoterpene Detection. *Forests* **2021**, *12*, 228. https://doi.org/10.3390/f12020228

Academic Editor: Angela Lo Monaco

Received: 15 January 2021
Accepted: 11 February 2021
Published: 16 February 2021

Publisher's Note: MDPI stays neutral with regard to jurisdictional claims in published maps and institutional affiliations.

1. Introduction

Climate change and the corresponding temperature increase have an impact on the stability of forest ecosystems, especially on the distribution and activity of xylophagous beetles [1,2]. The economic effect of a higher xylophagous beetle activity was quantified for Canada, with an average annual loss of 1274 million dollars of the net present value between 2009 and 2054 [3]. In the USA, the annual loss was 1500 million dollars between 1971 and 2000 (equivalent to 41% of total annual loss including fire, hurricane, tornado, ice, invasive species, landslides, and drought) [4]. In Germany, infestation from the last three years (2018–2020) led to wood loss of 178 mio m^3 [5], which is 4.9% of the wood stock calculated by the national forest inventory in 2012 [6].

The bark beetles *Ips typographus* and *Pityogenes chalcographus* show very high fertility in hot and dry seasons [2,7]. They can have up to four generations per year at an optimal temperature of 30 °C [8]. Detecting such a pest outbreak between four months and one year after the initial infestation, e.g., by alternating crown color [9], is far too late, as the bark beetle population multiplies by a factor of approximately 203–1003 per female individuum [8,10–12]. It was reported that the detection of beetle infestation at a late stage within the spectral range of 450–890 nm was possible with a prediction efficiency of 64%. This approach is very sensitive to interactions with direct and scattered sunlight [13]. Comparable results were presented for the utilization of a hyperspectral camera [14]. This is a limited approach for pest control of aggressive species, e.g., *Ips typographus*, because early detection of green stage infestation is needed to prevent a gradation. Detecting trees in the red stage of infestation at low detection rates cannot prevent a gradation, though it could indicate larger areas with high infestation risk. Another approach is the utilization of satellite data for infestation extrapolation based on the easily identifiable grey infestation stage [15]. Although early detection methods are necessary in order to reduce timber losses caused by frequent bark beetle gradations, holistic approaches can allow a prediction system to narrow down the possible areas of infestation to optimize the search for green stage infestation.

Monoterpenes and especially α-pinene are considered to have an impact on global air quality [16], with forests as a major emission source [16–18]. A model study of Guenther et al. estimated the annual global forest volatile organic compound (VOC) flux at 1150 Tg C, with a share of 11% monoterpenes. A study by Berg et al. modeled the bark beetle attack with a low impact on overall forest emissions, although the emission impact of bark beetles on the overall tree mortality emissions was very high [19]. Monoterpene emissions differed between forest plant species [17]. In healthy conifer trees, the monoterpene emission is a function of the temperature-dependent vapor pressure that can be explained by the Clausius Clapeyron equation [20–22]. α-pinene was found to be emitted in a diurnal cycle in forests [16]. This shows that, in healthy plants, monoterpenes are emitted passively through the stomata.

Monoterpenes belong to the group of plant VOCs that increase in concentration after plant damage (inducible VOCs) [23] or drought stress [22,24]. Coniferous trees can emit increasing amounts of monoterpenes after an attack of xylophagous beetles by induced host defense mechanisms [25–28]. These emissions are caused by leaking resin from physically damaged resin channels below the bark and by the general stress response of the tree. The resin consists of resin acids and monoterpenes. The monoterpenes are a nonpolar solvent for resin acids that decreases their viscosity and enables their flow into the bark hole. There, the solvent evaporates, the resin acids start to polymerize, and the hole closes. This mechanism enables the tree to defend itself by "resinating" the beetle out of its bore hole and by successively closing the bore hole. The general stress response comprises the increase in monoterpene cyclase levels after damage to a 5 to 15 times higher level. Therefore, trees produce monoterpenes after physical damage to restore their host defense capability [29] and to passively emit more monoterpenes through the stomata [20–22]. The major resin components in economically relevant coniferous trees in boreal and temperate forests are monoterpenes and, among the group of monoterpenes, especially α-pinene [11,26–28,30]. This compound is an attractant for bark beetles as well [11,31]. It is usually emitted by infested trees in the early or "green" stage of the pest attack [25,27]. However, an increased emission in the yellow stage was reported as well [26]. Drought stress also has a strong impact on α-pinene emission. The effect of draught is discussed controversially in the literature, as both increases and decreases in emission rates were detected [22,24]. Table 1 shows an overview on the literature that quantified emissions from infested forest trees or forest trees under drought stress.

Bark beetle pheromones are produced and dispersed by the beetles and can be a cue for individuals of the same species to gather or disperse according to the nutritional state of the host plant [12,31,32]. Interspecific pheromone interactions among bark beetle species were reported as well [33]. The chemical ecology of bark beetles was reviewed by Blomquist et al., by Tittinger and Bloquist [31,34], and by earlier sources [35–37]. The concentration of bark beetle pheromones is several orders of magnitude lower than the concentration of α-pinene in forests [38]. The beetles are comprised of a biological detection system that is based on odorant-binding proteins (OBPs). The OBPs allow the selective binding of VOCs and their transportation to neurons in the beetles antennae for information processing [39–41]. Binding sites of OBPs and especially of the subgroup pheromone-binding proteins (PBPs) can be highly selective and low VOC concentrations can elicit a neuronal response [41].

Table 1. Emission comparison between infested and not infested tree species (above) and emission comparison between tree species with and without drought stress (below) in the literature. [a] Emission ratio = emission (infested or drought stress)/emission (not infested or no drought stress). [b] The green infested stage always showed lower emission in comparison to the not infested stage.

Tree Species (Specification)	Compound (Unit)	Emission (Not Infested)	Emission (Infested)	Infestation Stage	Emission Ratio [a]	Source
Pinus contorta (canopy)	α-pinene ($ng \times L^{-1}$)	1.1 ± 0.56 2.6 ± 0.9	1.5 ± 0.6 2.7 ± 1.1	unknown	1.4 1.0	[28]
Pseudotsuga menziesii (lower branches)	α-pinene ($ng \times h^{-1} \times g_{FM}^{-1}$)	322 ± 166	813 ± 482	Green	2.5	[27]
Pseudotsuga menziesii (lower branches)	Total terpenes ($ng \times h^{-1} \times g_{FM}^{-1}$)	870 ± 417 1515 ± 737 1695 ± 797 571 ± 226 284 ± 59 557 ± 251 842 ± 268 948 ± 684	4472 ± 2759 1472 ± 834 5881 ± 3685 2480 ± 1094 462 ± 189 229 ± 68 724 ± 243 2124 ± 1782	Green	5.1 1.0 3.4 4.3 1.6 0.4 0.9 2.2	[27]
Pseudotsuga menziesii (lower branches)	α-pinene ($ng \times h^{-1} \times g_{FM}^{-1}$)	77 100 60 70	190 190 134 184	Yellow [b]	2.5 1.9 2.2 2.6	[26]

Tree Species	Compound	Emission (No Draught)	Emission (Draught)	Draught Intensity	Emission Ratio [a]	Source
Pinus halepensis (seedlings)	α-pinene ($\mu g \times h^{-1} \times g_{DM}^{-1}$)	23.9 ± 5.8 5.6 ± 0.7 8.8 ± 3.2 30.7 ± 4.5 10.9 ± 4.5 34.2 ± 7.5	17.9 ± 1.3 16.4 ± 5.3 4.0 ± 1.3 20.6 ± 4.8 21.2 ± 5.8 21.4 ± 4.8	2 weeks 4 weeks 6 weeks 8 weeks 10 weeks 12 weeks	0.7 2.9 0.5 0.7 1.9 0.6	[22] [22] [22] [22] [22] [22]
Pinus halepensis (seedlings)	α-pinene ($\%_{total\ VOC}$)	30.8 ± 10.6	60.0 ± 14.3	1 week	1.9	[24]

Semiconductor metal-oxide gas sensors are feasible detectors for volatile organic compound (VOCs) classes in air, like monoterpenes. A semiconductor surface, e.g., tungsten oxide, changes its conductance when it is heated. At a constant and high temperature, e.g., 275 °C, the oxidation state of the oxygen atoms in the molecular tungsten oxide grid determines the conductance. When VOCs interact with the heated semiconductor tungsten oxide surface, they are thermally oxidized. The oxygen atoms in the tungsten oxide grid can also change their oxidation state due to the electron exchange in the thermally induced conversion of the VOCs. This electron exchange can oxidize or reduce the oxygen in the metal oxide. This leads to a conductance decrease in the metal oxide. The intensity of this decrease is usually not VOC-specific but VOC class-specific [42]. The VOC class selectivity can be improved by using several metal oxide sensors that exhibit different reaction intensities, thereby creating a VOC class-specific signal pattern. Earlier studies used a gas sensor below a quadcopter in the rotor down winds to obtain signals while flying through a large smoke plume [43]. This study investigates the monoterpene detection performance of an quadcopter semiconductor gas sensor system using α-pinene as a model compound. The sensor setup is based on three semiconductor gas sensors with different surface properties and a sampling option to sample the undisturbed air outside the rotor down winds.

2. Materials and Methods

2.1. Sensor Calibration

Synthetic air (100 °C, $1.5 \text{ L} \times min^{-1}$, 5.5 purity, 80% N_2, 20% O_2, Messer Industriegase GmbH, Siegen, Germany) was mixed with a molar 10^{-2} α-pinene dilution at different flow rates. The α-pinene flow rate was controlled by a syringe pump (LA 30, Landgraf Laborsysteme HLL GmbH, Langenhagen, Germany) at pump rates between $1.5 \text{ }\mu L \times min^{-1}$ to $15 \text{ mL} \times min^{-1}$. Due to the constant synthetic air flow, final molar concentrations between 10^{-8} to 10^{-4} were mixed. This α-pinene gas flow was led over the surface of three heated metal oxide sensors (GGS1330, metal: SnO_2, 450 °C, 5 V, and 26 Ω; GGS2330, metal: SnO_3, 400 °C, 4.5 V, and 25 Ω; and GGS5330, metal: WO_2, 275 °C, 3 V, and 20.5 Ω; UST Umweltsensortechnik, Geschwenda, Germany) for three minutes, and the conductance was measured (signal processing technology by Cadmium GmbH, Regenstauf, Germany). Between stimuli, the mixing chamber, the heated connection tubes, and the sensor chamber were purged with synthetic air for three minutes.

Data Analysis of Sensor Calibration

Each α-pinene concentration was tested 13 times. The sensor signal decreased under the presence of the stimulus and reached an equilibrium state. This equilibrium state was maintained for three minutes. The mean sensor values of these three minutes were calculated. All mean values per concentration (10^{-8} to 10^{-4}) and sensor (GGS1330, GGS2330, and GGS5330) were grouped to calculate the mean, standard deviation, median, upper and lower quartiles, minimum and maximum, and the variation coefficient to obtain the indices for signal stability and repeatability of the sensor responses. The mean of the mean values of three-minute stimuli sensor values were used to calculate a linear dose–response correlation for each sensor. All mean values of the three-minute stimuli per sensor were used to calculate a second-grade polynomial function as a dose–response curve for each sensor. This data was also used to calculate the significant differences between adjacent dose–response reactions with the t-test for unpaired samples. The coefficients of determination (R^2) were calculated for all dose–response curves as a quality measure for the correlations.

2.2. Sensor Test

A custom-made wind tunnel was adjusted to a linear wind flow of 1 or 2 m $\times$ s^{-1}. The sensor, as schematically depicted in Figure 1, was placed at a distance of 1.5 or 3 m on the level of the wind tunnel exit. Either 0.5 or 0.1 mL α-pinene (CAS number: 80-56-8, 98% purity, Merck KgaA, Darmstadt, Germany) were dropped on a filter paper (ROTILABO® Typ 113A, Carl Roth, Stuttgart, Germany), put in a closed petri dish, and placed directly in the wind flow at the exit of the wind tunnel. The sensor was activated, and the petri dish was opened for 3 min and closed for 3 min. This procedure was repeated six times. The first stimulus was not considered for the successive data analysis, as the longer preconcentration time always led to higher sensor signals. After each factor combination (distance, wind flow, and α-pinene concentration), the air in the room was exchanged by a high-performance hood. The factor combinations were repeated at room temperatures of 15 °C, 20 °C, and 25 °C.

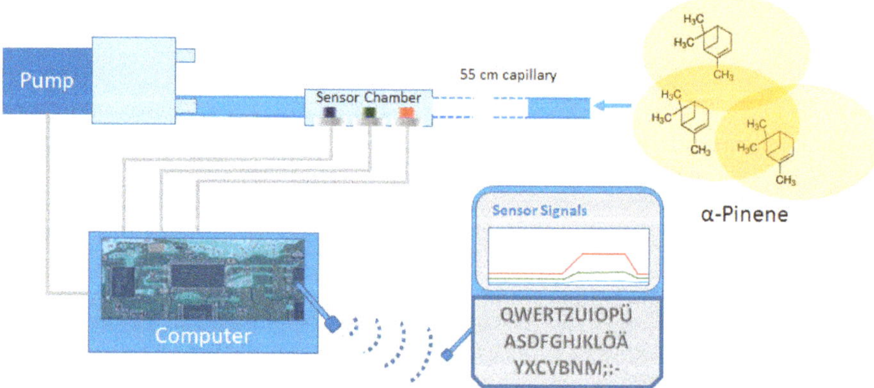

Figure 1. Schematic sensor setup that was used in the laboratory experiment (see Section 2.2) and that was attached to the drone for the field tests (see Sections 2.3 and 2.4).

Data Analysis of Sensor Test

The combination of the four selected factors resulted in 24 variations. Again, the three-minute equilibrium state of the sensor signals were used to calculate the mean sensor signals. All mean sensor signals were grouped to calculate mean, standard deviation, and correlation coefficient.

The sensor signals of GGS1330 and GGS2330 were related to the signal of GGS5330 in order to normalize the signals and, thus, to increase the comparability of the signals between different trials. This should compensate for the baseline shifts of the sensors and should reduce the effect of the baseline variation between the trials. Significant differences between distance, wind flow, and α-pinene concentration were tested with the *t*-test for unpaired samples, resulting in 72 pairwise comparisons. The effect of temperature on the sensor signals was analyzed with the Kruskal–Wallis test and the Conover Iman pairwise comparison with Bonferroni adaption, resulting in 16 triple comparisons. The level of significance was $p < 0.05$.

2.3. Sensor Field Test under Artificial Conditions

A pole that was 4 m high was placed in the botanical garden of the faculty of forest science and forest ecology, University of Göttingen, on 17 March 2020. The sky was clear, and the temperature was 15 °C, with wind speeds below $1 \text{ m} \times \text{s}^{-1}$ during the trials. A 15 × 20 cm tissue paper with approximately 10 mL of α-pinene was placed at the top of the pole. A drone (Hexacopter, Cadmium GmbH, Regenstauf, Germany) was equipped with the gas sensor system used for the lab experiments. A 55 cm long horizontal glass fiber capillary with a 0.8 mm diameter allowed for air sampling outside the rotor winds (Figure 2). The air was sucked by a pump (1420VP BLDC, Thomas Garden Denver, Fürstenfeldbruck, Germany) through this capillary at $1.5 \text{ L} \times \text{min}^{-1}$ and was led over the sensor surfaces. The drone was steered manually at approximately 4 m heights over a flat and rectangular area (approximately 40 m × 20 m), with the pole in the center. The minimum distance to the pole was 50 cm from the end of the horizontal capillary. The experiment was repeated one time (N = 1).

Figure 2. Picture of the PROTECT[FOREST] drone with the sensor system and capillary.

Data Analysis of Sensor Field Test under Artificial Conditions

The georeferenced resistance changes of the three sensors were inverted in order to display sensor reactions as upward peaks. The data were displayed as georeferenced dots, with the color indicating the concentration. The georeferenced points with sampled signal intensities were interpolated with the program ArcGIS Geostatistical Analyst (ESRI, Redlands, CA, USA) to a heat map by gradual point overlay and interpolated raster map generation.

2.4. Sensor Field Test above a Forest Stand

An approximately 40 m × 40 m test stand consisting of 100% spruce trees at different ages between 40 and 80 years at the Windgfällweiher lake, Baden-Württemberg, Germany, was assigned by the forest administration Fürst zu Fürstenberg. The test site was recorded with a multispectral imaging camera (Sequoia Parrot, Pix4D S.A., Prilly, Switzerland) at a height of 80 m with a 90% front and a 80% side overlap in order to create a digital surface model (DSM) with the heights above the ground (drone starting point as reference point) of the forest canopy.

One hundred waypoints were defined over the area, and the assigned height of the waypoints was calculated to be 10 m above the respective heights above the ground, with a lower limit of 20 m. The mikrocopter tool (V2.20, HiSystems GmbH, Moormerland, Germany) was used to transfer the waypoint data to the drone microcontroller.

The weather during the flight mission was slightly cloudy with temperature at 13 °C. The wind was measured with a DJI Mavic 2 Enterprise Zoom (Da-Jiang Innovations Science and Technology Co., Ltd., Nanshan, China) that flew at 27 m heights close to the stand. Wind speed and direction were calculated by a postprocessing IMU (internal measurement unit) data analysis algorithm (AirData, El Dorado Hills, CA, USA). The stand was searched for bark beetle infestation and trees with resin flow caused by physical damage. It was only possible to assess the status of the trees in the lower and middle stem compartments by sight; some trees with many branches in the lower stem compartment could not be completely assessed. All trees in the sampled area were georeferenced using the LogBuch GPS system (SDP Digitale Produkte GmbH, Waiblingen, Germany). The experiment was repeated one time (N = 1).

Data Analysis of Sensor Field Test above a Forest Stand

The DSM was generated by creating digital orthophotos (DOPs) using the software Metashape (V1.6.3, Agisoft, St. Petersburg, Russia). The resulting point cloud was used to calculate the height of each point with the reference of the drone starting position set to 0. In order to highlight damaged trees on the map, individual tree detection (ITD) was performed by identifying local minima in the inverted DSM using the watershed segmentation algorithm (ESRI, Redlands, CA, USA).

The recorded sensor data were displayed with geocoordinates as labels and normalized with the minimum and maximum sensor values for each sensor using R (V.4.0.3, R Core Team). The resulting sensor values were multiplied with each other to obtain a combined sensor response. The sensor values were canopy distance-corrected by assigning weights to the distance of the drone to the canopy in order to compensate for differences between programmed and actual flight paths. The values were normalized again to the mean of the minimum and maximum values to values between −0.75 and 0.26. The georeferenced sensor data were merged with the DSM using ArcGIS Geostatistical Analyst (ESRI, Redlands, CA, USA) to compute an interpolated heat map (universal kriging algorithm assuming autocorrelated errors to trend function): 80% of the sensor data was used to calculate the interpolation, and 20% was used to validate the accuracy of the interpolation.

3. Results

3.1. Sensor Calibration

The results of the sensor calibration are shown in Figure 3. The sensor conductance was increasingly reduced with increasing stimulus concentration, while concentrations below 10^{-6} did not show a clear and repeatable response and are, therefore, not displayed. At the exposition of the 10^{-6} concentration, only 7 out of 13 repetitions showed a clear response. The correlation between the stimuli means was very high (Table 2), but the variation in sensor values was high as well (Figure 3 and Table 2). A linear regression showed an acceptable correlation coefficient (Table 2). The GGS2330 showed the highest response to α-pinene.

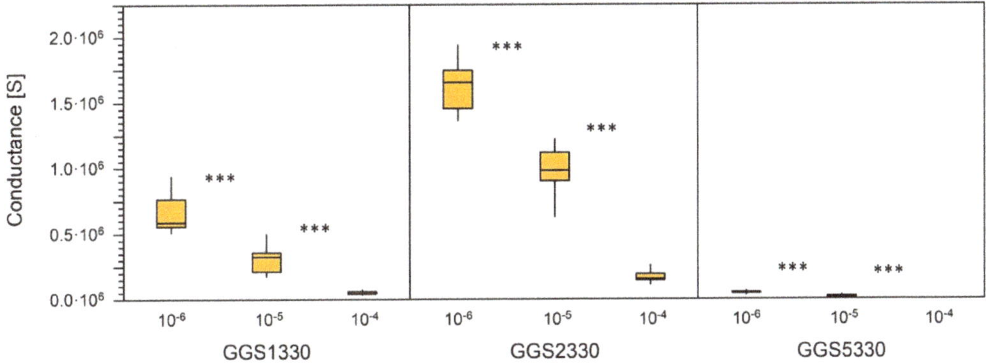

Figure 3. Sensor response (median, upper and lower quartiles, minimum, and maximum) to three concentrations of α-pinene (10^{-6}, 10^{-5}, and 10^{-4}) for each sensor (GGS1330, GGS2330, and GGS5330), with N = 13: the conductance was reduced with increasing stimulus concentration. Concentrations below 10^{-6} are not shown as they did not elicit a sensor response. Significant differences between responses to adjacent concentrations were calculated with the *t*-test for unpaired samples and are indicated by stars (*** $p < 0.001$).

Table 2. Descriptive statistics (mean, standard deviation, and variation coefficient) and dose–response curve regression analysis with coefficients of determination (R^2) of the sensor responses (mean value over a three-minute stimulus) to the selected α-pinene dilutions of 10^{-6}, 10^{-5}, and 10^{-4}.

	10^{-6}	10^{-5}	10^{-4}
	GGS1330, N = 13		
Mean (S)	669,022	308,851	51,870
Standard deviation (S)	147,007	114,605	11,260
Variation coefficient (%)	22	37	22
R^2 of mean		$y = 51{,}595x^2 - 514{,}955x + 1{,}132{,}382 \ R^2 = 1.00$	
R^2 of all values (linear)		$y = 1{,}171{,}683x - 302{,}126 \ R^2 = 0.83$	
	GGS2330, N = 13		
Mean (S)	1,622,037	990,640	171,820
Standard deviation (S)	193,071	15,4917	40,674
Variation coefficient (%)	12	16	24
Linear R^2 of mean		$y = -93{,}711x^2 - 350{,}263x + 2{,}066{,}011 \ R^2 = 1.00$	
R^2 of all values (linear)		$y = 2{,}748{,}138x - 736{,}822 \ R^2 = 0.94$	
	GGS5330, N = 13		
Mean (S)	47,119	16,323	1551
Standard deviation (S)	10,525	6814	538
Variation coefficient (%)	22	42	35
Linear R^2 of mean		$y = 8011x^2 - 54{,}830x + 93{,}938 \ R^2 = 1.00$	
R^2 of all values (linear)		$y = -87{,}914x + 21{,}782 \ R^2 = 0.83$	

Forests **2021**, *12*, 228

3.2. Sensor Test under Lab Condition

The sensor test under lab conditions showed low variations in the five stimuli for each parameter combination (Table 3). In general, an increasing temperature decreased the signal relation between GGS1330 and GGS2330 to GGS5330 (Figure 4). Most of these differences were significant or highly significant, with GGS1330 at 0.5 mL, 3 m, and 2 m $\times$ s^{-1} and GGS2330 at 0.5 mL, 3 m, and 2 m $\times$ s^{-1} as the only exceptions. The higher α-pinene amount (0.5 mL) caused a higher signal relation at all temperatures in comparison to 0.1 mL. Out of the 24 compared data sets, 15 data sets (60%) showed significant or highly significant differences.

When using 0.5 mL α-pinene, an increase in the wind flow caused an increased sensor signal relation at 15 °C, while at higher temperatures, the wind flow had a decreasing effect. However, all data sets showed no significant differences between the two wind flows. The distance to the source had an inconsistent effect at 20 °C for six out of eight data sets recorded with 0.5 mL α-pinene. Four data sets showed a significant increase in the values and two showed significant decreases, which showed that there was no repeatable correlation between the two selected distances.

When using 0.1 mL α-pinene, the influence of the wind flow was low. Only two data sets out of 12 showed a significant difference (both GGS1330 and GGS2330 at 20 °C, 0.1 mL, and 1.5 m). At short distance, there was a tendency for higher signal relations (Figure 4), which complemented the tendencies of the 0.5 mL α-pinene amount.

Table 3. The mean sensor response to the selected combinations of temperature (°C) and amount of α-pinene (mL) (row labels) and distance to α-pinene source (m) and wind flow (m/s) (column labels): for each parameter combination, five sensor signal means over a three-minute stimulus (S), standard deviation (S), and variation coefficient (%) are shown. The three data values are displayed above each other in the respective order for each data set. Low values correspond to high responses.

°C	Amount α-pinene (mL)	GGS1330 (S), N = 5				GGS2330 (S), N = 5				GGS5330 (S), N = 5			
		1.5 m; 1 m/s	1.5 m; 2 m/s	3 m; 1 m/s	3 m; 2 m/s	1.5 m; 1 m/s	1.5 m; 2 m/s	3 m; 1 m/s	3 m; 2 m/s	1.5 m; 1 m/s	1.5 m; 2 m/s	3 m; 1m/s	3 m; 2 m/s
15	0.1	78.383 3.648 5%	89.711 1.663 2%	79.732 1.669 2%	91.260 1.449 2%	73.774 4.160 6%	71.256 2.965 4%	77.092 1.121 1%	86.777 3.056 4%	9.107 711 8%	10.700 345 3%	9.567 321 3%	13.508 689 5%
	0.5	100.692 1.742 2%	96.687 2.850 3%	92.099 1.954 2%	89.899 2.556 3%	85.104 3.467 4%	61.596 3.668 6%	80.888 2.518 3%	77.253 2.263 3%	9.535 759 8%	6.773 249 4%	8.116 374 5%	8.857 520 6%
20	0.1	81.612 2.430 3%	80.852 2.567 3%	74.108 2.184 3%	88.336 1.241 1%	82.872 3.098 4%	76.040 6.387 8%	77.313 2.465 3%	93.394 2.877 3%	14.005 901 6%	11.515 482 4%	10.917 365 3%	16.927 538 3%
	0.5	64.598 2.165 3%	73.860 2.577 3%	77.323 3.070 4%	74.619 2.350 3%	69.516 3.243 5%	59.824 1.711 3%	78.921 2.986 4%	75.072 1.373 2%	7.755 323 4%	8.280 577 7%	9.721 278 3%	10.165 275 3%
25	0.1	57.741 787 1%	70.578 1.754 2%	77.092 2.472 3%	73.391 1.328 2%	69.281 3.098 4%	75.375 3.199 4%	92.740 6.931 7%	87.419 3.379 4%	10.042 1.030 10%	13.602 499 4%	18.530 1.654 9%	16.703 398 2%
	0.5	54.504 4.158 8%	63.156 3.247 5%	68.440 2.804 4%	66.394 2.112 3%	63.844 4.094 6%	62.421 3.408 5%	82.840 1.203 1%	79.421 987 1%	7.813 588 8%	8.494 514 6%	12.276 919 7%	11.734 737 6%

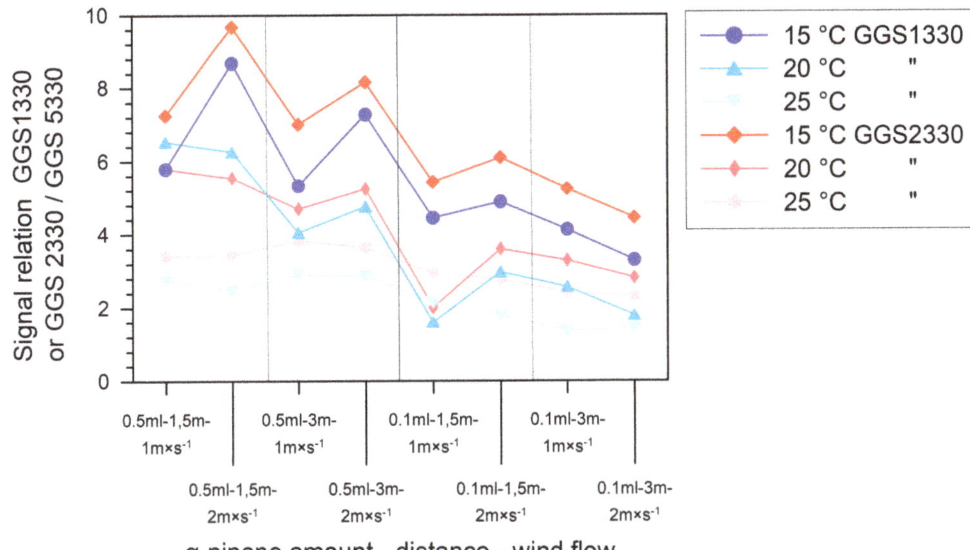

Figure 4. Sensor response of GGS1330 and GGS2330 in relation to GGS5330 (dimensionless) at different temperatures (°C) for the selected factor combinations of α-pinene amount (mL), distance to α-pinene source (m), and wind speed (m × s^{-1}). Lines between the points do not indicate interpolation but improve readability. N was 5 for all combinations.

3.3. Sensor Field Test under Artificial Conditions

The sensor field test under artificial conditions showed a response of all three sensors close to the pole with the strong α-pinene source (Figure 5). Sensor GGS1330 showed signals while flying north of the test field, and sensor GGS5330 showed signals scattered over the area. All three sensors responded in the northeast corner of the test field.

3.4. Sensor Field Test above a Forest Stand

Wind measurement with the drone resulted in an average wind speed of 4.5 m × s^{-1} from northwest. The heat map was calculated based on an interpolation of all three sensor signals (universal kriging, accuracy R^2 = 0.477) and is shown in Figure 6. The calculated tree crown shapes are indicated with black lines. The infested trees and the trees with physical damage, which showed resin flow, are marked with dots.

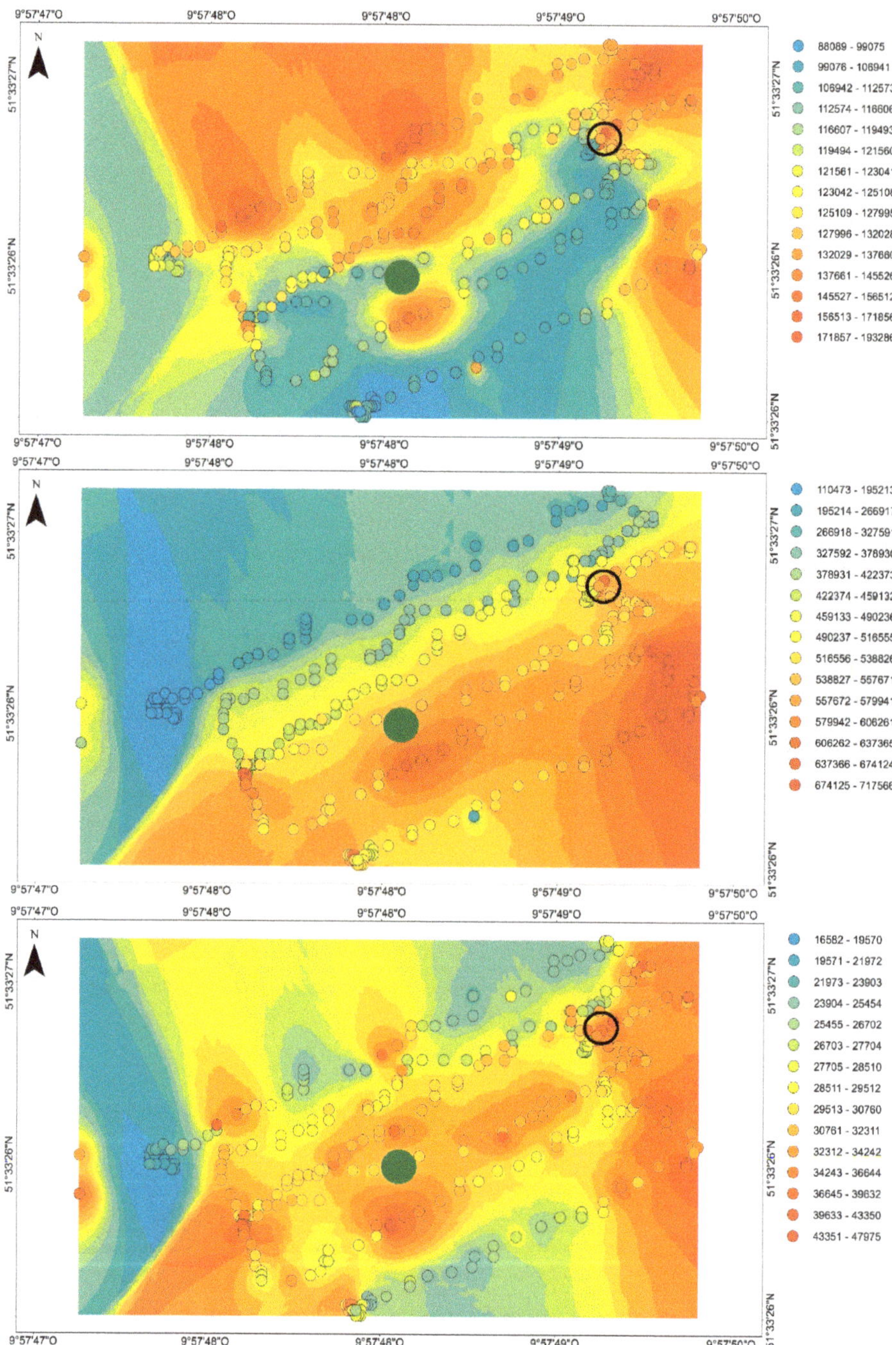

Figure 5. The heat maps generated from the inverted and interpolated sensor signals in conductance (S) (from top: GGS1330, GGS2330, and GGS5330, with N = 1): the sensors were attached to a drone that flew over a plane field with an α-pinene source held by a pole at 4 m heights (green dot). A *prunus domestica* tree is marked with a circle. Red indicates a high sensor response, and blue indicates a low sensor response.

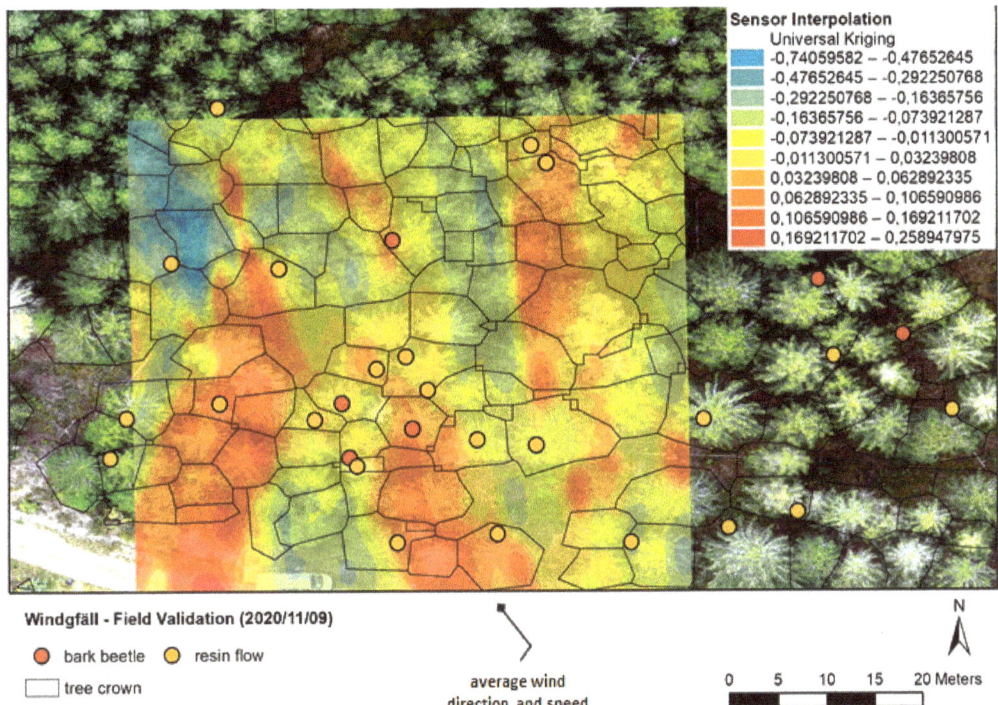

Figure 6. The sensor signal heat map (inverted, normalized, all three sensor values multiplied, and height above crown corrected and interpolated by universal kriging with $R^2 = 0.477$) with the calculated tree crown geometries (inverted DSM with the watershed algorithm) and manual assignment of the infested trees and trees that show resin flow from physical damage (visual stand assessment at the site). N = 1.

4. Discussion

The three sensors tested in this study show different response intensities towards α-pinene (Figure 3 and Table 2). The sensor GGS2330 has a SnO_3 surface, which allows for the detection of easily thermally oxidizable VOCs. The GGS1330 sensor has a SnO_2 surface, which responds to VOCs that are more stable towards thermal oxidization, and the GGS5330 sensor has a WO_3 surface, which responds to thermally reducible VOCs. α-Pinene is easily oxidizable at low temperatures [44], which explains the high response of the sensor GGS2330, the lower response of the sensor GGS1330, and the low response of the sensor GGS5330 in Figure 3. GGS2330 had the highest baseline and showed the steepest signal decrease at increasing concentrations, with the highest R^2 for the linear correlation of all datapoints (Figure 3 and Table 2). These results correspond to former studies on oak wood emission detection [45] and to a study on the detection of standard chemicals (e.g., 2,4-nonadienal) with SnO_2 and WO_3 gas sensors [42].

The calibration data were not transformed, so the variation displayed in Figure 1 shows the variation in the raw data and takes into account the differences in the zero-signal values of each sensor at each trial. As the variation coefficients are between 12% to 42%, a reliable field concentration determination can only be achieved by field calibration, e.g., with the method presented by Schultealbert et al. 2017 [46]. However, a relative data analysis as presented in Figure 4 can be sufficient to identify infested trees, which can be located by the relative signal increase. Such an approach was demonstrated by Sauerwald in Paczkowski et al. [30] for thermally treated wood flakes. The highly significant differences between the factor 10 dilutions in Figure 3 suggest that the differentiation capacities of all three sensors are below a factor 10 difference in the range between 10^{-4} and 10^{-6}. According to Kohl et al., a differentiation of a factor 2 difference is possible at a 10^{-6} concentration of 2,4-nonadienal detected with a ZnO/SnO_2 sensor [42]. A study by Schüler et al. suggests that, under lab conditions, a much higher sensitivity in the sub ppb range accompanied by a good selectivity can be possible [47].

As described in Section 2.1 of this article, the sensors were run with constant heater resistances both in the lab and field experiments. These resistances can be changed and adapted to the detection of monoterpenes, including α-pinene, in order to further improve both the selectivity and the sensitivity of the sensor system [48]. An example of the impact of the heater resistance on the sensor signal by induced sensor surface temperature alternation is given by Sauerwald in Paczkowski et al. 2013 [30]. Temperature cycles over the sensor surface increased the signal selectivity, as selected temperatures showed a compound-specific response pattern. This concept can be used to improve detection by increasing the selectivity towards monoterpenes, including α-pinene, in complex gas mixtures. However, such gas mixtures would be rare above conifer forests, as, e.g., Giunta et al. [27] and Page et al. [25,26] both only found terpenes in the VOC pattern of healthy and infested conifer trees.

It is not possible to detect bark beetle pheromones with semiconductor gas sensors. As shown in Erbilgin et al. [38] the pheromone concentrations are far below the expected α-pinene levels at ratios between 50:1 to 5000:1. Under such circumstances, the α-pinene oxidation on the sensor surface will strongly superimpose the pheromone-induced conductance change, which will disable pheromone detection. This is one of many examples that show the lack of selectivity of semiconductor gas sensors [42]. Bark beetles derive a sensitive and selective biological VOC detection system based on odor-specific odor-binding proteins [31,34,39,40,42]. This allows the beetles to communicate on the basis of low pheromone concentrations in complex odor mixtures [38]. At present, it is not possible to use OBPs on sensor surfaces for reliable odor detection under field conditions [49].

Six out of 12 data sets showed significant differences between the two distances in the lab experiment (Table 2 and Figure 2). In three cases, the correlations were positive, and in three cases, the correlations were negative. Therefore, these significances are considered nonexplanatory. Apparently, the difference in the distances between 1.5 m and 3 m is neglectable from a practical point of view. Probably, complex odor dispersion is responsible for the contradictive results, as plumes of odor show a high spatial concentration distribution that is weakly correlated to the distance from the source [50–53]. A significant dilution of α-pinene will only occur at larger distances, which corresponds to studies on VOC dispersion in nature [50,54]. This indicates that temperature and source concentration determine the emission rate at short ranges between 1.5 m and 3 m and low wind speeds of $1\ m \times s^{-1}$ and $2\ m \times s^{-1}$ (Figure 4).

Relating this information to field conditions, the limit of detecting the first generation of bark beetles in spring will be the environmental temperature at this time of the year. As the flight time of the bark beetles can start at low degrees, the first attack of the bark beetle might not be detectable at all. In this case, secondary plant stress emissions can serve as cues for infestation after an early initial infestation. Also, bark beetles attracted by the pheromones of the first successful bark beetles can cause additional resin flow after an early initial infestation.

VOC vapor pressure is positively correlated with temperature (Clausius Clapeyron relation), while bark beetle breeding cycle time is negatively correlated with temperature [8]. Therefore, a warm and dry spring season will facilitate both beetle development and the detection probability of monoterpenes, e.g., α-pinene. Increasing temperatures in the course of the season will further improve the detectability of bark beetle infestation at an early stage by means of monoterpene emissions [2,8]. This concept can be compromised by stands under initial drought stress, which will emit increased amounts of monoterpenes such as α-pinene even if they are not affected by bark beetle infestation [22,24] (Table 1). This emission increase in the initial drought stage is caused by a higher temperature and a successive increasing vapor pressure of α-pinene. In such cases, the background level of monoterpenes could be too high to differentiate between infested and not infested trees.

Table 1 shows that the overall emission differences between infested and healthy trees can be high. The data in the table also show few exceptions, which indicates a variable emission response to infestation. Further research has to address and clarify this issue, and the proposed drone system can help establish a large data basis, either with sensor data or, e.g., with a sorbent tube sampling method and a successive laboratory analysis.

Stands weakened by drought stress will have a higher probability of intense and uncontrollable bark beetle infestations, so the proposed drone-based sensor detection concept with single tree identification is not practical (Table 1). After a longer drought period, the trees will emit less terpenes [22] (Table 1) and start to increase the terpene levels inside the needle and wood tissue due to stomata closure [24]. Stands under strong and continuous drought stress lose their ability to resinate, as the trees are less physiological active [25–27] and lose their ability to produce monoterpene cyclase [29]. Such stands are usually severely attacked by bark beetles [28], and the proposed detection concept is not applicable under these extreme circumstances. The drone-based sensor concept is applicable for healthy trees under initial bark beetle attack [28].

Table 1 shows the literature values on monoterpene and α-pinene emission from conifer trees. In most cases of bark beetle infestation, the emission ratio increases between 1.4–5.1, which shows a higher emission rate after the infestation. However, there are some inconsistencies in the literature. Page et al. reported an increase in the emission rate in the yellow stage after infestation [26], while in an earlier study, applying the same method resulted in increasing emissions in the green infested stage [25] (Table 1). Ormeno et al. sampled drought stress emissions of *Pinus halepensis* seedlings, and the result was a fluctuation of the emission ratio throughout the experiment (Table 1). The absolute values given by the authors should not be compared between studies using different methods. For instance, Ormeno et al. [24] and Blanch et al. [22] used three- and two-year-old seedlings for their drought experiments, respectively, while Page et al. used the lower branches of adult trees [25,26]. Amin et al. [28] sampled in the crown and compared the α-pinene emissions with the trunk emissions. The authors of this study found that the emission difference between infested and not infested trees are higher in the trunk, as the bark beetles attack this tree compartment directly. The α-pinene emitted at the trunk diffuses to the crown, where it is present at lower concentrations. A general stress response can add to these emissions by stomata gas exchange of monoterpenes [20,21,29]. The studies summarized in Table 1 indicate that the emission dynamic of infested trees is complex. The comparison between infestation-induced and drought-induced terpene emission shows that the assumptions on the limitations of the drone-based early bark beetle infestation presented above are reasonable. Intense drought can disable the possibility to detect bark beetle infestation at an early stage.

The sensor calibration data shown in Figure 3 and Table 2 can be displayed as an emission rate. In a comparative approach, the resulting values match the proposed emission intensities in Table 1 (sensor response at 10^{-6} equals 90 ng $\times$ h^{-1} or 1 ng $\times$ L^{-1}). This does not take into consideration that the data in Table 1 are related to only one gram of dry or fresh biomass. Eighty-year-old spruce trees have a crown mass of approximately 40 kg [55]. Although these calculations are based on studies with different methodological approaches, there is a resemblance between the expected emissions from infested trees and sensor sensitivity. Additionally, there is a high order of magnitude between the sampled emissions from 1 g of biomass in the experiments listed in Table 1 and the expected amount of biomass in the field. This supports the feasibility of the drone-based sensor approach for early detection of bark beetle infestation.

Although the drone was able to locate the strong α-pinene source in Figure 5 (green dot), it also located another source (black circle). This source was a *prunus domestica* tree that caused a signal increase in the inverted signal. Trees emit isoprene and other volatile compounds during leave and bud development [56,57]. It is probable that the increasing physiological activity of the broad leave tree in march caused the weak signal in Figure 3 (black circle). These results confirm the measuring concept. However, false identifications are possible depending on the emission of monoterpenes or other VOCs. In cases where unspecific emissions of monoterpenes from plants are needed, e.g., for forest/atmosphere interactions [16–18,20], the sensor can provide valuable information for small-scale analysis or upscaling model approaches. It also allows us to investigate diurnal emission differences [18] or climate dependent emission dynamics [20,21,29]. In Figure 5, the interpolation of the sensor data led to artifact signals at larger distances to the sensor signal points, which did not correspond to high VOC concentrations.

Other sources of false positive sensor signals can be logging sites or trees broken by storm or ice, which emit very high concentrations of α-pinene. This suggests that the assessment of VOC emission is not enough to differentiate between false positive and positive signals. Optical assessments by drones, planes, or even satellite can add valuable information to identify the emission source. However, false-positive detections cannot be completely excluded.

The field trial was performed under unfavorable conditions, as the temperature was low and the wind was relatively strong (Ø: 4.5 m $\times$ s^{-1} or 8.7 knots), blowing northwest. Due to GPS inaccuracy the drone deviated from the programmed flight path and, therefore, maneuvered between 5 m and 15 m above the calculated DSM. This is both too high and too variable for consistent VOC detection performance. Therefore, the results in the heat map (Figure 6) have to be considered preliminary. It has to be mentioned that northwest of all bark beetle-infested trees there is a signal increase at a constant distance to the potential source, which could be caused by monoterpenes, especially α-pinene, that were emitted and distributed by the wind to the northwest. Due to the restrictions described above, there were many signal increases over the area. Therefore, consistency of the signal increases northwest of the sources does not confirm the measuring concept.

Bark beetle detection can only be performed during spring and summer, when the beetle is active and the temperatures are elevated. Therefore, no distinct correlation between the positions of the infested trees and the sensor signals are visible in Figure 6. It is possible that other sources, e.g., damaged twigs that were not visible from the ground or in the aerial survey, had an impact on the VOC distribution. The field trials have to be repeated at warmer temperatures, e.g., in springtime 2021, and under less wind influence. An improved GPS system, e.g., an RTK (real-time kinematic) GPS, has to ensure guidance of the drone along the precalculated flight path to maintain a constant distance above the canopy surface. Improved drone distance sensors that can detect small geometric structures, e.g., small branches and twigs, will allow the drone to fly closer to the canopies. This will improve the signal noise ratio of the monoterpene detection.

An economic feasibility study of the new bark beetle detection system (unpublished) showed that the drone sensor approach will be financially competitive against conventional early bark beetle detection by sight in steep areas, in areas with dense understory, and for large areas. The average costs per ha were estimated to be below 60 USD. Multiple flights throughout the vegetation period will be necessary in order to effectively prevent bark beetle gradation.

5. Conclusions

The drone-based semiconductor metal oxide sensor array is a promising new technology that allows for sampling of volatile emissions in the air. α-Pinene was detectable as an artificial source by a drone-based gas sensor, but the proof of principle above forest stands with trees in an early infestation stage still has to be performed. Technical barriers are the drone technology (close canopy flight by high-precision GPS and distance sensors) and the sensor selectivity. Further research will also address the VOC distribution dynamics in forest stands to understand the relation between detected VOCs above the crown and the position of infested trees under wind influence. The drone-based sensor can play an important role in understanding forest borne monoterpene emissions from an atmospheric research perspective. The approach can yield improved data for precision modeling, e.g., of drought stress-induced atmospheric emissions of forest stands.

Author Contributions: Conceptualization, S.P. (Sebastian Paczkowski) and P.D.; methodology, S.P. (Sebastian Paczkowski), P.D., M.P., T.H., H.I., and D.J.; software, P.D., H.I., and M.P.; validation, S.P. (Sebastian Paczkowski), T.H., and D.J.; formal analysis, S.P. (Sebastian Paczkowski), H.I., P.D., M.P., and T.H.; investigation, S.P. (Sebastian Paczkowski), P.D., H.I., and M.P.; resources, D.J. and S.P. (Sebastian Paczkowski); data curation, H.I. and M.P.; writing—original draft preparation, S.P. (Sebastian Paczkowski), P.D., and D.J.; writing—review and ed-iting, S.P. (Sebastian Paczkowski) and P.D.; visualization, S.P. (Sebastian Paczkowski), M.P., P.D., and H.I.; supervision, D.J. and S.P. (Sebastian Paczkowski); project ad-ministration, S.P. (Sebastian Paczkowski), P.D., D.J., and S.P. (Stefan Pelz); funding acquisition, S.P. (Sebastian Paczkowski), D.J., and S.P. (Stefan Pelz). All authors have read and agreed to the published version of the manuscript.

Funding: This research was funded by the FNR—Fachagentur für Nachwachsende Rohstoffe, Germany, under the grant number 22011018.

Institutional Review Board Statement: Not applicable.

Informed Consent Statement: Not applicable.

Data Availability Statement: The data presented in this study are available on request from the corresponding author.

Acknowledgments: The authors want to thank the forest owners Graf Ferdinand v. Drechsel, Christian Vogt, and Christian Fürst zu Fürstenberg and the forest manager Dietrich Nübling for their cooperation considering the field tests and the team of Cadmium GmbH for their technical support.

Conflicts of Interest: The authors declare that they have no competing financial interests or personal relationships that could have influenced the work reported in this paper.

References

1. De Grandpré, L.; Pureswaran, D.; Bouchard, M.; Kneeshaw, D. Climate-induced range shifts in boreal forest pests: Ecological, economic, and social consequences. *Can. J. For. Res.* **2018**, *48*, v–vi. [CrossRef]
2. Rouault, G.; Candau, J.-N.; Lieutier, F.; Nageleisen, L.-M.; Martin, J.-C.; Warzée, N. Effects of drought and heat on forest insect populations in relation to the 2003 drought in Western Europe. *Ann. For. Sci.* **2006**, *63*, 613–624. [CrossRef]
3. Corbett, L.J.; Withey, P.; Lantz, V.A.; Ochuodho, T.O. The economic impact of the mountain pine beetle infestation in British Columbia: Provincial estimates from a CGE analysis. *Forestry* **2016**, *89*, 100–105. [CrossRef]
4. Dale, V.; Joyce, A.L.; McNulty, S.; Neilson, R.P.; Ayres, M.P.; Flannigan, M.D.; Hanson, P.J.; Irland, L.C.; Lugo, A.E.; Peterson, C.J.; et al. Climate change and forest disturbances. *BioScience* **2001**, *51*, 723–734. [CrossRef]
5. Bundesministerium für Ernährung und Landwirtschaft. Waldschäden: Bundesministerium Veröffentlicht Aktuelle Zahlen. Available online: https://www.bmel.de/SharedDocs/Pressemitteilungen/DE/2020/040-waldschaeden.html;jsessionid=E322EBC9C439CDFD426657E41A3E9DC5.internet2851 (accessed on 1 July 2020).
6. Bundesministerium für Ernährung und Landwirtschaft. Bundeswaldinventur. Available online: https://bwi.info/inhalt1.3.aspx?Text=3.05%20Altersklasse&prRolle=public&prInv=BWI2012&prKapitel=3.05 (accessed on 10 October 2020).
7. McCollum, D.W.; Lundquist, J.E. Bark Beetle Infestation of Western US Forests: A Context for Assessing and Evaluating Impacts. *J. Forest.* **2019**, *117*, 171–177. [CrossRef]
8. Wermelinger, B.; Seifert, M. Temperature-dependent reproduction of the spruce bark beetle Ips typographus, and analysis of the potential population growth. *Ecol. Entomol.* **1999**, *24*, 103–110. [CrossRef]
9. Immitzer, M.; Einzmann, K.; Pinnel, N.; Seitz, R.; Atzberger, C. Vitaliltätserfassung von Fichten mittels Fernerkundung. *AFZ Wald* **2018**, *17*, 20–23.
10. Anderbrant, O. Reemergence and Second Brood in the Bark Beetle *Ips typographus. Holartic Ecol.* **1989**, *12*, 494–500. [CrossRef]
11. Thatcher, C.R. *The Southern Pine Beetle*; United State Department of Agriculture: Washington, DC, USA, 1981.
12. Schwerdtfeger, F. *Waldkrankheiten*; Paul Parey: Hamburg/Berlin, Germany, 1970.
13. Lausch, A.; Heurich, M.; Gordalla, D.; Dobner, H.-J.; Gwillym-Margianto, S.; Salbach, C. Forecasting potential bark beetle outbreaks based on spruce forest vitality using hyperspectral remote-sensing techniques at different scales. *Forest. Ecol. Manag.* **2013**, *308*, 76–89. [CrossRef]
14. Fassnacht, F.E.; Latifi, H.; Ghosh, A.; Joshi, P.K.; Koch, B. Assessing the potential of hyperspectral imagery to map bark beetle-induced tree mortality. *Remote Sens. Environ.* **2014**, *140*, 533–548. [CrossRef]
15. Hais, M.; Wild, J.; Berec, L.; Brůna, J.; Kennedy, R.; Braaten, J.; Brož, Z. Landsat Imagery Spectral Trajectories—Important Variables for Spatially Predicting the Risks of Bark Beetle Disturbance. *Remote Sens. Environ.* **2016**, *8*, 687. [CrossRef]

16. Guenther, A.; Hewitt, C.N.; Erickson, D.; Fall, R.; Geron, C.; Graedel, T.; Harley, P.; Klinger, L.; Lerdau, M.; Mckay, W.A.; et al. A global model of natural volatile organic compound emissions. *J. Geophys. Res.* **1995**, *100*, 8873. [CrossRef]

17. Isidorov, V.A.; Zenkevich, I.G.; Ioffe, B.V. Volatile organic compounds in the atmosphere of forests. *Atmos. Environ.* **1985**, *19*, 1–8. [CrossRef]

18. Enders, G.; Dlugi, R.; Steinbrecher, R.; Clement, B.; Daiber, R.; Euk, J.; Gäb, S.; Haziza, M.; Helas, G.; Herrmann, U.; et al. Biosphere/Atmosphere interactions: Integrated research in a European coniferous forest ecosystem. *Atmos. Environ.* **1992**, *26*, 171–189. [CrossRef]

19. Berg, A.R.; Heald, C.L.; Huff Hartz, K.E.; Hallar, A.G.; Meddens, A.J.H.; Hicke, J.A.; Lamarque, J.-F.; Tilmes, S. The impact of bark beetle infestations on monoterpene emissions and secondary organic aerosol formation in western North America. *Atmos. Chem. Phys.* **2013**, *13*, 3149–3161. [CrossRef]

20. Lerdau, M.; Dilts, S.B.; Westberg, H.; Lamb, B.K.; Allwine, E.J. Monoterpene emission from ponderosa pine. *J. Geophys. Res.* **1994**, *99*, 16609. [CrossRef]

21. Yokouchi, Y.; Ambe, Y. Factors Affecting the Emission of Monoterpenes from Red Pine (*Pinus densiflora*). *Plant. Physiol.* **1984**, *75*, 1009–1012. [CrossRef]

22. Blanch, J.-S.; Peñuelas, J.; Llusià, J. Sensitivity of terpene emissions to drought and fertilization in terpene-storing *Pinus halepensis* and non-storing Quercus ilex. *Physiol. Plant.* **2007**, *131*, 211–225. [CrossRef]

23. Holopainen, J.K. Multiple functions of inducible plant volatiles. *Trends Plant. Sci.* **2004**, *9*, 529–533. [CrossRef] [PubMed]

24. Ormeño, E.; Mévy, J.P.; Vila, B.; Bousquet-Mélou, A.; Greff, S.; Bonin, G.; Fernandez, C. Water deficit stress induces different monoterpene and sesquiterpene emission changes in Mediterranean species. Relationship between terpene emissions and plant water potential. *Chemosphere* **2007**, *67*, 276–284. [CrossRef]

25. Page, W.G.; Jenkins, M.J.; Runyon, J.B. Mountain pine beetle attack alters the chemistry and flammability of lodgepole pine foliage. *Can. J. For. Res.* **2012**, *42*, 1631–1647. [CrossRef]

26. Page, W.G.; Jenkins, M.J.; Runyon, J.B. Spruce Beetle-Induced Changes to Engelmann Spruce Foliage Flammability. *For. Sci.* **2014**, *60*, 691–702. [CrossRef]

27. Giunta, A.D.; Runyon, J.B.; Jenkins, M.J.; Teich, M. Volatile and Within-Needle Terpene Changes to Douglas-fir Trees Associated With Douglas-fir Beetle (Coleoptera: Curculionidae) Attack. *Environ. Entomol.* **2016**, *45*, 920–929. [CrossRef]

28. Amin, H.; Atkins, P.T.; Russo, R.S.; Brown, A.W.; Sive, B.; Hallar, A.G.; Huff Hartz, K.E. Effect of bark beetle infestation on secondary organic aerosol precursor emissions. *Environ. Sci. Technol.* **2012**, *46*, 5696–5703. [CrossRef] [PubMed]

29. Lewinsohn, E.; Gijzen, M.; Croteau, R. Defense Mechanisms of Conifers Differences in Constitutive and Wound-Induced Monoterpene Biosynthesis Among Species. *Plant. Physiol.* **1991**, *96*, 44–49. [CrossRef] [PubMed]

30. Paczkowski, S.; Paczkowska, M.; Dippel, S.; Schulze, N.; Schütz, S.; Sauerwald, T.; Weiß, A.; Bauer, M.; Gottschald, J.; Kohl, C.-D. The olfaction of a fire beetle leads to new concepts for early fire warning systems. *Sens. Actuat. B Chem.* **2013**, *183*, 273–282. [CrossRef]

31. Blomquist, G.J.; Figueroa-Teran, R.; Aw, M.; Song, M.; Gorzalski, A.; Abbott, N.L.; Chang, E.; Tittiger, C. Pheromone production in bark beetles. *Insect Biochem. Mol. Biol.* **2010**, *40*, 699–712. [CrossRef]

32. Hedgren, P.O. The bark beetle *Pityogenes chalcographus* (L.) (Scolytidae) in living trees: Reproductive success, tree mortality and interaction with *Ips typographus*. *J. Appl. Entomol.* **2004**, *128*, 161–166. [CrossRef]

33. Byers, J.A.; Wood, D.L. Interspecific effects of pheromones on the attraction of the bark beetles, *Dendroctonus brevicomis* and *Ips paraconfusus* in the laboratory. *J. Chem. Ecol.* **1981**, *7*, 9–18. [CrossRef] [PubMed]

34. Tittiger, C.; Blomquist, G.J. Pheromone biosynthesis in bark beetles. *Curr. Opin. Insect Sci.* **2017**, *24*, 68–74. [CrossRef]

35. Byers, J.A. Chemical ecology of bark beetles. *Experientia* **1989**, *45*, 271–283. [CrossRef]

36. Gitau, C.W.; Bashford, R.; Carnegie, A.J.; Gurr, G.M. A review of semiochemicals associated with bark beetle (Coleoptera: Curculionidae: Scolytinae) pests of coniferous trees: A focus on beetle interactions with other pests and their associates. *For. Ecol. Manag.* **2013**, *297*, 1–14. [CrossRef]

37. Progar, R.A.; Gillette, N.; Fettig, C.J.; Hrinkevich, K. Applied Chemical Ecology of the Mountain Pine Beetle. *For. Sci.* **2014**, *60*, 414–433. [CrossRef]

38. Erbilgin, N.; Powell, J.S.; Raffa, K.F. Effect of varying monoterpene concentrations on the response of Ips pini (Coleoptera: Scolytidae) to its aggregation pheromone: Implications for pest management and ecology of bark beetles. *Agric. For. Ent.* **2003**, *5*, 269–274. [CrossRef]

39. Andersson, M.N.; Grosse-Wilde, E.; Keeling, C.I.; Bengtsson, J.M.; Yuen, M.M.S.; Li, M.; Hillbur, Y.; Bohlmann, J.; Hansson, B.S.; Schlyter, F. Antennal transcriptome analysis of the chemosensory gene families in the tree killing bark beetles, *Ips typographus* and *Dendroctonus ponderosae* (Coleoptera: Curculionidae: Scolytinae). *BMC Genom.* **2013**, *14*, 198. [CrossRef]

40. Yuvaraj, J.K.; Roberts, R.E.; Sonntag, Y.; Hou, X.-Q.; Grosse-Wilde, E.; Machara, A.; Zhang, D.-D.; Hansson, B.S.; Johanson, U.; Löfstedt, C.; et al. Putative ligand binding sites of two functionally characterized bark beetle odorant receptors. *BMC Biol.* **2021**, *19*, 16. [CrossRef]

41. Pelosi, P. Odorant-binding proteins. *Crit. Rev. Biochem. Mol. Biol.* **1994**, *29*, 199–228. [CrossRef]

42. Kohl, D.; Heinert, L.; Bock, J.; Hofmann, T.; Schieberle, P. Systematic studies on responses of metal-oxide sensor surfaces to straight chain alkanes, alcohols, aldehydes, ketones, acids and esters using the SOMMSA approach. *Sens. Actuat. B Chem.* **2000**, *70*, 43–50. [CrossRef]

43. Krüll, W.; Tobera, R.; Willms, I.; Essen, H.; von Wahl, N. Early Forest Fire Detection and Verification using Optical Smoke, Gas and Microwave Sensors. *Procedia Eng.* **2012**, *45*, 584–594. [CrossRef]
44. Neuenschwander, U.; Guignard, F.; Hermans, I. Mechanism of the Aerobic Oxidation of α-Pinene. *ChemSusChem* **2010**, *3*, 75–84. [CrossRef]
45. Paczkowski, S.; Pelz, S.; Jaeger, D. Semi-conductor metal oxide gas sensors for online monitoring of oak wood VOC emissions during drying. *Dry. Technol.* **2019**, *37*, 1081–1086. [CrossRef]
46. Schultealbert, C.; Baur, T.; Schütze, A.; Böttcher, S.; Sauerwald, T. A novel approach towards calibrated measurement of trace gases using metal oxide semiconductor sensors. *Sens. Actuat. B Chem.* **2017**, *239*, 390–396. [CrossRef]
47. Schüler, M.; Helwig, N.; Ventura, G.; Schütze, A.; Sauerwald, T. *IEEE Sensors, Proceedings of the 12th IEEE Sensors Conference, Baltimore, Maryland, USA, 3–6 November 2013*; IEEE: Piscataway, NJ, USA, 2013; ISBN 9781467346405.
48. Leidinger, M.; Sauerwald, T.; Conrad, T.; Reimringer, W.; Ventura, G.; Schütze, A. Selective Detection of Hazardous Indoor VOCs Using Metal Oxide Gas Sensors. *Proc. Engin.* **2014**, *87*, 1449–1452. [CrossRef]
49. Bohbot, J.D.; Vernick, S. The Emergence of Insect Odorant Receptor-Based Biosensors. *Biosensors* **2020**, *10*, 26. [CrossRef] [PubMed]
50. Meng, Q.-H.; Yang, W.-X.; Wang, Y.; Li, F.; Zeng, M. Adapting an ant colony metaphor for multi-robot chemical plume tracing. *Sensors* **2012**, *12*, 4737–4763. [CrossRef] [PubMed]
51. de Croon, G.; O'Connor, L.M.; Nicol, C.; Izzo, D. Evolutionary robotics approach to odor source localization. *Neurocomputing* **2013**, *121*, 481–497. [CrossRef]
52. Monroy, J.; Hernandez-Bennets, V.; Fan, H.; Lilienthal, A.; Gonzalez-Jimenez, J. GADEN: A 3D Gas Dispersion Simulator for Mobile Robot Olfaction in Realistic Environments. *Sensors* **2017**, *17*, 1479. [CrossRef]
53. Ishida, H.; Wada, Y.; Matsukura, H. Chemical Sensing in Robotic Applications: A Review. *IEEE Sens. J.* **2012**, *12*, 3163–3173. [CrossRef]
54. Balkovsky, E.; Shraiman, B.I. Olfactory search at high Reynolds number. *Proc. Natl. Acad. Sci. USA* **2002**, *99*, 12589–12593. [CrossRef] [PubMed]
55. Turski, M.; Beker, C.; Kazmierczak, K.; Najgrakowski, T. Allometric equations for estimating the mass and volume of fresh assimilational apparatus of standing scots pine (*Pinus sylvestris* L.) trees. *For. Ecol. Manag.* **2008**, *255*, 2678–2687. [CrossRef]
56. Li, Y.; Ma, H.; Wan, Y.; Li, T.; Liu, X.; Sun, Z.; Li, Z. Volatile Organic Compounds Emissions from *Luculia pinceana* Flower and Its Changes at Different Stages of Flower Development. *Molecules* **2016**, *21*, 531. [CrossRef] [PubMed]
57. Kuhn, U.; Rottenberger, S.; Biesenthal, T.; Wolf, A.; Schebeske, G.; Ciccioli, P.; Kesselmeier, J. Strong correlation between isoprene emission and gross photosynthetic capacity during leaf phenology of the tropical tree species *Hymenaea courbaril* with fundamental changes in volatile organic compounds emission composition during early leaf development. *Plant. Cell Environ.* **2004**, *27*, 1469–1485. [CrossRef]

Article

Industrial Heat Treatment of Wood: Study of Induced Effects on Ayous Wood (*Triplochiton scleroxylon* K. Schum)

Emiliano Gennari, Rodolfo Picchio and Angela Lo Monaco *

Department of Agriculture and Forest Sciences (DAFNE), University of Tuscia, Via S. Camillo de Lellis, 01100 Viterbo, Italy; emiliano.gennari@unitus.it (E.G.); r.picchio@unitus.it (R.P.)
* Correspondence: lomonaco@unitus.it

Abstract: High-temperature treatment of wood is a useful method for improving certain physical characteristics, ensuring durability without biocides, and improving the performance of wood when exposed to degradation agents. This work aims to determine the effects induced by a heat treatment performed industrially on ayous wood (*Triplochiton scleroxylon* K. Schum) from Cameroon, through the study of the main physical and mechanical characteristics. The heat treatment at 215 °C for three hours with a slight initial vacuum determined a reduction of the mechanical characteristics (compression strength 26%, static bending 46%, Brinell hardness 32%) and some physical properties (dry density 11%, basic density 9%), while it improved the behaviour towards variations of environment moisture. The anti-shrinkage efficiency was 58.41 ± 5.86%, confirming the increase of the dimensional stability. The darkening (ΔE 34.76), clearly detectable (L* 39.69 ± 1.13; a* 10.59 ± 081; b* 18.73 ± 1.51), was supported almost equally by both the lightness parameter (L*) and the a* chromatic parameter. The data collected during the laboratory tests were then subjected to statistical analysis to verify correlations between the characteristics examined. Statistical differences were highlighted between each physical and mechanical properties of ayous wood modified or not.

Keywords: wood modification; mechanical properties; physical properties; anti shrinkage efficiency; colour; Cameroon

Citation: Gennari, E.; Picchio, R.; Lo Monaco, A. Industrial Heat Treatment of Wood: Study of Induced Effects on Ayous Wood (*Triplochiton scleroxylon* K. Schum). *Forests* **2021**, *12*, 730. https://doi.org/10.3390/f12060730

Academic Editor: Miha Humar

Received: 6 May 2021
Accepted: 1 June 2021
Published: 3 June 2021

Publisher's Note: MDPI stays neutral with regard to jurisdictional claims in published maps and institutional affiliations.

1. Introduction

The biological origin of woody materials makes them a valuable ally in reducing the environmental impact of human activities. Some intrinsic properties of wood, such as natural origin, nontoxicity, wide diffusion, and versatility of the material are the basis of the multitude of possible uses. In addition, the possibility of recycling primary wood products at the end of their life cycle, for example, to produce particle boards or similar secondary products, and finally, the valorisation of waste to produce energy from renewable sources (cascading), are characteristics that contribute to making wood an important resource against climate change and the rational and optimized exploitation of natural resources [1]. However, the natural origin of wood is associated with some drawbacks, among these the susceptibility to be altered by degradation factors is of great importance; in general, the most important degradation factors are UV, water and moisture, and biological organisms. The durability of wood is a characteristic variable among species that expresses the intrinsic ability to resist the action of degradation and decay agents. Often it is necessary to improve the natural durability of wood, particularly when it is used for outdoor purposes. Improving wood durability is achieved by treating wood with preservatives, however, these composites are often toxic, with a high environmental impact and high cost. There are alternatives to prevent degradation and decay; these are different modification processes classified into: chemical processes, thermo–hydro–mechanical processes, electromagnetic processes, and other types less common and developed. Among them, wood thermal modification processes have been extensively studied by researchers over the past decades, an activity that has been confirmed in the market with an increasing volume of heat-treated wood produced annually [2]. Thermal modification of wood is a process that involves exposing the material to a temperature between 100 and 300 °C for periods from 15 min to 24 h. Depending on the treatment cycle, it is possible to obtain different characteristics in the treated materials. For example, the purpose of the treatment may be to soften the wood, using heat and steam, to release internal tension and facilitate subsequent manufacturing processes; or to achieve controlled degradation of the material to improve dimensional stability and durability. For this purpose, the temperature range is generally between 150 and 260 °C [2]. Thermal modifications of wood are performed at temperatures between 180 and 260 °C; no significant variation of the material properties occurs at temperatures below 140 °C [3,4], while exceeding 260 °C excessive degradation occurs. Modification cycles with temperature over 300 °C make no sense for excessive degradation of the material [5]. The improved durability of heat-treated wood makes it suitable for outdoor use, where greater exposure to moisture generally contributes to more rapid alteration of the material, so reducing water affinity is an advantage. In fact, heat-treated wood is widely used as a building material, including ayous, especially for the construction of building cladding and exterior flooring. Previous studies have evaluated the moisture behaviour of heat-treated wood used for building facades, concluding that heat-treated wood performs better than untreated wood [6,7]. In the literature, however, few works are still available on heat-treated ayous, a wood that is not very durable in its natural state [8,9]. To date, studies conducted on the subject are limited to the analysis of the chemical composition of wood and the analysis with dynamic methods of the mechanical properties of samples subjected to different cycles of heat treatment [10]; on the study of the physical properties of ayous wood subjected to heat treatment with boiling oil by analysing the effect of the treatment on wood hygroscopicity, stability and colour [11]; and on the water resistance of the mechanical properties of joints made of heat-treated wood [12].

Thermally modified wood is a non-biocidal alternative to the classical techniques of extending the natural durability of wood [13] and an effective product with better behaviour in outdoors than natural wood due to a better dimensional stability [6].

Despite market interest, too little is known about natural ayous wood and even less about its characteristics after heat treatment. This study aims to present the first set of data discussed at the First International Electronic Conference on Forests—IECF 2020—[14], allowing the evaluation of industrially thermally modified ayous wood (*Triplochiton scleroxylon* K. Schum) to highlight the influence of heat treatment at 215 °C on the selected physical and mechanical characteristics; to add further data and make a comparison with untreated wood from the same area (Cameroon); and to report the results published by other authors to compare the results obtained with the data literature.

2. Materials and Methods

The ayous (*Triplochiton scleroxylon* K. Schum) planks derive from timber taken in a natural forest, FSC (Forest Stewardship Council) certified for forest management and chain of custody, from Cameroon Department of Boumba et Ngoko. The thermal treatment was industrially performed on ayous planks in an autoclave (Model TVS 6000 WDE Maspell srl, Terni, Italy) at a treatment temperature of 215 °C for three hours with a slight initial vacuum.

The samples were prepared according to the general requirements of the standard for physical and mechanical tests [15] and preserved at laboratory conditions at 65% relative humidity and 20 °C. Moisture content was determined according to the standard UNI ISO 13061-1 [16]. A ventilated oven was used at 103 ± 2 °C for 24 + 6 h until the mass was constant. The properties were assessed at a moisture content of 12% when required, for comparison with literature data. The property adjustment was carried out by placing the samples, before each test, inside a conditioning chamber for 168 h (or in any case, when the moisture content of 12% ± 0.3% was reached). This operation was necessary because the correction factor (α) used for normal wood is not applicable to heat-treated wood (result assessed after some preliminary tests). In this sense, further research is underway in order to formulate α suitable for this type of wood.

Sample dimensions were measured with a digital caliper (± 0.01 mm), and the mass was recorded at a precision scale $\pm$ 0.001 g. Demineralized water was used to reach the maximum swelling.

The basic and dry densities were calculated on a set of specimens (20 $\times$ 20 $\times$ 30) mm according to UNI ISO 13061-2 [17]. The tangential, radial, and volumetric shrinkage were determined by considering the total dimensional change from the fully swollen to the oven-dry condition, according to the reference standard ISO 13061-13 and 14 [18]. The shrinkage anisotropy factor was calculated as the ratio between tangential shrinkage and radial shrinkage [19], and the anti-shrink efficiency (ASE) was determined for volumetric shrinkage and cross-section shrinkage using the following formula:

$$\text{ASE} = \frac{\beta_{max} - \beta_{min}}{\beta_{max}} \times 100 \tag{1}$$

where:

β_{max} is the shrinkage of untreated wood in percent;
β_{min} is the shrinkage of heat-treated wood in percent.

The cross-section shrinkage in percent was calculated using the following formula:

$$\beta_{cs} = \left[1 - \left(1 - \frac{\beta_t}{100} \right) \times \left(1 - \frac{\beta_r}{100} \right) \right] \times 100 \tag{2}$$

where:

β_{cs} is the cross-section shrinkage in percent;
β_t is the tangential direction shrinkage;
β_r is the radial direction shrinkage.

A reflectance spectrophotometer X-Rite CA 22 (X-Rite Inc., MI, USA) was used to quantitatively characterise the colour in the CIELAB colour system. The characteristics of the instrument are as follows: colour scale CIEL*a*b*; illuminant D65; standard observer 10°; geometry of measurement 45°/0°; spectral range 400–700 nm; spectral resolution 10 nm; measurement diameter 4 mm; and white reference supplied with the instrument. 60 points were selected to consider the variability of the wood colour, as already discussed [20]. For each point, three measures were made; so, 180 measures were collected. To obtain the colour difference, a comparison was performed with natural wood. The differences in lightness (L*), chromatic coordinates (a* and b*), and the total colour difference (ΔE*) were then calculated using these parameters according to UNI EN-15886 [21], as previously described [22]. The "*" designates the lightness and colorimetric parameters in the CIELab colour space [21].

Mechanical tests, axial compression strength, static bending strength, and Brinell hardness were performed at the equilibrium moisture content (EMC) of the laboratory conditions and then adjusted at moisture content 12%, when required.

The ultimate axial compression strength was determined following the reference standard UNI ISO 3787 [23] on a set of specimens (20 $\times$ 20 $\times$ 30) mm. The load reached the maximum axial compression strength in 1.5 to 2 min.

The standard ISO 13061-3 [24] was applied to determine the ultimate tensile strength in static bending on a set of specimens (20 $\times$ 20 $\times$ 300) mm. The test was carried out with a span of 260 mm, and the transverse load was applied at mid-span to the radial surface of the specimens; width and height were measured at mid-span with 0.01 mm of tolerance. The test piece was broken in 1.5–2 min from the start of the applied load.

The resistance to indentation, formerly Brinell hardness, was quantified by UNI EN 1534 [25] on a set of specimens (50 $\times$ 300 $\times$ 20) mm. The maximum load of 1 kN was reached in 15 s from the start and maintained for 25 s. Two diameters of the residual indentation were measured: one parallel to the fibre direction and the other perpendicular to the fibre direction, at least three minutes after load application.

Statistical analyses were carried out with StatisticaTM version 7.1 (TIBCO Software Inc., Palo Alto, CA, USA). Data distribution was plotted and checked for normality and homogeneity of variance using the Lilliefors and Levene tests, respectively. To check differences between treatments, t-test for independent samples was applied to assess possible differences among the data of this study and those of a specific research on untreated ayous wood [14,26] originating from the same wood macro sample. Non-linear regression analysis was conducted for all the mechanical variables in relation to the wood density.

3. Results

3.1. Physical Properties

Table 1 provides an overview of the data from the physical characterization of heat-treated ayous wood. The density of this material, referred to as 12% moisture content, was 0.33 ± 0.02 g/cm^3, while in anhydrous condition, it was 0.31 ± 0.02 g/cm^3. The obtained value of basic density was 0.30 ± 0.02 g/cm^3. The linear shrinkages were $1.27 \pm 0.26\%$ in the radial direction, and $1.92 \pm 0.25\%$ in the tangential direction. The observed volumetric shrinkage was equal to $3.16 \pm 0.44\%$, and the cross-section shrinkage was $1.89 \pm 0.24\%$. The shrinkage anisotropy factor was 1.55. The percentage of reduction in the physical properties between treated and untreated wood was 15% and 11%, respectively. The density was at a moisture content of 12% and in anhydrous state, 9% for basic density. In shrinkages, the reduction was 55%, 62%, 59%, and 61%, respectively, for radial, tangential, volumetric, and cross-section shrinkage. The reduction of the shrinkage anisotropy factor was 15% [14,26].

Table 1. Physical properties of heat-treated ayous wood.

Properties	Samples N.	Mean Value	St. Dev.
Density 12% MC (g/cm^3)	30	0.33	0.02
Dry density (g/cm^3)	30	0.31	0.02
Basic density (g/cm^3)	30	0.30	0.02
Radial shrinkage βr (%)	30	1.27	0.26
Tangential shrinkage βt (%)	30	1.92	0.25
Axial shrinkage βa (%)	30	0.16	0.07
Volumetric shrinkage βv (%)	30	3.16	0.44
Cross-section shrinkage βcs (%)	30	1.89	0.24
Shrinkage anisotropy factor	30	1.55	0.27

MC = moisture content.

Regarding the anti-shrinkage efficiency (ASE) of heat-treated ayous wood, Table 2 summarises the obtained results. ASE referred to as volumetric shrinkage was $58.41 \pm 5.86\%$, while ASE referred to as the cross-section was $61.01 \pm 4.82\%$.

Table 2. Anti-Shrinkage Efficiency (ASE) of heat-treated ayous wood referred to as volumetric and cross-section shrinkage.

ASE (%)	Samples N.	Mean Value	St. Dev.
Volumetric	30	58.41	5.86
Cross-section	30	61.01	4.82

3.2. Colour

Ayous wood is an undifferentiated wood and in its natural state has a light colour, often described as yellowish or creamy [27–29]. This nonobjective representation is better described in the CIELAB colour space, in which this wood has a lightness of just over 70, with a yellow component due to the colorimetric parameter b* near to 28, higher than the component a*, which concerns the component that gives the reddish perception [26]. Heat treatment induced a notability darker colouring throughout the entire thickness of the assortment (Table 3). The percentage variation of these parameters between untreated and heat-treated wood was much greater in L* and a* (45% and 43% respectively) than in the b* parameter (33%).

Table 3. Lightness and colorimetric parameters in CIELAB colour space.

	Sampling Point [1] N.	L*	a*	b*
Mean value	60	39.69	10.59	18.73
Standard deviation	60	1.13	0.81	1.51

[1] Note that three measures were made in each sampling point, 180 measures were collected.

However, hue (colour in colloquial terms) is defined by the variation of the parameters a* and b*, which in our case indicated a substantial increase in the red component (a*) and a decrease in the yellow component (b*) (Table 4). Although the colour change was in this case easily appreciable to the naked eye, the colorimetric parameter that unequivocally indicated the darkening was the colorimetric difference ΔE* (Table 4).

Table 4. Lightness and chromatic parameters variation due to industrial heat treatment.

ΔL*	Δa*	Δb*	ΔE*
−33.38	3.19	−9.15	34.76

3.3. Mechanical Properties

The results of these tests were presented in Table 5. The mechanical characterisation of heat-treated ayous wood showed a compression strength parallel to the grain equal to 26.98 ± 1.99 MPa. The ultimate strength in static bending was equal to 32.90 ± 3.09 MPa; and the Brinell hardness was 8.30 ± 1.05 N/mm^2. The reported values for compression strength parallel to grain and ultimate strength in static bending were both referred to 12% of moisture content. The percentage of reduction of the mechanical properties between untreated and heat-treated ayous wood were 26% in compression strength, 46% in static bending strength, and 32% in Brinell hardness [14,26].

Table 5. Mechanical properties of heat-treated ayous wood.

Properties	Samples N.	Mean Value	St. Dev.
Compression strength 12% MC (MPa)	35	26.98	1.99
Static bending strength 12% MC (MPa)	40	32.90	3.09
Brinell hardness (N/mm^2)	68	8.30	1.05

Regarding the relation of the mechanical characteristics with wood density, the results of the regression analysis did not show statistical significance (p-value > 0.05) and the R^2 values were very low (<0.1), except for the compression strength. Figure 1 shows the correlation between compression strength parallel to grain and wood density. Even if, as stated above, the regression analysis did not show statistical significance, there is a positive correlation between these parameters and in this specific case, the R^2 value could be considered appreciable at 0.517.

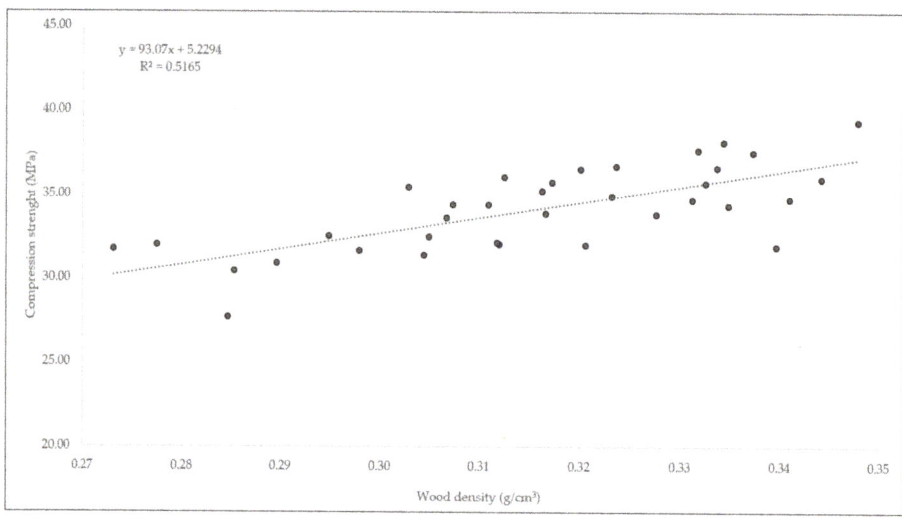

Figure 1. Compression strength of heat-treated ayous wood in function of density referred to MC 12%.

3.4. Comparison with Untreated Ayous Wood

Industrial heat treatment statistically significantly modified both the physical and mechanical properties of ayous wood when compared to natural wood [14,26] (Table 6).

Table 6. *T*-test for independent sample results, for physical and mechanical properties, compared between heat-treated wood and normal wood [14,26].

Parameter	*p*-Value
Density 12% MC	<0.001
Basic density	<0.001
Radial shrinkage βr	<0.001
Tangential shrinkage βt	<0.001
Volumetric shrinkage βv	<0.001
Compression strength	<0.001
Static bending strength	<0.001
Brinell hardness	<0.001

All the characteristics showed a decrease ranged between 8.2% and 61.6% (Figure 2) with high statistical significance.

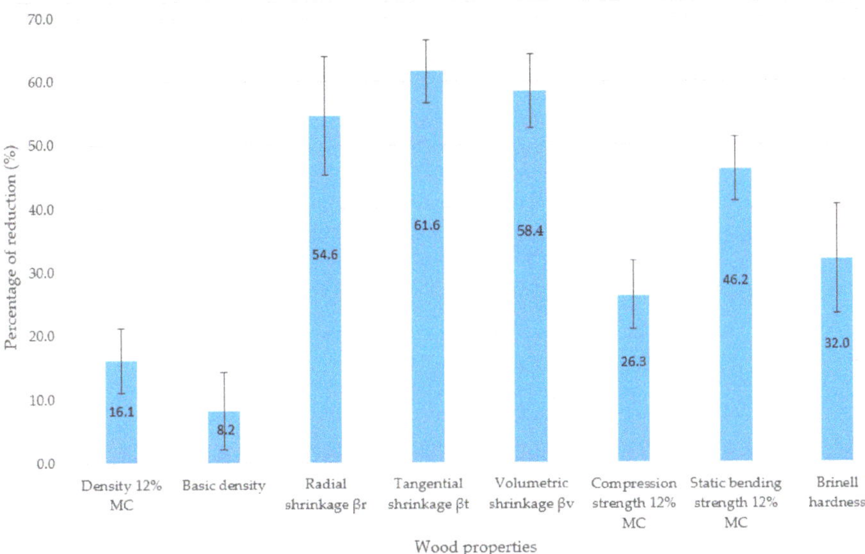

Figure 2. Percentage of variation (reduction) of the mean values, for the heat-treated wood with respect to the normal wood [14,26]. The bars showed the standard deviation.

4. Discussion

The consequences of heat treatment on ayous planks affected the mechanical and physical properties of the wood, producing a statistically significant reduction in strength, shrinkage, and swelling, as well as in densities. The mechanical properties generally showed a reduction between 10 and 50% after the thermal modification; on the other hand, for physical properties, the most important consequences are related to the alteration of the material behaviour with moisture and water. Heat treatment has exacerbated the position of ayous wood among light-density timbers. The reported density values make heat-treated ayous a medium-low density wood material.

One of the main effects of the thermal modification on the physical properties of the wood is related to the behaviour towards water. As a consequence of the treatment, the shrinkage values were rather low, indicating an appreciable dimensional stability of the material, also expressed by the shrinkage anisotropy factor, which was close to that of the wood of species considered to have a low propensity to shape modification [19]. The anti-shrinkage efficiency (ASE) also confirms this improvement; however, the ASE results, referred as to volumetric and cross-section shrinkage, are both lower than 75%, which is considered a threshold in terms of treatment effectiveness for improving the dimensional stability of the material. Reducing the water reactivity of wood leads to improved dimensional stability due to a reduction in swelling; it also leads to a reduction in equilibrium moisture content and, consequently, wood density. Given the effects of moisture as a decay factor in wood, the consequences induced by heat treatment cause an overall increase in the service life of the material [13].

Colour is often a decisive element as it influences the aesthetic of products [30,31]. The effect of heat treatment on wood colour consisted of darkening. The darkening, clearly visible even to the naked eye, was supported almost equally by both the lightness parameter (L*) and the a* parameter, which indicates the red component of the CIELAB colour space. Our eye begins to perceive colour differences with ΔE* greater than three, therefore, the value obtained indicated a considerable variation. This empirical observation was confirmed by the chromatic coordinates' values. Lightness of treated ayous was recorded to be like that found in chestnut thermally treated at 200 °C [4]. The wood colour is determined by its chemical components and the colour variation indicates chemical modifications [32–34]. In fact, the products of lignin degradation, including phenolic substances and quinones, contribute to the darkening of wood [35,36].

As expected, the mechanical properties of the heat-treated wood showed low values. These findings confirm those of earlier studies, where a general reduction of the mechanical properties was described after the thermal modification of wood obtained from other species [5,37]. This trend was highlighted in the comparison with untreated ayous wood.

In detail, the shrinkages were the features' high percentage of reduction, and these wood characteristics were then improved, while the other wood features analysed showed modest percentages of reduction but with a completely opposite meaning: these wood characteristics were negatively affected by heat treatment.

The explanation for the change in physical, colorimetric, and mechanical characteristics induced by heat treatment is related to the chemical modification of the molecular components of the cell wall. As Tjeerdsma et al. [36] observed, the chemical composition of wood after heat treatment is mainly related to the thermal degradation of hemicelluloses, which initially leads to the formation of acetic acid, which in turn catalyses the degradation of carbohydrates into simple molecules. Generally, there is also an increase in lignin content, this is due to both the apparent increase in lignin content related to the loss of other components subject to more rapid degradation caused by heat, but also, as reported by Tjeerdsma et al. [36], to the self-condensation phenomenon of lignin that is thought to occur due to the formation of methylene bridges connecting aromatic rings of various organic compounds to some lignin molecules that have available reactive sites [36]. The degradation of hemicelluloses occurs as soon as the temperature reaches 180–200 °C, earlier and faster than cellulose because of their lower molecular weight and amorphous structure [38]; even the amorphous regions of cellulose are structurally modified, obtaining less hygroscopic compounds, while crystalline cellulose is less affected by the changes related to heat treatment up to 300 °C [37,39]. In addition, the reduction of hygroscopicity is also related to the formation of lignin–carbohydrate compounds induced by the exposure to high temperature. This leads to the surrounding of the cellulose microfibrils by a network of more rigid compounds reducing their expansion possibilities and, as a consequence, their absorption capacity. This results in reduced cell wall swelling and lower fibre saturation point of the wood [36]. Another effect of heat treatment, as reported by [40], is the improvement of the wood permeability that allows a better penetration of polar and nonpolar compounds into the texture of the wood. This phenomenon is related to the reduction of the wettability of the material, due to the lower availability of free polar sites on the cell wall compounds, and this can increase the permeability of both water-based and oil-based preservative compounds, which can contribute to a further improvement of the material durability [40].

5. Conclusions

In this study, the modifications induced on the mechanical, physical, and colorimetric characteristics were evaluated. Statistically significant differences were found in physical and mechanical properties. The heat treatment resulted in a reduction in compression strength (26.98 ± 1.99 MPa) of 26%, static bending strength (32.90 ± 3.09 MPa) of 46%, and Brinell hardness (8.30 ± 1.05 N/mm^2) of 32%. A light ayous wood was confirmed since dry density was found to be 0.31 ± 0.02 g/cm^3. The densities were reduced between 15% and 9%. The positive effect of heat treatment was detected in shrinkage and swelling. Shrinkages were rather low and ayous showed an appreciable dimensional stability, also expressed by the shrinkage anisotropy factor (1.55), and the anti-shrinkage efficiency (58.41 ± 5.86). The darkening (ΔE^* 34.76) was clearly detected (L* 39.69 ± 1.13; a* 10.59 ± 081; b* 18.73 ± 1.51), and it was supported almost equally by both the lightness parameter (L*) and the a* chromatic parameter.

The rapid growth and easy processing of *Triplochiton scleroxylon* wood has made this species increasingly attractive on the international market. A serious concern for outdoor use is its poor durability, and heat treatment can be a processing method that allows ayous wood to improve its stability, durability, and prolong its service life, with reduced environmental impact in service, and after dismantling.

Author Contributions: Conceptualization, A.L.M. and R.P.; methodology, A.L.M., R.P., and E.G.; validation, A.L.M., E.G., and R.P.; formal analysis, A.L.M.; investigation, A.L.M., R.P., and E.G.; resources, A.L.M.; data curation, A.L.M., R.P., and E.G.; writing—original draft preparation, A.L.M., E.G.; writing—review and editing, A.L.M., R.P., and E.G.; visualization, E.G.; supervision, A.L.M.; project administration, A.L.M.; funding acquisition, A.L.M. and R.P. All authors have read and agreed to the published version of the manuscript.

Funding: This research did not receive any specific grant.

Acknowledgments: Authors are grateful to Vasto Legno spa, who donated the industrially heat-treated and untreated wooden planks, used in this study. This work was supported by the Italian Ministry for education, University and Research (MIUR) for financial support (Law 232/2016, Italian University Departments of excellence)—UNITUS -DAFNE WP3 and WP4 (Rodolfo Picchio and Angela Lo Monaco) and for basic research activities of Angela Lo Monaco.

Conflicts of Interest: The authors declare no conflict of interest. The funders had no role in the design of the study; in the collection, analyses, or interpretation of data; in the writing of the manuscript, or in the decision to publish the results.

References

1. Höglmeier, K.; Steubing, B.; Weber-Blaschke, G.; Richter, K. LCA-based optimization of wood utilization under special consideration of a cascading use of wood. *J. Environ. Manag.* **2015**, *152*, 158–170. [CrossRef]
2. Jones, D.; Sandberg, D.; Goli, G.; Todaro, L. *Wood Modification in Europe: A State-of-the-Art about Processes, Products and Applications*; Firenze University Press: Firenze, Italy, 2019.
3. Esteves, B.M.; Pereira, H.M. Wood modification by heat treatment: A review. *BioResource* **2009**, *4*, 370–404. [CrossRef]
4. Lo Monaco, A.; Pelosi, C.; Agresti, G.; Picchio, R.; Rubino, G. Influence of thermal treatment on selected properties of chestnut wood and full range of its visual features. *Drewno* **2020**, *63*, 1–20.
5. Hill, C.A.S. *Wood Modification: Chemical, Thermal and Other Processes*; John Wiley and Sons, Ltd.: Chichester, UK, 2006.
6. Humar, M.; Lesar, B.; Kržišnik, D. Moisture Performance of Façade Elements Made of Thermally Modified Norway Spruce Wood. *Forests* **2020**, *11*, 348. [CrossRef]
7. Alao, P.; Visnapuu, K.; Kallakas, H.; Poltimäe, T.; Kers, J. Natural Weathering of Bio-Based Façade Materials. *Forests* **2020**, *11*, 642. [CrossRef]
8. Scheffer, T.C.; Morrell, J.J. *Natural Durability of Wood: A Worldwide Checklist of Species. Forest Research Laboratory*; Oregon State University; Research Contribution: Corvallis, OR, USA, 1998; Volume 22, p. 58.
9. Kukachka, B.F. *Characteristics of Some Imported Woods. Foreign Wood*; Series Report 2242; USDA Forest Service, Forest Products Laboratory: Madison, WI, USA, 1962.
10. Fabiyi, J.S.; Ogunleye, B.M. Mid-infrared spectroscopy and dynamic mechanical analysis of heat-treated obeche (*Triplochiton scleroxylon*) wood. *Maderas-Cienc. Tecnol.* **2015**, *17*, 5–16. [CrossRef]
11. Fotsing, J.A.M.; Fokoua, A.D.S. Effects of thermal modification by the hot oil treatment process on some physical properties of two Cameroonian hardwood species. *Int. J. Heat. Technol.* **2012**, *30*, 43–50. [CrossRef]
12. Zor, M. Water Resistance of Heat-treated Welded Iroko, Ash, Tulip, and Ayous Wood. *BioResources* **2020**, *15*, 9584–9595.
13. Kamdem, D.P.; Pizzi, A.; Jermannaud, A. Durability of heat-treated wood. *Wood Raw Mater.* **2002**, *60*. [CrossRef]
14. Gennari, E.; Picchio, R.; Tocci, D.; Lo Monaco, A. Modifications of Physical and Mechanical Characteristics Induced by Heat Treatment: Case Study on Ayous Wood (*Triplochiton scleroxylon* K. Schum). *Environ. Sci. Proc.* **2021**, *3*, 27. [CrossRef]
15. International Organization for Standardization. *ISO 3129 Wood—Sampling Methods and General Requirements for Physical and Mechanical Testing of Small Clear Wood Specimens*; International Organization for Standardization: Geneve, Switzerland, 2019.
16. Ente Nazionale Italiano di Unificazione. *UNI ISO 13061-1 Physical and Mechanical Properties of Wood—Test Methods for Small Clear Wood Specimens Determination of Moisture Content for Physical and Mechanical Tests*; Ente Nazionale Italiano di Unificazione: Milano, Italy, 2017.
17. Ente Nazionale Italiano di Unificazione. *UNI ISO 13061-2 Physical and Mechanical Properties of Wood—Test Methods for Small Clear Wood Specimens Determination of Density for Physical and Mechanical Tests*; Ente Nazionale Italiano di Unificazione: Milano, Italy, 2017.
18. International Organization for Standardization. *ISO 13061-14 Physical and Mechanical Properties of Wood—Test Methods for Small Clear Wood Specimens Determination of Volumetric Shrinkage*; International Organization for Standardization: Geneve, Switzerland, 2016.
19. Ferreira, R.C.; Monaco, A.L.; Picchio, R.; Schirone, A.; Vessella, F.; Schirone, B. Wood anatomy and technological properties of an endangered species: *Picconia azorica* (*Oleaceae*). *IAWA J.* **2012**, *33*, 375–390. [CrossRef]

20. Lo Monaco, A.; Marabelli, M.; Pelosi, C.; Picchio, R. Color measurements of surfaces to evaluate the restoration materials. In *O3A: Proc Spie*; Pezzati, L., Salimbeni, R., Eds.; SPIE: Washington, DC, USA, 2011; Volume 8084, pp. 1–14.
21. Ente Nazionale Italiano di Unificazione. *UNI-EN-15886. Conservation of Cultural Property—Test Methods—Colour Measurement of Surfaces*; Ente Nazionale Italiano di Unificazione: Milano, Italy, 2010.
22. Lo Monaco, A.; Todaro, L.; Sarlatto, M.; Spina, R.; Calienno, L.; Picchio, R. Effect of moisture on physical parameters of timber from Turkey oak (*Quercus cerris* L.) coppice in Central Italy. *For. Stud. China* **2011**, *13*, 276–284. [CrossRef]
23. Ente Nazionale Italiano di Unificazione. *UNI ISO 3787 Wood—Test Methods—Determination of Ultimate Stress in Compression Parallel to Grain*; Ente Nazionale Italiano di Unificazione: Milano, Italy, 1985.
24. International Organization for Standardization. *ISO 13061-3 Determination of Ultimate Tensile Strength in Static Bending—Physical and Mechanical Properties of Wood—Test Methods for Small Clear Wood Specimens*; International Organization for Standardization: Geneve, Switzerland, 2014.
25. Ente Nazionale Italiano di Unificazione. *UNI EN 1534 Wood Flooring and Parquet—Determination of Resistance to Indentation—Test Method*; Ente Nazionale Italiano di Unificazione: Milano, Italy, 2020.
26. Gennari, E.; Picchio, R.; Lo Monaco, A. Physical and Mechanical Properties of *Triplochiton scleroxylon* K. Schum from Cameroon. *Drewno* **2021**. under review.
27. Giordano, G. *Tecnologia del Legno*; UTET: Torino, Italy, 1988; Volume 3/II, pp. 1091–1093.
28. Berti, R.N.; Abbate, M.L.E. *Legnami Tropicali Importati in Italia: Anatomia ed Identificazione*, 1st ed.; Ribera Editore: Milan, Italy, 1988; pp. 200–201.
29. Olorunnisola, A.O. *Anatomy and Physical Properties of Tropical Woods, Design of Structural Elements with Tropical Hardwoods*; Springer International Publishing: Cham, Switzerland, 2018; Chapter 2; pp. 7–29.
30. Calienno, L.; Pelosi, C.; Picchio, R.; Agresti, G.; Santamaria, U.; Balletti, F.; Lo Monaco, A. Light-induced color changes and chemical modification of treated and untreated chestnut wood surface. *Stud. Conserv.* **2015**, *60*, 131–139. [CrossRef]
31. Sedliačiková, M.; Moresová, M.; Aláč, P.; Malá, D. What Is the Supply and Demand for Coloured Wood Products? An Empirical Study in Slovakian Practice. *Forests* **2021**, *12*, 530. [CrossRef]
32. Agresti, G.; Bonifazi, G.; Calienno, L.; Capobianco, G.; Lo Monaco, A.; Pelosi, C.; Picchio, R.; Serranti, S. Surface investigation of photo-degraded wood by colour monitoring, infrared spectroscopy and hyperspectral imaging. *J. Spectrosc.* **2013**. [CrossRef]
33. Pelosi, C.; Agresti, G.; Calienno, L.; Lo Monaco, A.; Picchio, R.; Santamaria, U.; Vinciguerra, V. Application of spectroscopic techniques for the study of the surface changes in poplar wood and possible implications in conservation of wooden artefacts. *Proc. SPIE* **2013**, *8790*, 1–14.
34. Calienno, L.; Lo Monaco, A.; Pelosi, C.; Picchio, R. Colour and chemical changes on photodegraded beech wood with or without red heartwood. *Wood Sci. Technol.* **2014**, *48*. [CrossRef]
35. González-Peña, M.M.; Hale, M.D.C. Colour in thermally modified wood of beech, Norway spruce and Scots pine. I. Colour evolution and colour changes. *Holzforschung* **2009**, *63*, 385–393. [CrossRef]
36. Tjeerdsma, B.F.; Boonstra, M.; Pizzi, A.; Tekely, P.; Militz, H. Characterisation of thermally modified wood: Molecular reasons for wood performance improvement. *Wood Raw Mater.* **1998**, *56*, 149–153. [CrossRef]
37. Yildiz, S.; Gezer, E.D.; Yildiz, U.C. Mechanical and chemical behavior of spruce wood modified by heat. *Build. Environ.* **2006**, *41*, 1762–1766. [CrossRef]
38. Lewin, M.; Goldstein, I.S. Overview of the chemical composition of wood. In *Wood Structure and Composition*; Marcel Dekker Inc.: New York, NY, USA, 1991; pp. 1–6.
39. Yildiz, S.; Gümüskaya, E. The effects of thermal modification on crystalline structure of cellulose in soft and hardwood. *Build. Environ.* **2007**, *42*, 62–67. [CrossRef]
40. Taghiyari, H.R.; Abbasi, H.; Militz, H.; Papadopoulos, A.N. Fluid Flow of Polar and Less Polar Liquids through Modified Poplar Wood. *Forests* **2021**, *12*, 482. [CrossRef]

Article

Trichoderma spp. from Pine Bark and Pine Bark Extracts: Potent Biocontrol Agents against Botryosphaeriaceae

Vera Karličić [1,*], Milica Zlatković [2], Jelena Jovičić-Petrović [1], Milan P. Nikolić [3], Saša Orlović [2] and Vera Raičević [1]

[1] Faculty of Agriculture, University of Belgrade, 11000 Belgrade, Serbia; jelenap@agrif.bg.ac.rs (J.J.-P.); verar@agrif.bg.ac.rs (V.R.)

[2] Institute of Lowland Forestry and Environment (ILFE), University of Novi Sad, 21102 Novi Sad, Serbia; milica.zlatkovic@uns.ac.rs (M.Z.); sasao@uns.ac.rs (S.O.)

[3] Faculty of Agronomy, University of Kragujevac, 32000 Čačak, Serbia; milanik@kg.ac.rs

* Correspondence: vera.karlicic@agrif.bg.ac.rs; Tel.: +381-6428-17485

Abstract: *Pinus sylvestris* bark represents a rich source of active compounds with antifungal, antibacterial, and antioxidant properties. The current study aimed to evaluate the antifungal potential of *P. sylvestris* bark against *Botryosphaeria dothidea*, *Dothiorella sarmentorum*, and *Neofusicoccum parvum* (Botryosphaeriaceae) through its chemical (water extracts) and biological (*Trichoderma* spp. isolated from the bark) components. The water bark extracts were prepared at two temperatures (80 and 120 °C) and pH regimes (7 and 9). The presence of bark extracts (30%) caused inhibition of mycelial growth of *B. dothidea* and *D. sarmentorum* for 39 to 44% and 53 to 60%, respectively. Moreover, we studied the antagonistic effect of three *Trichoderma* isolates originating from the pine bark. *Trichoderma* spp. reduced growth of *B. dothidea* by 67%–85%, *D. sarmentorum* by 63%–75% and *N. parvum* by 55%–62%. Microscopic examination confirmed typical mycoparasitism manifestations (coiling, parallel growth, hook-like structures). The isolates produced cellulase, β-glucosidase and N-acetyl-β-glucosaminidase. The volatile blend detected the emission of several volatile compounds with antimicrobial activity, including nonanoic acid, cubenene, cis-α-bergamotene, hexanedioic acid, and verticillol. The present study confirmed in vitro potential of *P. sylvestris* bark extracts and *Trichoderma* spp. against the Botryosphaeriaceae. The study is an important step towards the use of environmentally friendly methods of Botryosphaeriaceae disease control.

Keywords: Botryosphaeriaceae; biocontrol; pine bark extracts; *Trichoderma citrinoviride*; VOCs; lytic enzymes

Citation: Karličić, V.; Zlatković, M.; Jovičić-Petrović, J.; Nikolić, M.P.; Orlović, S.; Raičević, V. *Trichoderma* spp. from Pine Bark and Pine Bark Extracts: Potent Biocontrol Agents against Botryosphaeriaceae. *Forests* **2021**, *12*, 1731. https://doi.org/10.3390/f12121731

Academic Editors: Angela Lo Monaco, Cate Macinnis-Ng and Om P. Rajora

Received: 29 October 2021
Accepted: 3 December 2021
Published: 9 December 2021

Publisher's Note: MDPI stays neutral with regard to jurisdictional claims in published maps and institutional affiliations.

1. Introduction

Pesticides are generally considered a quick, easy, and inexpensive solution against plant pathogens. However, constant reliance on chemicals has led to the emergence of more virulent strains with higher resistance to active compounds [1,2]. Understanding the seriousness of this problem has triggered an intense search for alternative solutions, among which "naturally-based" products have attracted special attention. For example, a collaboration between sawmill industries and plant health experts has led to multiple solutions for plant disease control. Sawmill industries generate enormous amounts of bark waste which are mostly burned or disposed to landfills [3,4]. Removed bark represents raw material for substrate formulations, soil conditioners, a variety of human health and industrial products, and bioremediation agents [3,5]. Moreover, it also exhibits antifungal, antibacterial, and insecticidal properties which main carriers are compounds such as terpenes, phenolics, flavonoids, tannins, and pinosylvins [4–6]. In addition, a bark represents a habitat with a complex set of niches available to various microorganisms and communities [7]. Species of the Botryosphaeriaceae (Ascomycota: Botryosphaeriales) are important pathogens of forest, ornamental, fruit trees, and agricultural plants. These fungi are distributed worldwide causing a variety of symptoms such as crown die-back, cankers, leaf blights, and shot hole disease [8–11]. The Botryosphaeriaceae are difficult to control as they reside as endophytes and latent pathogens in wood, and the disease symptoms appear when host is under stress [12]. Moreover, these fungi colonize the xylem tissue, and most species are known as generalist pathogens, able to infect different taxonomically unrelated hosts [12–14]. This is of particular concern in urban environments which represent mixtures of native and non-native trees, conifers and broadleaves, and where environmental conditions are such that promote stress on the trees [9].

Fungi of the genus *Trichoderma* (Ascomycota: Hypocreales: Hypocreaceae) represent a large, ecologically diverse group of well-known biocontrol agents (BCA) [1]. Worldwide, more than 60% of registered biopesticides are *Trichoderma*-based [15]. The modes of biocontrol action are various and include competition, mycoparasitism, antibiosis, inactivation of the pathogen's enzymes, and induction of plant disease resistance [2,16]. Moreover, these fungi produce powerful secondary metabolites such as cell-wall degrading enzymes (cellulase, protease, chitinase), volatile organic compounds (VOCs) and non-volatile compounds. Many VOCs have been associated with *Trichoderma* spp., including sesquiterpenes, alcohols, ketones, lactones, esters, thioalcohols, thioesters, and cyclohexenes [15]. Some VOCs are directly involved in communication of *Trichoderma* spp. with their co-habitants and in antibiosis [15]. Moreover, the VOCs are usually involved in indirect antagonistic actions, by diffusing and affecting distant opponents [17].

The research of *Trichoderma* fungi as BCA has been directed towards diseases of agricultural plants and *Trichoderma* spp. proved to be effective antagonists of numerous pathogens, including *Armillaria* spp., *Botrytis* spp., *Fusarium* spp., *Phytophthora* spp., *Pythium* spp., *Rhizoctonia* spp., *Sclerotinia* spp., and *Verticillium* spp. [18,19]. Moreover, pine bark extracts have been shown to inhibit growth of *Coniophera puteana* (Schumach.) P.Karst., *Trametes versicolor* (L.ex. Fr.) Pilát, *Botrytis cinerea* Pers., *Colletotrichum acutatum* J.H. Simmonds, *Phytophthora cactorum* (Lebert and Cohn) J. Schröt., and *Mycosphaerella fragariae* (Tul.) Lindau [6,20]. However, the research on *Trichoderma* spp. as BCA of Botryosphaeriaceae has been limited and mostly focused on grapevine trunk diseases [21]. Moreover, except for a preliminary study [22], there has been no previous research related to pine bark extracts as a biological control option for fungi classified in the Botryosphaeriaceae.

In this regard, the aims of this study were: (1) to conduct in vitro evaluation of the antifungal activity of *P. sylvestris* bark extracts and *Trichoderma* spp. against *Botryosphaeria dothidea* (Moug. ex. Fr.) Ces. et de Not., *Dothiorella sarmentorum* (Fr.) AJL Phillips, A. Alves and J. Luque, and *Neofusicoccum parvum* (Pennycook and Samuels) Crous, Slippers and AJL Phillips (Ascomycota: Botryosphaeriaceae); (2) to conduct in vitro evaluation of antifungal effects of *Trichoderma*-VOCs on Botryosphaeriaceae mycelial growth; and (3) to detect effective VOCs and cell-wall degrading enzymes produced by *Trichoderma* spp. when confronted with the Botryosphaeriaceae.

2. Materials and Methods

2.1. Preparation of Pinus Sylvestris Bark Extracts

Pinus sylvestris bark extracts were prepared, and extraction yields calculated as described in Karličić et al. [22]. The water extracts were obtained at two temperatures (80 and 120 °C) and pH regimes (7 and 9) in three repetitions. The extraction yields represented a percentage of soluble bark powder and were 4% for extracts prepared at 80 °C and 5% for extracts prepared at 120 °C [22].

2.2. Fungal Isolates

Phytopathogenic isolates of *B. dothidea* (CMW 39314), *D. sarmentorum* (CMW 39365), and *N. parvum* (BOT 275) were isolated from *Picea abies* (L.) H. Karst., *Thuja occidentalis* L., and *Prunus laurocerasus* L., respectively. The identity and pathogenicity of the isolates was determined in previous studies [8,13].

The putative biocontrol fungal agents *Trichoderma* spp. were isolated from *P. sylvestris* bark using a serial dilution in a previous study by Karličić et al. [22] and kept in 20% glycerol at −80 °C until use. Morphological identification of the fungal isolates based on colony appearance and microscopic examination (Leica DMLS, Leica Microsystems GmbH, Wetzlar, Germany) was determined after 3-day incubation on PDA at 25 °C in darkness. Genomic DNA was extracted from five days-old cultures grown on sterile cellophane circles pierced with a sterile hypodermic needle by following the manufacturer's protocol for the ZR Soil Microbe DNA Kit (Zymo Research, Irvine, CA, USA). The ITS region of the rDNA was amplified using primers ITS1F [23] and ITS 4 [24]. Part of the *tef 1-α* gene and part of the *RPB2* gene were amplified using primers EF1 and EF2 and fRPB2-5f and fRPB2-7cr, respectively [25,26]. The 25 μL PCR reaction mixtures contained 2.5 μL of $10\times$ Taq buffer with $(NH_4)SO_4$ (Thermo Scientific, Waltham, MA, USA), 3 μL of 25 mM $MgCl_2$ (Thermo Scientific, Waltham, MA, USA), 1 μL of 100 mM of each dNTPs (Thermo Scientific, Waltham, MA, USA), 0.5 μL of 10 μM of each primer (Invitrogen, Paisley, UK), 2 μL (40 ng) of genomic DNA, 0.3 μL (1.5 U) of Taq DNA polymerase (Thermo Scientific, Waltham, MA, USA) and 15.2 μL of autoclave-sterilized ultra-pure water The PCR was performed in an Eppendorf Mastercycler epgradient S thermal cycler (Eppendorf AG, Hamburg, Germany) under the conditions described in Kovač et al. [27]. However, the *tef 1-α* region was amplified using annealing temperature of 60 °C instead of 55 °C.

The PCR products were purified and sequenced by Macrogen Europe (Amsterdam, The Netherlands).

Nucleotide sequences were examined for sequencing errors, edited, and assembled using BioEdit v. 7.2.5, and MEGA X, whereas sequence alignments were carried out using MEGA X and MAFFT v. 7 (on-line) as described in Zlatković et al. [28]. Sequences of the three loci (*ITS, tef 1-α, RPB2*) were analysed individually and in combination following the GCPSR concept [29]. The phylogenetic analyses of the individual genes were carried out using Maximum Likelihood (ML) analyses, whereas analyses of the combined datasets (*tef 1-α + RPB2*), and (*ITS + tef 1-α + RPB2*) were performed using ML, Maximum parsimony (MP) and Bayesian analyses (BI). The ML and MP analyses were conducted using PhyML v. 3.0 (on-line) and PAUP v. 40b10 as described by Zlatković et al. [8,28]. The BI analyses were carried out in MrBayes v. 3.0b4 as described in Kovač et al. [27]. The DNA sequences generated in this study were deposited in the GenBank genetic sequence database (Table S1 in Supplementary Materials).

2.3. In Vitro Assessment of Antifungal Activity of Pinus Sylvestris Bark Extracts

The antifungal activity of *P. sylvestris* bark extracts towards *B. dothidea*, *D. sarmentorum* and *N. parvum* was assessed, mycelial growth inhibition was calculated and the degree of toxicity was estimated as described in Karličić et al. [22]. The experiment was repeated three times.

2.4. In Vitro Assay of Trichoderma spp. Antifungal Activity

The antifungal activity of *Trichoderma* strains towards *B. dothidea*, *D. sarmentorum* and *N. parvum* was assessed, growth inhibition calculated, and the degree of antagonistic activity estimated as described by Karličić et al. [22]. The experiment was repeated three times. At the same time, microscopic observations of the fungal interactions were performed. Small amounts of mycelia from the plates with individual fungi (negative controls) and from the interaction zones of plates with dual cultures were collected using a sterile hypodermic needle, mounted in distilled water on microscope slides and examined using an Olympus BX53F light microscope (Olympus Co., Tokyo, Japan) with Nomarski differential interference contrast (DIC) equipped with an Olympus SC50 digital camera and accompanying software.

2.5. Biochemical Characterization of Trichoderma spp. Antifungal Activity

The biochemical characterization of *Trichoderma* spp. antifungal activity included the determination of cell wall-degrading enzymes and the production of siderophores. A semiquantitative determination of cell-wall degrading enzymes (lipase, esterase-lipase, N-acetyl-β-glucosaminidase and β-glucosidase) was performed using an API ZYM kit according to the manufacturer's protocol (BioMereux, Craponne, France). The presence of cellulase was determined using carboxymethyl cellulose (CMC) agar method in three repetitions [30]. Siderophore production was detected on the Chrome azurol S (Sigma-Aldrich, St. Louis, USA) agar medium in three repetitions [31]. The chrome azurol S (CAS) agar plates were inoculated with 5-mm-diameter mycelia discs of three *Trichoderma* isolates and incubated at 28 °C for 72 h. The appearance of yellow-orange halo zones around colonies was considered as a positive result.

2.6. The Effect of Trichoderma VOCs on Mycelial Growth of Botryosphaeriaceae

The effect of VOCs produced by *Trichoderma* strains on the mycelial growth of Botryosphaeriaceae was tested using the method of confronted cultures without contacts of the two mycelia [32]. The two Petri dishes containing 20 mL of potato dextrose agar (PDA, Himedia, India) were individually inoculated with 5-mm-diameter mycelia discs of a pathogen (Botryosphaeriaceae) and an antagonist (*Trichoderma* spp.). Inoculated plates were sealed with Parafilm®, arranged to face each other and incubated at 25 °C in a microbiological incubator (Binder, Tuttlingen, Germany) in the dark until fungi in the control plates (plates with individual fungi, negative control) reached edges of plates. The experiment was repeated three times. The effects of volatile metabolites were estimated as percentage of mycelial growth inhibition (MGI) calculated using the following equation:

$$MGI\ (\%) = ((DC - DT)/DC)) \times 100, \tag{1}$$

where MGI is mycelial growth inhibition, DC is the average diameter of a fungal colony of the control group, and DT is the average diameter of a fungal colony of the treatment group [33].

The antagonistic levels were estimated as described in Ruiz-Gómez et al. [34] and classified as low (MGI 50%); medium (50% < MGI ≤ 60%); high (60% < MGI ≤ 75%); and very high (MGI > 75%).

2.7. Collection of VOCs and GC−MS Analysis

To determine VOCs emitted by *Trichoderma* strains penicillin bottles containing 5ml PDA were inoculated with *Trichoderma* spp. The VOCs were collected six days after inoculation using headspace solid-phase micro-extraction followed by gas chromatography (Agilent Technologies 7890 B GC System, AIM, Littleton, CO, USA) coupled with mass spectrometry (Agilent Technologies 5977A MSD, AIM, Littleton, CO, USA). Briefly, the 0.2 g of the sample (PDA + *Trichoderma* strain) was placed in a headspace vial, followed by the addition of 0.5 mL of sterilized distilled water. Bottles containing only sterile PDA served as negative controls. Each vial was sealed using a cap with PTFE/silicone septa and incubated at 70 °C. The solid phase microextraction fiber (Polydimethylsiloxane (PDMS) 100 μm, Agilent Technologies, AIM, Littleton, CO, USA) was inserted into the head space of the vial containing the sample solution. The extraction was carried out at 70 °C with 90 min of fiber-exposed time. After sampling, the SPME fiber was withdrawn into the needle, removed from the tube, and inserted into the hot injector port (270 °C) of the GC system where the extracted analyte was desorbed and transferred to the analytical column (HP-5, Agilent Technologies, AIM, Littleton, CO, USA). A relatively long desorption time in the injector (10 min) was selected to avoid carryover between runs to ensure full desorption of analyte from the fiber. Ultra-high purity 5.0 grade helium (Messer Tehnogas AD, Belgrade, Serbia) was used as a carrier gas at a flow rate at 1.2 mL/min along with the spitless injection. The oven temperature was programmed for an initial 50 °C for 2 min and was then increased in two steps: 50–80 °C at a rate of 20 °C/min and held for 6 min at this temperature; 80–280 °C at a rate of 15 °C/min and 240–280 °C and held for 6 min at this temperature. During the analysis, data acquisition was carried out in full scan mode (*m/z* 27–350) operated in the electron ionization mode at 70 eV with a source temperature of 230 °C. Volatile compounds were identified by comparison with the National Institute of Standards and Technology (NIST) database. The VOCs that showed mass spectra with match factor ≥80% were considered as identified substances.

2.8. Statistical Analyses

The data were subjected to ANOVA followed by Tukey's HSD post-hoc comparison tests to determine if there were statistically significant differences between the means ($p = 0.05$). All statistical analyses were performed using Statistica 12.0 (StatSoft, Tulsa, OK, USA).

3. Results

3.1. In Vitro Assessment of Antifungal Activity of Pinus Sylvestris Bark Extracts

The antifungal properties of pine bark extracts were tested at two concentrations, i.e., 20% and 30% (Table 1). The concentration of 20% inhibited growth of *B. dothidea* for 34%–39% while the concentration of 30% increased the level of inhibition to 39%–44%. The pine bark extracts inhibited *D. sarmentorum* mycelia for 48%–66%. The lowest inhibition (48%) was obtained using neutral water extracts prepared at 120 °C, while alkaline water extract prepared at 120 °C showed the highest inhibition rate (66%). Pine bark extracts showed no signs of inhibition of radial growth of *N. parvum* (Table 1). However, the aerial mycelium of this fungus was sparse compared to the control plate.

Table 1. Mycelial growth inhibition (%) by pine bark water extracts.

Pine Bark Water Extracts		Water pH7/80 °C	Water pH9/80 °C	Water pH7/120 °C	Water pH9/120 °C
B. dothidea CMW 39314	20%	39 [bA]	36 [abA]	35 [aA]	34 [aA]
	30%	44 [aB]	43 [aB]	39 [aB]	40 [aB]
D. sarmentorum CMW 39365	20%	56 [aA]	56 [aA]	48 [aA]	60 [aA]
	30%	60 [abA]	62 [abB]	53 [aA]	66 [bA]
N. parvum BOT 275	20%	NI	NI	NI	NI
	30%	NI	NI	NI	NI

NI—no inhibition; mean values in the same row with different lowercase letters are significantly different according to the Tukey test ($p = 0.05$); mean values in the same column and same pathogen with different uppercase letters are significantly different according to the Tukey test ($p = 0.05$).

3.2. Identification of Trichoderma spp.

Morphology of the three isolates from the *P. sylvestris* bark was characterized by scarce mycelium producing typical diffusing yellow pigment which was particularly pronounced in isolate *T. citrinoviride* DEMf TR4. Conidiophores were sparsely branched, representing a long strongly developed central axis from which lageniform and mostly solitary phialides arise, bearing smooth-walled subglobose to ellipsoidal conidia. These morphological features corresponded to common features of the *Longibrachiatum* section of the genus *Trichoderma* [35]. Considering the difficulties differentiating *Trichoderma* species belonging to this section, the additional molecular analyses were performed. *Trichoderma* spp. were identified using phylogenetic analyses of the *ITS rDNA, TEF 1-α,* and *RPB2* genes. The concatenated datasets with the two loci (*tef 1-α* + *RPB2*) and three loci (*ITS* + *tef 1-α* + *RPB2*) had 1532 characters of which 577 characters were parsimony informative and 2164 characters of which 623 characters were parsimony informative, respectively. The PHT test showed that the loci can be combined ($p = 0.01$ for both concatenated datasets). The MP analyses of the concatenated dataset of the two loci produced a tree with CI = 0.6, RI = 0.7, TL = 1850, whereas the analyses of the three loci resulted in two equally most parsimonious trees with CI = 0.6, RI = 0.7, TL = 1948. The ML, BI and MP analyses of each concatenated dataset yielded trees with the similar topology (Figures 1, S1 and S2).

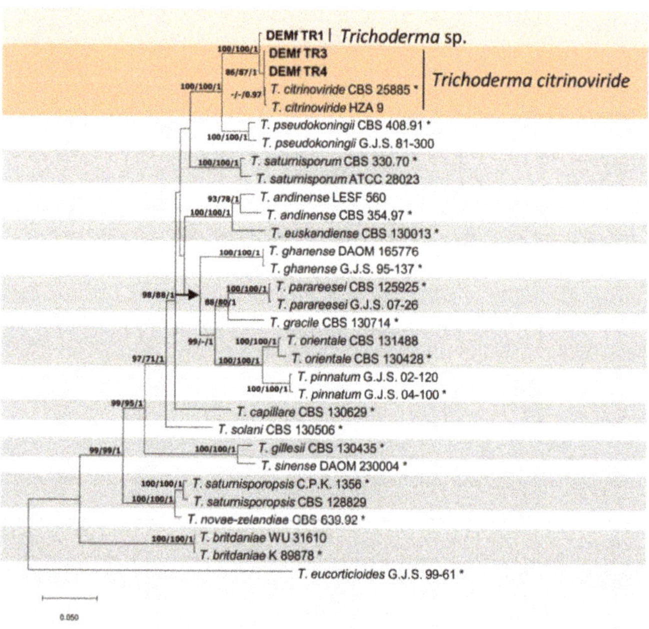

Figure 1. Phylogenetic tree generated from maximum likelihood analysis (ML) based on concatenated alignments of ITS, tef 1-α and RPB2 sequence data showing the position of *Trichoderma citrinoviride* in relation to its closely related species belonging to the *Longibranchiatum* clade. ML and maximum parsimony (MP) bootstrap support values greater than 70% and Bayesian posterior probability values (PP) greater than 0.95 are indicated at the tree nodes (ML/MP/PP). The type strains are marked with an asterisk and isolates sequenced in this study are shown in bold. *T. eucorticioides* G.J.S. 99-61 (clade Hypocreanum) is included as an outgroup. Scale bar indicates expected number of substitutions per site.

In the phylogenetic analyses of both single and combined loci isolates sequenced in this study clustered within a clade strongly supported in the analyses of the *RPB2* gene and fully supported in the analyses of the *tef 1-α* and combined analyses of the two and three loci (Figures 1 and S2). Isolates DEMf TR3 and DEMf TR4 from this study clustered within a sub-clade corresponding to *T. citrinoviride* (Figures 1 and S2). These isolates had only two single nucleotide polymorphisms (SNPs) that differentiated them from the type strain of *T. citrinoviride* CBS 25885 (Table S2). The sub-clade corresponding to *T. citrinoviride* was well-supported in all three analyses (86/87% ML, MP bootstrap support; posterior probability: 1).

Isolate DEMf TR1 clustered within a sub-clade corresponding to *T. citrinoviride* only in the individual analyses of the *ITS* and *tef 1-α* (Figure S1). In analyses of the *RPB2*, and combined analyses of *tef 1-α/RPB2* and *ITS/tef 1-α/RPB2* genes, the isolate clustered within a sister clade to *T. citrinoviride* (Figures 1 and S2). There were 14 bp differences that differentiated this isolate from the type strain of its phylogenetically closest relative *T. citrinoviride* (Table S2). Based on phylogenetic analyses, isolates from this study were identified as *T. citrinoviride* and *Trichoderma* sp.

3.3. In Vitro Assay of Trichoderma spp. Antifungal Activity

The *Trichoderma* isolates caused inhibition of Botryosphaeriaceae in confrontation test (Table 2). The highest percentage of *B. dothidea* growth inhibition (85%) was achieved by *T. citrinoviride* DEMf TR4, followed by *Trichoderma* sp. DEMf TR1. This level of antagonistic activity is characterized as very strong. Moreover, *T. citrinoviride* DEMf TR4 and *Trichoderma* sp. DEMf TR1 showed very high antagonistic activity towards D. sarmentorum. *T. citrinoviride* DEMf TR4 was the only *Trichoderma* isolate to show high antagonistic activity against *N. parvum*.

Table 2. Mycelial growth inhibition (%) of *B. dothidea*, *D. sarmentorum* and *N. parvum* by *Trichoderma* spp. isolated from *P. sylvestris* bark.

Pathogen	*Trichoderma* sp. DEMf TR 1	*T. citrinoviride* DEMf TR3	*T. citrinoviride* DEMf TR4
B. dothidea CMW 39314	76 [ab]	67 [a]	85 [b]
D. sarmentorum CMW 39365	75 [b]	63 [a]	75 [b]
N. parvum BOT 275	59 [b]	55 [a]	62 [c]

Mean values in the same row with different lowercase letters are significantly different according to Tukey test ($p = 0.05$).

The confrontation test revealed different competition strategies. Interaction of *Trichoderma* strains with *B. dothidea* and *D. sarmentorum* resulted in overgrowth with replacement [36]. Interaction of *Trichoderma* strains with *N. parvum* (Figure 2) was labeled as a deadlock at distance [36] manifested through the presence of an inhibition zone. Moreover, the co-inoculation of these fungi resulted in enhanced production of dark pigmentation in *N. parvum*.

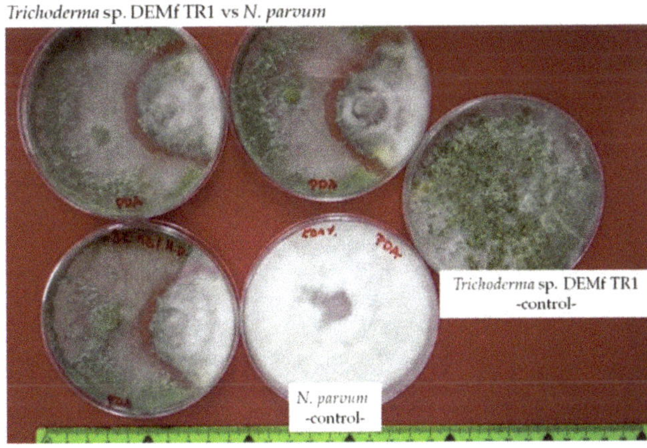

Figure 2. Plate confrontation assay of *Trichoderma* sp. DEMf TR 1 and *Neofusicoccum parvum*.

Microscopic observations of the *Trichoderma*-pathogen meeting zone (Figure 3) revealed different manifestations of BCA action and suggested mycoparasitism as a mode of action. Briefly, the hyphae of *Trichoderma* sp. DEMf TR1 grew alongside and coiled compactly around the *B. dothidea* hypha. The hook-like structures of *Trichoderma* spp. were formed around *B. dothidea* and *D. sarmentorum* hyphae. Moreover, several other morphological alternations were observed in *B. dothidea*, *D. sarmentorum* and *N. parvum* hyphae, such as vesicle-like structures, vacuolation and cytoplasmatic coagulation as a response to *Trichoderma* spp. presence. In addition, the micrograph of the meeting zone showed that both *Trichoderma* spp. and Botryosphaeriaceae isolates, reacted to mutual recognition by intensive production of chalmydospores whereas chlamydospores were rarely formed in the control Petri dishes.

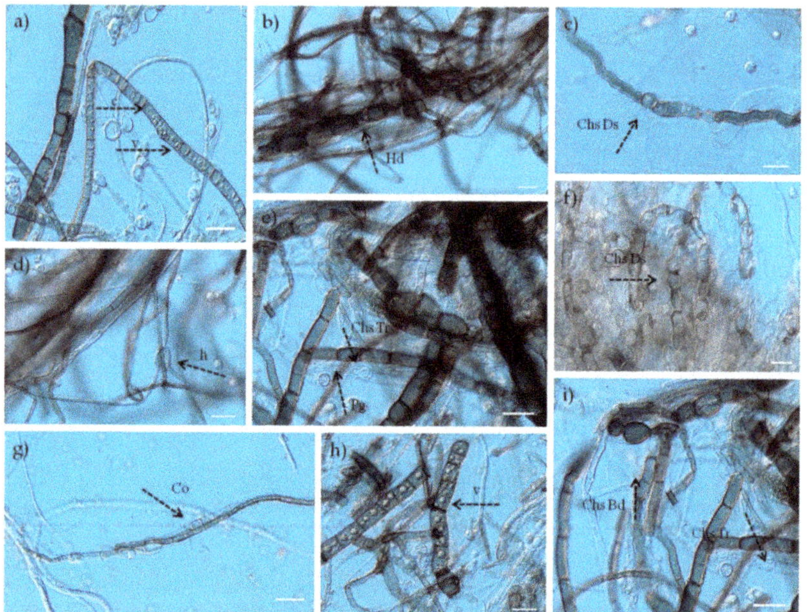

Figure 3. Observations of mycelial interactions between *Trichoderma* spp. and *B. dothidea* and *D. sarmentorum*; (**a**): c—cytoplasmatic coagulation, v—vacuolation; (**b**): Hd—hyphal desintegration; (**c**): Chs Ds—chlamydospores of *D. sarmentorum*; (**d**): h—hook-like structures; (**e**): ChsTr—chlamydospores of *T. citrinoviride* DEMf TR4, Pg—parallel growth; (**f**): Chs Ds—chlamydospores of *D. sarmentorum*; (**g**): Co—coiling; (**h**): v—vacuolation; (**i**): Chs Bd—chlamydospores of *B. dothidea*, ChsTr-chlamydospores of *T. citrinoviride* DEMf TR4; scale bar 20 µm.

3.4. Biochemical Characterization of Trichoderma spp. Antifungal Activity

Semiquantitative analyses of *Trichoderma* spp. enzymatic profiles showed the ability of these fungi to produce lipase and esterase-lipase at a moderate level as well as high amounts of N-acetyl-β-glucosaminidase. *T. citrinoviride* DEMf TR3 produced high amounts of β-glucosidase; *Trichoderma* sp. DEMf TR1 was marked as a moderate producer, while *T. citrinoviride* DEMf TR4 produced low amounts of the enzyme (Table 3). The three *Trichoderma* isolates were capable of producing cellulase. The CAS assay confirmed the ability of *Trichoderma* spp. to produce siderophores.

Table 3. Biochemical characteristics of *Trichoderma* spp. used in this study.

Strains	Cell-Wall Degrading Enzymes					Sid
	Lipase	Est-Lip	Na-β	β-Glu	Cell	
Trichoderma sp. DEMf TR 1	2	2	3	2	+	+
T. citrinoviride DEMf TR3	2	2	3	3	+	+
T. citrinoviride DEMf TR4	2	2	3	1	+	+

(est-lip)—esterase-lipase, (Na-β)—N-acetyl-β-glucosaminidase, (β-glu)—β-glucosidase, (cell)—cellulase, (sid)—siderophores; 1—low production, 2—moderate production, and 3—high production according to the API ZYM reading color scale; + positive reaction.

3.5. Effect of Trichoderma VOCs on Mycelial Growth of Botryosphaeriaceae

The VOCs emitted by *Trichoderma* sp. DEMf TR 1, *T. citrinoviride* DEMf TR3 and *T. citrinoviride* DEMf TR4 inhibited mycelial growth of *D. sarmentorum* for 35%, 40%, and 41%, respectively. Moreover, the VOCs induced loss of mycelia pigmentation in *B. dothidea* (Figure S3). The *N. parvum* growth was not inhibited. The GC-MS analysis of *Trichoderma* isolates used in this study confirmed equal profiles of the volatile compounds emitted by *Trichoderma* sp. DEMf TR1 and *T. citrinoviride* DEMf TR4. *T. citinoviride* DEMf TR3 produced six compounds and showed different spectra of VOCs. The exception was trichoacorenol which was produced by all three *Trichoderma* isolates (Figure 4).

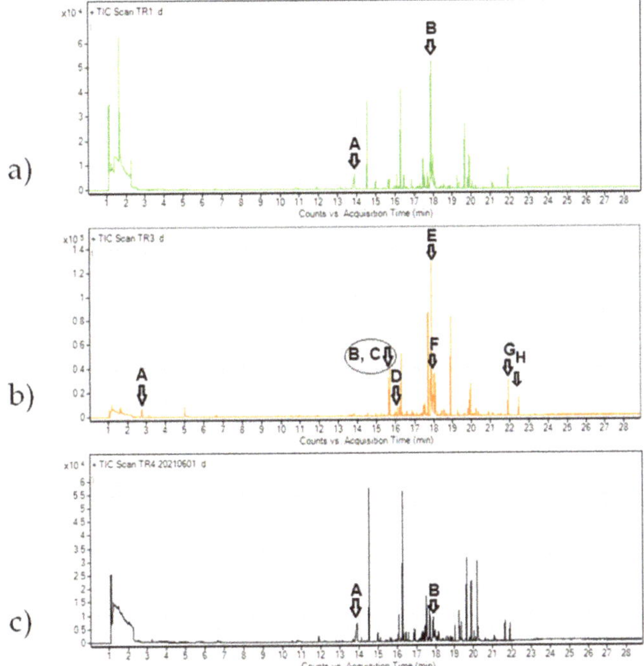

Figure 4. HS-GC-MS profiles of VOCs emitted by *Trichoderma* isolates: (**a**) *Trichoderma* sp. DEMf TR1: A: Nonanoic acid, B: Trichoacorenol; (**b**) *T. citrinoviride* DEMf TR3: A: Acetic acid, B: Cubenene, C: Cubenene, D: cis-α-bergamotene, E: Trichoacorenol, F: Trichoacorenol, G: Verticillol, H: Hexanedioic acid, bis(2-ethylhexyl) ester; (**c**) *T. citrinoviride* DEMf TR4: A: Nonanoic acid, B: Trichoacorenol.

Two compounds were detected in *Trichoderma* sp. DEMf TR1 and *T. citrinoviride* DEMf TR4 VOCs spectrum, i.e., trichoacorenol and nonanoic acid. An acetic acid, cubenene, cis-α-bergamotene, trichoacorenol, verticillol, hexanedioic acid, and bis(2-ethylhexyl) ester were detected in *T. citrinoviride* DEMf TR3 spectrum (Table 4). The amount of trichoacorenol in *Trichoderma* sp. DEMf TR1 and *T. citrinoviride* DEMf TR3 was similar and higher compared to the amount of this compund detected in *T. citrinoviride* DEMf TR4. Nonanoic acid emitted by *Trichoderma* sp. DEMf TR1 and *T. citrinoviride* DEMf TR4 were absent in *T. citrinoviride* DEMf TR3 VOCs profile, but another fatty acid, namely hexanedioic acid, was detected in VOCs spectra of this isolate. Cubenene (syn. naphthalene) was the second most abundant compund in *T. citrinoviride* DEMf TR3 VOCs blend with 8.25 relative abundance.

Table 4. VOCs emitted by *Trichoderma* sp. DEMf TR1, *T. citrinoviride* DEMf TR3 and *T. citrinoviride* DEMf TR4 identified by HS-GC-MS.

Isolate	Retention Time(min)	Peak	Volatile Compound	Relative Abundance
Trichoderma sp.	13.8756	A	Nonanoic acid	4.84
DEMf TR 1	17.8761	B	Trichoacorenol	17.68
	2.7580	A	Acetic acid	1.19
	15.6375	B	Cubenene	3.75
	15.7034	C	Cubenene	4.54
T. citrinoviride	16.2008	D	cis-α-bergamotene	0.93
DEMf TR3	17.8873	E	Trichoacorenol	18.88
	17.9652	F	Trichoacorenol	1.54
	21.9053	G	Verticillol	3.43
	22.4483	H	Hexanedioic acid	1.64
T. citrinoviride	13.9	A	Nonanoic acid	5.38
DEMf TR4	17.8773	B	Trichoacorenol	3.78

4. Discussion

The present study showed biocontrol potential of *P. sylvestris* bark estimated through chemical (water extracts) and biological (*T. citrinoviride* and *Trichoderma* sp.) components against three Botryospaheriaceae species, i.e., *B. dothidea*, *D. sarmentorum* and *N. parvum*. The *Trichoderma* isolates showed the ability to use multiple biocontrol mechanisms and to produce various antifungal substances from cell-wall degrading enzymes, siderophores to VOCs. The VOCs spectra analyses revealed presence of many volatile compounds among which nonanoic acid, cubenene, cis-α-bergamotene, hexanedioic acid, and verticillol were previously unknown as *T. citrinoviride* metabolites.

In this study, *P. sylvestris* bark extracts showed antifungal effects against *B. dothidea*, *D. sarmentorum* and *N. parvum*. The effects on *B. dothidea* mycelium growth were characterized as little toxic following the classification of Mori et al. [37]. Moreover, the bark extracts showed moderately toxic effect on *D. sarmentorum* while extracts were non-toxic towards *N. parvum*. These results are in accordance with the research of Alfredsen et al. [38] who reported slightly toxic effects of Scots pine extracts toward *Heterobasidion annosum* (Fr.) Bref., *Nectria ditissima* Tul. and C.Tul., *Ceratocystis polonica* Siem. (C. Moreau) and moderate level of its toxicity towards *Phacidium coniferarum* (Hahn) DiCosmo, Nag Raj and W.B. Kendr. Similarly, Vek et al. [39] used 5% pine bark extracts and reported moderate level of its toxicity towards several fungi, including *Schizophyllum commune* Fr., *Gloeophyllum trabeum* (Pers.) Murrill, and *Fibroporia vaillantii* (DC.) Parmasto. Several studies indicate the high toxicity of the extract to phytopathogens. In a study by Minova et al. [6] pine and spruce bark extracts (2%) showed high effectiveness in growth inhibition of several strawberry pathogens, including *B. cinerea*, *C. acutatum*, and *P. cactorum*.

Lomeli-Ramírez et al. [40] reported toxic effects of extracts from *P. strobus*, *P. douglasiana*, *P. caribea and P. leiophylla* toward *T. versicolor*. Similarly, Maritime pine extracts exhibited a significant antifungal potency against *C. puteana* and *T. versicolor* which mycelia growth was reduced by 89% and 87%, respectively [20]. In the study, even though water extracts were prepared on different temperature, duration and pH regimes, in most cases, the method of preparation did not influence the antifungal effects on *B. dothidea*, *D. sarmentorum* and *N. parvum*. However, chemical composition of the extracts was not determined and future work should also include further optimization of the extraction conditions. Nevertheless, this is the first study to show that *P. sylvestris* bark extracts could serve as biocontrol agents against Botryosphaeriaceae. Further studies should include *in planta* screening of the bicontrol potential of the extracts. The representatives of *Trichoderma* gender are well-known as antifungal agents with a long history of usage as BCA against a wide range of pathogens [21]. Several products based on *Trichoderma* are already used as protectants against grapevine trunk diseases caused by botryosphaeriaceous fungi [21].

The assay of *Trichoderma* spp. antagonistic capabilities toward three representatives of Botryospheriaceae showed that all three isolates obtained from the pine bark are effective in reducing mycelial growth. The antagonistic activity of *Trichoderma* spp. was marked as high to very high in the case of *B. dothidea* and *D. sarmentorum* and the isolates inhibited the mycelial growth of the pathogens by up to 85% and 75%, respectively. The *Trichoderma* isolates expressed moderate to high antagonistic activity against *N. parvum* by inhibiting mycelial growth by up to 62%. These results are consistent with several in vitro experiments conducted with botryosphaeriaceous fungi [21,41]. Úrbez-Torres [21] reported high antagonistic activity of *T. atroviride* who inhibited the growth of *Diplodia seriata* De Not. by 70%, and *T. koningiopsis* who inhibited the growth of *N. parvum* by 74%. Mondello et al. [41] reviewed *T. atroviride* and *T. harzianum* as highly efficient against *D. seriata*, *Lasiodiplodia theobromae* (Pat.) Griffon and Maubl, *Neofusicoccum australe* Slippers, Crous and M.J. Wingf., *N. parvum* and *T. longibrachiatum* as highly efficient against *D. seriata*. Marraschi et al. [42] reported *T. asperelloides*, *T. asperellum*, and *T. koningiopsis* as antagonists of *L. theobromae*. Moreover, *T. citrinoviride* expressed high antagonistic activity toward *Rhizoctonia solani* J.G. Kühn, *B. cinerea*, *Alternaria panax* Whetzel, *Cylindrocarpon destructans* Gerlach and L. Nilsson, *P. cactorum* and *Pythium* spp. by inhibiting mycelial growth up to 90% [43]. Similarly, Kuzmanovska et al. [44] confirmed strong antagonistic properties of *T. asperellum* and *T. harzianum* against *B. cinerea* whose growth was inhibited 77%–97% and 71%–94%, respectively.

In vitro confrontations of three *Trichoderma* isolates and representatives of Botryosphaeriaceae revealed two antagonistic strategies, i.e., overgrowth and production of antifungal compounds. This is in accordance with the work of Kotze et al. [45] who reported formation of inhibition zones between *T. atroviride* and *D. seriata*, *L. theobromae*, and *N. australe*., and *T. atroviride* overgrowth of *N. parvum*. According to Pellan et al. [46] overgrowth of the pathogen suggests stabile interaction of BCA with pathogen metabolism which indicates mycoparasitism as the main mechanism. Moreover, microscopical observations (coiling, parallel growth, hook-like structures) suggested mycoparasitic nature of *Trichoderma* isolates. Similarly, Kotze et al. [45] reported coiling and hyphal adhesion during interactions of *T. atroviride* with *N. parvum*. Moreover, in a study by Park et al. [43] *T. citrinoviride* showed ability to stick closely and coil around hyphae of *B. cinerea* and *R. solani*. Similar microscopic manifestations of direct mycoparasitism were observed during *T. asperellum* and *T. harzianum* confrontation with *B. cinerea* [44]. Microscopic observations conducted in this study revealed changes in the cytoplasm (coagulation, vacuolation) which are of great importance since they lead to hyphae disintegration and death [47]. According to Krause et al. [36] those changes indicate that antagonistic action is a result of biocontrol mechanisms such as production of antibiotic compounds or competition for nutrients.

In this study, enzymatic profiles of *Trichoderma* spp. confirmed the ability of these fungi to produce cell-wall degrading enzymes such as esterase-lipase, N-acetyl-β-glucosaminidase, β-glucosidase, and cellulase. The common constituents of fungal cell-walls are chitin and glucan susceptible to the presence of the extracellular enzymes such as lipases, chitinase, N-acetyl-β-glucosaminidase, β-glucosidase and protease [48,49]. *Trichoderma* isolates expressed high production of N-acetyl-β-glucosaminidase which is already recognized by Lorito et al. [50] as a metabolite of *T. harzianum* and an effective inhibitor of *B. cinerea* spore germination. Similarly, Park et al. [43] reported a low level of β-glucosidase activity of *T. citrinoviride* which is consistent with enzymatic activity of *Trichoderma* isolates in this study. The three isolates of *Trichoderma* spp. produced cellulase and this coincides with previous studies which described *T. citrinoviride* as a producer of strong cellulases [43]. Moreover, Nidhina et al. [48] showed that this enzyme inhibited mycelia growth of *Phytophthora* spp. Furthermore, *Trichoderma* isolates from this study expressed another significant biocontrol mechanism, i.e., production of siderophores. This is in accordance with the study of Chen et al. [51] who reported siderophore production by *T. citrinoviride* and *T. atroviride*.

The results of in vitro confrontations from this study conducted to estimate VOCs revealed inhibitory effect of volatile metabolites on growth of *D. sarmentorum*. Similarly, in the study of Chen et al. [52] *T. koningiopsis* VOCs showed inhibitory effects on Epicoccum nigrum Link growth. *B. dothidea* and *N. parvum* growth was not inhibited and this is in accordance with the study of Stracquadanio et al. [17] who reported the absence of inhibition of *Neofusicoccum batangarum* Begoude, Jol. Roux and Slippers and *N. parvum* by *T. asperellum* and *T. atroviride* VOCs.

The volatile compounds detected in this study ae already recognized as part of *Trichoderma* spp. VOCs spectra [1,15]. However this study represents the first report of *T. citrinoviride* as producer of nonanoic acid, cubenene, cis-α-bergamotene, hexanedioic acid, and verticillol. *Trichoderma* sp. DEMf TR1 and *T. citrinoviride* DEMf TR4 produced nonanoic acid. Similarly, in a study by Aneja et al. [53] nonanoic acid was detected as part of *T. harzianum* VOCs spectra and showed inhibitory effects on fungal growth and spore germination of *Crinipellis perniciosa* Stahel and *Moniliophthora roreri* Cif. H.C. Evans. However, in this study we focused only on the effect of VOCs on the mycelial growth of Botryosphaeriaceae, and the effect of nonanoic acid on spore germination of these fungi will remain as a task for a future study. Moreover, verticillol was detected in *T. citrinoviride* DEMf TR3 VOCs spectra. Similarly, in a study of Zhang et al. [19] verticillol was one of the main components of *T. harzianum* volatile metabolites that caused growth inhibition of *Fusarium oxysporum* Schlecht. Emend. Snyder and Hansen. *T. citrinoviride* DEMf TR3 produced cubenene and hexanedioic acid which are an antioxidant and antibacterial compounds [54,55]. Considering that trichoacorenol was produced by all three *Trichoderma* isolates, and that similar levels of *D. sarmentorum* inhibition were observed when co-inoculated with each of the three *Trichoderma* spp., we hypothesize that trichoacorenol could represent the antifungal component of VOCs. Trichoacorenol together with cis-α-bergamotene belongs to the group of volatile sesquiterpenes with biological potential in suppression of microbial growth [15].

5. Conclusions

The present study showed biocontrol potential of *P. sylvestris* bark through its chemical (water extracts) and biological (*Trichoderma* spp.) components against three Botryosphaeriaceae species, i.e., *B. dothidea*, *D. sarmentorum* and *N. parvum*.

Trichoderma spp. were capable to activate multiple antifungal mechanisms, from mycoparasitism, production of cell-wall degrading enzymes, competition for nutrients (siderophores) to antifungal volatile metabolites. The study showed the ability of *T. citrinoviride* to produce nonanoic acid, cubenene, cis-α-bergamotene, hexanedioic acid and verticillol, which represent volatile compounds with antimicrobial activity. Moreover, it represents the first report of inhibitory effect of *Trichoderma* spp. on mycelial growth of *D. sarmentorum*.

Biological control is an important component of an integrated pathogen management, and this in vitro study is the first and promising step towards the introduction of *P. sylvestris* bark extracts and *Trichoderma* spp. in biological control programs for landscape pathogens such as Botryosphaeriaceae. Future work should examine the in planta activity of these potential biocontrol agents.

Supplementary Materials: The following are available https://www.mdpi.com/article/10.3390/f12121731/s1, Figure S1: Phylogenetic trees generated from Maximum likelihood (ML) analyses based on a single gene alignment of RPB2, tef 1-α and ITS sequence data, Figure S2: Phylogenetic tree generated from a maximum likelihood analysis (ML) based on a concatenated alignment of tef 1-α and RPB2 sequence data showing the position of *Trichoderma citrinoviride* in relation to its closely related species belonging to the *Longibranchiatum* clade, Figure S3: Effects of VOCs of *Trichoderma citrinoviride* DEMf TR4 on mycelial growth of *D. sarmentorum* (a) and *B. dothidea* (b), Table S1: Sequences used in the phylogenetic analyses, Table S2. Nucleotide differences between *Trichoderma* sp. and its closest phylogenetic relative *T. citrinoviride*.

Author Contributions: Conceptualization, V.K., M.Z. and J.J.-P.; funding acquisition, S.O. and V.R.; methodology, V.K., M.Z., J.J.-P. and M.P.N.; visualization, M.Z. and J.J.-P.; writing—original draft, V.K. and M.Z.; writing—review and editing, V.K., M.Z., J.J.-P., S.O. and V.R. All authors have read and agreed to the published version of the manuscript.

Funding: This research was funded by The Ministry of Education, Science, and Technological Development of the Republic of Serbia, grant number 451-03-9/2021-14/200116 and 451-03-9/2021-14/200197.

Institutional Review Board Statement: Not applicable.

Informed Consent Statement: Not applicable.

Conflicts of Interest: The authors declare no conflict of interest.

References

1. Elsherbiny, E.A.; Amin, B.H.; Aleem, B.; Kingsley, K.L.; Bennett, K.K. *Trichoderma* volatile organic compounds as a biofumigation tool against late blight pathogen *Phytophthora infestans* in post harvest potato tubers. *J. Agric. Food Chem.* **2020**, *68*, 8163–8171. [CrossRef] [PubMed]
2. Aarti, T.; Meenu, S. Role of volatile metabolites from *T. citrinoviride* in biocontrol of phytopathogens. *Int. J. Res. Chem. Environ.* **2015**, *5*, 86–95.
3. Pásztory, Z.; Mohácsiné, I.R.; Gorbacheva, G.; Börcsök, Z. The utilization of tree bark. *BioResources* **2016**, *11*, 7859–7888. [CrossRef]
4. Bianchi, S. Extraction and Characterization of Bark Tannins from Domestic Softwood Species. Ph.D. Thesis, University of Hamburg, Hamburg, Germany, 13 January 2017.
5. Fregoso-Madueño, J.N.; Goche-Télles, J.R.; Rutiaga-Quiñones, J.G.; González-Laredo, R.F.; Bocanegra-Salazar, M.; Chávez-Simental, J.A. Alternative uses of sawmill industry waste. *Rev. Chapingo Ser. Cienc For. Ambient* **2017**, *23*, 243–260. [CrossRef]
6. Minova, S.; Seškēna, R.; Voitkâne, S.; Metla, Z.; Daugavietis, M.; Jankevica, L. Impact of pine (*Pinus sylvestris* L.) and spruce (*Picea abies* (L.) Karst.) bark extracts on important strawberry pathogens. *Proc. Latv. Acad. Sci. Sect. B Nat. Exact Appl. Sci.* **2015**, *69*, 62–67. [CrossRef]
7. Hagge, J.; Bässler, C.; Gruppe, A.; Hoppe, B.; Kellner, H.; Krah, F.-S.; Müller, J.; Seibold, S.; Stengel, E.; Thorn, S. Bark coverage shifts assembly processesof microbial decomposer communitiesin dead wood. *Proc. R. Soc. B* **2019**, *286*, 1–9. [CrossRef] [PubMed]
8. Zlatković, M.; Keča, N.; Wingfield, M.J.; Jami, F.; Slippers, B. Shot hole disease on *Prunus laurocerasus* caused by *Neofusicoccum parvum* in Serbia. *For. Pathol.* **2016**, *46*, 666–669. [CrossRef]
9. Zlatković, M.; Keča, N.; Wingfield, M.J.; Jami, F.; Slippers, B. Botryosphaeriaceae associated with the die-back of ornamental trees in the Western Balkans. *Antonie Leeuwenhoek* **2016**, *109*, 543–564. [CrossRef]
10. Zlatković, M. Botryosphaeriaceae species associated with canker and die-back disease of conifers in urban environments in Serbia. Topola/Poplar. *Inst. Lowl. For. Univ. Novi Sad.* **2017**, *199*, 55–75. (In Serbian)
11. Popović, T.; Blagojević, J.; Aleksić, G.; Jelušić, A.; Krnjajić, S.; Milovanović, P. A blight disease on highbush blueberry associated with *Macrophomina phaseolina* in Serbia. *Can. J. Plant Pathol.* **2018**, *40*, 121–127. [CrossRef]
12. Slippers, B.; Wingfield, M.J. Botryosphaeriaceae as endophytes and latent pathogens of woody plants: Diversity, ecology and impact. *Fung. Biol. Rev.* **2007**, *21*, 90–106. [CrossRef]
13. Zlatković, M.; Wingfield, M.J.; Jami, F.; Slippers, B. Host specificity of co-infecting Botryosphaeriaceae on ornamental and forest trees in the Western Balkans. *For. Pathol.* **2018**, *48*, e12410. [CrossRef]
14. Zlatković, M.; Wingfield, M.J.; Jami, F.; Slippers, B. Genetic uniformity characterizes the invasive spread of *Neofusicoccum parvum* and *Diplodia sapinea* in the Western Balkans. *For. Pathol.* **2019**, *49*, e12491. [CrossRef]
15. Moya, P.; Girotti, J.R.; Toledo, A.V.; Sisterna, M.N. Antifungal activity of *Trichoderma* VOCs against *Pyrenophora teres*, the causal agent of barley net blotch. *J. Plant Prot. Res.* **2018**, *58*, 45–53. [CrossRef]
16. Shah, M.M.; Afiya, H. *Introductory Chapter: Identification and Isolation of Trichoderma spp.—Their Significance in Agriculture, Human Health, Industrial and Environmental Application. Trichoderma—The Most Widely Used Fungicide*; Shah, M.M., Sharif, U., Buhari, T.R., Eds.; IntechOpen: London, UK, 2019; pp. 1–13. [CrossRef]
17. Stracquadanio, C.; Quiles, J.M.; Meca, G.; Cacciola, S.O. Antifungal activity of bioactive metabolites produced by *Trichoderma asperellum* and *Trichoderma atroviride* in liquid medium. *J. Fungi* **2020**, *6*, 263. [CrossRef] [PubMed]
18. Guo, Y.; Ghirardo, A.; Weber, B.; Schnitzler, J.-P.; Benz, J.P.; Rosenkranz, M. *Trichoderma* species differ in their volatile profiles and in antagonism toward ectomycorrhiza *Laccaria bicolor*. *Front. Microbiol.* **2019**, *10*, 1–15. [CrossRef]
19. Zhang, F.; Yang, X.; Ran, W.; Shen, Q. *Fusarium oxysporum* induces the production of proteins and volatile organic compounds by *Trichoderma harzianum* T-E5. *FEMS Microbiol. Lett.* **2014**, *359*, 116–123. [CrossRef] [PubMed]
20. Özgenç, Ö.; Durmaz, S.; Yildiz, Ü.C.; Erişir, E. A comparison between some wood bark extracts: Antifungal activity. *Kastamonu Univ. Orman. Derg.* **2017**, *17*, 502–508. [CrossRef]
21. Úrbez-Torres, J.R.; Tomaselli, E.; Pollard-Flamand, J.; Boulé, J.; Gerin, D.; Pollastro, S. Characterization of *Trichoderma* isolates from southern Italy, and their potential biocontrol activity against grapevine trunk disease fungi. *Phytopathol. Mediterr.* **2020**, *59*, 425–439. [CrossRef]

22. Karličić, V.; Jovičić-Petrović, J.; Marojević, V.; Zlatković, M.; Orlović, S.; Raičević, V. Potential of *Trichoderma* spp. and *Pinus sylvestris* bark extracts as biocontrol agents against fungal pathogens residing in the Botryosphaeriales. *Environ. Sci. Proc.* **2021**, *3*, 99. [CrossRef]

23. Gardes, M.; Bruns, T.D. ITS primers with enhanced specificity for basidiomycetes-application to the identification of mycorrhizae and rusts. *Mol. Ecol.* **1993**, *2*, 113–118. [CrossRef] [PubMed]

24. White, T.J.; Bruns, T.; Lee, S.; Taylor, J.; Innis, M.A.; Gelfand, D.H.; Sninsky, J.J.; White, T.J. Amplification and direct sequencing of fungal ribosomal RNA genes for phylogenetics. In *PCR Protocols: A Guide to Methods and Applications*; Innis, M.A., Gelfand, D.H., Sninsky, J.J., White, T.J., Eds.; Academic Press: San Diego, CA, USA, 1990; pp. 315–322.

25. O'Donnell, K.; Kistler, H.C.; Cigelnik, E.; Ploetz, R.C. Multiple evolutionary origins of the fungus causing Panama disease of banana: Concordant evidence from nuclear and mitochondrial gene genealogies. *Proc. Natl. Acad. Sci. USA* **1998**, *95*, 2044–2049. [CrossRef] [PubMed]

26. Liu, Y.L.; Whelen, S.; Hall, B.D. Phylogenetic relationships among ascomycetes: Evidence from an RNA polymerase II subunit. *Mol. Biol. Evolut.* **1999**, *16*, 1799–1808. [CrossRef] [PubMed]

27. Kovač, M.; Diminić, D.; Orlović, S.; Zlatković, M. *Botryosphaeria dothidea* and *Neofusicoccum yunnanense* causing canker and die-back of *Sequoiadendron giganteum* in Croatia. *Forests* **2021**, *12*, 695. [CrossRef]

28. Zlatković, M.; Tenorio-Baigorria, I.; Lakatos, T.; Tóth, T.; Koltay, A.; Pap, P.; Marković, M.; Orlović, S. Bacterial canker disease on *Populus* x *euramericana* caused by *Lonsdalea populi* in Serbia. *Forests* **2020**, *11*, 1080. [CrossRef]

29. Taylor, J.W.; Jacobson, D.J.; Kroken, S.; Kasuga, T.; Geiser, D.M.; Hibbett, D.S.; Fisher, M.C. Phylogenetic species recognition and species concepts in fungi. *Fungal Genet. Biol.* **2000**, *31*, 21–32. [CrossRef]

30. Romsaiyud, A.; Singkasiri, W.; Nopharatana, A.; Chaiprasert, P. Combination effect of pH and acetate on enzymatic cellulose hydrolysis. *J. Environ. Sci.* **2009**, *21*, 965–970. [CrossRef]

31. Alexander, D.B.; Zuberer, D.A. Use of chrome azurol S reagents to evaluate siderophore production by rhizosphere bacteria. *Biol. Fert. Soils* **1991**, *12*, 39–45. [CrossRef]

32. Dennis, C.; Webster, J. Antagonistic properties of species groups of *Trichoderma*-II. Production of volatile antibiotics. *Trans. Br. Mycol Soc.* **1971**, *57*, 47–48. [CrossRef]

33. Mohareb, A.S.; Kherallah, I.E.A.; Badawy, M.E.; Salem, M.Z.M.; Yousef, H.A. Chemical composition and activity of bark and leaf extracts of *Pinus halepensis* and *Olea europaea* grown in AL-Jabel AL-Akhdar region, Libya against some plant phytopathogens. *J. Appl. Biotechnol. Bioeng.* **2017**, *3*, 331–342. [CrossRef]

34. Ruiz-Gómez, F.J.; Miguel-Rojas, C. Antagonistic potential of native *Trichoderma* spp. against *Phytophthora cinnamomiin* the control of Holm Oak decline in Dehesas ecosystems. *Forests* **2021**, *12*, 945. [CrossRef]

35. Samuels, G.; Petrini, O.; Kuhls, K.; Lieckfeldt, E.; Kubicek, C.P. The *Hypocrea schweinitzii* complex and *Trichoderma* sect. *Longibrachiatum*. *Stud. Mycol.* **1998**, *41*, 1–54.

36. Krause, K.; Jung, E.-M.; Lindner, J.; Hardiman, I.; Poetschner, J.; Madhavan, S.; Mtthäus, C.; Kai, M.; Menezes, R.C.; Popp, J.; et al. Responseof the wood-decay fungus *Schizophyllum commune* to co-occurring microorganisms. *PLoS ONE* **2020**, *15*, e0232145. [CrossRef] [PubMed]

37. Mori, M.; Aoyama, M.; Doi, S.; Kanetoshi, A.; Hayashi, T. Antifungal activity of bark extracts of conifers. *Holz Roh Werkst* **1995**, *53*, 81–82. [CrossRef]

38. Alfredsen, G.; Solheim, H.; Slimestad, R. Antifungal effect of bark extracts from some European tree species. *Eur. J. Forest Res.* **2008**, *127*, 387–393. [CrossRef]

39. Vek, V.; Poljanšek, I.; Humar, M.; Willför, S.; Oven, P. In vitro inhibition of extractives from knotwood of *P. sylvestris* (*Pinus sylvestris*) and black pine (*Pinus nigra*) on growth of *Schizophyllum commune, Trametes versicolor, Gloeophyllum trabeum* and *Fibroporia vaillantii*. *Wood Sci. Technol.* **2020**, *54*, 1645–1662. [CrossRef]

40. Lomeli-Ramírez, M.G.; Dávila-Soto, H.; Silva-Guzmán, J.A.; Ruiz, H.G.O.; García-Enriquez, S. Fungitoxic potential of extracts of four *Pinus* spp. bark to inhibit fungus Trametes versicolor (L.ex. Fr.) Pilát. *BioResources* **2016**, *11*, 10575–10584. [CrossRef]

41. Mondello, V.; Songy, A.; Battiston, E.; Pinto, C.; Coppin, C.; Trotel-Aziz, P.; Clément, C.; Mugnai, L.; Fontaine, F. Grapevine trunk diseases: A review of fifteen years of trials for their control with chemicals and biocontrol agents. *Plant Dis.* **2018**, *102*, 1189–1217. [CrossRef] [PubMed]

42. Marraschi, R.; Ferreira, A.B.M.; da Silva Bueno, R.N.; Leite, J.A.B.P.; Lucon, C.M.M.; Harakava, R.; Leite, L.G.; Padovani, C.R.; Bueno, C.J. A protocol for selection of *Trichoderma* spp. to protect grapevine pruning wounds against *Lasiodiplodia theobromae*. *Braz. J. Microbiol.* **2019**, *50*, 213–221. [CrossRef]

43. Park, Y.-H.; Mishra, R.C.; Yoon, S.; Kim, H.; Park, C.; Seo, S.-T.; Bae, H. Endophytic *Trichoderma citrinoviride* isolated from mountain-cultivated ginseng (*Panax ginseng*) has great potential as a biocontrol agent against ginseng pathogens. *J. Ginseng Res.* **2019**, *43*, 408–420. [CrossRef] [PubMed]

44. Kuzmanovska, B.; Rusevski, R.; Jankulovska, M.; Oreshkovikj, K.B. Antagonistic activity of *Trichoderma asperellum* and *Trichoderma harzianum* against genetically diverse *Botrytis cinerea* isolates. *Chil. J. Agric. Res.* **2018**, *78*, 391–399. [CrossRef]

45. Kotze, C.; Van Niekerk, J.; Mostert, L.; Halleen, F.; Fourie, P. Evaluation of biocontrol agents for grapevine pruning wound protection against trunk pathogen infection. *Phytopathol. Mediterr.* **2011**, *50*, S247–S263.

46. Pellan, L.; Durand, N.; Martinez, V.; Fontana, A.; Schorr-Galindo, S.; Strub, C. Commercial biocontrol agents reveal contrasting comportments against two mycotoxigenic fungi in cereals: *Fusarium graminearum* and *Fusarium verticillioides*. *Toxins* **2020**, *12*, 152. [CrossRef] [PubMed]

47. Maheshwari, D.K.; Dubey, R.C.; Sharma, V.K. Biocontrol effects *of Trichoderma virens* on *Macrophomina phaseolina* causing charcoal rot of peanut. *Ind. J. Microbiol.* **2001**, *41*, 251–256.

48. Nidhina, K.; Sharadraj, K.M.; Prathibha, V.H.; Hegde, V.; Gangaraj, K.P. Antagonistic activity of *Trichoderma* spp. to *Phytophthora* infecting plantation crops and its beneficial effect on germination and plant growth promotion. *Vegetos* **2016**, *29*, 19–26. [CrossRef]

49. Danilović, G.; Radić, D.; Raičević, V.; Jovanović, L.J.; Kredics, L.; Panković, D. Extracellular enzyme activity of *Trichoderma* strains isolated from different soil types. In Proceedings of the 2nd International Symposium for Agriculture and Food, Ohrid, North Macedonia, 7–9 October 2015.

50. Lorito, M.; Hayes, C.K.; Di Pietro, A.; Woo, S.L.; Harman, G.E. Prification, characterization, andsynergistic activity of glucan 1,3-β-glucosidse and an N-cetyl-β-glucosaminidase from *Trichoderma harzianum*. *Mol. Plant Pathol.* **1994**, *84*, 398–405.

51. Chen, D.; Hou, Q.; Jia, L.; Sun, K. Combined use of two *Trichoderma* strains to promote growth of pakchoi (*Brassica chinensis* L.). *Agronomy* **2021**, *11*, 726. [CrossRef]

52. Chen, J.-L.; Liu, K.; Miao, C.-P.; Sun, S.-Z.; Chen, Y.-W.; Xu, L.-H.; Guan, H.-L.; Zhao, L.-X. Salt tolerance of endophytic *Trichoderma koningiopsis* YIM PH30002 and its volatile organic compounds (VOCs) allelopathicactivity against phytopathogens associated with Panax notoginseng. *Ann. Microbiol.* **2016**, *66*, 981–990. [CrossRef]

53. Aneja, M.; Gianfagna, T.J.; Hebbar, P.K. *Trichoderma harzianum* produces nonanoic acid, an inhibitor of spore germination and mycelial growth of two cacao pathogens. *Physiol. Mol. Plant Pathol.* **2005**, *67*, 304–307. [CrossRef]

54. Shareef, H.K.; Muhammed, H.J.; Hussein, H.M.; Hameed, I.H. Antibacterial effect of ginger (*Zingiber officinale*) roscoe and bioactive chemical analysis using Gas chromatography mass spectrum. *Orient. J. Chem.* **2016**, *32*, 817–837. [CrossRef]

55. Kim, J.-E.; Seo, J.-H.; Bae, M.-S.; Bae, C.-S.; Yoo, J.-C.; Bang, M.-A.; Cho, S.-S.; Park, D.-H. Antimicrobial constituents from *Allium hookeri* Root. *Nat. Prod. Commun.* **2016**, *11*, 237–238. [CrossRef]

Article

Bark Thickness and Heights of the Bark Transition Area of Scots Pine

Florian Wilms, Nils Duppel, Tobias Cremer and Ferréol Berendt *

Department of Forest Utilization and Timber Markets, Faculty of Forest and Environment, Eberswalde University for Sustainable Development, 16225 Eberswalde, Germany; Florian.Wilms@hnee.de (F.W.); Nils.Duppel@hnee.de (N.D.); Tobias.Cremer@hnee.de (T.C.)
* Correspondence: Ferreol.Berendt@hnee.de

Abstract: The estimation of forest biomass is gaining interest not only for calculating harvesting volumes but also for carbon storage estimation. However, bark (and carbon) compounds are not distributed equally along the stem. Particularly when looking at Scots pine, a radical change in the structure of the bark along the stem can be noted. At the bark transition area, the bark changes from thick and rough to thin and smooth. The aim of our study was (1) to analyze the height of the bark transition area where the bark structure changes and (2) to analyze the effect of cardinal direction on the bark thickness. Regression analyses and forward selection were performed including measured tree height, DBH, bark thickness, crown base height and upper and lower heights of the bark transition areas of 375 trees. While the cardinal direction had no effect on bark thickness, DBH was found to have a significant effect on the heights of the bark transition areas, with stand density and tree height having a minor additional effect. These variables can be used to estimate timber volume (without bark) with higher accuracy and to predict the carbon storage potential of forest biomass according to different tree compartments and compounds.

Keywords: bark structure; *Pinus sylvestris*; forward selection; bark types

Citation: Wilms, F.; Duppel, N.; Cremer, T.; Berendt, F. Bark Thickness and Heights of the Bark Transition Area of Scots Pine. *Forests* **2021**, *12*, 1386. https://doi.org/10.3390/f12101386

Academic Editor: Blas Mola-Yudego

Received: 10 September 2021
Accepted: 5 October 2021
Published: 11 October 2021

Publisher's Note: MDPI stays neutral with regard to jurisdictional claims in published maps and institutional affiliations.

1. Introduction

Scots pine (*Pinus sylvestris*, L.) is one of the main tree species in Central and Northern European forests. In addition, it is often planted in pure stands due to its short rotation time [1] and its low site and climate requirements. In the total forest area of Germany, the proportion (by area) of pine stands amounts to 22%. In the federal state of Brandenburg, this proportion increases to up to 70% according to the last Federal Forest Inventory in 2012 [2]. Thus, Scots pine is the second most represented tree species in Germany after Norway spruce (*Picea abies*, L.) [2]. Suitable for use as construction timber and sawlogs, as well as for pulp and chipboard, this tree species is very important for forestry, especially in Eastern Europe [3]. Currently, the focus of the timber industry lies mostly on the processing of wood. The bark is often considered as an industrial by-product and is primarily used for energy production [4]. However, due to its unique chemical composition, bark can be used to produce biomaterials with the potential to replace fossil fuel-based products. Interest in these new value-added products made from the valuable compounds of tree bark is increasing steadily [5–7].

Regardless of whether bark is being used as high-value raw material or treated as a by-product, there is a real need to determine the proportions of bark accurately. The bark of most woody plants is defined as all tissues located outside the vascular cambium, divided into the inner (phloem) and the outer bark [8]. As the correct determination of the bark proportion is of great importance when calculating the volume of wood without bark, it has a substantial economic impact [9,10]. The accurate recording of the ratio between wood and bark biomass can also be used for other purposes such as carbon inventory [11] or to determine the fire-adaptedness of tree species [12,13]. Generally, bark factors based on bark thickness models are used to predict the diameter under bark in order to further estimate the wood volume [14,15]. Since bark percentage varies among different tree species from 4% to 30% [16,17], bark thickness has to be determined at the species level. Additionally, the respective growing region has to be considered [18]. Several bark types (e.g., smooth bark, white bark, fissured bark and scaly bark) with different bark rugosities (ridges and furrows) increase the complexity of precise and accurate estimations of tree bark allometry or volume [11,19–21].

The bark of Scots pine poses additional challenges to the development of bark deduction factors, as its thickness varies greatly across different stem segments: (i) rough, coarse and furrowed bark that is brown to gray-brown, (ii) a transitional phase, and (iii) smooth bark with a distinct reddish-orange color [22–25]. In addition to the difference in structure and color (Figure 1), these various bark types within a tree have different thicknesses and proportions [26–29]. Thus, the heights of the bark transition area from one bark type to another have to be determined in order to relate these to other growth parameters.

Figure 1. Bark transition phases along the stem of Scots pine, with HT1: bark transition height 1 and HT2: bark transition height 2.

Therefore, the purpose of this study was to determine the tree height at which Scots pine bark changes from thick and rough to thin and smooth. Several growth parameters were analyzed in order to reliably predict the bark type proportion of standing Scots pines. Consequently, it could be expected that the heights of the bark transition area would increase with increasing DBH, increasing tree height and decreasing stand density. A second purpose was to analyze the influence of the cardinal direction on the bark thickness. We hypothesized that bark thickness would increase with increasing DBH and tree height and that the bark thickness would be affected by the cardinal directions. Precise estimations on the heights of the bark transition area and of bark thickness may further be used for a wood volume estimation of standing trees, carbon inventories of forests and for developing new bark deduction factor algorithms, e.g., for harvesters.

2. Materials and Methods

2.1. Site Description

The research area is situated near Eberswalde in the German federal state of Brandenburg. The analyzed forest stands were all located within the forest district Heegermuehle. The mean precipitation and the mean annual temperature (for the period between 2010 and 2020) recorded by the German weather service were 498 mm/a and 10.0 °C, respectively [30]. Across the research area, cambisols are prevalent, with podzols being present in the vicinity of stand 3 [31].

2.2. Research Design

The measurements were carried out during November 2020. In total, 375 Scots pines (*Pinus sylvestris*, L.) originating from three single-aged, monoculture stands were measured. The three stands had different ages and diameters at breast height (DBH): the age of stand 1 was 64 years and the trees had a mean DBH of 18.5 ± 5.2 cm; DBH of stand 2 (69 years) was higher at 27.8 ± 4.9 cm; and stand 3 (84 years) had a mean DBH of 34.5 ± 5.3 cm (Table 1). In each stand, three distinct plots were chosen with approximately 70–100 m distance between the plot centers. Within a specific radius around the center of the plot, all Scots pine trees were measured. To ensure a minimum sample size of 30 trees per plot, the plots' radii (r) were 12.62 m (corresponding to an area (A) of 500 m^2), 15.45 m (A = 750 m^2) and 17.84 m (A = 1000 m^2) for stands 1 to 3, respectively. The radii were increased according to stand age in order to balance out the lower stand density. In stand 1, the densities were 1020, 1300 and 800 trees/ha for the three plots, in stand 2 they were 480, 507 and 587, and in stand 3 the densities were 330, 360 and 320 trees/ha.

Table 1. Characteristics of the forest stands, with DBH: diameter at breast height, BT: bark thickness, H: tree height, HC: crown base height, HT1: transition height 1 and HT2: transition height 2.

Characteristic	Stand 1	Stand 2	Stand 3	All
WGS84 Coordinates	13.762056, 52.818503	13.752174, 52.82378	13.744799, 52.815309	
Plot radius (in m)	12.62	15.45	17.84	
Cumulative plot area (in m^2)	1500	2250	3000	
Stand age (in years)	64	69	84	
Trees measured (N)	156	118	101	375
Stand density (N/ha)	1040	524	337	
Mean DBH ± SD (in cm)	18.5 ± 5.2	27.8 ± 4.9	34.5 ± 5.3	25.7 ± 8.4
Mean BT ± SD (in cm)	1.24 ± 0.41	1.68 ± 0.39	1.99 ± 0.39	1.58 ± 0.50
Mean H ± SD (in m)	18.9 ± 2.8	25.5 ± 2.0	26.8 ± 1.9	23.1 ± 4.3
Mean HC ± SD (in m)	14.1 ± 2.2	19.2 ± 2.2	19.1 ± 2.3	17.0 ± 3.3
Mean HT 1 ± SD (in m)	3.0 ± 0.9	4.3 ± 1.5	5.8 ± 1.7	4.2 ± 1.8
Mean HT2 ± SD (in m)	5.6 ± 1.8	7.1 ± 2.2	9.1 ± 2.5	7.0 ± 2.5

The DBH was measured at 1.3 m tree height with a slide caliper and two measurements were taken at a 90° angle for every analyzed tree. The bark thickness was measured with the Swedish bark gauge (SUUNTO, 60 mm, Vantaa, Finland) at 1.3 m tree height. As proposed by Stängle et al. [32] five bark thickness measurements were taken to reduce errors. The first bark thickness measurement was always taken from the cardinal position south, with the following measurements moving clockwise around the stem.

The height parameters tree height (H), crown base height (HC) and both heights of the bark transition area (transition height 1 and 2: HT1 and HT2) were measured using a VertexIV hypsometer (Haglöf AB, Bromma, Sweden) to a precision of 0.1 m. HC was defined as the height of the lowest knot with three or more living branches and without any dead branches upwards, which is also known as the analytical crown base height [33–38]. The height of the first bark transition (HT1) was defined as the lowest occurrence where smooth bark occupied at least half of the stem circumference (Figure 1). Smaller single patches of smooth bark farther downwards were not considered as representative for HT1. For the second bark transition height (HT2), the height measurement was performed at the uppermost end of rough bark that is connected to the stem base (Figure 1). Freshly formed patches of gray or rough bark around knots, which were separate from the interconnected rough bark portion, were not considered as the upper bark transition point HT2.

2.3. Data Analysis and Statistics

To identify the best model to determine HT1 and HT2 for several variables, a forward selection was performed [39]. The analyzed response variables were: DBH, H, HC, stand density (StD) and plot density (PD). While PD is the number of trees/ha in a single circle plot, StD is the average of the three respective PD figures in the stand. The forward regression analysis was carried out with the R-packages *caret* [40] and *MASS* [41]. The forward selection selects the best model for a certain amount of n predictors out of a set pool of variables [39]. The model whose single predictor represents the best fit (with lowest RMSE and highest R^2) was defined as the base linear model. Subsequently, the next model was formed with an additional variable, keeping the first predictor. This was performed using k-fold cross-validation, meaning that the total sample was divided into 10 subsets (folds), as far as possible of the same size. Out of these ten subsets, nine were then used to build the model and one was used as a test set in order to minimize RMSE. Prior to regression analysis, some trees were removed from the analysis due to obvious measurement errors in the height measurements or because DBH and/or H of trees were significantly different from the respective stand, resulting in N = 366 trees being analyzed for HT1 and HT2 and N = 373 for bark thickness.

As the Shapiro–Wilk Test did not confirm a normal distribution ($p \leq 0.05$) for DBH, H and HC, all vectors were transformed to their natural logarithm. As a result, both transformed response variables ln(HT1) and ln(HT2) followed a normal distribution. The correlation of each predictor to ln(HT2) followed the same order as with ln(HT1), but with a generally weaker Spearman-ρ by ca. 0.1. Regarding collinearity, a Spearman-rank-correlation test was executed for all transformed predictor variables (Table 2). The highest collinearity of predictors was found between ln(DBH) and ln(H) with a Spearman-ρ of 0.8486. ln(HC) was also found to correlate strongly with ln(H) as both are height parameters. ln(StD) and ln(PD) naturally correlate highly as they provide the same information but on different scales. Moreover, for comparability reasons, StD and PD [N/ha] were divided by 100 to have a similar scale as DBH, HC and H. Consequently, a model with multiple predictors including StD and/or PD can be plotted without the point cloud being separated by stands with high x-axis difference.

Table 2. Combined correlation matrix (Spearman-ρ) of logarithm-transformed variables, with HT1: transition height1, HT2: transition height2, DBH: diameter at breast height, H: tree height, HC: crown base height, StD: stand density and PD: plot density.

Variables	ln(HT1)	ln(DBH)	ln(H)	ln(HC)	ln(StD)	ln(PD)	Variables
	1	0.7293	0.6391	0.4975	−0.6675	−0.6668	**ln(HT1)**

Table 2. *Cont.*

Variables	ln(HT1)	ln(DBH)	ln(H)	ln(HC)	ln(StD)	ln(PD)	Variables
ln(HT2)	1	1	0.8486	0.6238	−0.8104	−0.8135	ln(DBH)
ln(DBH)	0.6553	1	1	0.8393	−0.8250	−0.8120	ln(H)
ln(H)	0.5532	0.8486	1	1	−0.7063	−0.6960	ln(HC)
ln(HC)	0.4157	0.6238	0.8393	1	1	0.9439	ln(StD)
ln(StD)	−0.5409	−0.8104	−0.8250	−0.7063	1	1	ln(PD)
ln(PD)	−0.5461	−0.8135	−0.8120	−0.6960	0.9439	1	
	ln(HT2)	ln(DBH)	ln(H)	ln(HC)	ln(StD)	ln(PD)	

All statistical computations were executed with R 4.0.3 [42] using the interface RStudio [43]. The quality of linear regression models was assessed using the adjusted coefficient of determination, Adjusted-R^2 (adj-R^2). Whereas R^2 of a linear regression model increases just by adding predictors [44], adj-R^2 compensates by including the degrees of freedom into the equation. The root mean square error (RMSE) was used to calculate the mean square deviation of residuals from the regression line. By subtracting the number of predictors p in the denominator of the RMSE function, the residual standard error (RSE) accounts for additional predictors in multiple regression analysis [44]. After confirming DBH as the single best predictor, a non-linear regression analysis for HT1 and HT2~DBH was performed with the R-package *minpack.LM* [45].

The significance level (α) was set to 5 %, with $\alpha < 1\%$ considered as highly significant.

3. Results

3.1. Bark Thickness

The mean bark thickness increased from stand 1 to stand 3. The mean bark thickness was significantly different with 1.2 + 0.4 cm, 1.7 ± 0.4 cm and 2.0 ± 0.4 cm for stands 1, 2 and 3, respectively. The mean proportion of bark (as the ratio between the double bark thickness and DBH) was 12.5% ± 2.3%. The bark proportion decreased with increasing age from 13.4% ± 2.4% in stand 1 (64 years) to 11.7% ± 1.83% in stand 3 (84 years).

The analysis showed that the two independent variables DBH and H had a significant effect on the bark thickness with $p = 2.2 \times 10^{-16}$ and $p = 0.036$, respectively. The bark thickness model using a multiple linear regression showed an adj-R^2 of 0.705 when considering DBH and H. According to the following equation, for a given DBH, the bark thickness decreases with increasing tree height:

$$y = 0.0564 \times \text{DBH} - 0.0134 \times \text{H} + 0.439 \tag{1}$$

where y is the bark thickness in cm, DBH the diameter at breast height in cm and H the tree height in m.

Through an ANOVA, the influence of the cardinal direction on the bark thickness was tested. The result showed that no significant effect of the cardinal direction was observed when considering all stands with a *p*-value of 0.538 (Figure 2). Similarly, when looking at single stands, no significance of the cardinal direction was found for stand 1 (*p*-value = 0.444), stand 2 (*p*-value = 0.64) or stand 3 (*p*-value = 0.328).

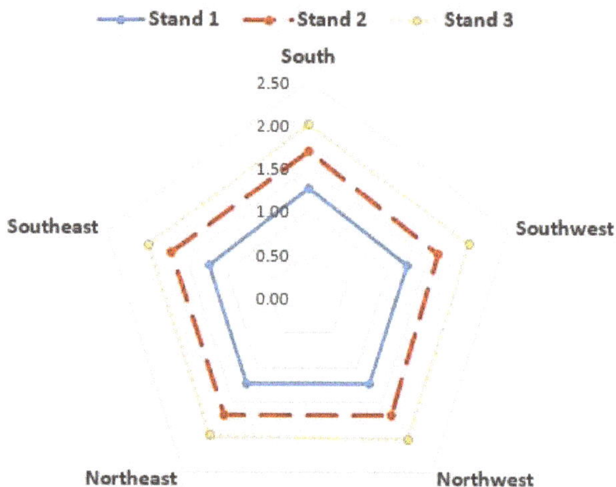

Figure 2. Bark thickness for different cardinal directions for the three stands.

3.2. Bark Transition Heights

A multiple linear regression analysis was carried out separately for ln(HT1) and ln(HT2) by using a forward selection algorithm. The variable ln(DBH) was identified as the strongest single predictor for both ln(HT1) and ln(HT2) with adj-R^2 values of 0.4833 and 0.3935, respectively. By adding ln(StD) as the second predictor to the regression, an additional 1.5% of explained variance was observed for ln(HT1). When including the third predictor ln(H) in the model, adj-R^2 reached 0.5153 for ln(HT1). In the case of ln(HT2), only ln(DBH) had a significant effect upon the regression. As ln(PD) showed a lower adj-R^2 than ln(StD), ln(PD) was not further analyzed. Moreover, the variable ln(HC) was found to be not significant. Table 3 provides the estimates and model statistics. According to the Shapiro–Wilk Test, the residuals of both models are normally distributed.

Table 3. Model statistics for bark transition heights 1 (HT1) and 2 (HT2), with DBH: diameter at breast height, StD: stand density and H: tree height.

Variable	Estimate	SE	*p*-Value	Adj-R^2	RSE
ln(HT1)				0.5135	0.2965
Intercept	0.7427	0.5469	0.175		
Coef. ln(DBH)	0.7618	0.0999	2.11×10^{-13}		
Coef. ln(StD)	−0.2909	0.1985	0.0355		
Coef. ln(H)	−0.4189	0.0586	1.07×10^{-6}		
ln(HT2)				0.3935	0.2781
Intercept	−0.2233	0.1378	0.106		
Coef. ln(DBH)	0.6598	0.0428	1.19×10^{-41}		

The scatter plots after back transformation into metric values are displayed in Figure 3. The regression of the transformed log values for HT2 shows a similar image to HT1, with a metric regression curve which does not follow the exponentially shaped scatter. Naturally, the measured values and the value range of HT2 are higher than those of HT1, resulting in a worse regression according to the lower R^2 and higher RSE.

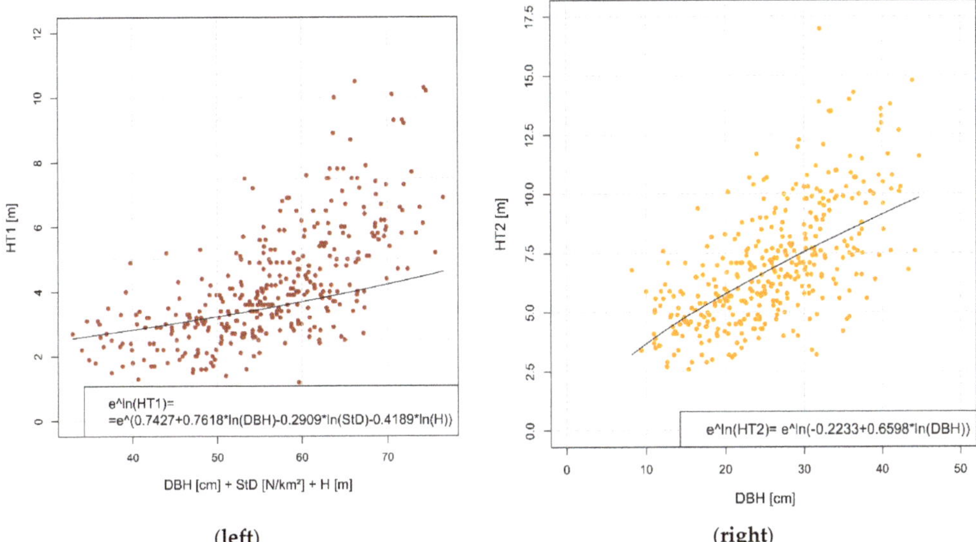

Figure 3. Scatter plot of the bark transition heights (HT) after back transformation into metric value for the HT1 regression (**left**) and the HT2 regression (**right**).

Since only a minimal gain in explained variance and decrease in error was achieved by adding independent variables to the linear regression of logarithmically transformed values, DBH was considered the main predictor for HT1 and HT2. In order to fit the non-linear relationship of HT1 or HT2 to DBH, a non-linear regression fit was applied. While HT1 $\sim$ a + (b $\times$ DBH)^c reached an RSE of 1.2527, the RSE of HT2 $\sim$ a + (b $\times$ DBH)^c increased by about 54% to 1.9323 (Table 4 and Figure 4).

Table 4. Model statistics for bark transition heights 1 (HT1) and 2 (HT2) with the non-linear regression.

Variable	Estimate	SE	*p*-Value	RSE
HT1				1.2527
Intercept a	1.8376	0.3766	1.59×10^{-6}	
b	0.0586	0.0102	1.85×10^{-8}	
c	1.8532	0.3121	6.79×10^{-9}	
HT2				1.9323
Intercept a	3.6428	0.7179	6.23×10^{-7}	
b	0.0797	0.0247	1.38×10^{-3}	
c	1.6199	0.3613	9.84×10^{-6}	

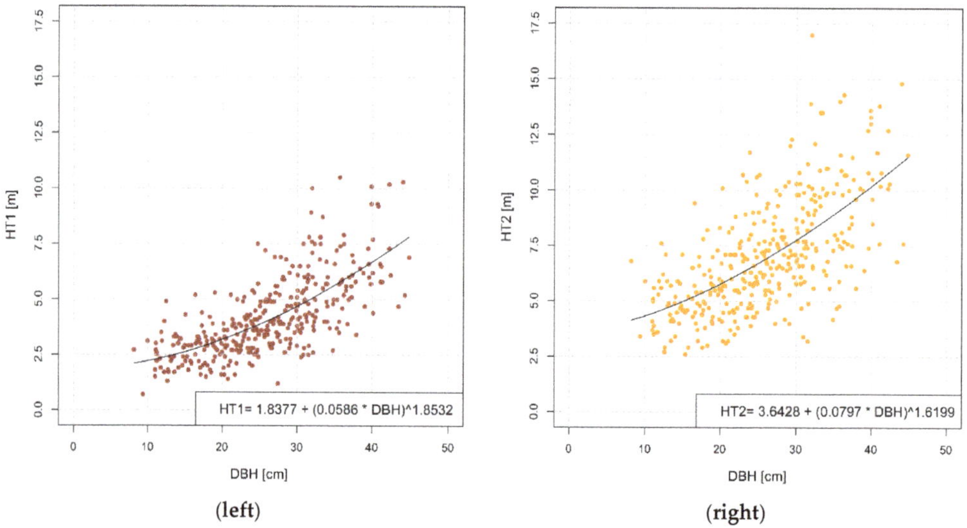

Figure 4. Scatter plots of the non-linear regressions of the bark transition heights (HT) with HT1 regression (**left**) and HT2 regression (**right**).

4. Discussion

4.1. Bark Thickness

The analysis showed that DBH and tree height significantly affect bark thickness at a height of 1.3 m in Scots pine trees. The strong influence of DBH and the weak influence of tree height on bark thickness for Scots pine were in line with earlier studies [46,47] which stated that DBH was the most relevant independent variable for estimating bark thickness at 1.3 m for Norway spruce (*Picea abies*, H. Karst) and Turkish pine (*Pinus brutia*, Ten.), respectively. Moreover, a strong correlation between DBH and bark thickness was also found for Oriental spruce (*Picea orientalis*, Link) [48] and Radiata pine (*Pinus radiata*, D.Don) [49]. While for Norway spruce the tree height did not have an effect on bark thickness [46], smaller trees had a thicker bark when looking at Radiata pine. The regression analysis showed that tree height was negatively correlated with bark thickness. Thus, similar to the results of Gorden [49], for Scots pine higher trees had a thinner bark than smaller trees. This could be because smaller trees use more energy on bark production compared to height growth, or there may be genetic or site influences [12,21,46]. The measurement showed a mean bark thickness of 1.2 cm for a mean DBH of 18.5 cm, and a bark thickness of 2.0 cm for trees with a mean DBH of 34.5 cm. Compared with available data on Scots pine bark thickness from the literature, the results from this study were of similar dimensions and they were only slightly higher compared to bark thicknesses from 0.85 cm to 1.35 cm [26], between 0.75 cm and 1.4 cm [27] and between 1.14 cm and 1.54 cm [28]. However, even if the measurements from both Gorden [49] and Kahriman [47] were performed on logs with thick and gray bark from the lower part of the tree, the results are only partially comparable with the bark thickness at DBH. Another study found that, at a relative height of 5%, the bark proportion is 14.7% for Scots pine [50]. A relative height of 5% corresponds to a height of 1.3 for a 26 m tall tree and, thus, is comparable with the bark proportions found in this study, which were between 11.7% and 13.4%.

The bark thickness measurements were performed with a bark gauge. Different authors note that the use of a bark gauge may lead to biased measurements [32,46,49]. Moreover, the time of year affects the measurements. Penetration of the bark gauge into the growing annual rings was observed for the relatively soft early growth, especially during spring and summer [46]. As the field work occurred during November and was performed by the same person, and as five bark thickness measurements were taken for each tree, the possible risks of measurement errors were minimized. This method is non-destructive as opposed to more precise methods such as measuring the diameters over and under bark [51] or computed tomography for highly accurate measurements [32], which can only be achieved by analyzing logs or wood discs. When analyzing the influence of the cardinal direction on bark thickness, no significant differences were found. Nevertheless, some factors may influence the bark thickness around the stem, as shown with sun radiation on the bark thickness of Oriental spruce [48].

4.2. Bark Transition Heights

All analyzed trees were from homogeneous pure Scots pine stands. An original study on smooth bark proportions on Scots pine was described by Dengler [22] in 1937, noting a positive relation between the smooth bark onset height and greater DBH. Based on the above-mentioned initial findings of Dengler, Wagenknecht [23] carried out similar measurements two years later. Furthermore, in 1963, Erteld [24] directly referred to Wagenknecht's findings. His criticism was that Wagenknecht only listed mean values of the diameter steps, which Erteld took as an opportunity to establish, among other things, linear regression functions between the smooth bark onset height and DBH. These three authors also confirmed additional positive influences on the smooth bark onset height: age, lower stand density, branchiness and tree height.

The main result of HT1 and HT2 regression analysis is that DBH has the strongest significant influence among all the investigated parameters. This is underlined by the small gain of the explained variance when adding additional predictors. The variance of the back-transformed values corresponds to $e^{RSE} = e^{0.30} \approx 1.35$. Thus, there is already an average range of at least ±1.35 m for the explained variance. While the RSE values of non-linear regression analysis are higher compared to the linear equations for transformed parameters, the $a + (b \times DBH)^c$ regression curve clearly shows a better fit to the exponential relationship of HT1 and HT2~DBH. Due to a possible over- or underestimation of the location of the bark transition area, these models give an estimation of both bark transition heights. However, it should be mentioned that the mean length of the bark transition areas was 2.87 m $\pm$ 1.42 m over all trees. Thus, it could be expected that the bark transition area can be accurately located even if this was not possible for the two distinct bark transition heights. For an accurate prediction of the tree height where the bark structure is changing, the model should be improved. In order to improve the models, further data acquisition on the Scots pine bark transition area should include a wider range of plots with a more incremental variety of stand characteristics, such as DBH, heights or tree age classes.

Erteld's findings of the linear regression functions of a single bark transition height correlating to DBH could not be replicated in this study [24]. On visual inspection of the scatter plots, HT1 and HT2 seem to correlate linearly with DBH on a stand scale. Nevertheless, a linear regression of HT $= a + b \times$ DBH was not possible due to residuals not being normally distributed. Another drawback of the more complex models is the negative correlation between DBH and stand density. In other words, stand density depends on thinning measures, which aim to improve the diameter and crown growth of the remaining stand. In addition, DBH and tree height are correlated, so the combined use of both variables as predictors does not necessarily improve the model. An older stand has the largest DBH but has most often been thinned accordingly. In summary, additional collinear predictors have only minuscule influences supplementing DBH.

5. Conclusions

The unusual bark structure of Scots pine along the stem, divided in two distinct bark types and a transition area, poses a key challenge when the ratio between wood and bark is calculated, or when looking at bark deduction factors. Such factors are used to predict the diameter under the bark in order to estimate the timber volume without bark. For Scots pine, some general bark functions are available for lower and for upper stem parts. However, by modelling the respective heights of the bark transition area, the bark proportion and thus also the bark volume of standing trees can be determined. This non-destructive method may be useful for both timber and carbon inventory purposes: For estimating Scots pine timber volume, new algorithms can be developed according to our findings and, thus, result in more differentiated bark factor models for the two distinct bark structures. Such models may be included in the analysis of forest inventory data to provide a more accurate estimation of, e.g., carbon sequestered by trees. As well, our results can be used as a basis for a more correct estimation of the bark proportion of harvested trees, or even in wood stacks. Therewith, a better economic valuation of logs and stacks is possible, according to the purchaser's demands. Furthermore, the valuable raw material bark that is important for a developing bioeconomy can be valued on a more reliable basis, to support a shift from fossil to renewable raw materials in the coming years.

Author Contributions: F.W., N.D. and F.B. conceived and designed the study, with contributions from T.C.; F.W. and N.D. collected the data; F.W. and N.D. analyzed the data, with contributions from F.B.; F.W., N.D., F.B. and T.C. wrote the manuscript. All authors have read and agreed to the published version of the manuscript.

Funding: This research was undertaken within the framework of the project "HoBeOpt", which was funded by the Fachagentur Nachwachsende Rohstoffe e.V. (FNR), grant number 22008518.

Institutional Review Board Statement: Not applicable.

Informed Consent Statement: Not applicable.

Data Availability Statement: The datasets analyzed for this study are available from the corresponding author on reasonable request.

Acknowledgments: The authors want to thank Evlyn Wallor for her support as well as the Landesbetrieb Forst Brandenburg, in particular, Lubomir Blasko from the Landeswaldoberförsterei Chorin.

Conflicts of Interest: The authors declare no conflict of interest.

References

1. FVA. Baumartenporträt: Die Waldkiefer. Available online: https://www.waldwissen.net/de/lebensraum-wald/baeume-und-waldpflanzen/nadelbaeume/die-waldkiefer (accessed on 14 December 2020).
2. BWI³. Dritte Bundeswaldinventur—Ergebnisdatenbank. Available online: https://bwi.info/ (accessed on 12 January 2020).
3. Schütt, P.; Stimm, B. Pinus sylvestris. In *Enzyklopädie der Holzgewächse: Handbuch und Atlas der Dendrologie/begründet von Peter Schütt*; Roloff, A., Weisgerber, H., Lang, U.M., Stimm, B., Eds.; Wiley-VCH: Weinheim, Germany, 2007; ISBN 9783527321414.
4. Routa, J.; Brännström, H.; Hellström, J.; Laitila, J. Influence of storage on the physical and chemical properties of Scots pine bark. *Bioenerg. Res.* **2020**, *29*, 53. [CrossRef]
5. Bauhus, J.; Kouki, J.; Paillet, Y.; Asbeck, T.; Marchetti, M. How does the forest-based bioeconomy impact forest biodiversity? In *Towards a Sustainable European Forest-Based Bioeconomy: Assessment and the Way Forward*; Winkel, G., Ed.; European Forest Institute: Joensuu, Finland, 2017; pp. 67–76. ISBN 9789525980417.
6. Jansone, Z.; Muizniece, I.; Blumberga, D. Analysis of wood bark use opportunities. *Energy Procedia* **2017**, *128*, 268–274. [CrossRef]
7. Eberhardt, T.L. Longleaf Pine Inner Bark and Outer Bark Thicknesses: Measurement and Relevance. *South. J. Appl. For.* **2013**, *37*, 177–180. [CrossRef]
8. Romero, C. Bark: Structure and functional ecology. *Adv. Econ. Bot.* **2014**, *17*, 5–25.
9. Diamantopoulou, M.J.; Özçelik, R.; Yavuz, H. Tree-bark volume prediction via machine learning: A case study based on black alder's tree-bark production. *Comput. Electron. Agric.* **2018**, *151*, 431–440. [CrossRef]
10. Marshall, H.D.; Murphy, G.E.; Lachenbruch, B. Effects of bark thickness estimates on optimal log merchandising. *For. Prod. J.* **2006**, 87–92.
11. Neumann, M.; Lawes, M.J. Quantifying carbon in tree bark: The importance of bark morphology and tree size. *Methods Ecol. Evol.* **2021**, *12*, 646–654. [CrossRef]
12. Hammond, D.H.; Varner, J.M.; Kush, J.S.; Fan, Z. Contrasting sapling bark allocation of five southeastern USA hardwood tree species in a fire prone ecosystem. *Ecosphere* **2015**, *6*, art112. [CrossRef]
13. Dantas, V.d.L.; Pausas, J.G. The lanky and the corky: Fire-escape strategies in savanna woody species. *J. Ecol.* **2013**, *101*, 1265–1272. [CrossRef]
14. Diamantopoulou, M.J. Artificial neural networks as an alternative tool in pine bark volume estimation. *Comput. Electron. Agric.* **2005**, *48*, 235–244. [CrossRef]

15. Murphy, G.; Cown, D. Within-tree, between-tree, and geospatial variation in estimated Pinus radiata bark volume and weight in New Zealand. *N. Z. J. Sci.* **2015**, *45*, 55. [CrossRef]
16. Volz, K.-R. Untersuchung über die Elgenschaften der Rinde von Ficht, Kiefer und Buche und ihr Eignung als Rohstoff für Flachpressplatten: Dissertation zur Erlangung des Doktorgrades der Forstlichen Fakultät der Georg-August-Universität Göttingen. Ph.D. Thesis, Georg-August-Universität Göttingen, Göttingen, Germany, 1974.
17. UNECE. Forest product Conversion Factors for the UNECE Region: Geneva Timber and Forest; Discussion Paper 49. Available online: https://unece.org/fileadmin/DAM/timber/publications/DP-49.pdf (accessed on 8 October 2021).
18. Stängle, S.M.; Sauter, U.H.; Dormann, C.F. Comparison of models for estimating bark thickness of Picea abies in southwest Germany: The role of tree, stand, and environmental factors. *Ann. For. Sci.* **2017**, *74*, 49. [CrossRef]
19. Adams, D.C.; Jackson, J.F. Estimating the Allometry of Tree Bark. *Am. Midl. Nat.* **1995**, *134*, 99. [CrossRef]
20. Nicolai, V. The bark of trees: Thermal properties, microclimate and fauna. *Oecologia* **1986**, *69*, 148–160. [CrossRef]
21. Shearman, T.M.; Varner, J.M. Variation in Bark Allocation and Rugosity Across Seven Co-occurring Southeastern US Tree Species. *Front. For. Glob. Change* **2021**, *4*, 346. [CrossRef]
22. Dengler, A. 52jährige finnische und märkische Kiefern im Forstamt Eberswalde: Untersuchungen aus dem Waldbau-Institut der Forstlichen Hochschule Eberswalde. *Z. Forst Jagdwes.* **1937**, *69*, 555–566.
23. Wagenknecht, E. Untersuchungen über den Spiegelrindenanteil verschiedener Kiefernrassen im Zusammenhang mit der Ästigkeit. *Z. Forst Jagdwes.* **1939**, *21*, 505–526.
24. Erteld, W. Die Bedeutung der Spiegelrinde an der Kiefer. *Sozial. Forstwirtsch.* **1963**, *13*, 331–335.
25. Houston Durrant, T.; de Rido, D.; Caudullo, G. Pinus sylvestris in Europe: Distribution, habitat, usage and threats. In *European Atlas of Forest Tree Species*; European Commission: Luxembourg, 2016; pp. 132–133.
26. Altherr, E.; Unfried, P.; Hradetzky, V. *Statistische Rindenbeziehungen als Hilfsmittel zur Ausformung und Aufmessung unentrindeten Stammholzes: Teil 1: Kiefer, Buche, Hainbuche, Esche und Roterle*; Forstlichen Versuchs- und Forschungsanstalt Baden-Württemberg: Freiburg, Germany, 1974; Volume 61.
27. Hamilton, G.J. *Forest Mensuration Handbook*; Her Majesty's Stationery Office: London, UK, 1985.
28. Sedmíková, M.; Löwe, R.; Jankovský, M.; Natov, P.; Linda, R.; Dvořák, J. Estimation of Over- and Under-Bark Volume of Scots Pine Timber Produced by Harvesters. *Forests* **2020**, *11*, 626. [CrossRef]
29. Berendt, F.; Pegel, E.; Blasko, L.; Cremer, T. Bark proportion of Scots pine industrial wood. *Eur. J. Wood Wood Prod. (Holz Roh Werkst.)* **2021**, *128*, 268. [CrossRef]
30. DWD. CDC Open Data. Available online: https://opendata.dwd.de/climate_environment/CDC/observations_germany/climate/annual/more_precip/recent/ (accessed on 18 April 2021).
31. LFE. Forstliche Standortskartierung Brandenburg. 2021. Available online: http://www.brandenburg-forst.de:80/geoserver/wms?request=GetCapabilities& (accessed on 29 April 2021).
32. Stängle, S.M.; Weiskittel, A.R.; Dormann, C.F.; Brüchert, F. Measurement and prediction of bark thickness in Picea abies: Assessment of accuracy, precision, and sample size requirements. *Can. J. For. Res.* **2016**, *46*, 39–47. [CrossRef]
33. Freise, C. Die relative Kronenlänge als Steuerungsparameter des Einzelbaumwachstums der Fichte. Ph.D. Thesis, Albert-Ludwigs-Universität Freiburg, Freiburg, Germany, 2005.
34. Epp, P. *Zur Abschätzung des Zuwachspotenzials von Fichten und Tannen nach starker Freistellung in Plenterüberführungsbeständen*; Albert Ludwigs Universität Freiburg: Freiburg, Germany, 2003.
35. Gerecke, K.-L. Herleitung und Anwendung von "Referenzbäumen" zur Beschreibung des Wachstumsganges vorherrschender Tannen. Ph.D. Thesis, Albert-Ludwigs-Universität Freiburg, Freiburg, Germany, 1988.
36. Kramer, H. Kronenaufbau und Kronenentwicklung gleichalter Fichtenpflanzbestände. *Allg. Forst Jagdztg.* **1962**, *133*, 249–256.
37. Kramer, H.; Dong, P.H. Kronenanalyse für Zuwachsuntersuchungen in immissionsgeschädigten Nadelholzbeständen. *Forst Holz* **1985**, *40*, 115–118.
38. Spathelf, P. Orientierungshilfe zur Prognose und Steuerung des Wachstums von Fichten (*Picea abies* (L.) Karst.) und Tannen (*Abies alba* Mill.) in Überführungswäldern mit Hilfe der relativen Kronenlänge. Ph.D. Thesis, Albert-Ludwigs-Universität Freiburg, Freiburg, Germany, 1999.
39. R Core Team. *The R Project for Statistical Computing*; The R Foundation for Statistical Computing: Vienna, Austria, 2002; Available online: https://www.R-project.org/ (accessed on 18 April 2021).
40. RStudio Team. *R Studio: Integrated Development for R*; RStudio, PBC: Boston, MA, USA, 2020; Available online: http://www.rstudio.com/ (accessed on 18 April 2021).
41. Wooldridge, J.M. *Introductory Econometrics: A Modern Approach*, 7th ed.; Cengage Learning South-Western: Cincinnati, OH, USA, 2020; ISBN 9781337558860.
42. James, G.; Witten, D.; Hastie, T.; Tibshirani, R. *An Introduction to Statistical Learning*; Springer: New York, NY, USA, 2013; ISBN 978-1-4614-7137-0.
43. Kuhn, M. Caret: Classification and Regression Training. 2020. Available online: https://CRAN.R-project.org/package=caret (accessed on 26 September 2021).
44. Venables, W.N.; Ripley, B.D. *Modern Applied Statistics with S*, 4th ed.; Springer: New York, NY, USA, 2002.

45. Elzhov, T.V.; Mullen, K.M.; Spiess, A.-N.; Bolker, B. minpack.lm: R Interface to the Levenberg-Marquardt Nonlinear Least-Squares Algorithm Found in MINPACK, Plus Support for Bounds. 2016. Available online: https://CRAN.R-project.org/package=minpack.lm (accessed on 26 September 2021).
46. Laasasenaho, J.; Melkas, T.; Aldén, S. Modelling bark thickness of Picea abies with taper curves. *For. Ecol. Manag.* **2005**, *206*, 35–47. [CrossRef]
47. Kahriman, A. A bark thickness model vor calabrian pine in Turkey. In Proceedings of the 2nd International Conference on Science, Ecology and Technology, Barcelona, Spain, 14–16 October 2016; pp. 661–670.
48. Sonmez, T.; Keles, S.; Tilki, F. Effect of aspect, tree age and tree diameter on bark thickness of Picea orientalis. *Scand. J. For. Res.* **2007**, *22*, 193–197. [CrossRef]
49. Gordon, A. Estimating bark thickness of Pinus Radiata. *N. Z. J. For. Sci.* **1983**, *13*, 340–353.
50. Liepins, J.; Liepins, K. Evaluation of bark volume of four tree species in Latvia. *Res. Rural. Dev.* **2015**, *2*, 22–28.
51. Berendt, F.; de Miguel-Diez, F.; Wallor, E.; Blasko, L.; Cremer, T. Comparison of different approaches to estimate bark volume of industrial wood at disc and log scale. *Sci. Rep.* **2021**, *11*, 15630. [CrossRef] [PubMed]

Article

Soil Temperature in Disturbed Ecosystems of Central Siberia: Remote Sensing Data and Numerical Simulation

Tatiana V. Ponomareva [1,2], Kirill Yu. Litvintsev [3], Konstantin A. Finnikov [4], Nikita D. Yakimov [2], Andrey V. Sentyabov [3] and Evgenii I. Ponomarev [1,2,*]

1 V.N. Sukachev Institute of Forest SB RAS of the Federal Research Center "Krasnoyarsk Science Center, SB RAS", 660036 Krasnoyarsk, Russia; bashkova_t@mail.ru
2 Institute of Ecology and Geography, Siberian Federal University, 660041 Krasnoyarsk, Russia; nyakimov96@mail.ru
3 Kutateladze Institute of Thermophysics, Siberian Branch, Russian Academy of Sciences, 630090 Novosibirsk, Russia; sttupick@yandex.ru (K.Y.L.); sentyabov_a_v@mail.ru (A.V.S.)
4 Institute of Engineering Physics and Radioelectronics, Siberian Federal University, 660041 Krasnoyarsk, Russia; f_const@mail.ru
* Correspondence: evg@ksc.krasn.ru; Tel.: +7-391-249-4092

Abstract: We investigated changes in the temperature regime of post-fire and post-technogenic cryogenic soils of Central Siberia using remote sensing data and results of numerical simulation. We have selected the time series of satellite data for two variants of plots with disturbed vegetation and on-ground cover: natural ecosystems of post-fire plots and post-technogenic plots with reclamation as well as dumps without reclamation. Surface thermal anomalies and temperature in soil horizons were evaluated from remote data and numerical simulation and compared with summarized experimental data. We estimated the influence of soil profile disturbances on the temperature anomalies forming on the surface and in soil horizons based on the results of heat transfer modeling in the soil profile. According to remote sensing data, within 20 years, the thermal insulation properties of the vegetation cover restore in the post-fire areas, and the relative temperature anomaly reaches the level of background values. In post-technogenic plots, conditions are more "contrast" comparing to the background, and the process of the thermal regime restoration takes a longer time (>60 years). Forming "neo-technogenic ecosystems" are distinct in special thermal regimes of soils that differ from the background ones both in reclamated and in non-reclamated plots. An assumption was made of the changes in the moisture content regime as the main factor causing the long-term existence of thermal anomalies in the upper soil horizons of disturbed plots. In addition, we discussed the formation of transition zones ("ecotones") along the periphery of the disturbed plots due to horizontal heat transfer.

Keywords: disturbances; wildfires; natural and technogenic ecosystems; permafrost; thermal anomaly; soil; numerical simulation; remote sensing; Siberia

Citation: Ponomareva, T.V.; Litvintsev, K.Y.; Finnikov, K.A.; Yakimov, N.D.; Sentyabov, A.V.; Ponomarev, E.I. Soil Temperature in Disturbed Ecosystems of Central Siberia: Remote Sensing Data and Numerical Simulation. *Forests* **2021**, *12*, 994. https://doi.org/10.3390/f12080994

Academic Editor: Rosemary Sherriff

Received: 4 June 2021
Accepted: 22 July 2021
Published: 27 July 2021

Publisher's Note: MDPI stays neutral with regard to jurisdictional claims in published maps and institutional affiliations.

1. Introduction

Regimes of soil functioning, dynamics, and rate of soil processes are subjects of physical characteristics of soils (temperature and moisture content). Analysis and modeling of temperature distribution in horizons of cryolithozone soils, including the seasonally thawed layer (STL) of soil, are urgent and widely discussed problems [1–7]. Central Siberia is an important subject of these studies due to that 50% of forest ecosystems (about 3 mln km^2 of 6 mln km^2 of forested area) is in the permafrost zone [8].

Changes in thermal regimes of soil reveal themselves in the root-inhabited soil layer (active layer of soil). Possible changes in the state of permafrost soils are a significant factor affecting eco-systems [7,9–13] and even on global climate [14–16].

The heat exchange in the soil profile is a subject of the temperature gradient, thermal radiation balance, turbulent heat exchange in air, and evaporation and soil moisture.

Temperature gradients determine the heterogeneity of the soil cover functioning [17]. The vegetation cover causes the most significant influence on the soil thermal regime, moss–lichen layer, and organic soil layer [18–21]. The temperature of the upper soil layer (0–0.4 m of depth) depends on the thermal insulating properties of moss and lichen covers and on the thickness of the organic soil horizon [21–23], as well as on the tree stand crown canopy [24]. Heat exchange in soils depends also on the granulometric composition of mineral horizons [18].

Under conditions of loss of thermal insulating organic layer, soil heating causes changes in ecosystems functioning according to the seasonal variations of the active layer of soil comparing to the statistical norm [3,9,13]. Such a situation may arise as a result of a number of destructive natural or human-made factors (wildfires, logging, and activities of the mining complex).

Disturbances may have a larger and immediate impact on Low Arctic ecosystems of Siberia than temperature warming alone. While soil conditions strongly affect vegetation patterns, plants may also have strong feedback effects on soil thermal and hydrological properties [25].

Destructive impacts affect surface albedo, emissivity, moisture, and water regimes of the upper soil horizons. As a result, there is a significant change in the temperature regime of soils comparing to the background. It is estimated that the post-fire effects have an "accumulative" character [26,27] on the area up to 20% of boreal forests of Siberian and can stay significant for 15 years [11,27]. This factor is likely to increase in the future [28]. Anomalous heating of the surface is also observed for post-technogenic areas (open pit mining, quarries, overburden dumps, logging, etc.), characterized by the intense mechanical impact on the vegetation cover and soil [29,30]. This issue is relevant for the zone of the resource-mining complex of Siberia and, in particular, for the Arctic zone (>65° N).

In the context of the diversity of soil properties, the issue of mathematical modeling based on ground-based and remote sensing data is urgent [7,27,31–35]. Available models of heat and moisture transfer in soil under freezing and thawing conditions take into account the moisture conductivity of soil [36–39]. An obligatory element of models of moisture transfer in freezing soil is taking into account cryosuction, i.e., capillary force associated with the presence of a phase transition [40–42], natural convection in water that fills soil pores [43], and vapor transfer in water-free soil pores and snow cover [44–46]. After all, it is important to account correctly for solar radiation, which depends on the location and state of the atmosphere [47,48].

The main aim of this work was to assess the influence of the transformation of the soil profile on the formation of temperature anomalies in the surface and in the soil horizons in disturbed plots based on numerical simulation of heat transfer in the soil profile. The following aspects of the issue were considered:

(1) remote sensing of the state of disturbed natural and technogenic ecosystems based on satellite monitoring data and subsequent numerical assessment of the influence of changes in the optical properties of the surface (spectral albedo and surface emissivity) and variability of soil moisture content on the formation of thermal anomalies in disturbed natural and technogenic ecosystems;

(2) modeling of conditions for abnormal heating of soil horizons in disturbed areas during the growing and non-growing seasons of a year;

(3) time-lag of recovery of the thermal state in disturbed ecosystems and the level of stabilization of the temperature characteristics of the surface under conditions of long-term post-pyrogenic and post-technogenic recovery;

(4) analysis of heat transfer at the boundary of disturbed plots/background areas and the peculiarities of the temperature regime of soils in the formed transition zones ("ecotones") along the periphery of the disturbed plots due to horizontal heat transfer.

In the current work, it will be demonstrated that the characteristic features of the temperature effect of the damaging impact, revealed during field observations, are reproducible in a mathematical model that takes into account the differences in the structure and moisture content of disturbed and undisturbed soil. We intend to verify, by means of the numerical simulation, that not only the change in the optical properties of the surface, but also the disturbance of the structure and properties of soil horizons are the factor providing the long-term existence of temperature anomalies.

2. Area of Interest and Research Objects

In the current research, we took into account the properties of cryogenic soils, which are typical for the taiga region of Central Siberia (59–66° N, 90–107° E). The forest stands of the region are dominated by larch (*Larix sibirica* Lebed., *Larix gmelinii* Rupr. Rupr.) and sparse larch stands, accounting for >50% of the total forested area [26,49,50]. Herb-shrub layer, mosses, and lichens represent the vegetation cover as well as litter loads [51].

The continental climate of the permafrost zone and characteristics of parent rocks determine the genetic specificity of the soils in the region. The soils are characterized by low thickness (0.20–0.50 m) and weak differentiation of the mineral part of the soil profile. The permafrost level (<1.0 m) depends significantly on the topography [11]. Permafrost is widespread on flat areas and is not found in the soil profile on drained areas of slopes and valleys. Periodic freezing and thawing of the soils determine the soil structure, the low rate of humus accumulation in the soil, the peculiarities of moisture transport, and waterlogging. In addition, freezing and thawing processes affect slightly acidic or neutral reaction of the soil, as well as loamy or clayey granulometric composition with a high gravel content in the lower part of the soil profile [49,50]. Finally, cryoturbation, solifluction, cryogenic heaving, thermokarst, etc. characterize such soils.

According to the Russian soil classification of 2004 [52], the soil cover is represented by cryozems (In World Reference Base for Soil Resources (WRB, 2014 [53]) equivalents is Histic Cryosols), podburs (WRB equivalents is Cambic Crysols), and peat soils (WRB equivalents is Cryic Histosols). Cryozems are formed in the lower parts of slopes and microdepressions on rocks of heavy granulometric composition. The soil profile contains a peat litter (thickness of 0.05–0.10 m), a thin humus or coarse humus horizon, and a uniformly mixed cryoturbated mineral horizon with inclusions of plant residues of varying degrees of decomposition. A high (10–14%) humus content characterizes such soils, as the consequence of the low rate of organic matter decomposition. The lower supra-permafrost part of the soil profile is saturated with moisture, structureless, and is more dense and homogeneous. Podburs are formed on fine-earth-clastic products of destruction of metamorphic rocks, which leads to good drainage and to large thickness of the active soil layer. The profile consists of a peat litter, under which lies the Al-Fe-humus horizon, which gradually turns into the parent rock. Peat-cryozems are present in combination with cryozems, located in low relief areas, and are diagnosed by the presence of peat (T) and cryoturbated (CR) horizons.

Background cryogenic soils typically are full-profile, and according to the modern classification [50] have the formula T(Oao)-O(F+H)-CR-C$_\perp$ (Figure 1).

Wildfire impact changes the structural organization of the soil profile. After a wildfire the soil loses its upper horizon, the level of permafrost occurrence decreases, and the soil formula transforms to Opir-O(F+H)pir-CR-C. Under the conditions of high-intensity burning, the profile structure can transform more strongly. Depending on the degree of litter burnout the soil formula changes to O(F+H)pir-CRpir-C or O(H)pir-CRpir-C. Locally, complete burnout of the upper organic horizons of the soil is possible [54]. Upper part of the soil profile, as a rule, have a thin humus-accumulative (underdeveloped) horizon W or a fragmented coarse humus horizon Oao both in case of post-fire mineralization of cryozems and in case of technogenic transforming of soils. Such soils formula is Oao(W)-C.

Evolution of soils after destructive impacts on thermal insulating covers includes an increase in thawing depth, an increase in soil moisture due to thawing of schlieren and wedge ice, an increase in gleying, and cryoturbation processes, and it proceeds under strong influence of succession.

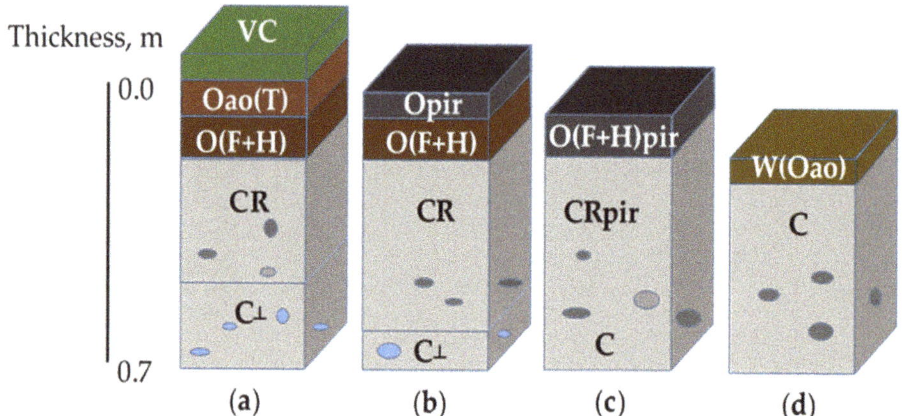

Figure 1. Models of soil profiles of non-disturbed cryozems (**a**), post-pyrogenic cryozems (**b**,**c**), and technogenic soils (**d**), which are typical for the study area in the permafrost zone of Siberia with marks of peat (T) and cryoturbated (CR) horizons, permafrost level (C⊥), postfire litter O(F+H)pir and mineral horizon (CRpir), coarse humus horizon (Oao), vegetation cover (VC), and underdeveloped organic horizon (W).

For instance, the effect of post-fire rapid increase in the thawing depth (by 30–100% in the first year after the fire) [34] leads to a unique path of soil evolution, when the underlying gleyed rock layers are included in soil formation for a significant period [55]. This type of evolution is possible only in the permafrost zone. Moreover, an increase in the STL thickness leads to the thawing of schlieren (sometimes wedge ice). An excessive amount of moisture under conditions of difficult drainage stimulates the gleying process [55].

The objects of modeling were (i) native variant, which is undisturbed natural state of cryozem, and (ii) variants of soils with the topsoil layer disturbed as a result of a destructive factor.

The set of input parameters (Tables 1 and 2) for the numerical simulation of the soil temperature regime was summarized from the data of field studies of the morphological and physical properties of soils carried out in the region [49,50,56–59].

Table 1. Input parameters used for numerical simulation of the temperature regime in the soil profile. Physical characteristics of non-disturbed soils, summarized from the works in [49,50,56–59].

Horizon	Specific Heat, J/(kg·K)	Volumetric Water/Ice Content (Q_w) *	Porosity	Bulk Density of Dry Soil, ρ_b, kg/m^3	Depth, m
Oao	1500	0.3/0.8	0.91	60	0.0–0.12
O(F+H)	1880	0.35/0.79	0.81	70	0.12–0.17
CR	750	0.275/0.3	0.48	1100	0.17–0.35
C	920	0.3/0.33	0.36	1500	0.35–Z_{end} **

* Constant values of the volumetric water/ice content. ** Z_{end}—maximum depth of computational domains, m.

Table 2. Input parameters used for numerical simulation of the temperature regime in the soil profile. Physical characteristics of disturbed soils (post-fire and technogenic impact), summarized from the works in [49,50,56–59].

Horizon	Specific Heat, J/(kg·K)	Volumetric Water/Ice Content *	Porosity	Bulk Density of Dry Soil, ρ_b, kg/m^3	Depth, m
O(F+H)pir	1880	0.35/0.79	0.81	70	0.12–0.17
CR (CRpir)	750	0.275/0.3	0.48	1100	0.17–0.35
C	920	0.3/0.33	0.36	1500	0.35–Z_{end} **

* Constant values of the volumetric water/ice content. ** Z_{end}—maximum depth of computational domains, m.

According to field experiments in the permafrost zone [50], "thermal inversion" is possible in different layers of cryogenic soils of disturbed plots. Thus, there is stronger heating of the surface in summer (due to lower thermal resistance). Next, there is a higher temperature in the soil horizons during the non-growing period (early spring, winter, autumn) (due to the higher thermal resistance) in comparison with the distribution in the background plots. During the growing season (summer), such differences are explained by the state (or loss) of disturbed upper soil horizons and the reduced albedo value. The relative change in thermal resistance during the non-growing season can only be explained by a higher ice content in the soils of non-damaged plots, if the snow cover is equal.

As the model used did not consider the transfer of moisture within the soil, values of the volumetric water/ice content were tabulated as constants for the summer and winter seasons (Tables 1 and 2).

3. Materials and Methods

3.1. Multispectral and Infrared Survey of Research Objects

We used satellite images of average spatial resolution (15–30 m) from Landsat-5/7/8 for 1975–2020, which are freely available in The United States Geological Survey (USGS) database (https://earthexplorer.usgs.gov/, accessed on 21 May 2021). The surface temperature was evaluated from the calibrated B6 channel (λ = 10.4–12.5 μm, Landsat-5/TM—Thematic Mapper), B6/1 channel (Landsat-7/ETM—Enhanced Thematic Mapper), and B10 channel (λ = 10.6–11.9 μm, Landsat-8/OLI—Operational Land Imager). We implemented radiometric correction method to the initial data using calibration constants from the metadata files [60,61]. First, we analyzed the availability of the imagery for selected experimental post-technogenic and post-fire plots to obtain time-series for each plot for 1975–2020. The only criteria were the percentage of cloudiness, the binding of the survey dates to the middle of the growing season (July), and the regularity of survey covering of selected objects. Daytime imagery was used only. Thus, we did not implement any special procedure for the selection of Landsat images.

Additionally, we used low spatial resolution survey data (250–1000 m) from Terra/MODIS (Moderate Resolution Imaging Spectroradiometer) for 2002–2020. Standard products L2G and L3 used (free USGS database, https://lpdaac.usgs.gov/dataset_discovery/modis, accessed on 21 May 2021). We operated with reflectances (albedo) measured in MODIS band #1 (λ = 0.620–0.670 μm) and band #2 (λ = 0.841–0.876 μm) (product MOD09GQ), and to analyze surface temperature anomalies we used the MOD11A1 standard product. We used 10-day averaged data across the entire set of initial data, with a special focus on the disturbed ecosystem's recovery succession stages. A monthly data average procedure, as well as a procedure of spatial averaging within each natural and technogenic disturbed ecosystem plots, were applied.

We have selected the time series of satellite data for two variants of plots with disturbed vegetation and on-ground cover: (1) natural ecosystems of post-fire plots (PF) for 1996–2018 and (2) post-technogenic plots for 1975–2019.

Using time series of satellite data, we performed indirect assessments of the state of disturbed ecosystems (disturbed plots) at different periods of recovery successions based on changes in the anomalies of spectral features comparing to characteristics of the background areas. We analyzed data for 30 post-fire plots recorded in Siberia in 1996, 2006, and 2017 (Figure 2a), and the dynamics of spectral anomalies for two technogenic plots: Borodinsky Coal Mine (BCM) for 1975–2018 and Olimpiada Mining Plant (OMP) for 2000, 2011, 2019 (Figure 2b,c).

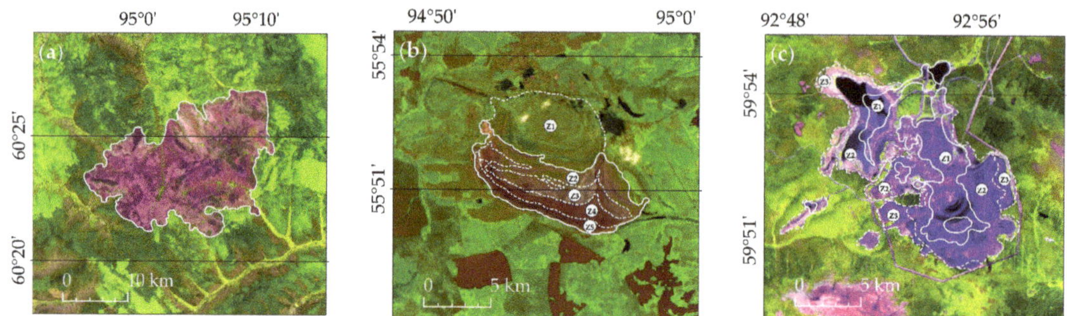

Figure 2. Initial images on natural and technogenic disturbed plots: (**a**) an example of a post-fire plot (burned 2 July–24 July 2018), image from Landsat-8 OLI/TIRS, 12 July 2019; (**b**) technogenic disturbances at the Borodinsky Coal Mine (BCM), Landsat-8 OLI/TIRS, 02 July 2018; and (**c**) technogenic disturbances at the Olimpiada Mining Plant (OMP), 12 July 2019. Plots of different ages (from 1970th up to 2018) are marked as Z1–Z5 polygons.

We analyzed the dynamics of albedo and temperature anomalies during 1, 5, 10 (or 12), and 20 years of post-fire vegetation and on-ground cover restoration. For technogenic disturbances, we analyzed albedo and surface temperatures during 1, 10, 20, 40, and 60 years after impact a destructive factor on soil covers.

To determine the relative anomalies of the surface temperature of disturbed plots to the characteristics of background territory ($\Delta T/T_{bg}$ (°C/°C), in %), we compared averaging over 10 measurements for each period of recovery:

$$\Delta T/T_{bg} = 100\% \cdot (T_{tg} - T_{bg})/T_{bg}, \tag{1}$$

where T_{tg} is surface temperature of target (disturbed plot), °C, and T_{bg} is surface temperature of background (non-disturbed) plot, °C.

3.2. Mathematical Model of Heat Transfer in the Soil

Mathematical models of heat transfer in soils taking into account water-ice phase transformation are quite diverse [62]. The choice of the model depends on the specifics of the problem being solved. Climatological studies often impose severe constraints on the computational resources that can be assigned to calculations of soil heat transfer, therefore analytical and semi-analytical methods are widely used [63,64]. These methods are based on constant or piecewise constant approximations of the main thermophysical characteristics (density, specific heat capacity, and thermal conductivity coefficients) in their dependence on the depth and on the assumption that the temperature depends on the vertical coordinate only.

The formulation of the issue of heat transfer in disturbed cryogenic soil took into account the following factors.

(1) The multidimensional character of the temperature distribution, being a result of vicinity of disturbed and non-disturbed plots and, accordingly, the presence of a significant difference in the state of the permafrost layer. Under these conditions, significant heat transfer in the horizontal direction can arise.

(2) Inhomogeneity of the thermophysical properties of the soil, due to the vertical structure of the soil profile and its differences in disturbed and background plots.

(3) The presence of a transition zone with a mixed phase state of water and ice. Such areas inevitably arise due to local heterogeneities of the soil, its composition, and moisture content. For the used scale of grid sampling in the considered problem, the transition zone is described in terms of the average content of the liquid and solid phases of water.

In a mathematical model of heat transfer in soil, either the temperature field or the specific enthalpy field can be considered as a sought field [65]. The choice of temperature as a sought field is more expedient for the considered problem. This facilitates greatly the process of composition of the thermophysical properties of soil. Thus, the following formulation for the heat transport equation was chosen:

$$c_p(\mathbf{x}, T)\rho(\mathbf{x})\frac{\partial T}{\partial t} - \frac{\partial}{\partial x_i}\left(\lambda(\mathbf{x}, T)\frac{\partial T}{\partial x_i}\right) = S, \tag{2}$$

where T is the temperature (K), $c_p(T)$ is the thermal capacity (J/(kg·K)), ρ is the density (kg/m^3), λ (T) is the thermal conductivity coefficient (W/(m·K)), and S is the heat source associated with the phase change latent heat (W/m^3).

Formulating the model of the heat source S, we consider the liquid and solid phases as separate components of the medium. They can present at the same place (i.e., in the same finite volume of the discrete mathematical model) simultaneously. This approach allows taking into account a fine structure of intermitting water and ice that is typical for freezing and thawing soils [66] but cannot be resolved in a discrete model due to the restrictions for a grid dimension. The mass sources of solid and liquid phases are opposite and proportional to the temperature on the Celsius scale [66]. In accordance with the chosen approach, the source term S in Equation (2) has the following form:

$$S = -y_w(\mathbf{x}, T)\rho_w L\frac{\partial \alpha_l}{\partial t}, \tag{3}$$

where y_w is the soil volumetric content of water regardless to the state (m^3/m^3), α_l is the liquid water fraction, and L is the latent heat (J/kg).

The phase transition rate is determined by the heat transfer flux to the contacts between the phases. This flux is proportional to the difference between the mean temperature of the soil and the phase change temperature, in other words, to the mean temperature of the soil measured in Celsius degrees. Considering the phase transition process as the exponential decay of the diminishing phase, we can write the following equation for the liquid water fraction:

$$\frac{\partial \alpha_l}{\partial t} = k(\max(T\ [^{\circ}C],\ 0)(1 - \alpha_l) + \min(T[^{\circ}C],\ 0)\alpha_l), \tag{4}$$

where k is a coefficient connecting the characteristic timescale of the exponential decay of the diminishing phase and temperature.

Equation (2) was solved in a two-dimensional formulation, in which the vector of coordinate $\mathbf{x}$ has vertical and horizontal components (to take into account the effects of the transition region between disturbed and undisturbed soil areas). Large-scale variability of the soil structure and its moisture content, as well as the presence of dynamically changing depth of snow cover, are taken into account in the expressions for the coefficients of the Equation (2): $c_p(\mathbf{x}, T)$, $\rho(\mathbf{x})$, $\lambda(\mathbf{x}, T)$, $y_w(\mathbf{x}, T)$.

Equation (2) was solved with the use of the finite volume method in that the equation domain is divided into a large number of cells that have four faces as the case is two-dimensional. The sought temperature fields are determined as a set of temperature values at the centers of the cells and the differential Equation (2) is replaced by a system of linear algebraic equations (grid equations) connecting the values at the centers of the cells [67]. Integration of Equation (2) along the cell volumes leads to the algebraic grid equations in that the sought variables are the temperature values in the cell centers [68].

Available models [1,4,5,69,70] were used to derive expressions for the thermal conductivity depending on the vertical coordinate. Empirical models by Pavlov [69], Anisimov et al. [70] are results of the generalization of the great amount of field measurements. In these models, the thermal conductivity coefficient is a function of the dry soil density and the volume fraction of water or ice. The models proposed in the works [1,4,5] are based on the supposition [71] that the thermal conductivity of a partially water-saturated soil correlates linearly with the thermal conductivities of a dry and a water-saturated soil:

$$\lambda = Ke\lambda_{sat} + (1 - Ke)\lambda_d, \tag{5}$$

where λ_{sat} and λ_d are the thermal conductivity coefficients of a soil in the water-saturated and dry states (W/(m·K)), and Ke is Kersten number.

We used the model by Porada et al. [5] for the moss and lichen layer (Oao horizon). For the horizon O(F+H) that consists mainly of medium and well decomposed organic residues, the model by Anisimov et al. was used [70]. For the following two mineral horizons CR and C two types of models were considered [4,69] (Table 3). In calculations relating to periods with stable snow cover the thermal conductivity of snow was admitted to be a constant, $\lambda_{snow} = 0.23$ W/(m·K).

Table 3. Models of thermal conductivity coefficients used for different soil horizons.

Horizon	Models of Thermal Conductivity Coefficients	Parameters Used	Author, Reference
Oao	Equation (5) with $\lambda_d = 0.05$ and $\lambda_{sat} = \lambda_o{}^{(1-(w_f+w_s))} \cdot \lambda_l{}^{w_l} \cdot \lambda_s{}^{w_s}$ $\lambda_o = 0.25;\ \lambda_l = 0.56;\ \lambda_s = 2.25$	λ_d is the thermal conductivity of dry organic substance, λ_o is the thermal conductivity of the organic substance in the water-saturated condition, w is the volumetric fraction of a certain state of water, s is the solid state of water, l is the liquid state of water.	Porada et al., 2016 [5]
O(F+H)	$\lambda_t = 0,08 \cdot e^{3.88w_f}$ $\lambda_{f_r} = (615w_s + 22.2) \cdot 10^{-3}$ $\lambda = k(0.001 \cdot \rho_b + 10^4 \cdot w/\rho_b - 1.1) - 1.16 \cdot 10^4 \cdot w/\rho_b$	t is the thawed soil, fr is the frozen soil	Anisimov et al., 2012 [70]
CR	Equation (5) was used, where $\lambda_{sat} = \lambda_o{}^{(1-(w_f+w_s))} \cdot \lambda_f{}^{w_f} \cdot \lambda_i{}^{w_i}$ $\lambda_d = (A \cdot \rho_b + B)/(\rho_{soil} + C \cdot \rho_b)$ Coefficients A, B, C, D summarized from [1,4]	ρ_b is the bulk density of dry soil; ρ_{soil} is the density of the material of soil particles; w is the volumetric water/ice fraction; k is the empiric coefficient specific for the type and condition of soil	Pavlov, 1979 [69], Ekici et al., 2014 [4], Tarnawski et al., 2000 [1]

Other thermophysical properties of snow (density, heat capacity, and albedo) were also assumed to be constant, but the depth of the snow cover was considered as alternating according to meteorological data.

The boundary conditions at the upper boundary of a soil or snow take into account the presence of solar (shortwave) and atmospheric (longwave) radiation and convective heat transfer by air [72,73]:

$$-\lambda\left(\frac{\partial T}{\partial x}\right) = q_c + (1 - \alpha_s) \cdot q_{Sr} + q_{Lr}, \tag{6}$$

where q_c is the convective heat flux, q_{Sr} and q_{Lr} are the shortwave and longwave radiation heat fluxes, and α_s is the surface albedo.

The surface convective heat flux is calculated with the use of the following expression [69]:

$$q_c = a_s \cdot (T_{atm} - T_s),$$
$$\alpha_s = u_1^{0.5} \cdot (6 + 3.1 \cdot \Delta T/u_1^2) \tag{7}$$

where α_s is the heat transfer coefficient, W/(K·m²); u_1 is the wind velocity at the altitude 1 m; and ΔT is the temperature difference between the solid surface and atmosphere.

The shortwave radiation heat flux divides into the direct solar radiation $\left(I_{Si}^{Dir}\right)$ and the scattered one $\left(I_{Si}^{Dif}\right)$:

$$q_{Sr} = q_{Sr}^{Dir} + q_{Sr}^{Dif}, \tag{8}$$

The direct solar heat flux on a horizontal surface is found as [47]

$$q_{Sr}^{Dir} = q_{Sr0} \cdot \cos Z \cdot \prod_i \tau_i,$$ (9)

where Z is the reduced incidence angle, τ_i are the atmosphere transmittance factors that characterize absorption by aerosols, water vapor, ozone, other gases, and Rayleigh scattering; q_{Sr0} (W/m^2), is the Solar heat flux reaching the Earth atmosphere. The value of q_{Sr0} depends on the position of the Earth relative to the Sun [73]. The factors τ_i in (9) are calculated according to [47].

Contributions to the scattered radiation flux are made by the Rayleigh and aerosol scattering and multiple reflection of the direct solar radiation. The meteorological data for cloud cover were used to calculate the change in relative insolation.

Longwave radiation heat exchange between the solid surface and the atmosphere is calculated as

$$q_{Lr} = \varepsilon_s \cdot \left(\varepsilon_{atm} \sigma \cdot T_{atm}^4 - \sigma \cdot T_s^4 \right),$$ (10)

where ε_s is the surface emissivity and ε_{atm} is the atmosphere emissivity. For the surface of the damaged and undamaged soil, the values of the integral albedo were taken equal to 0.13 and 0.18 (a decrease by 40% for the case of a significant change in the state of the ground cover and the upper organic horizon), the emissivity values are 0.9 and 0.98, respectively [74]. The atmosphere emissivity ε_{atm} was calculated as a function of cloud cover and humidity [75,76].

3.3. Meteorological Data

We used long-term series of meteorological information from open data banks for the period 1990–2020: Climatic Research Unit (http://www.cru.uea.ac.uk/, accessed on 25 March 2021), Weather archive (http://rp5.ru, accessed on 25 March 2021), and the National Climatic Data Center (NCDC) (http://www7.ncdc.noaa.gov/CDO/cdo, accessed on 25 March 2021).

To calculate solar heat flux (4), we used the geographic coordinates of Tura (64° N, 100° E) (Evenkia, Krasnoyarsk region, Siberia, Russia). The data on the depth of snow cover (mm), air temperature (°C), wind speed (m/s), and cloudiness (oktas) were taken from the annual series of meteorological data. The observations have a regularity of 4 h (6 times a day). Numerical simulations are performed for 2018 data (Table S1, Supplementary Materials).

3.4. Main Features and Assumptions of the Mathematical Model of Heat Transfer in the Soil

The governing equations of the mathematical model are (1) and (4), that describe heat transfer in soil in presence of water-ice phase transition neglecting the moisture transfer processes. The main thermophysical properties, in addition to thermal conductivity, were set constant for different soil horizons, but different for temperatures above and below 0 °C. The calculations were performed for a time interval up to three years required to establish a quasiperiodic solution.

The result of the numerical solution of the model equations was the dynamics of two-dimensional temperature fields and phase transition boundaries.

In the mathematical model, the discretization of the computational domain was built based on a Cartesian (orthogonal, structured) grid. The vertical grid step was 0.01 m, the horizontal one was from 0.25 m, and the time step was 15 min.

In the mathematical model, the boundaries of the soil horizons were assumed constant and dynamic change of the snow cover depth according to the data of meteorological observations was taken into account. A linear temperature distribution changing from the air temperature at the surface to the specified temperature at the lower boundary (Z_{end}) (see Tables 1 and 2) of the equations domain was assumed as the initial temperature distribution in the soil.

4. Results

4.1. Surface Temperature Anomalies of Disturbed Plots on Satellite Data in IR Range

Time series of satellite imagery reflect the change in spectral characteristics and temperature anomalies of disturbed areas caused by recovery of vegetation and of a ground cover (Figure 3). For PF plots it was evaluated for the period of 20 years, for both BCM and OMP plots it was analyzed for the period of >60 years. The peculiarity of plots after fire disturbances is that natural recovery processes occur there. The peculiarity of post-technogenic plots is that there are two possible options for dynamics of the ecosystem: (1) restoration after reclamation or (2) long-term condition in the format of non-reclaimed lands (dumps, mineralized surfaces). Each of these options is of interest from the point of view of the formation of surface temperature anomalies and its effect on the properties of soil horizons.

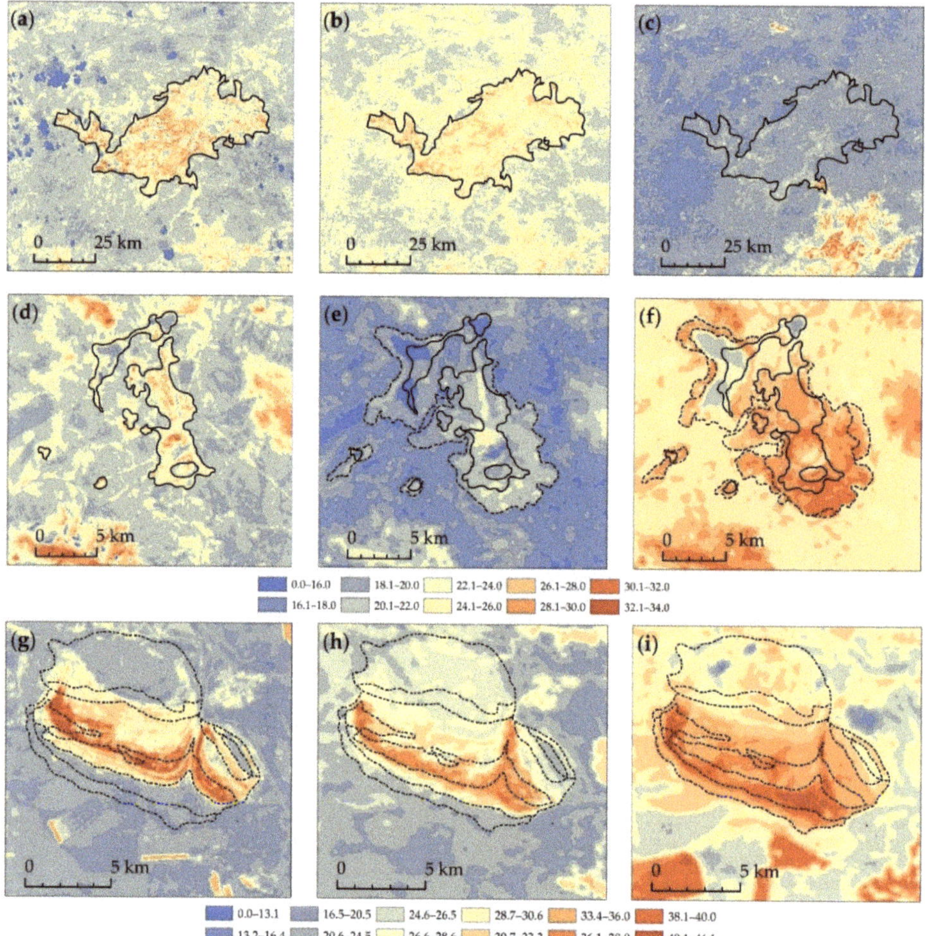

Figure 3. Time series of satellite data on surface temperature in July for (1) post-fire plot in 1997 after 1-years recovery (**a**), 10-years recovery in 2006 (**b**), and 20-years recovery in 2016 (**c**); (2) OMP plot with the identification of zones of different ages (since 2000) formation of a technogenic ecosystem on dumps in (**d**) 2000, (**e**) 2011, and (**f**) 2019; and (3) BCM plot with the identification of zones of different ages (since 1960–1970th) formation of a technogenic ecosystem with reclamation in (**g**) 2000, (**h**) 2010, and (**i**) 2018.

The long-term dynamics of spectral albedo (α) and surface temperature anomalies in disturbed plots of natural and technogenic landscapes in comparison with the background data are summarized in Table 4.

Table 4. Changes in the spectral characteristics of disturbed areas in % to background values for the options PF, BUR (reclamation), and OMP (without reclamation). Mean value for July $\pm$ SD, $p < 0.05$.

Recovery Time, Years	Albedo Decreasing (%) $\lambda = 0.620$–0.670 µm		Albedo Decreasing (%) $\lambda = 0.841$–0.876 µm		$\Delta T/T_{bg}$ (%) $\lambda = 10$–12 µm	
	PF/BCM (Reclamation)	OMP (without Reclamation)	PF/BCM (Reclamation)	OMP (without Reclamation)	PF	BCM (Reclamation)/OMP (without Reclamation)
1	17.5 ± 4.8	53.1 ± 11.3	48.5 ± 1.5	46.8 ± 5.2	28.5 ± 3.4	77.9 ± 10.4
5	16.9 ± 5.6	–	27.5 ± 3.6	–	15.0 ± 2.5	–
10–12	15.4 ± 5.5	50.6 ± 8.1	25.0 ± 3.7	51.4 ± 5.6	12.5 ± 1.1	55.0 ± 8.3
22	3.1 ± 0.2	62.9 ± 13.9	12.5 ± 1.1	50.2 ± 5.5	3.6 ± 0.6	43.0 ± 6.2
30	–	–	–	–	–	32.4 ± 4.5
>40	–	–	–	–	–	18.7 ± 0.3

The data obtained for the two considered channels of satellite equipment (two spectral ranges) (Table 4) correspond to the change in the integral albedo of the disturbed plots [74] incorporated in the model of current research as one of the input parameter.

4.2. The Results of Numerical Simulation

The direct results of numerical simulation are dynamics of fields of temperature and fraction of liquid and solid states of water, radiation, and convective heat flux on the surface (Table S2, Supplementary Materials). The phase transition boundary and the magnitude of the relative surface temperature anomaly for disturbed areas in comparison with the background ones are the indicators most important for analysis.

Figure 4a shows the formation of a periodic solution during a three-year calculation period. Figure 4b demonstrates the differences in the dynamics of the phase transition boundary for damaged and undamaged soils. Figure 4c shows the dynamics of the surface temperature anomaly in the summer months through the values of the relative temperature deviation ($\Delta T/T_{bg}$, °C/°C) from background values. Figure 4d reveals the influence of the season and optical properties of the surface (soil or snow) on the annual dynamics of the total radiation flux.

The influence of changes in the optical properties and moisture content of disturbed soil areas on the formation of thermal anomalies in the numerical simulation results are shown in Table 5. Several variants of the set of soil conditions were examined. One of them was considered as the basic set of conditions (Tables 1 and 2) and others differed from it in a certain parameter value in order to reveal the influence of this parameter. We considered the following variants: (i) basic set of conditions, (ii) no change in the moisture content for the horizon O(F+H), (iii) equal moisture content in the upper soil horizons, and (iv) moisture content similar to var. 3 with greater change in albedo. In the cases considered, $\Delta T/T_{bg}$ varied from 5 to 25%.

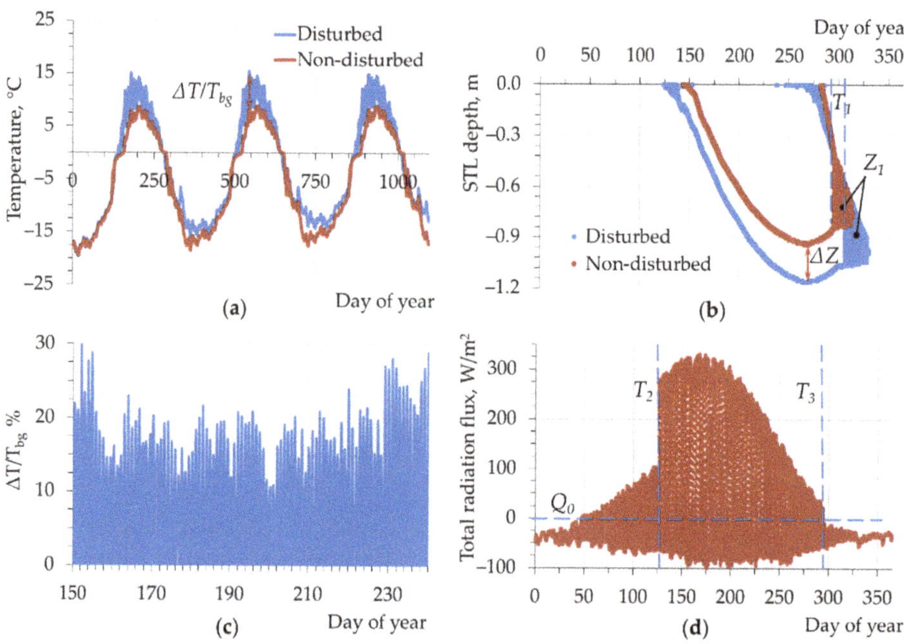

Figure 4. Results of numerical simulation of the heat transfer in soils: (**a**) temperature at the depth of 0.1 m, (**b**) the vertical coordinate of the phase change boundary, (**c**) relative temperature anomaly at the soil surface during summer $\Delta T/T_{bg}$, °C/°C (%), and (**d**) annual cycle of the total heat flux on the damaged soil surface (W/m^2).

Table 5. Moisture content and albedo in the variants of model problem statement and calculated values of $\Delta T/T_{bg}$ for summer. Moisture content and albedo are given for disturbed/non-disturbed plots.

Horizon	Volumetric Water Content	Albedo for Disturbed Plot/Background	$\Delta T/T_{bg}$, %
	Var. 1. Basic set of conditions		
Oao	–/0.6	0.13/0.18	15–25
O(F+H)	0.35/0.14		
	Var. 2. No change in the moisture content for the horizon O(F+H)		
Oao	–/0.6	0.13/0.18	7–12
O(F+H)	0.35/0.35		
	Var. 3. Equal moisture content in the upper soil horizons		
Oao	–/0.6	0.13/0.18	<5
O(F+H)	0.35/0.6		
	Var. 4. Moisture content similar to Var. 3 with greater change in albedo		
Oao	–/0.6	0.07/0.28	10–25 (mean ~15%)
O(F+H)	0.35/0.6		

5. Discussion

5.1. Long-Term Surface Temperature Anomalies

Temperature anomalies in the post-fire areas are typical for permafrost conditions of Siberia [11,12,56].

A decrease in the spectral and broadband albedo (α, %) of the surface in post-fire areas, due to partial or complete combustion of the ground cover, provokes excessive heating of the surface. As shown by the remote measurements (Table 4), the spectral albedo in the short-wavelength bands (MODIS band #1 and #2) is reduced by 20–48% relative to the background immediately after the fire, by 15–25% after 10 years of recovery, and by 3–12% after 20 years of vegetation regeneration. For the variant of post-technogenic plot BCM (with reclamation), the albedo values correspond to the dynamics in PF, while for OMP (dumps without reclamation) the level of the albedo reduction remained 45–60% throughout the observation period (Table 4).

Under conditions of positive air temperatures in summer, as well as in spring and early-autumn, the effect of significant surface temperature anomalies ($\Delta T/T_{bg}$, %) was recorded in all variants of the disturbed plots (PF, BCM, OMP) in comparison with background surface temperature.

In PF plot $\Delta T/T_{bg}$ on average reaches values of $33 \pm 6\%$ with maxima of ~46%, which were recorded immediately after fire impact. During 10–12 years, on average, the value of the temperature anomaly decreases to ~$12.5 \pm 1\%$ (sporadic maxima ~20%). During at least 20 years (the stabilization time lag T_{stab}, Figure 5a), the anomaly amplitudes decrease to the level of background values and the measurement error of $3 \pm 1\%$ (Table 4, Figure 5a). Thus, during the period under consideration, stabilization of temperature parameters is observed in PF plot, which is determined by the thermal insulation properties of the surface layer recovering to the pre-fire state under conditions of a successful vegetation regeneration [11,34,56,77].

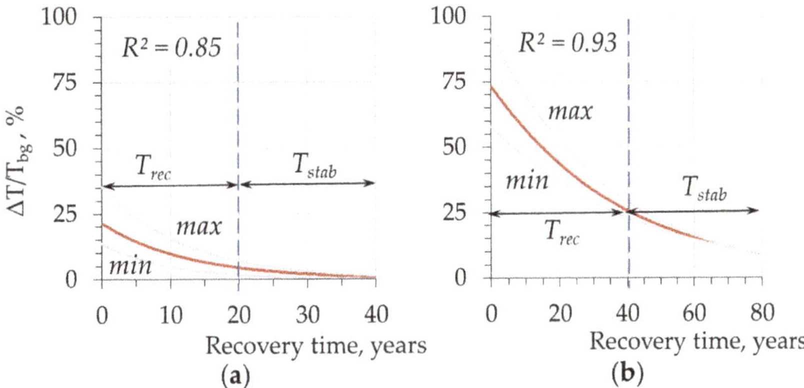

Figure 5. Long-term dynamics of surface temperature anomalies ($\Delta T/T_{bg}$, %) marked with the period of intensive recovery (T_{rec}) and stabilization time lag (T_{stab}) averaged for post-fire plots (**a**) and for post-technogenic plots (**b**).

The considered post-technogenic plots were characterized by a higher level of initial relative surface temperature anomalies (up to 100% of the background). Further, an exponential decrease was observed, similar to PF plot ($R^2 = 0.87$–0.92) (Figure 5). However, the stabilization time lag (T_{stab}) is 40–60 years, which is characterized with $\Delta T/T_{bg}$ of 18–20% under the conditions of recovery (Figure 5b).

A significantly high level of $\Delta T/T_{bg}$ in technogenic areas remains twice as long as in PF plot. In disturbed ecosystems, the morphological and physical properties of the upper soil horizons are not restored to the background level for a long time [56,78]. Thermal regimes remain significantly anomalous for up to 60–80 years (Figure 5b). Even after the time of stabilization ($T_{stab} > 60$ years), the level of relative anomaly overestimated at least by 15–20% in relation to the background values. Probably, we can talk about "neo-technogenic ecosystems" forming, which are characterized by special thermal regimes of soils that differ from the background ones both for reclamated (BCM) and for non-reclamated (OMP) plots. The main reason for such changes is a significant mechanical effect on the soil and/or complete destruction of its structure, which restoration requires much longer time.

These results can be explained by the fact that mechanical reclamation cannot provide an increase in the rate of restoration of the state of the litter and upper soil horizons to the level of the background territories. Although, the restoration of vegetation (as a factor that changes the albedo) proceeds faster in reclaimed areas, where successions acquire a character closer to natural restoration processes.

The formation of thermal anomalies in disturbed plots can be caused by a sharp change in three factors (i) surface emissivity, (ii) albedo, and (iii) moisture content in the upper soil horizons.

Surface emissivity determines the heat loss due to the re-emission of absorbed short-wave solar radiation into the atmosphere. The coefficient of surface emissivity decreases in the disturbed plots [74,77]. However, heat losses from the surface with damaged vegetation and soil cover at significant $\Delta T / T_{bg}$ should be higher than from the undisturbed surface, due to the strong nonlinear dependence of thermal radiation on temperature (according to Stefan–Boltzmann law). Consequently, heat losses contribute to a decrease in $\Delta T / T_{bg}$ values. Thus, the surface emissivity does not fully determine the formation of thermal anomalies.

According to field observations conducted in the study area [7,50], changes in albedo and moisture content conditions can also contribute to the formation of thermal anomalies.

The spectral albedo of disturbed areas in all variants (PF, BCM, OMP) significantly differs from the background values (Table 4). At the initial stages (during the first years after the destructive factor impact), this factor significantly affects the absorption of solar radiation. However, it is numerically shown (Table 5) that the albedo reduction by 45–55% leads to a decrease in the proportion of absorbed solar radiation by <10%, and only by 6% for the model of "native" variant (see Table 5).

In addition to changing the optical properties in disturbed areas, conditions are formed for changing the water content in the near-surface soil horizons [35]. Such changes in moisture content directly depend on the degree of disturbance of the vegetation cover and soil, including the disturbance of its structural organization. Probably, it is this parameter that determines thermal anomalies and their presence at later stages of recovery processes (10–20 years in PF plots or >60 years in BCM and OMP), when the spectral albedo and the surface emissivity are leveled as a result of the restoration of vegetation and litter cover. It was previously shown [27,79] that vegetation signs are restored no later than 7 years after a wildfire. Nevertheless, a change in the soil moisture regime in disturbed plots provokes a decrease in the thermal conductivity of the upper soil horizons. We believe that this is the main reason for the existence of significant temperature anomalies even after the normalization of the optical properties (albedo, surface emissivity) of the disturbed ecosystems.

5.2. Results of Numerical Simulation of the Annual Dynamics of Heat Transfer

As the results of the numerical simulation show, the initial temperature distribution is changed by a periodically time-dependent one starting from the second year. At a depth of 0.1 m, this change occurs after five months (Figure 4a).

At the beginning of the negative air temperatures period a thawed cavity is formed, surrounded by frozen soil in the area above the permafrost zone (Figure 4b). It decreases on both upper and lower sides. This phenomenon exists for 14 days in the non-disturbed soil, and up to 21 days in the soil of the damaged plot. Freezing in the intact soil starts two weeks later than in the soil of the damaged plot.

The maximum inflow and losses of heat due to longwave and shortwave radiation occur in summer and amount to about 350 and 100 W/m^2, respectively (Figure 4d). Heat loss in summer occurs mostly during nights. In winter, with a high snow albedo, the soil loses 30–50 W/m^2 due to radiation (Figure 4d). Abrupt changes in the profile of the radiation flux are associated with snowfalls and melting of snow, which affects the surface albedo sharply.

The numerical solution shows a positive temperature anomaly $\Delta T / T_{bg} = 15–25\%$ in the damaged areas in summer (Table 5, var. 1), which correlates with the data of satellite observations (Table 4, Figure 5), in particular for the post-fire areas (PF). Such values of

$\Delta T/T_{bg}$ are achieved in the case of a simultaneous significant decrease in moisture content and a change in the optical characteristics of damaged areas.

The simulation results demonstrate that a significant role in the formation of the temperature anomaly belongs to the change in moisture content. When the influence of the moisture content factor is excluded, the effect of changing in the radiant flux does not exceed $\Delta T/T_{bg} = 5\%$ (Table 5, var. 3) that is 4–5 times less than data from remote sensing. With a decrease in moisture content for the upper soil horizon to the level of the second soil horizon of the non-disturbed soil, $\Delta T/T_{bg}$ becomes two times higher but remains below the observed values (Table 5, var. 2). Only with further 2.5 times decrease in the moisture content (Table 5, var. 1), the simulation results correlate with the data of satellite observations. To reproduce the observed thermal anomaly in the mathematical model changing only the amount of the absorbed radiation, the albedo in the disturbed area must be several times less than in the undisturbed areas (Table 5, var. 4).

5.3. Thermal Inversion

Thermal inversion in soil temperature values in comparison of disturbed and undisturbed areas was observed both in numerical simulation (Figure 6a) and in field measurements (Figure 6b) [50]. The results of numerical simulation allow us to state that the condition for this effect is a higher ice content in the soils of the background areas, in comparison with the variants of disturbed territories.

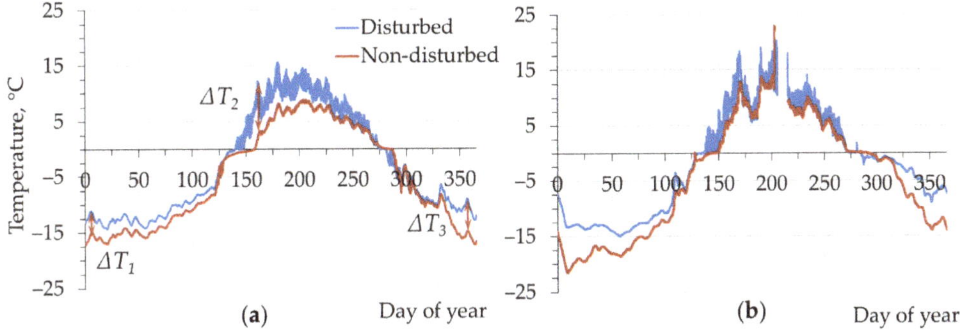

Figure 6. Annual temperature dynamics at a depth of 0.1 m: (**a**) simulation results for the basic variant of model parameters set (see Table 5) and (**b**) data of field observations in the research area (summarized experimental data from the work in [50]).

Due to the higher ice content, despite the presence of the heat-insulating soil horizon Oao, the temperature in the soil of non-disturbed plots decreases faster (Table 3 [5,70]). In the summer, the opposite situation is realized. Greater heating of soils in disturbed plots causes an increase in the depth of the active layer by $\Delta Z = 29\%$ according to the results of numerical modeling (Figure 4b) in comparison with the background areas, both due to the absence of an additional thermal insulating horizon and due to lower albedo and emissivity values. The calculated increase in thawing depth for disturbed soil is consistent with observational data, which show the magnitude of the relative increase in thawing depth in the range of 25–40% [34].

Comparison of the results of modeling and field observations (Figure 6a,b) shows the similarity of heat transfer processes: thermal inversion, similar nature of temperature dynamics at a depth of 0.1 m for soils of disturbed and background areas. The most significant differences are observed in the summer period (in general, the maxima are higher by about 5 °C), which can be explained by differences in meteorological conditions and model assumptions of soil characteristics (Tables 1 and 2) that differ from natural ones. In particular, in the presented results for the mineral horizon CR clayey granulometric composition, the model from [69] was used, as the model from the works in [1,4] led to a significant overestimation of the depth of soil thawing [34]. The necessity to refine the model coefficients in accordance with the characteristics of natural soils is obvious.

5.4. Transition Zone

The problem of heat transfer simulation is two-dimensional due to boundary effects at the periphery of the damaged plots. The calculated two-dimensional distribution of the temperature field in the summer and autumn months (Figure 7) indicates that at a depth of up to 0.5 m (within the root-inhabited layer), the horizontal propagation of thermal disturbances from plots with damaged soil cover to a non-damaged area can reach ~5.5 m.

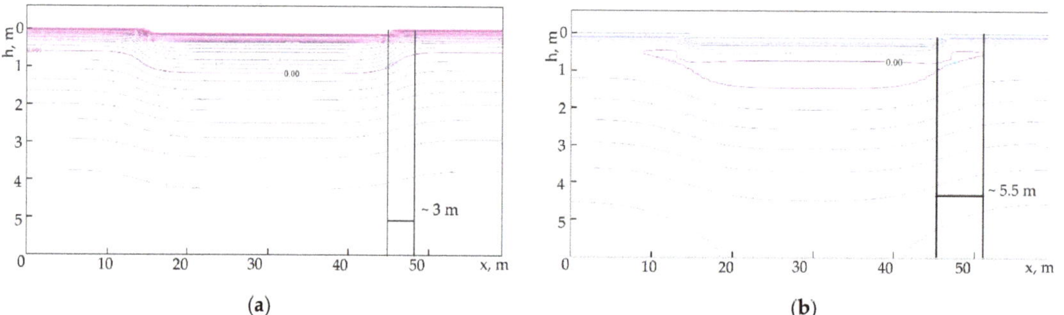

(a) (b)

Figure 7. Results of numerical modeling of a two-dimensional temperature field (°C) for 200th day (a) and for 294th day (b).

Transferring the modeling results to post-fire and post-technogenic disturbed ecosystems, it should be taken into account that along the boundary of disturbed plots, transitional zones (or "ecotones") (in terms of thermal and moisture conditions) are formed. Their characteristic width is up to ~5 m and area can be up to 6% of the area of the disturbed plots (Figure 8).

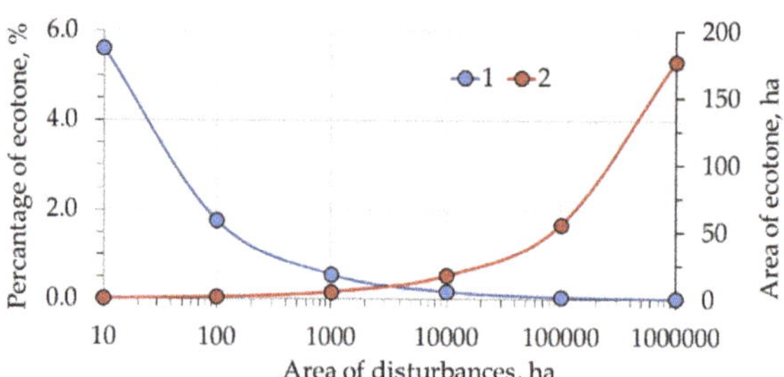

Figure 8. Areas and percentage of the emerging transitional zone ("ecotone") versus the total area of the disturbed plots. In the calculations, we used the approximation of the circular shape of the disturbed plots. 1 is the percentage of ecotone; 2 is the area of ecotone.

The zones formed along the border of disturbed areas should also be considered as transitional zones or "ecotones" characterized by "edge effect", where conditions favorable for vegetation and biota can develop in the root-inhabited layer of the soil. This factor can be favorable for conditions of vegetation and biota earlier exit from the winter state due to increased temperatures in the root-inhabited layer, which is confirmed for Siberia by field studies [9,10,13,56]. As an ecological factor, this can determine an increase in the density of some populations and be a zone of additional introduction of plant, microbial, and animal communities into the disturbed plots territory [80].

6. Conclusions

Under conditions of the same insolation, plots with disturbances in the upper soil horizons and ground cover are accompanied by the formation of long-term surface temperature anomalies. Similar processes were recorded for both natural (post-fire) and post-technogenic landscapes. Within 20 years, the thermal insulation properties of the vegetation recover in the post-fire areas. Thus, the relative temperature anomaly reaches the level of the mathematical error of measurement. In post-technogenic plots, conditions are more "contrast" compared to the background, and the processes of restoration of the thermal regime take a longer period (>60 years). "Neo-technogenic ecosystems" are formed, characterized with special thermal regimes of soils that differ from the background ones both for reclamated (BCM) and for non-reclamated (OMP) plots.

Along with a change in the optical properties of the surface (the spectral and the broadband albedo, the emissivity coefficient), the main reason for the increased surface and soil temperatures of disturbed plots is a decrease in moisture content in the upper soil horizons. Probably, it is this parameter that determines thermal anomalies in disturbed plots as well as their presence at later stages of recovery processes for 10–15 years after a fires impact and for >60 years in post-technogenic ecosystems of Siberia. Thus, monitoring of surface thermal anomalies can be used as a diagnostic criterion of post-technogenic ecosystems state in the context of territories development after technogenic impacts. The data on the moisture content change in the damaged areas are crucially important for the numerical studies of the temperature profiles evolution. However, these data cannot be obtained directly by the remote observation. Further work on this topic involves not only field measurements of moisture content, but also taking into account moisture transfer in a mathematical model.

The destruction of the upper soil layer and the complete degradation of the moss–lichen cover leads not only to increased surface temperatures in summer, but also causes thermal inversion in soil temperature in winter when the thermal resistance in the upper organic layer of the disturbed plots becomes less than in the upper organic horizons of the background soil. This factor can be favorable for the formation of conditions for an earlier exit from the winter state of vegetation and biota due to the increased temperatures of the root-inhabited layer, which is confirmed for Siberia by field studies [9,10,13,56].

The simulation results show the formation of transition zones ("ecotones") along the periphery of the disturbed plots due to horizontal heat transfer. The typical size of such a zone is ~5.5 m, and the area is up to 6% of the area of disturbed plots. Probably, zones of rapid recovery successions can form along the boundaries of disturbed areas, which makes further study of their properties urgent.

Supplementary Materials: The following are available online at https://www.mdpi.com/article/10.3390/f12080994/s1, Table S1: weather data for 2018 in Tura (Evenkia region, Central Siberia, Russia); Table S2: results of numerical simulation.

Author Contributions: Conceptualization, K.Y.L., T.V.P., E.I.P. and K.A.F.; methodology, K.Y.L. and E.I.P.; software, K.Y.L., K.A.F., N.D.Y. and A.V.S.; validation, T.V.P., E.I.P. and K.Y.L.; resources, E.I.P. and T.V.P.; data curation, T.V.P.; writing—original draft preparation, E.I.P., T.V.P. and K.Y.L.; writing—review and editing, E.I.P., K.A.F.; visualization, E.I.P., N.D.Y., A.V.S. and K.Y.L.; supervision, T.V.P., E.I.P.; funding acquisition, T.V.P. All authors have read and agreed to the published version of the manuscript.

Funding: This work was performed using the subject of project of IF SB RAS no. 0287-2021-0010. The study of heat transfer in soil was performed using the subject of project of IT SB RAS no. 0257-2021-0001. This research was partly funded by the Russian Foundation for Basic Research (RFBR) and Government of the Krasnoyarsk krai, and Krasnoyarsk krai Foundation for Research and Development Support, no. 20-44-242002 ("Instrumental monitoring of physical properties and numerical modeling of the state of technogenically disturbed soils in Siberia"), and by Siberian Federal University and Government of the Krasnoyarsk krai, and Krasnoyarsk krai Foundation for

Research and Development Support, 2020, no. KF-782 49/20 ("Long-term consequences of extreme fires in the permafrost zone of Siberia by the materials of satellite monitoring").

Institutional Review Board Statement: Not applicable.

Informed Consent Statement: Not applicable.

Data Availability Statement: Publicly available datasets were analyzed in this study. This data can be found here: https://ladsweb.modaps.eosdis.nasa.gov/ (accessed on 25 March 2021), The United States Geological Survey (USGS) database (https://earthexplorer.usgs.gov/, accessed on 21 May 2021), Climatic Research Unit (http://www.cru.uea.ac.uk, accessed on 25 March 2021), Weather archive (http://rp5.ru) (accessed on 25 March 2021), and the National Climatic Data Center (NCDC) (http://www7.ncdc.noaa.gov/CDO/cdo, accessed on 25 March 2021), and https://worldview.earthdata.nasa.gov (accessed on 13 April 2021).

Acknowledgments: The satellite data-receiving equipment used was provided by the Centre of Collective Usage of Federal Research Center "Krasnoyarsk Science Center, Siberian Branch of Russian Academy of Sciences", Krasnoyarsk, Russia.

Conflicts of Interest: The authors declare no conflict of interest.

References

1. Tarnawski, V.R.; Leong, W.H.; Bristow, K.L. Developing a temperature-dependent Kersten function for soil thermal conductivity. *Int. J. Energy Res.* **2000**, *24*, 1335–1350. [CrossRef]
2. Dankers, R.; Burke, E.J.; Price, J. Simulation of permafrost and seasonal thaw depth in the JULES land surface scheme. *Cryosphere* **2011**, *5*, 773–790. [CrossRef]
3. Desyatkin, R.V.; Desyatkin, A.R.; Fedorov, P.P. Temperature regime of taiga cryosoils of Central Yakutia. *Kriosf. Zemli Earth Cryosphere* **2012**, *16*, 70–78. (In Russian)
4. Ekici, A.; Beer, C.; Hagemann, S.; Boike, J.; Langer, M.; Hauck, C. Simulating high-latitude permafrost regions by the JSBACH terrestrial ecosystem model. *Geosci. Model Dev.* **2014**, *7*, 631–647. [CrossRef]
5. Porada, P.; Ekici, A.; Beer, C. Effects of bryophyte and lichen cover on permafrost soil temperature at large scale. *Cryosphere* **2016**, *10*, 2291–2315. [CrossRef]
6. Pavlov, A.V. *Cryolithozone Monitoring*; The Academic Publishing House "GEO": Novosibirsk, Russia, 2008; p. 229. (In Russian)
7. Orgogozo, L.; Prokushkin, A.S.; Pokrovsky, O.S.; Grenier, C.; Quintard, M.; Viers, J.; Audry, S. Water and energy transfer modeling in a permafrost-dominated, forested catchment of Central Siberia: The key role of rooting depth. *Permafr. Periglac. Process.* **2019**, *30*, 75–89. [CrossRef]
8. Brown, J.; Ferrians, O.J.; Heginbottom, J.A.; Melnikov, E.S. *Circum-Arctic Map of Permafrost and Ground Ice Conditions*; National Snow and Ice Data Center, Digital Media: Boulder, CO, USA, 2002. Available online: https://nsidc.org/data/ggd318 (accessed on 1 June 2021).
9. Knorre, A.A.; Kirdyanov, A.V.; Prokushkin, A.S.; Krusic, P.J.; Büntgen, U. Tree ring-based reconstruction of the long-term influence of wildfires on permafrost active layer dynamics in Central Siberia. *Sci. Total Environ.* **2019**, *652*, 314–319. [CrossRef] [PubMed]
10. Kirdyanov, A.V.; Prokushkin, A.S.; Tabakova, M.A. Tree-ring growth of Gmelin larch in the north of Central Siberia under contrasting soil conditions. *Dendrochronologia* **2013**, *31*, 114–119. [CrossRef]
11. Dymov, A.; Abakumov, E.; Bezkorovaynaya, I.; Prokushkin, A.; Kuzyakov, Y.; Milanovsky, E. Impact of forest fire on soil properties (Review). *Theor. Ecol.* **2018**, *4*, 13–23. [CrossRef]
12. Zhang-Turpeinen, H.; Kivimäenpää, M.; Aaltonen, H.; Berninger, F.; Köster, E.; Köster, K.; Menyailo, O.; Prokushkin, A.; Pumpanen, J. Wildfire effects on BVOC emissions from boreal forest floor on permafrost soil in Siberia. *Sci. Total Environ.* **2020**, *711*, 134851. [CrossRef] [PubMed]
13. Kirdyanov, A.; Saurer, M.; Siegwolf, R.; Knorre, A.; Prokushkin, A.S.; Churakova, O.; Fonti, M.V.; Büntgen, U. Long-term ecological consequences of forest fires in the continuous permafrost zone of Siberia. *Environ. Res. Lett.* **2020**, *15*, 034061. [CrossRef]
14. Anisimov, O.A. Potential feedback of thawing permafrost to the global climate system through methane emission. *Environ. Res. Lett.* **2007**, *2*. [CrossRef]
15. Anisimov, O.A.; Sherstiukov, A.B. Evaluating the effect of environmental factors on permafrost in Russia. *Kriosf. Zemli. Earth Cryosphere* **2016**, *2*, 78–86. (In Russian)
16. Masyagina, O.V.; Menyailo, O.V. The impact of permafrost on carbon dioxide and methane fluxes in Siberia: A meta-analysis. *Environ. Res.* **2020**, *182*, 109096. [CrossRef] [PubMed]
17. Arkhangelskaya, T.A. *The Thermal Regime of Soils*; GEOS: Moscow, Russia, 2012; p. 282. (In Russian)
18. Trofimova, I.E.; Balybina, A.S. Regionalization of the West Siberian Plain from thermal regime of soils. *Geogr. Nat. Resour.* **2015**, *3*, 27–38. [CrossRef]
19. Goncharova, O.Y.; Matyshak, G.V.; Bobrik, A.A.; Moskalenko, N.G.; Ponomareva, O.E. Temperature regimes of northern taiga soils in the isolated permafrost zone of Western Siberia. *Eurasian Soil Sci.* **2015**, *48*, 1329–1340. [CrossRef]

20. Koronatova, N.G.; Mironycheva-Tokareva, N.P.; Solomin, Y.R. Thermal regime of peat deposits of palsas and hollows of peat plateaus in Western Siberia. *Kriosf. Zemli. Earth Cryosphere* **2018**, *6*, 15–23. (In Russian) [CrossRef]

21. Park, H.; Launiainen, S.; Konstantinov, P.Y.; Iijima, Y.; Fedorov, A.N. Modeling the Effect of Moss Cover on Soil Temperature and Carbon Fluxes at a Tundra Site in Northeastern Siberia. *J. Geophys. Res. Biogeosci.* **2018**, *123*, 3028–3044. [CrossRef]

22. Krasnoshchekov, Y.N.; Sorokin, N.D.; Bezkorovainaya, I.N.; Yashikhin, G.I. Genetic peculiarities of Northern Taiga soils in the Yenisei region of Siberia. *Eurasian Soil Sci.* **2001**, *34*, 12–20. (In Russian)

23. Prokushkin, S.G.; Abaimov, A.P.; Prokushkin, A.S. *Structural and Functional Characteristics of the Gmelin Larch in the Permafrost Zone of Central Evenkia*; IL SB RAS: Krasnoyarsk, Russia, 2008; p. 164, ISBN 978-5-903055-13-5. (In Russian)

24. Galenko, E.P. Thermal regime formation of soils in coniferous ecosystems of boreal zone in reference to dominating tree species and forest type. *Proc. Komi Sci. Cent. Ural. Div. Russ. Acad. Sci.* **2013**, *1*, 32–37. (In Russian)

25. Ehrenfeld, J.G.; Ravit, B.; Elgersma, K. Feedbacks in the plant-soil system. *Annu. Rev. Environ. Resour.* **2005**, *30*, 75–115. [CrossRef]

26. Kharuk, V.I.; Ponomarev, E.I. Spatiotemporal characteristics of wildfire frequency and relative area burned in larch-dominated forests of Central Siberia. *Russ. J. Ecol.* **2017**, *48*, 507–512. [CrossRef]

27. Ponomarev, E.I.; Ponomareva, T.V. The Effect of Postfire Temperature Anomalies on Seasonal Soil Thawing in the Permafrost Zone of Central Siberia Evaluated Using Remote Data. *Contemp. Probl. Ecol.* **2018**, *11*, 420–427. [CrossRef]

28. Kharuk, V.I.; Ponomarev, E.I.; Ivanova, G.A.; Dvinskaya, M.L.; Coogan, S.C.; Flannigan, M.D. Wildfires in the Siberian taiga. *Ambio* **2021**, 1–22. [CrossRef]

29. Krasnoshchekov, K.V.; Dergunov, A.V.; Ponomarev, E.I. Evaluation of underlying surface temperature maps on logging sites using Landsat data. *Sovrem. Probl. Distantsionnogo Zondirovaniya Zemli Iz Kosm.* **2019**, *16*, 87–97. [CrossRef]

30. Uzarowicz, Ł.; Charzyński, P.; Greinert, A.; Hulisz, P.; Kabała, C.; Kusza, G.; Kwasowski, W.; Pędziwiatr, A. Studies of technogenic soils in Poland: Past, present, and future perspectives. *Soil Sci. Annu.* **2020**, *71*, 281–299. [CrossRef]

31. Rodionova, N. Sentinel 1 Radar Data Correlation WITH GROUND Measurements of Soil Moisture and Temperature. *Issled. Zemli kosmosa* **2018**, *4*, 32–42. [CrossRef]

32. Vinogradov, Y.B.; Semenova, O.M.; Vinogradova, T.A. Hydrological modeling: Calculation of the dynamics of thermal energy in the soil profile. *Kriosf. Zemli Earth Cryosphere* **2015**, *19*, 11–21. (In Russian)

33. Zwieback, S.; Westermann, S.; Langer, M.; Boike, J.; Marsh, P.; Berg, A. Improving Permafrost Modeling by Assimilating Remotely Sensed Soil Moisture. *Water Resour. Res.* **2019**, *55*, 1814–1832. [CrossRef]

34. Ponomarev, E.; Masyagina, O.; Litvintsev, K.; Ponomareva, T.; Shvetsov, E.; Finnikov, K. The effect of post-fire disturbances on a seasonally thawed layer in the permafrost larch forests of Central Siberia. *Forests* **2020**, *11*, 790. [CrossRef]

35. Arkhangelskaya, T.A. Parameters of the Thermal Diffusivity vs. Water Content Function for Mineral Soils of Different Textural Classes. *Eurasian Soil Sci.* **2020**, *53*, 39–49. [CrossRef]

36. Deng, X.; Pan, S.; Wang, Z.; Ke, K.; Zhang, J. Application of the Darcy-Stefan model to investigate the thawing subsidence around the wellbore in the permafrost region. *Appl. Therm. Eng.* **2019**, *156*, 392–401. [CrossRef]

37. Cohen, D.; Zwinger, T.; Koskinen, L.; Karvonen, T. Long-term coupled permafrost-groundwater interactions at Olkiluoto, Finland. In Proceedings of the 22nd EGU General Assembly, Online, 4–8 May 2020; p. 8972. [CrossRef]

38. Swenson, S.C.; Lawrence, D.M.; Lee, H. Improved simulation of the terrestrial hydrological cycle in permafrost regions by the Community Land Model. *J. Adv. Model. Earth Syst.* **2012**, *4*. [CrossRef]

39. Thomas, H.R.; Cleall, P.; Li, Y.-C.; Harris, C.; Kern-Luetschg, M. Modelling of cryogenic processes in permafrost and seasonally frozen soils. *Geotechnique* **2009**, *59*, 173–184. [CrossRef]

40. Zhou, M.M.; Meschke, G. A three-phase thermo-hydro-mechanical finite element model for freezing soils. *Int. J. Numer. Anal. Methods Géoméch.* **2013**, *37*, 3173–3193. [CrossRef]

41. Painter, S.L.; Karra, S. Constitutive model for unfrozen water content in subfreezing unsaturated soils. *Vadose Zone J.* **2014**, *13*, 1–8. [CrossRef]

42. Semin, M.A.; Levin, L.Y.; Zhelnin, M.S.; Plekhov, O.A. Natural convection in Water-saturated rock mass under artificial freezing. *J. Min. Sci.* **2020**, *56*, 297–308. [CrossRef]

43. Gong, F.; Jacobsen, S. Modeling of water transport in highly saturated concrete with wet surface during freeze/thaw. *Cem. Concr. Res.* **2019**, *115*, 294–307. [CrossRef]

44. Addassi, M.; Schreyer, L.; Johannesson, B.; Lin, H. Pore-scale modeling of vapor transport in partially saturated capillary tube with variable area using chemical potential. *Water Resour. Res.* **2016**, *52*, 7023–7035. [CrossRef]

45. Wen, W.; Lai, Y.; You, Z. Numerical modeling of water–heat–vapor–salt transport in unsaturated soil under evaporation. *Int. J. Heat Mass Transf.* **2020**, *159*, 120114. [CrossRef]

46. Jafari, M.; Gouttevin, I.; Couttet, M.; Wever, N.; Michel, A.; Sharma, V.; Rossmann, L.; Maass, N.; Nicolaus, M.; Lehning, M. The Impact of Diffusive Water Vapor Transport on Snow Profiles in Deep and Shallow Snow Covers and on Sea Ice. *Front. Earth Sci.* **2020**, *8*, 249. [CrossRef]

47. Psiloglou, B.E.; Santamouris, M.; Asimakopoulos, D.N. Atmospheric broadband model for computation of solar radiation at the Earth's Surface. Application to Mediterranean Climate. *Pure Appl. Geophys.* **2000**, *157*, 829–860. [CrossRef]

48. Kambezidis, H.D.; Psiloglou, B.E.; Karagiannis, D.; Dumka, U.C.; Kaskaoutis, D.G. Meteorological Radiation Model (MRM v6. 1): Improvements in diffuse radiation estimates and a new approach for implementation of cloud products. *Renew. Sustain. Energy Rev.* **2017**, *74*, 616–637. [CrossRef]

49. Abaimov, A.P.; Bondarev, A.I.; Zyryanova, O.A.; Shitov, S.A. *Polar Forests of Krasnoyarsk Region*; Nauka Publisher: Novosibirsk, Russia, 1997; p. 208. (In Russian)

50. Osawa, A.; Zyryanova, O.A.; Matsuura, Y.; Kajimoto, T.; Wein, R.W. (Eds.) *Permafrost Ecosystems*; Siberian Larch Forests; Springer: Berlin/Heidelberg, Germany, 2010; p. 528. [CrossRef]

51. Masyagina, O.V.; Evgrafova, S.Y.; Titov, S.V.; Prokushkin, A.S. Dynamics of soil respiration at different stages of pyrogenic restoration succession with different-aged burns in Evenkia as an example. *Russ. J. Ecol.* **2015**, *46*, 27–35. [CrossRef]

52. Shishov, L.L.; Tonkonogov, V.D.; Lebedeva, I.I.; Gerasimova, I.I. *Classification Diagnosis of Russian Soils*; Oikumena: Smolensk, Russia, 2004; p. 341. (In Russian)

53. World Reference Base for Soil Resources. 2014. Available online: https://www.isric.org/explore/wrb (accessed on 10 May 2021).

54. Dymov, A.A.; Dubrovsky, Y.A.; Gabov, D.N. Pyrogenic changes in iron-illuvial podzols in the middle taiga of the Komi Republic. *Eurasian Soil Sci.* **2014**, *47*, 47–56. [CrossRef]

55. Mergelov, N.S. Post-pirogenic Transformation of Soils and Soil Carbon Stocks in Sub-Tundra Woodlands of Kolyma Lowland: A Cascading Effect and Feedbacks. *Izv. Ross. Akad. Nauk. Seriya Geogr.* **2015**, 129–140. [CrossRef]

56. Bezkorovaynaya, I.N.; Borisova, I.V.; Klimchenko, A.V.; Shabalina, O.M.; Zakharchenko, L.P.; Il'in, A.A.; Beskrovny, A.K. Influence of the pyrogenic factor on the biological activity of the soil under permafrost conditions (Central Evenkia). *Vestnik KrasGAU* **2017**, *9*, 181–189. (In Russian)

57. Startsev, V.V.; Dymov, A.A.; Prokushkin, A.S. Soils of postpyrogenic larch stands in Central Siberia: Morphology, physicochemical properties, and specificity of soil organic matter. *Eurasian Soil Sci.* **2017**, *50*, 885–897. [CrossRef]

58. Gavriliev, R.I. *Catalog of Thermophysical Properties of Rocks in the North-East of Russia*; Publishing house of the Institute of Permafrost. P.I. Melnikov SO RAN: Yakutsk, Russia, 2013; p. 174. (In Russian)

59. Gubin, S.V.; Lupachev, A.V. Suprapermafrost horizons of the accumulation of raw organic matter in cryozems of Northen Yakutia. *Eurasian Soil Sci.* **2018**, *51*, 772–781. [CrossRef]

60. Mishra, N.; Haque, M.O.; Leigh, L.; Aaron, D.; Helder, D.; Markham, B.L. Radiometric Cross Calibration of Landsat 8 Operational Land Imager (OLI) and Landsat 7 Enhanced Thematic Mapper Plus (ETM+). *Remote Sens.* **2014**, *6*, 12619–12638. [CrossRef]

61. Zanter, K. Landsat 8 (L8) data users handbook // U.S. Geological Survey. Available online: https://prd-wret.s3-us-west-2.amazonaws.com/assets/palladium/production/s3fs-public/atoms/files/LSDS-1574_L8_Data_Users_Handbook.pdf (accessed on 1 June 2021).

62. Van Huissteden, J. *Thawing Permafrost: Permafrost Carbon in A Warming Arctic*; Springer: Berlin/Heidelberg, Germany, 2020; p. 508. [CrossRef]

63. Kurylyk, B.L.; MacQuarrie, K.T.; McKenzie, J.M. Climate change impacts on groundwater and soil temperatures in cold and temperate regions: Implications, mathematical theory, and emerging simulation tools. *Earth-Sci. Rev.* **2014**, *138*, 313–334. [CrossRef]

64. Riseborough, D.; Shiklomanov, N.; Etzelmüller, B.; Gruber, S.; Marchenko, S. Recent advances in permafrost modelling. *Permafr. Periglac. Process.* **2008**, *19*, 137–156. [CrossRef]

65. Lunardini, V.J. *Heat Transfer with Freezing and Thawing*; Elsevier: Amsterdam, The Netherlands, 1991; p. 65.

66. Zhang, Y.; Carey, S.K.; Quinton, W.L. Evaluation of the algorithms and parameterizations for ground thawing and freezing simulation in permafrost regions. *J. Geophys. Res. Space Phys.* **2008**, *113*. [CrossRef]

67. Ferziger, J.H.; Peric, M. *Computational Methods for Fluid Dynamics*; Springer: Berlin, Germany, 2002; p. 423.

68. Patankar, S. *Numerical Heat Transfer and Fluid Flow*; Hemisphere Publishing Corporation: New York, NY, USA, 1980; p. 197.

69. Pavlov, A.V. *Thermal Physics of Landscapes*; Nauka: Novosibirsk, Russia, 1979; p. 285. (In Russian)

70. Anisimov, O.A.; Borshch, S.V.; Georgiyevskiy, V.Y. *Methods for Assessing the Effects of Climate Change on Physical and Biological Systems, Chapter 8. Continental Permafrost*; Planeta: Moscow, Russia, 2012; p. 509.

71. Johansen, O. Thermal Conductivity of Soils. Ph.D. Thesis, Norwegian University of Science and Technology, Trondheim, Norway, 1975. (CRREL draft transl. 637, 1977) ADA 044002. p. 291.

72. Oke, T.R. *Boundary Layer Climates*, 2nd ed.; Routledge: London, UK, 1987; p. 435.

73. Paltridge, G.; Platt, C. *Radiative Processes in Meteorology and Climatology*; Elsevier Scientific Pub, Co.: Amsterdam, The Netherlands, 1976; p. 318.

74. Bezkorovainaya, I.N.; Ivanova, G.A.; Tarasov, P.A.; Sorokin, N.D.; Bogorodskaya, A.V.; Ivanov, V.A.; Makrae, D.D. Pyrogenic transformation of soils in pine forests in the middle taiga of the Krasnoyarsk Territory. *Sib. Ecol. J.* **2005**, *12*, 143–152. (In Russian)

75. Prata, A.J. A new long-wave formula for estimating downward clear-sky radiation at the surface. *Q. J. R. Meteorol. Soc.* **1996**, *122*, 1127–1151. [CrossRef]

76. Herrero, J.; Polo, M.J. Parameterization of atmospheric longwave emissivity in a mountainous site for all sky conditions. *Hydrol. Earth Syst. Sci.* **2012**, *16*, 3139–3147. [CrossRef]

77. Kornienko, S.G. Transformation of tundra land cover at the sites of pyrogenic disturbance: Studies based on LANDSAT satellite data. *Kriosf. Zemli Earth's Cryosphere* **2017**, *21*, 93–104.

78. Ponomareva, T.V. Assessment of the structural organization of soils of technogenic landscapes on the basis of radiometric survey in the thermal range. In *Soils in the Biosphere, Proceedings of the All-Russian Conf. Dedicated to the 50th Anniversary of the Institute of Soil Science and Agrochemistry SB RAS, Novosibirsk, Russia, 10 September–14 October 2018*; Syso, A.I., Ed.; National Research Tomsk State University Publisher: Tomsk, Russia, 2018; pp. 410–413.

79. Yakimov, N.; Ponomarev, E. Dynamics of Post-Fire Effects in Larch Forests of Central Siberia Based on Satellite Data. *E3S Web Conf.* **2020**, *149*, 03008. [CrossRef]
80. Dabrowska-Prot, E.; Wasiłowska, A. The role of ecotones in man-disturbed landscape: Boundaries between mixed forest and adjacent man-made ecosystems in the Kampinos National Park, Poland. *Pol. J. Ecol.* **2012**, *60*, 677–698.

Article

Tree Species Mapping on Sentinel-2 Satellite Imagery with Weakly Supervised Classification and Object-Wise Sampling

Svetlana Illarionova *, Alexey Trekin, Vladimir Ignatiev and Ivan Oseledets

Skolkovo Institute of Science and Technology, 143026 Moscow, Russia; a.trekin@skoltech.ru (A.T.);
V.Ignatiev@skoltech.ru (V.I.); i.oseledets@skoltech.ru (I.O.)
* Correspondence: s.illarionova@skoltech.ru

Citation: Illarionova, S.; Trekin, A.; Ignatiev, V.; Oseledets, I. Tree Species Mapping on Sentinel-2 Satellite Imagery with Weakly Supervised Classification and Object-Wise Sampling. *Forests* **2021**, *12*, 1413. https://doi.org/10.3390/f12101413

Academic Editor: Gang Chen

Received: 10 September 2021
Accepted: 12 October 2021
Published: 16 October 2021

Publisher's Note: MDPI stays neutral with regard to jurisdictional claims in published maps and institutional affiliations.

Abstract: Information on forest composition, specifically tree types and their distribution, aids in timber stock calculation and can help to better understand the biodiversity in a particular region. Automatic satellite imagery analysis can significantly accelerate the process of tree type classification, which is traditionally carried out by ground-based observation. Although computer vision methods have proven their efficiency in remote sensing tasks, specific challenges arise in forestry applications. The forest inventory data often contain the tree type composition but do not describe their spatial distribution within each individual stand. Therefore, some pixels can be assigned a wrong label in the semantic segmentation task if we consider each stand to be homogeneously populated by its dominant species. Another challenge is the spatial distribution of individual stands within the study area. Classes are usually imbalanced and distributed nonuniformly that makes sampling choice more critical. This study aims to enhance tree species classification based on a neural network approach providing automatic markup adjustment and improving sampling technique. For forest species markup adjustment, we propose using a weakly supervised learning approach based on the knowledge of dominant species content within each stand. We also propose substituting the commonly used CNN sampling approach with the object-wise one to reduce the effect of the spatial distribution of forest stands. We consider four species commonly found in Russian boreal forests: birch, aspen, pine, and spruce. We use imagery from the Sentinel-2 satellite, which has multiple bands (in the visible and infrared spectra) and a spatial resolution of up to 10 meters. A data set of images for Leningrad Oblast of Russia is used to assess the methods. We demonstrate how to modify the training strategy to outperform a basic CNN approach from F1-score 0.68 to 0.76. This approach is promising for future studies to obtain more specific information about stands composition even using incomplete data.

Keywords: deep learning; remote sensing; tree species; classification

1. Introduction

Many ecological and forest management studies are based on knowledge about tree species within a region of interest. Such knowledge can be used for the precise analysis of natural conditions [1], the development of ecological models [2], and for conservation and restoration decision-making [3]. Accompanied by other characteristics, such as tree age and height, crown width, and tree species information can be leveraged for timber volume [4,5] and biomass estimation [6].

A commonly used approach for forest type data gathering is field-based measurement, which has the obvious drawbacks of acquisition cost and difficulty. Many studies are now focused on the automatization of land-cover survey through the use of remote sensing-derived data. This approach is more preferable when analysing vast territories. For instance, the creation of large-scale maps has been described in [7,8]. For such tasks, both low spatial resolution and high resolution data can be used. Examples of frequently leveraged data sources with resolution lower than 30 m is Landsat satellite imagery [9,10]. Promising results have been shown in studies, both for single image and time-series data [11–13]. Nevertheless, some tasks require more precise data with higher resolution. Multispectral images with high resolution strive to provide more thorough land-cover analysis. Many studies have performed forest survey based on Sentinel images with spatial resolution adjusted to 10 m [14]. For instance, one of the open source packages for Sentinel data analysis is eo-learn project [15].

Recently, image classification algorithms have demonstrated high prediction accuracy in a variety of applied tasks. Algorithms based on machine learning methods are now commonly used for land-cover mapping—particularly for forest species prediction—using satellite imagery. Classical methods, such as Random Forest [16], Support Vector Machine [17], and Linear Regression, usually work with feature vectors, where each value corresponds to some spectral band or combination of bands (in the case of vegetation indices) [18,19]. Deep neural network approaches have proved to be more capable for many land-cover tasks [20–22]. In [23], a CNN was compared with XGBoost [24]. In [25], a CNN approach was examined for tree mapping, through the use of airplane-based RGB and LiDAR data. In [26], neural-based hierarchical approach was implemented to improve forest species classification.

In contrast with typical image classification tasks (such as in the Imagenet data set), land-cover tasks involve spatial data. Vast study regions are usually supplied, with a reference map covering the entire area. Classes within this area may not be evenly distributed in many cases [27]. Moreover, classes of vegetation types of land-cover are often imbalanced within the study region. In many works, the analysed territory can be covered by a single satellite tile (e.g., the size for Sentinel-2 is 100×100 km^2). Therefore, researchers must choose both how to select the training and validation regions and how to organize the training procedure to deal with imbalanced classes and a spatial distribution that is usually far from uniform. Sampling approach is vital for the remote sensing domain as simple image partition into tiles is ineffective for vast territories [28]. The training procedure depends on whether we use a pixel-wise [29] or object-wise approach [10,18]. In a pixel-wise approach, each pixel is ascribed a particular class label and the goal is to predict this label using a feature description of the pixel. In an object-wise approach, a set of pixels is considered as a single object. In some classical machine learning methods, a combination of the two approaches has also been considered [19]. An alternative approach to classical pixel- or object-wise has been provided in [25] for a CNN tree classification task using airplane-based data. During the described patch-wise training procedure, the model strove to predict one label for a whole input image of size 64×64 pixels. However, for some semantic segmentation tasks with lower spatial resolution, the input image can include pixels with different labels and, therefore, the aforementioned approach is not always applicable. The same issue was faced in [21], where patch-wise approach was implemented for CNN for a land-cover classification task using RapidEye satellite imagery. Some patches with mixed labels were excluded, in order to solve the problem. In our study, we aim to provide sampling approach for medium resolution satellite imagery for forest species classification. In contrast to [25], we focus on the particular area within a patch and do not exclude from training patches with mixed labels as in [21].

Another important issue is markup limitations. Field-based measurements are commonly used as reference data. Vast territories are often split into small aggregated areas comprised of groups of trees called individual stands [30]. These stands are not necessarily homogeneous but, in some cases, the percentages of different tree species within the stand is available [26]. The location of the non-dominant trees is unknown. In such cases, machine learning algorithms are often trained to predict the dominant class even for regions with mixed forest species [31], or just areas with a single dominant tree species are selected [32]. This raises the issue of weak markup adjustment. Among weakly supervision tasks, this one belongs to inexact supervision when only coarse-grained labels are given [33]. Weakly supervised images occur both in the general domain [34,35] and in specific tasks such as medical images segmentation [36]. These studies involve new neural network architectures or frameworks development to decrease requirements for labor-intensive data labeling. In the remote sensing domain weakly supervised learning was also considered in different tasks such as cropland segmentation using low spatial resolution satellite data [37], cloud detection through high resolution images [38], and detection of red-attacked trees with very high resolution areal images [39]. However, in the field of forest species classification, the weak markup problem requires additional analysis according to data specificity (both satellite and field-based). In this study, we propose a CNN-based approach to extract more homogeneous areas from the traditional forest inventory data that includes only species' content within stands and does not provide each species' location. We focus on semantic segmentation problem using high resolution multispectral satellite data. The approach is particularly based on the Co-teaching paradigm presented in [40] where two neural networks are trained, and small-loss instances are selected as clean data for image classification task. In contrast, we split the data adjustment and training process into two separate stages and implement this pipeline for the semantic segmentation task.

In this study, we aim to explore a deep neural network approach for forest type classification in Russian boreal forests using Sentinel-2 images. We set the following objectives:

- to develop a novel approach for forest species classification using convolutional neural networks (CNN) combining pixel- and object-wise approaches during the training procedure, and compare it with a typically used approach for semantic segmentation; and
- to provide a strategy for weak markup improvements and examine forest type classification both as a problem of (a) dominant class estimation for non-homogeneous individual stands and (b) more precise homogeneous classification.

2. Materials and methods

2.1. Study Site

The study was conducted in the Russian boreal forests of Leningrad Oblast. The coordinates of these regions are between 33°42′ and 33°76′ longitude and between 60°78′ and 61°01′ latitude (Figure 1). The vegetation cover is mixed and includes deciduous and conifer tree species. The main species are pine, spruce, aspen, and birch. The climate in the region is humid. An average daily high temperature in the vegetation period (from May to August) is above 15 °C. The rain period usually lasts for 7 months (from April to November). From September to May, it is snowy (or rain mixed with snow). Throughout the course of the year, the region is generally cloudy (with the clearer periods during the summer time, when the probability of a clear sky is about 20%) [41].

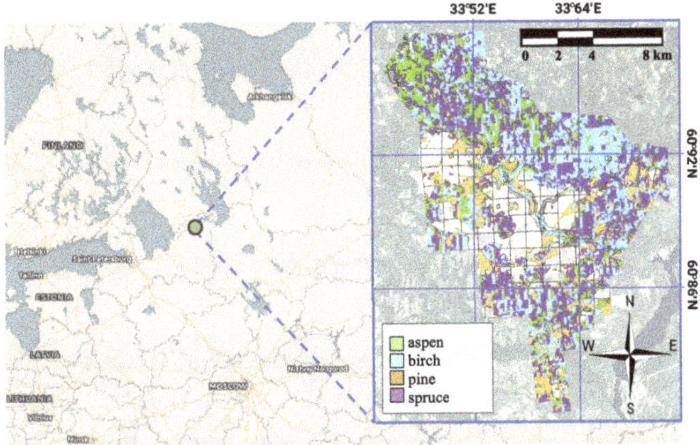

Figure 1. Region of interest. Enhanced RGB bands of Sentinel-2 image (tile id is L2A_T36VWN_A010343_20170615T090713) are shown.

2.2. Reference Data

Reference data was previously reported in [26]. It was collected by field-based measurements carried out in July-August 2018. The methodology of data gathering corresponded to the official Russian inventory regulation [30]. In accordance, the study area was split into individual stands with the following characteristics: polygonal coordinates, a certain percentage of each tree species, average age, and height within the stand. The distribution of stand sizes is presented in Figure 2. The majority of polygons had their longest side length between 100 and 600 m. Although the percentage for each stand was defined, the spatial distribution within the stand was unknown. The number of individual stands with particular dominant tree species (larger than 50% within the stand) is shown in Figure 3 and in Table 1. The vast majority of individual stands consisted of mixed species; for instance, there were less than 100 stands of pure (not mixed) birch type. Example of mixed individual stands are presented in Figure 4.

Table 1. Dataset statistics.

	Training Individual Stands	Test Individual Stands	All Individual Stands	Area (ha)
aspen	520	205	725	2298
birch	1143	501	1644	4165
pine	1569	726	2295	3620
spruce	1087	450	1537	6315

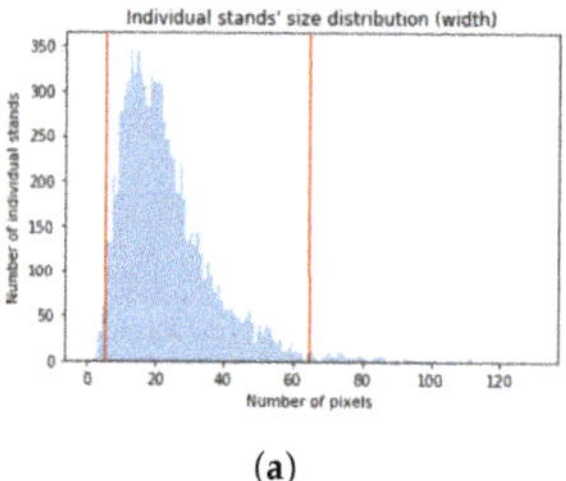

(a)

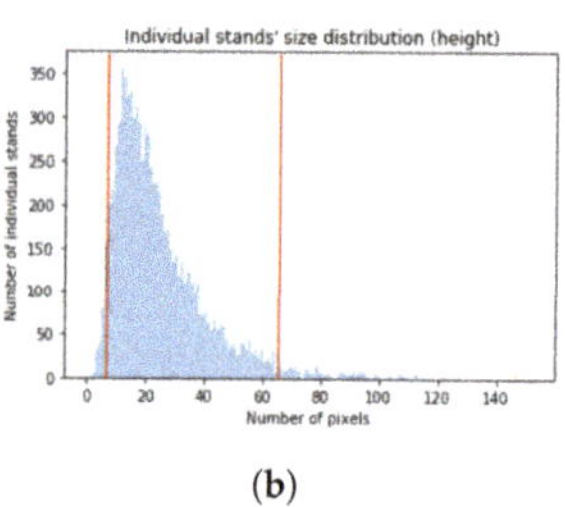

(b)

Figure 2. Size distribution of individual stands within the study area. Polygons with a side larger than 64 pixels or smaller than 8 pixels were eliminated.

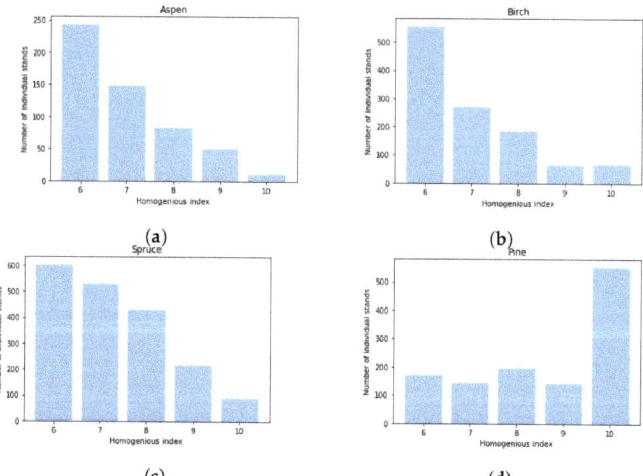

(a)

(b)

(c)

(d)

Figure 3. Distribution of classes.

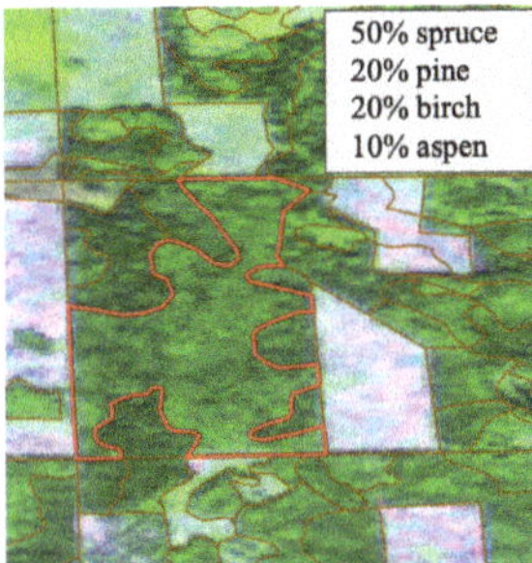

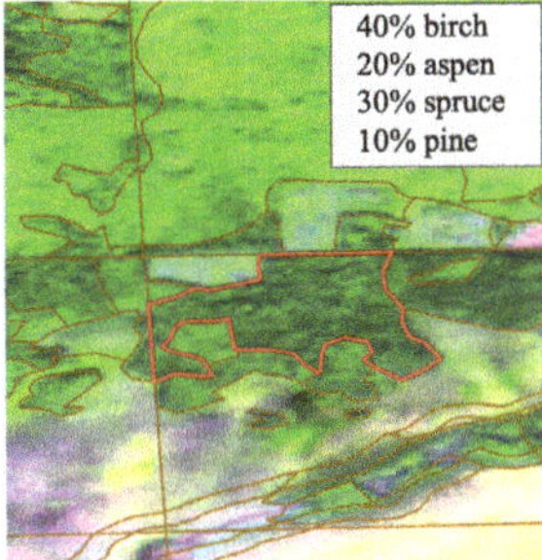

Figure 4. Composite of B12, B08, B04 Sentinel-2 bands (tile id is L2A_T36VWN_A010343_20170615T090713). Example of mixed individual stands (red polygon) with percentages of species.

2.3. Satellite Data

For optical multispectral imagery, we acquired Sentinel-2 data. This data is available for free download in L1C format from EarthExplorer USGS [42]. Tiles IDs and acquisition dates are presented in Table 2. In this study, we considered only summer images. High cloud cover imposes limits on data for this northern region. Therefore, only two summer images from different years but of the comparable summer period were used to create the training dataset. Images acquired in other summer dates did not provide a sufficient amount of clear areas without clouds. There were no significant forest cover changes between survey time and image acquisition time; therefore, both images are relevant for the study. 10 bands of the following wavelengths were used: Band 2: Blue, 458–523 nm; Band 3: Green, 543–578 nm; Band 4: Red, 650–680 nm; Band 5: Red-edge I (R-edge I), 698–713 nm; Band 6: Red-edge II (R-edge II), 733–748 nm; Band 7: Red-edge III (R-edge III), 773–793 nm; Band 8: Near infrared (NIR), 785–900 nm; Band 8A: Narrow Near infrared (NNIR), 855–875 nm; Band 11: Shortwave infrared-1 (SWIR1), 1566–1651 nm; Band 12: Shortwave infrared-2 (SWIR2), 2100–2280 nm). Images were pre-processed with the Sen2Cor package [43] for atmospheric correction. Although, Sen2Cor package provides a cloud and shadow map, which can be used to eliminate irrelevant pixels, we selected cloudless images for the study. The obtained data were in L2A format, including values of Bottom-Of-Atmosphere (BOA) reflectances. For CNN-based tasks, image values are often brought to the interval from 0 to 1 [44,45]. Therefore, pixel values were mapped to the interval $[0, 1]$ through division by 10000 (the maximum physical surface reflectance value for Sentinel-2 in level L2A) and clipping to 0 and 1. We used bands with a spatial resolution of 10 m per pixel (*B02, B03, B04, B08* bands) and 20 m per pixel (*B05, B06, B07, B11, B12, B8A* bands), adjusted to 10 m by Nearest Neighbor interpolation [46]. Each image covered the entire study area, and images were considered separately without any spatial averaging (the same as in [47]).

Table 2. Sentinel-2 images from USGS. Wavelength values corresponding to each band: Band 2: Blue, 458–523 nm; Band 3: Green, 543–578 nm; Band 4: Red, 650–680 nm; Band 5: Red-edge I (R-edge I), 698–713 nm; Band 6: Red-edge II (R-edge II), 733–748 nm; Band 7: Red-edge III (R-edge III), 773–793 nm; Band 8: Near infrared (NIR), 785–900 nm; Band 8A: Narrow Near infrared (NNIR), 855–875 nm; Band 11: Shortwave infrared-1 (SWIR1), 1566–1651 nm; Band 12: Shortwave infrared-2 (SWIR2), 2100–2280 nm).

Tile ID	Date	Cloud Coverage	10 m Bands	20 m Bands	Level of Processing
L1C_T36VWN_A010343_20170615T090713	2017.06.15	0	2, 3, 4, 8	5, 6, 7, 8A, 11, 12	L1C
L1C_T36VWN_A016206_20180730T090554	2018.07.30	0	2, 3, 4, 8	5, 6, 7, 8A, 11, 12	L1C

2.4. Organizing Samples for Classification

Four tree species were considered: aspen, birch, spruce, and pine. We also considered the 'conifer' class as a combination of spruce and pine, and the 'deciduous' class as a combination of aspen and birch. As a sample for the further analysis, we chose individual stands. There was no information on the spatial distribution of tree species within an individual stand. Therefore, we defined the label for each stand as the dominant tree species within it, if the stand contained more than 50% of this forest type (the same approach was described in [31]). For conifer and deciduous classes, we summed the percentages for spruce and pine, and for aspen and birch, respectively. The described sample definition assumed that the markup had some pre-defined uncertainty for non-homogeneous stands. However, it provided information necessary to the dominant species classification task. Thus, for each sample in the data set, we know the label of the dominant forest type, the percentage of secondary types (if any), and an ascribed polygon in a multispectral satellite image.

For the experiment of training procedure adjustment, we selected 8 test regions of about 450 ha each (Figure 5). For the experiment of weak markup improvement, 30% of samples were selected randomly for test. Samples outside test regions were split into train and validation sets randomly, in a ratio of 7:3, following the constraint of no occurrence of the same individual stand in both validation and training sets. For each polygon it can be more than one sample depending on the images' number covering the polygon. Non-overlapping parts of the same satellite image could appear in both the training and test sets.

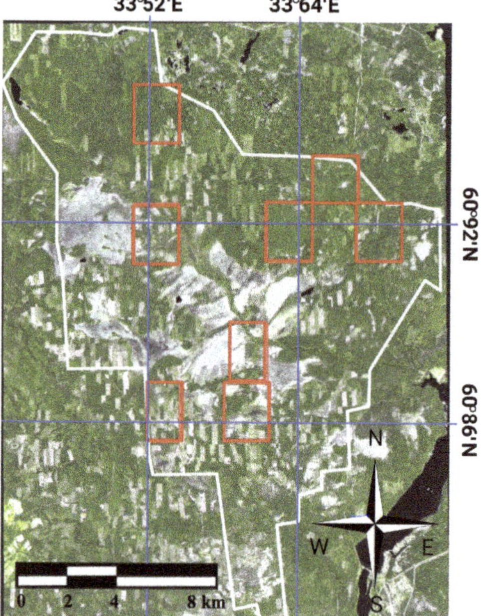

Figure 5. The whole study area (white polygon). Test regions (red polygons). Enhanced RGB bands of Sentinel-2 image (tile id is L2A_T36VWN_A010343_20170615T090713) are shown.

2.5. Forest Species Classification

Instead of typical multi-class classification, we used an hierarchical approach described in [26]. The task of four-species prediction was split into three tasks: (a) classification of conifer and deciduous; (b) classification of birch and aspen; and (c) classification of spruce and pine. The final results followed from the intersection between the predicted mask of birch and aspen and the predicted deciduous mask (with a similar approach followed in the conifer case). Such an hierarchical approach allows for solving each task independently and ensuring greater control over experiment at each step.

For the forest type classification, we implemented a deep neural network approach, which have been widely used for image classification and segmentation tasks when spatial characteristics are important in the remote sensing domain [48–50]. At the input of such a neural network, there is usually a combination of spectral bands. The output of the semantic segmentation model is a map, where each pixel is ascribed a particular class label. During the training procedure, a model is forced to correctly predict as many pixel labels as possible by observing random image patches with pre-defined size. This is achieved through the implementation of a particular loss function. The loss is computed for each step of neural network training, when all images patches from one batch have been processed. For our study, we implemented the categorical cross entropy per-pixel loss function:

$$\text{Loss} = -\frac{\sum_{i=1}^{N}\sum_{k=1}^{C}(y_{ik} \times \log \hat{y}_{ik})}{N},\tag{1}$$

where $\hat{y}_{ik}$—predicted probability of the ith pixel to belong to the kth class, y_{ik}—ground truth value for the ith pixel (1 if the pixel belongs to the kth class), N—number of not masked pixels, C—number of classes.

In this loss, all pixels in the scene are taken into account. Therefore, if the classes are highly unbalanced, a model rarely observes pixels labeled as the smaller class. This results in poor performance of the model for a less represented class. A common solution is using a larger penalty for errors on the smaller class samples, such as in the weighted categorical cross entropy:

$$\text{Weighted Loss} = -\frac{\sum_{i=1}^{N}\sum_{k=1}^{C}(y_{ik} \times \log \hat{y}_{ik}) \times weights(y_{ik})}{N},\tag{2}$$

where $\hat{y}_{ik}$—predicted probability of the ith pixel to belong to the kth class, y_{ik}—ground truth value for the ith pixel (1 if the pixel belongs to the kth class), N—number of not masked pixels, C—number of classes.

Another issue that should be taken into account is that samples of particular classes may not be evenly distributed across the study region. This means that random selection of image patches in batch can lead to a situation where samples concentrated in one area may be seldom observed.

To tackle this problem, we modified the classical sampling approach for semantic segmentation with CNN, as described in the next section.

2.6. Object-Wise Sampling Approach

We replaced the commonly used batch creation approach. The sample content was taken into account, instead of simply using random patch selection. The choice of patch size was governed by the relevant size of polygons. As we eliminated polygons with sides less than 80 m and larger than 640 m, the patch size was selected as 64 × 64 pixels. The number of patches per batch was set to 128. Although we considered two classes, the general approach is also applicable for more classes. For each class, we picked the same number of polygons and cut patches around these polygons to create the batch. As the polygon size could vary in the defined range, the patch crop could also differ for the same polygon. The only demand was that the polygon's bounding box should be within the patch boundary. The patch was also geometrically augmented, in order to provide more variability during the training procedure. We implemented random rotate, mirror, and flip operations. The general approach for batch creation is described in Algorithm 1.

Algorithm 1: Batch Creation

$N \leftarrow$ Batch size;
$M \leftarrow$ Number of classes;
$Pol\_set_0 \leftarrow$ Set of polygons of the class 0;
...;
$Pol\_set_{M-1} \leftarrow$ Set of polygons of the class M-1;
$Batch \leftarrow \varnothing$;
$Polygons\_masks \leftarrow \varnothing$;
$cl \leftarrow 0$;
while $cl \neq M$ **do**
    $patch\_ind \leftarrow 0$;
    while $patch\_ind \neq N/M$ **do**
        $pol \leftarrow SelectPolygon(Pol\_set_{cl})$;
        $patch \leftarrow CropPatch(pol)$;
        $patch\_mask \leftarrow ExtractPolLabel(pol)$;
        $Batch \leftarrow Augment(patch)$;
        $Polygons\_masks \leftarrow Augment(patch\_mask)$;
        $patch\_ind \leftarrow patch\_ind + 1$;
    end while
    $cl \leftarrow cl + 1$;
end while
return $Batch, Polygons\_masks$;

The next step was loss computation. The approach is described in the Figure 6. For this purpose, we used polygon mask. Patch has dimension $Patch\_Rows, Patch\_Columns, Number\_of\_classes$. The patch mask contains non-zero values for pixels within the polygon's area and for the appropriate correct class. Despite the fact that individual stands are not often homogeneous, all pixels within one stand were ascribed the same label. The loss was computed for this area. There can be an available markup for other pixels within the patch, but this was not considered. The main reason for this is that it can affect the balance of classes.

We compared this approach with the commonly used per-pixel semantic segmentation approach, for which the batch was randomly formed and an extra penalty for mistakes in the smaller class was added (Figure 7). In this approach, for calculation of the weighted categorical cross-entropy loss, all pixels within the patch were considered. The weights were set proportionally to the amount of each class represented.

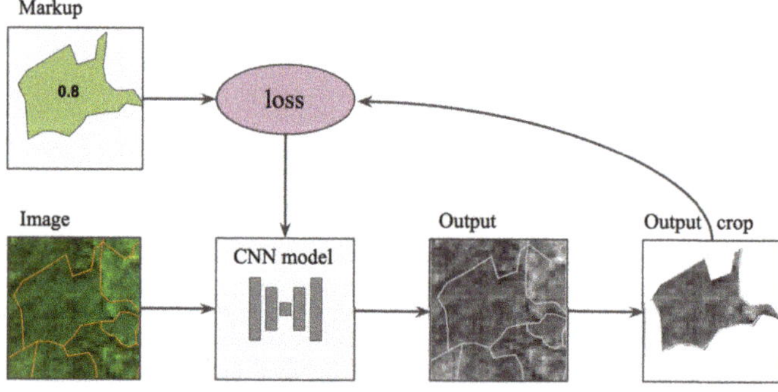

Figure 6. The object-wise semantic segmentation approach. The model produces the map where the probability of a class is recorded at each pixel. Loss is computed just for masked area of the polygon. The percentage of dominant class is also can be taken into consideration (in the example, the dominant species percentage for the individual stand is 0.8).

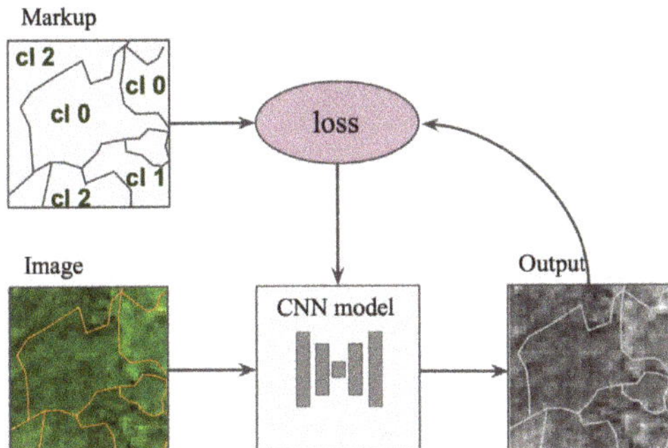

Figure 7. The commonly used per-pixel semantic segmentation approach. The model produces the map where the probability of a class is recorded at each pixel. Loss is computed for the entire patch. The patch includes stands with different dominant species (class 0, class 1, etc.).

2.7. Weak Markup

Another adjustment was aimed at addressing weak markup. It includes two stages, as shown in the Figure 8. The first stage was as follows. The aforementioned reference data consisted of the percentage of each class within the individual stand. We took this knowledge during the loss computation. The loss was calculated for each individual stand and multiplied by the dominant species percentage. For example, for a stand that consisted 60% of conifers and 40% of deciduous trees, the penalty will be 0.6. If the percentage is higher, then the penalty becomes stricter. For a homogeneous stand, all pixels have the maximum loss weights. Therefore, in Equation (2), *weights* were defined as the dominant species percentage. When the learning curve started to change less rapidly and could achieve sufficient results on the validation set (after about 15 epochs), we changed the training data set. We eliminated all individual stands with percentage less than 90%. Thus, for a few epochs (about 2 epochs), the model observes more pure data. Obviously, such a model will perform poorly, in terms of the initial dominant species problem statement. However, at the same time, it will not strive to label deciduous trees within a conifer individual stand as conifer trees (as for case with 60% conifer and 40% deciduous). Then, we used this model to predict conifer and deciduous species both for training and validation regions. The first stage of markup adjustment results was the intersection between the predicted mask and initial dominant species markup. It was assumed that the map acquired in this way contained less pixels of minor (i.e., non-dominant) classes.

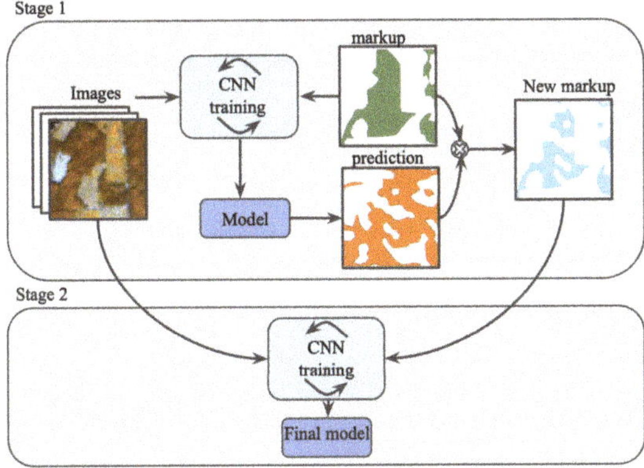

Figure 8. Markup adjustment strategy.

The next stage of the weak markup study was the implementation of the newly obtained markup in further training. We intersected the new conifer mask with the initial spruce and pine dominant masks, and the same for the deciduous classes. The goal of this intersection was to reduce the number of deciduous pixels within individual stands dominated by pine and spruce, and vice versa. For this study, we created the validation data set only from homogeneous individual stands.

2.8. Experimental Setup

For all experiments, the U-Net architecture [51] with ResNet [52] encoder was used, as it has been shown to successfully perform in popular image classification tasks both in general and remote sensing domains [50]. The model implementation referred to [53]. It used Keras [54] with Tensorflow [55] backend. For model training, a PC with GTX-1080Ti GPUs was used. The batch size was 128 patches, where each patch had size of 64×64 pixels. The batch size was chosen accord-ing to GPU memory limitations. There were 100–200 steps per epoch and the number of epochs varied from 10 to 30, depending on the size of classes. Similar results reproduction was achieved by fixing a random seed for pseudo-random number generator for all training methods.

2.9. Validation Methods

To assess the classification quality, we considered F1-score (Equation (5)). It is a commonly used score for semantic segmentation tasks [56], in particular in cases of unbalanced datasets. F1-score is also often considered in the remote sensing domain [50].

$$P = \frac{TP}{TP + FP} \tag{3}$$

$$R = \frac{TP}{TP + FN} \tag{4}$$

$$F1 = \frac{2 \times P \times R}{P + R} \tag{5}$$

where P is precision, R is recall, TP is True Positive (the number of correctly classified pixels/stands of a given class), FP is False Positive (the number of pixels/stands classified as a given class while being of another class), and FN is False Negative (the number of pixels/stands of a given class missed by the model).

In the one case, we evaluated the number of correctly predicted individual stands. To this end, per-pixel predictions within stands were aggregated and the dominant class was defined for each stand. Based on reference and predicted stand labels, the amounts of true positive, false positive, and false negative samples were estimated. In the second case, we evaluated the F1-score in a per-pixel manner.

A CNN model for each experiment was trained five times with different random seeds, and then results were averaged. Standard deviation was computed.

3. Results

3.1. Sampling Approach For Species Classification

We compared a typical sampling procedure for forest species semantic segmentation with our modified one. The results are presented in Table 3. For all classes, the object-wise sampling approach performed better. The average F1-score before aggregation was improved from 0.8 to 0.85. The final aggregated results were obtained by multiplying the predicted conifer binary mask with spruce and pine masks and multiplying the predicted deciduous mask with aspen and birch masks. Aggregated results for four forest classes are shown in Table 6. The object-wise sampling approach allows us to improve segmentation's F1-score from 0.68 to 0.74. The larger difference between the two methods was for the birch and aspen classes. The reason for this is that these classes were the most difficult to distinguish due to imbalance. The proposed approach leads to a more balanced training samples choice.

Standard deviation was computed for averaged F1-score of different model training running. It shows that achieved results are relevant for further forest analysis studies.

Table 3. Forest types classification using different sampling procedure (per-pixel F1-score).

	Aspen/Birch	Pine/Spruce	Conifer/Deciduous	Average
Simple sampling procedure	0.48/0.88	0.91/0.88	0.81/0.85	0.8 ± 0.003
Modified sampling procedure	0.63/0.91	0.94/0.88	0.85/0.87	0.85 ± 0.004

3.2. Markup Adjustment

We conducted experiments aimed to improve conifer and deciduous markup. Some areas were eliminated by the model predictions intersected with the initial dominant species map. It aims to leave only homogeneous areas with conifer or deciduous trees. The per-pixel metric is intended to label all pixels even within inhomogeneous individual stand as the dominant class type. Therefore, at this stage of the task, the goal was not to improve the per-pixel score. The average per-pixel F1-score for conifer and deciduous classification became 0.76, in comparison with the previously achieved 0.82 (Table 4). However, we aimed to preserve the score per individual stands than the per-pixel one. The score of dominant classification per individual stands was still approximately at the same level (F1-score 0.85). It means that the model was trained to ignore pixels of non-dominant classes within the stand. For the further assessment, homogeneous stands were considered.

Table 4. Conifer and deciduous classification (average score) using source markup and updated markup.

	Per-Pixel F1-Score	Per-Stand F1-Score
Source markup	0.827	0.851
Updated markup	0.769	0.854

The obtained map was then used for species classification. We compared the model trained on the source markup and that trained on updated one. Their performances were assessed on homogeneous individual stands for four species from the test set. The results are presented in Table 5. Although we eliminated pixels from the (non-homogeneous) training set, the model performed better than when using the larger training data of weaker quality. It allowed us to improve the average F1-score for four species from 0.74 to 0.76 compared with initial markup usage (Table 6). The results confirmed the benefit of the proposed approach.

Table 5. Forest types classification for more homogeneous individual stands (per-pixel F1-metric) using source markup and updated markup. Results on test samples.

	Aspen/Birch	Pine/Spruce	Average
Source markup	0.77/0.9	0.94/0.88	0.87 ± 0.003
Updated markup	0.79/0.91	0.95/0.9	0.89 ± 0.002

Table 6. Final aggregated results for forest types classification using modified sampling procedure and markup adjustment (F1-score).

	Aspen	Birch	Pine	Spruce	Average
Simple sampling procedure	0.42	0.72	0.84	0.74	0.68 ± 0.007
Modified sampling procedure	0.6	0.8	0.81	0.74	0.74 ± 0.004
Modified sampling procedure with new markup	0.62	0.83	0.82	0.76	0.76 ± 0.005

Example of the final predictions using both modified sampling approach and adjusted markup is presented in Figure 9.

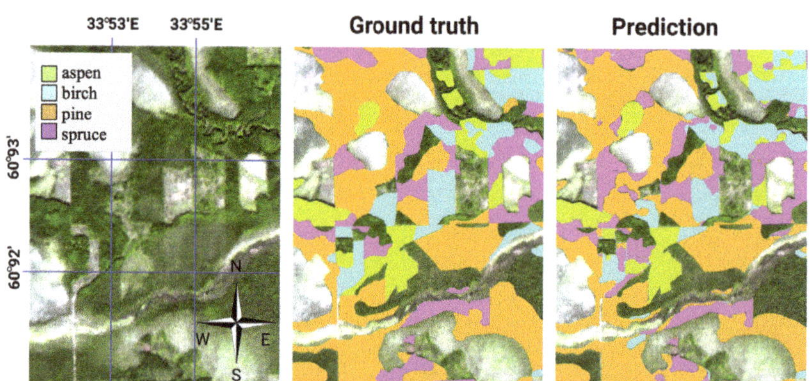

Figure 9. Sentinel-2 RGB image (tile id is L2A_T36VWN_A010343_20170615T090713). Final predictions using modified sampling approach and adjustment markup.

4. Discussion

4.1. Sampling Approach for Species Classification

The analysis showed that the sampling procedure is highly essential for the forest species classification task. The same approach can be implemented for other problems where maps of vast territories are used and some classes are distributed not uniformly. The proposed object-wise sampling approach for CNN leads to better results than the commonly used approach where patches are chosen randomly within the entire satellite image.

It is worth mentioning the reason why a classical patch-wise approach was not considered suitable for our problem. It implies that we can choose the patch size small enough to include just the pixels of one class. However, in our case, there are two obstacles to implement this. The first being that individual stands are not of rectangular shape and, therefore, the patch size must be rather small. The other point is that individual stands are not homogeneous and we do not know the spatial distributions within stands. Therefore, a random small patch within an individual stand may turn out to, in fact, be a set of pixels of a minor class. This makes the approach described in [21] inappropriate in the presented case.

Another alternative approach to classical pixel- and object-wise classification for remote sensing applications (e.g., airplane-based) has been discussed in [25]. It should be noted that, despite the apparent similarity of airplane and satellite-derived remote sensing data, they have substantive differences. The main difference is spatial resolution. The relevant observation field can vary by 100 times (e.g., 0.1 m for UAV and 10 m for satellite images). Thus, the approach have to be modified.

4.2. Markup Adjustment

Clear markup is essential for remote sensing tasks. In some cases, non-homogeneous areas are excluded from training set [32]. Another approach is to use plots with different species and ascribed it by the dominant species class [31]. It is reasonable to move further in the direction of an automatic markup adjustment, in order to make the data clearer without extra manual labeling. The next step of the study can be label adjustment for all classes, not only for conifer and deciduous. The weighted loss function adjustment can also be considered to improve homogeneous areas detection.

Weakly supervised learning is now applied in different remote sensing tasks. They vary by the target objects and remote sensing data properties such as spatial resolution and spectral bands number. In our study, we focus on 10 m spatial resolution and 10 multispectral bands. In cases of very high spatial resolution and just RGB bands such as in [39] markup constraints differ significantly. Particular tasks also pose some limitations and additional opportunities for a weakly supervised learning approach [38]. Therefore, remote sensing datasets can differ drastically from such datasets as MNIST or CIFAR considered in [40]. Another difference is that the forest species classification is considered as a semantic segmentation task instead of an image classification task, such as in the case of noisy labels problem in [40].

Markup adjustment can be also studied in the case with machine learning algorithms instead of neural network based such as methods described in [57–59].

The main error source in such land cover tasks is diversity within each forest species. Spectral characteristics vary drastically for different tree age and depend on environmental conditions. Therefore, markup adjustment and optimal sampling choice are promising approaches to improve model performance. Another error source is mixed border pixels of neighboring individual stands. In the case of 10 m spatial resolution, even for homogeneous forest stands, spectral characteristics on the border can be affected by other species outside this stand. A possible approach to address this problem for homogeneous stands is to consider just inner pixels remote from the border.

One of the potential limitations is the time and computational cost for markup adjustment model training. In our case, we used the same CNN architecture to perform this stage. We trained the model for markup adjustment and the final segmentation model sequentially. In future studies, an alternative approach can be developed and implemented to perform markup adjustment on the fly for remote sensing tasks.

In this study, we considered forest species classification. However, the proposed approach can be transferred in future studies for other tasks where samples are grouped, and for a group, label distribution is known. The described approach is also applicable for other neural network architectures. Therefore, experiments with new state-of-the-art architectures can be conducted using the same method. Both the sampling and markup adjustment approaches are transferable to new satellite data sources. We considered multispectral Sentinel-2 imagery with a spatial resolution of 10 m. However, it can also be implemented for high-resolution multispectral data such as WorldView or just RGB images such as base maps.

Vegetation indices are significant for environmental tasks as they provide relevant surface characteristics. Therefore, they are widely used as features for classical machine learning methods. However, in the case of deep neural networks, it is assumed that neural networks can learn non-linear connections between raw input data and use prior information for more general characteristics extraction. In our study, we considered only multispectral satellite bands. However, future studies might include vegetation indices or supplementary materials such as digital elevation or canopy height models to achieve higher results and reduce training time.

It is promising to study different augmentation techniques combined with improved markup and the object-wise sampling approach. For example, the object-based augmentation described in [60] can further be implemented to create more variable training samples with different homogeneous stands.

Precise forest species classification can also be implemented in ecological and environmental studies, as large forest patches have been proved to affect human health [61]. Detailed forest characteristics can be helpful for such analysis.

5. Conclusions

The sampling approach and ground truth markup quality are crucial in forestry tasks involving remote sensing data. In this study, we analyzed the potential of combining CNN and Sentinel-2 images for the task of forest species classification using weak markup with non-homogeneous individual stands. During the first stage, a CNN was trained to find the homogeneous areas within each stand, providing a more accurate markup. During the second stage, the final model was trained to predict four forest species. This markup adjustment allows us to increase F1-score from 0.74 to 0.76 compared to the initial markup. The experiment confirms the opportunity of finding weak labels and shows promising results for further classification enhancement. We also proposed the CNN-based sampling approach for spatial data in forest species classification. The proposed modification outperformed the prediction quality of a commonly used per-pixel semantic segmentation model (the average F1-metric was increased from 0.68 to 0.74). The described pipeline helps to address the issue of highly imbalanced and not evenly distributed classes. The provided training strategy can help solve forest species classification tasks more precisely, even when the reference data has significant limitations. Further study for other vast territories is promising, and the proposed sampling technique seems to be beneficial in such spatial studies.

Author Contributions: Conceptualization, S.I. and V.I.; methodology, S.I. and V.I.; software, S.I.; validation, S.I.; formal analysis, S.I. and A.T.; investigation, S.I.; data curation, V.I. and A.T.; writing—original draft preparation, S.I.; visualization, S.I.; supervision, V.I. and I.O. All authors have read and agreed to the published version of the manuscript.

Funding: This research received no external funding.

Conflicts of Interest: The authors declare no conflict of interest.

References

1. Lindenmayer, D.B.; Margules, C.R.; Botkin, D.B. Indicators of biodiversity for ecologically sustainable forest management. *Conserv. Biol.* **2000**, *14*, 941–950. [CrossRef]
2. Franklin, J.; Andrade, R.; Daniels, M.L.; Fairbairn, P.; Fandino, M.C.; Gillespie, T.W.; González, G.; Gonzalez, O.; Imbert, D.; Kapos, V.; et al. Geographical ecology of dry forest tree communities in the West Indies. *J. Biogeogr.* **2018**, *45*, 1168–1181. [CrossRef]
3. Wallace, K.J.; Clarkson, B.D. Urban forest restoration ecology: A review from Hamilton, New Zealand. *J. R. Soc. N. Z.* **2019**, *49*, 347–369. [CrossRef]
4. Hill, A.; Buddenbaum, H.; Mandallaz, D. Combining canopy height and tree species map information for large-scale timber volume estimations under strong heterogeneity of auxiliary data and variable sample plot sizes. *Eur. J. For. Res.* **2018**, *137*, 489–505. [CrossRef]
5. Bont, L.G.; Hill, A.; Waser, L.T.; Bürgi, A.; Ginzler, C.; Blattert, C. Airborne-laser-scanning-derived auxiliary information discriminating between broadleaf and conifer trees improves the accuracy of models for predicting timber volume in mixed and heterogeneously structured forests. *For. Ecol. Manag.* **2020**, *459*, 117856. [CrossRef]
6. Pandey, P.C.; Anand, A.; Srivastava, P.K. Spatial distribution of mangrove forest species and biomass assessment using field inventory and earth observation hyperspectral data. *Biodivers. Conserv.* **2019**, *28*, 2143–2162. [CrossRef]
7. Persson, H.J.; Olsson, H.; Soja, M.J.; Ulander, L.M.; Fransson, J.E. Experiences from large-scale forest mapping of Sweden using TanDEM-X data. *Remote Sens.* **2017**, *9*, 1253. [CrossRef]
8. Lei, Y.; Siqueira, P.; Chowdhury, D.; Torbick, N. Generation of large-scale forest height mosaic and forest disturbance map through the combination of spaceborne repeat-pass InSAR coherence and airborne lidar. In Proceedings of the 2016 IEEE International Geoscience and Remote Sensing Symposium (IGARSS), Beijing, China, 10–15 July 2016; pp. 5342–5345.
9. Pasquarella, V.J.; Holden, C.E.; Woodcock, C.E. Improved mapping of forest type using spectral-temporal Landsat features. *Remote Sens. Environ.* **2018**, *210*, 193–207. [CrossRef]
10. Gudex-Cross, D.; Pontius, J.; Adams, A. Enhanced forest cover mapping using spectral unmixing and object-based classification of multi-temporal Landsat imagery. *Remote Sens. Environ.* **2017**, *196*, 193–204. [CrossRef]
11. Stoian, A.; Poulain, V.; Inglada, J.; Poughon, V.; Derksen, D. Land cover maps production with high resolution satellite image time series and convolutional neural networks: Adaptations and limits for operational systems. *Remote Sens.* **2019**, *11*, 1986. [CrossRef]
12. Nguyen, T.H.; Jones, S.D.; Soto-Berelov, M.; Haywood, A.; Hislop, S. A spatial and temporal analysis of forest dynamics using Landsat time-series. *Remote Sens. Environ.* **2018**, *217*, 461–475. [CrossRef]
13. Campos-Taberner, M.; García-Haro, F.J.; Martínez, B.; Izquierdo-Verdiguier, E.; Atzberger, C.; Camps-Valls, G.; Gilabert, M.A. Understanding deep learning in land use classification based on Sentinel-2 time series. *Sci. Rep.* **2020**, *10*, 1–12. [CrossRef] [PubMed]
14. Illarionova, S.; Nesteruk, S.; Shadrin, D.; Ignatiev, V.; Pukalchik, M.; Oseledets, I. MixChannel: Advanced Augmentation for Multispectral Satellite Images. *Remote Sens.* **2021**, *13*, 2181. [CrossRef]
15. Eo-Learn. 2020. Available online: https://github.com/sentinel-hub/eo-learn (accessed on 20 August 2020).
16. Breiman, L. Random forests. *Mach. Learn.* **2001**, *45*, 5–32. [CrossRef]
17. Cortes, C.; Vapnik, V. Support-vector networks. *Mach. Learn.* **1995**, *20*, 273–297. [CrossRef]
18. Hamedianfar, A.; Barakat A.; Gibril, M. Large-scale urban mapping using integrated geographic object-based image analysis and artificial bee colony optimization from worldview-3 data. *Int. J. Remote Sens.* **2019**, *40*, 6796–6821. [CrossRef]

19. Chen, Y.; Zhou, Y.; Ge, Y.; An, R.; Chen, Y. Enhancing land cover mapping through integration of pixel-based and object-based classifications from remotely sensed imagery. *Remote Sens.* **2018**, *10*, 77. [CrossRef]

20. Kussul, N.; Shelestov, A.; Lavreniuk, M.; Butko, I.; Skakun, S. Deep learning approach for large scale land cover mapping based on remote sensing data fusion. In Proceedings of the 2016 IEEE International Geoscience and Remote Sensing Symposium (IGARSS), Beijing, China, 10–15 July 2016; pp. 198–201.

21. Mahdianpari, M.; Salehi, B.; Rezaee, M.; Mohammadimanesh, F.; Zhang, Y. Very deep convolutional neural networks for complex land cover mapping using multispectral remote sensing imagery. *Remote Sens.* **2018**, *10*, 1119. [CrossRef]

22. Illarionova, S.; Shadrin, D.; Trekin, A.; Ignatiev, V.; Oseledets, I. Generation of the NIR spectral Band for Satellite Images with Convolutional Neural Networks. *Sensors* **2021**, *21*, 5646. [CrossRef]

23. DeLancey, E.R.; Simms, J.F.; Mahdianpari, M.; Brisco, B.; Mahoney, C.; Kariyeva, J. Comparing deep learning and shallow learning for large-scale wetland classification in Alberta, Canada. *Remote Sens.* **2020**, *12*, 2. [CrossRef]

24. Chen, T.; Guestrin, C. Xgboost: A scalable tree boosting system. In Proceedings of the 22nd ACM Sigkdd International Conference On Knowledge Discovery and Data Mining, San Francisco, CA, USA, 13–17 August 2016; pp. 785–794.

25. Sun, Y.; Huang, J.; Ao, Z.; Lao, D.; Xin, Q. Deep Learning Approaches for the Mapping of Tree Species Diversity in a Tropical Wetland Using Airborne LiDAR and High-Spatial-Resolution Remote Sensing Images. *Forests* **2019**, *10*, 1047. [CrossRef]

26. Illarionova, S.; Trekin, A.; Ignatiev, V.; Oseledets, I. Neural-Based Hierarchical Approach for Detailed Dominant Forest Species Classification by Multispectral Satellite Imagery. *IEEE J. Sel. Top. Appl. Earth Obs. Remote Sens.* **2020**, *14*, 1810–1820. [CrossRef]

27. Xia, G.S.; Bai, X.; Ding, J.; Zhu, Z.; Belongie, S.; Luo, J.; Datcu, M.; Pelillo, M.; Zhang, L. DOTA: A large-scale dataset for object detection in aerial images. In Proceedings of the IEEE Conference on Computer Vision and Pattern Recognition, Salt Lake City, UT, USA, 18–22 June 2018; pp. 3974–3983.

28. Xu, G.; Zhu, X.; Tapper, N. Using convolutional neural networks incorporating hierarchical active learning for target-searching in large-scale remote sensing images. *Int. J. Remote Sens.* **2020**, *41*, 4057–4079. [CrossRef]

29. Trisasongko, B.H.; Panuju, D.R.; Paull, D.J.; Jia, X.; Griffin, A.L. Comparing six pixel-wise classifiers for tropical rural land cover mapping using four forms of fully polarimetric SAR data. *Int. J. Remote Sens.* **2017**, *38*, 3274–3293. [CrossRef]

30. Order of the Federal Forestry Agency (Rosleskhoz) of 12 December 2011 N 516 Moscow "On approval of the Forest Inventory Instruction" Prikaz Federal'nogo Agentstva Lesnogo Hozyajstva (Rosleskhoz) ot 12 Dekabrya 2011 g. N 516 g. Moskva "Ob Utverzhdenii Lesoustroitel'noj Instrukcii". Available online: https://rulaws.ru/acts/Prikaz-Rosleshoza-ot-12.12.2011-N-516/ (accessed on 20 August 2020).

31. Abdollahnejad, A.; Panagiotidis, D.; Shataee Joybari, S.; Surovỳ, P. Prediction of dominant forest tree species using quickbird and environmental data. *Forests* **2017**, *8*, 42. [CrossRef]

32. Knauer, U.; von Rekowski, C.S.; Stecklina, M.; Krokotsch, T.; Pham Minh, T.; Hauffe, V.; Kilias, D.; Ehrhardt, I.; Sagischewski, H.; Chmara, S.; et al. Tree species classification based on hybrid ensembles of a convolutional neural network (CNN) and random forest classifiers. *Remote Sens.* **2019**, *11*, 2788. [CrossRef]

33. Zhou, Z.H. A brief introduction to weakly supervised learning. *Natl. Sci. Rev.* **2017**, *5*, 44–53. [CrossRef]

34. Guo, S.; Huang, W.; Zhang, H.; Zhuang, C.; Dong, D.; Scott, M.R.; Huang, D. CurriculumNet: Weakly Supervised Learning from Large-Scale Web Images. In Proceedings of the European Conference on Computer Vision (ECCV), Munich, Germany, 8–14 September 2018.

35. Ahn, J.; Cho, S.; Kwak, S. Weakly Supervised Learning of Instance Segmentation With Inter-Pixel Relations. In Proceedings of the IEEE/CVF Conference on Computer Vision and Pattern Recognition (CVPR), Long Beach, CA, USA, 16–20 June 2019.

36. Xu, G.; Song, Z.; Sun, Z.; Ku, C.; Yang, Z.; Liu, C.; Wang, S.; Ma, J.; Xu, W. CAMEL: A Weakly Supervised Learning Framework for Histopathology Image Segmentation. In Proceedings of the IEEE/CVF International Conference on Computer Vision (ICCV), Seoul, Korea, 27–28 October 2019.

37. Wang, S.; Chen, W.; Xie, S.M.; Azzari, G.; Lobell, D.B. Weakly Supervised Deep Learning for Segmentation of Remote Sensing Imagery. *Remote Sens.* **2020**, *12*, 207. [CrossRef]

38. Li, Y.; Chen, W.; Zhang, Y.; Tao, C.; Xiao, R.; Tan, Y. Accurate cloud detection in high-resolution remote sensing imagery by weakly supervised deep learning. *Remote Sens. Environ.* **2020**, *250*, 112045. [CrossRef]

39. Qiao, R.; Ghodsi, A.; Wu, H.; Chang, Y.; Wang, C. Simple weakly supervised deep learning pipeline for detecting individual red-attacked trees in VHR remote sensing images. *Remote Sens. Lett.* **2020**, *11*, 650–658. [CrossRef]

40. Han, B.; Yao, Q.; Yu, X.; Niu, G.; Xu, M.; Hu, W.; Tsang, I.; Sugiyama, M. Co-teaching: Robust training of deep neural networks with extremely noisy labels. *arXiv* **2018**, arXiv:1804.06872.

41. Weather Spark. 2020. Available online: https://weatherspark.com/ (accessed on 20 August 2020).

42. EarthExplorer USGS. Available online: https://earthexplorer.usgs.gov/ (accessed on 12 August 2020).

43. Sen2Cor. Available online: https://step.esa.int/main/third-party-plugins-2/sen2cor/ (accessed on 12 August 2020).

44. Vaddi, R.; Manoharan, P. Hyperspectral image classification using CNN with spectral and spatial features integration. *Infrared Phys. Technol.* **2020**, *107*, 103296. [CrossRef]

45. Debella-Gilo, M.; Gjertsen, A.K. Mapping Seasonal Agricultural Land Use Types Using Deep Learning on Sentinel-2 Image Time Series. *Remote Sens.* **2021**, *13*, 289. [CrossRef]

46. Persson, M.; Lindberg, E.; Reese, H. Tree species classification with multi-temporal Sentinel-2 data. *Remote Sens.* **2018**, *10*, 1794. [CrossRef]

47. Astola, H.; Häme, T.; Sirro, L.; Molinier, M.; Kilpi, J. Comparison of Sentinel-2 and Landsat 8 imagery for forest variable prediction in boreal region. *Remote Sens. Environ.* **2019**, *223*, 257–273. [CrossRef]

48. Zhang, W.; Tang, P.; Zhao, L. Remote sensing image scene classification using CNN-CapsNet. *Remote Sens.* **2019**, *11*, 494. [CrossRef]

49. Song, J.; Gao, S.; Zhu, Y.; Ma, C. A survey of remote sensing image classification based on CNNs. *Big Earth Data* **2019**, *3*, 232–254. [CrossRef]

50. Kattenborn, T.; Leitloff, J.; Schiefer, F.; Hinz, S. Review on Convolutional Neural Networks (CNN) in vegetation remote sensing. *ISPRS J. Photogramm. Remote Sens.* **2021**, *173*, 24–49. [CrossRef]

51. Ronneberger, O.; Fischer, P.; Brox, T. U-net: Convolutional networks for biomedical image segmentation. In Proceedings of the International Conference on Medical Image Computing And Computer-Assisted Intervention, Munich, Germany, 5–9 October 2015; pp. 234–241.

52. He, K.; Zhang, X.; Ren, S.; Sun, J. Deep residual learning for image recognition. In Proceedings of the IEEE Conference on Computer Vision and Pattern Recognition, Las Vegas, NV, USA, 27–30 June 2016; pp. 770–778.

53. Yakubovskiy, P. Segmentation Models. 2019. Available online: https://github.com/qubvel/segmentation_models (accessed on 20 August 2020).

54. Keras. 2019–2020. Available online: https://keras.io/ (accessed on 20 August 2020).

55. TensorFlow. 2019–2020. Available online: https://github.com/tensorflow/tensorflow (accessed on 20 August 2020).

56. Csurka, G.; Larlus, D.; Perronnin, F.; Meylan, F. What is a good evaluation measure for semantic segmentation? In Proceedings of the British Machine Vision Conference BMVC, Bristol, UK, 9–13 September 2013; Volume 27, pp. 10–5244.

57. Xia, J.; Yokoya, N.; Pham, T.D. Probabilistic mangrove species mapping with multiple-source remote-sensing datasets using label distribution learning in Xuan Thuy National Park, Vietnam. *Remote Sens.* **2020**, *12*, 3834. [CrossRef]

58. Ha, N.T.; Manley-Harris, M.; Pham, T.D.; Hawes, I. A comparative assessment of ensemble-based machine learning and maximum likelihood methods for mapping seagrass using sentinel-2 imagery in Tauranga Harbor, New Zealand. *Remote Sens.* **2020**, *12*, 355. [CrossRef]

59. Pham, T.D.; Bui, D.T.; Yoshino, K.; Le, N.N. Optimized rule-based logistic model tree algorithm for mapping mangrove species using ALOS PALSAR imagery and GIS in the tropical region. *Environ. Earth Sci.* **2018**, *77*, 1–13. [CrossRef]

60. Illarionova, S.; Nesteruk, S.; Shadrin, D.; Ignatiev, V.; Pukalchik, M.; Oseledets, I. Object-Based Augmentation Improves Quality of Remote Sensing Semantic Segmentation. *arXiv* **2021**, arXiv:2105.05516.

61. Kim, J.; Park, D.B.; Seo, J.I. Exploring the Relationship between Forest Structure and Health. *Forests* **2020**, *11*, 1264. [CrossRef]

forests

Article

MDIR Monthly Ignition Risk Maps, an Integrated Open-Source Strategy for Wildfire Prevention

Luis Santos [1,2,3,*], Vasco Lopes [1] and Cecília Baptista [1,3]

1   School of Technology, Polytechnic Institute of Tomar, 2300-313 Tomar, Portugal;
     estt11828@ipt.pt (V.L.); cecilia@ipt.pt (C.B.)
2   Geosciences Research Center, Coimbra University, 3030-790 Coimbra, Portugal
3   Centre for Technology, Restoration and Art Enhancement (Techn&Art), 2300-313 Tomar, Portugal
*   Correspondence: lsantos@ipt.pt; Tel.: +351-967-743-365

Abstract: Countries unaccustomed to wildfires are currently experiencing wildfire as a new climate-change reality. Understanding how fire ignition and propagation are correlated with temperature, orography, humidity, wind, and the mixture and age of individual plants must be considered when designing prevention strategies. While wildfire prevention focuses on fire ignition avoidance, firefighting success depends on early ignition detection, meaning that, in either case, ignition plays a major role. The current case study considered three Portuguese municipalities that annually observe frequent fire ignitions (Tomar, Ourém, and Ferreira do Zêzere) as the testing ground for the Modernized Dynamic Ignition Risk (MDIR) strategy, thus evaluating the efficiency of MDIR and the efficacy of the variables used. This methodology uses geographic information systems technology sustained by open-source satellite imagery, along with the Habitat Risk Assessment model from the InVEST software package, as drivers for the MDIR application. The MDIR approach grants frequent update capabilities and fully open-sourced high ignition risk area identification, producing monthly ignition risk maps. The advantage of using this method is the ease of adaptation to any current monitoring strategy, awarding further efficiency and efficacy in reducing ignitions. The approach delivered adequate results in estimating ignitions for the three Portuguese municipalities, achieving, for several months, prediction accuracy percentages of over 70%. For the studied area, MDIR clearly identifies areas of high ignition risk and delivers an average of 62% success in predicting ignitions, thus showing potential for analyzing the impact of policy implementation and monitoring through the strategy design.

Keywords: fire ignition; fire hazard; QGIS; InVEST; NDVI; S2 NDWI; risk

Citation: Santos, L.; Lopes, V.; Baptista, C. MDIR Monthly Ignition Risk Maps, an Integrated Open-Source Strategy for Wildfire Prevention. *Forests* **2022**, *13*, 408. https://doi.org/10.3390/f13030408

Academic Editor: Costantino Sirca

Received: 14 December 2021
Accepted: 1 March 2022
Published: 3 March 2022

Publisher's Note: MDPI stays neutral with regard to jurisdictional claims in published maps and institutional affiliations.

1. Introduction

Certain biogeographic regions are historically prone to wildfires; the Mediterranean in Europe, Fynbos and Savannah in Africa, Malee in Australia, Matorral in Chile, and Chaparral in California, are among the most well-known biomes for common wildfire occurrence [1]. Corresponding ecosystems experience cyclic recurrence of wildfires, where ignition and spreading are driven by temperature, orography, humidity, wind, the mixture of plants, and the age of the individual plants [2,3]. The characteristic patchy habitats of such ecosystems exhibit a ratio of dead to live fuel, which varies from species to species, with ratios being much greater in higher succession stages as a result of flammable leaf litter deposition [4,5]. Despite ignition and spread tendencies, hot and dry climate ecosystems require fire: species have evolved to resist wildfires, and in some extreme cases, fire is a mandatory cue for species development [6].

Nonetheless, circumstances are changing, and countries unaccustomed to wildfires are currently experiencing this new phenomenon. In Europe, countries such as Sweden, Germany, Poland, and Slovakia, among others, are experiencing increased burnt area, while wildfire hotspot countries, such as Portugal, Spain, Greece, Italy, and France have experienced unusually severe fires, claiming hundreds of lives and resulting in serious economic loss [7,8]. Many argue that this is an indicator of increasing wildfire activity, resulting in destructive megafires under the current climate change scenario [8,9].

Europe's extensive urbanization and expansion trends are blurring the line between urban and rural areas, which under current climate change predictions, will increase fire risk connected with the urban-wildland interface [10,11]. Portugal follows the general European trend, with a turn in the opposite direction. The main culprits are the demographic shift from rural to urban areas, the changes in land use with evident abandonment, a fragmented land ownership that discourages investment in forest management, and poor fire planning strategies, resulting in the increased connectivity of unmanaged forested areas, consequently increasing wildfire severity [12,13].

Although the causes of most wildfires are proven to be accidental or criminally driven [14,15], wildfires can arise naturally from sources such as the spontaneous combustion of dry matter under high temperatures and wind conditions [16], or most commonly, lightning strikes [17]. Ignitions, no matter what underlies their origin, are the first and most important factor in fire prevention [18,19].

Wildfire risk is mainly associated with the number of ignitions that potentially result in large fires, considering that risk is a factor of hazard by the number of elements at risk, ignitions add to hazard, even though most don't result in large fires [20,21]. Ignition prevention associated with an adequate monitoring strategy represents a key element in fire risk reduction [22,23]. Several risk projection maps and techniques are based mostly on meteorological characteristics such as wind, humidity, and temperature; however, vegetation density and leaf litter, along with chlorophyll content and dryness, are factors of major importance, as has been thoroughly addressed in the literature [11,21,24–26].

Wildfire mitigation analysis has evolved significantly since the turn of the century with the development of satellite imagery and remote sensing (RS) analysis, mostly combined with geographic information systems (GIS). GIS offer a panoply of software and applications, both commercial and open-source, widely used to detect and monitor the behavior of operational firefighting activities and facilitate burnt area mapping. These technologies are proven to confer further efficiency and precision when compared with traditional surveying methods, reducing risk, burnt area, and saving human lives [27,28].

While most high-quality satellite imagery is still fee-based, new free satellite technology, providing good resolution imagery through a wide range of cameras, is currently widely available. One of these satellites is the Sentinel-2, which offers a wide wavelength range, 10 m resolution imagery, and free access [29,30].

Fire occurrence, prediction, and propagation models are widely tested and compared in various research papers and literature reviews [18,19,31], offering an indisputable myriad of solutions to fire problems, from statistical methods [18,31] to remote sensing [29,32] and machine learning [31,33]. Despite the many advances in scientific knowledge, such solutions are partly restricted to specialists and are not as readily accessible to the operational agents that play an important role in fire prevention and firefighting.

Adapting and mitigating wildfire risk requires a thorough understanding of all variables involved, both natural and anthropic, thus considering their intricate relationship. The development of suitable planning and fire management policy will only be possible by considering such variables along with the operational firefighting activities [34].

The current study aims to create dynamic monthly risk maps, designed to optimize monitoring strategies as part of the integrated dynamic process involved in fire prevention. The main output is monthly updateable ignition risk maps, which in combination with the current meteorological-based techniques, will enhance firefighting, conferring a more efficient monitoring strategy and enabling its use by both researchers and field operational staff.

2. Materials and Methods

Following the trends within geographic information technology, aided by the plethora of free satellite imagery, the current study applied a combined methodology which used the Habitat Risk Assessment model, one of the InVEST (Integrated Valuation of Ecosystem Services and Tradeoffs) models. The InVEST Habitat and Species Risk Assessment (HRA) model allows the assessment of cumulative risks posed to habitats by human activities and habitat-specific consequences, delivering a final ecosystem risk map for each individual habitat resulting from the contribution of exposure and consequence to overall risk [35].

Satellite imagery was obtained from the European Space Agency Sentinel-2 satellite, equipped with the MultiSpectral Instrument (MSI) sensor including 13 spectral bands with different resolutions covering the spectral regions of the visible (VIS), near infrared (NIR), and shortwave infrared (SWIR) ranges (https://sentinel.esa.int/web/sentinel/missions, accessed on 1 November 2020).

The fire ignition risk potential maps were created using the previously published Modernized Dynamic Ignition Risk (MDIR) strategy, which combines the InVEST HRA model engine and the variables of road network, historic ignitions, the Digital Terrain Model (DTM), visibility, the Normalized Difference Vegetation Index (NDVI), and population to produce risk maps. This study improves the previous strategy by including the dynamic variable of the Normalized Difference Water Index (S2-NDWI) [36]. The new MDIR approach uses as dynamic variables the Normalized Difference Vegetation Index (NDVI) and the Normalized Difference Water Index (S2-NDWI), which in combination with static variables, land occupation, slope, forest road network, historical fire ignitions, and visualization basins, enabled the new fire ignition prevention methodology.

All geographic variables within the habitat risk assessment model (HRA) were correlated with each other, with different weights assigned to each distinct land occupation habitat class, whereas the remaining variables were assigned as stressors contributing to fire risk. All weights were attributed based on the research noted in the bibliography [37].

2.1. Study Area

The study area belongs to the biogeographical transitional region situated between the floodplains of the Tagus River and the mountain range of Montejunto-Estrela in central Portugal (Figure 1).

Administratively, the region comprises the municipalities of Tomar, Ourém, and Ferreira do Zêzere belonging to the Santarém district, Nomenclature of Territorial Units for Statistical Purposes (NUTS) level II and III; the municipalities are inserted in the Central region and in the sub-region of the Middle Tagus. Biogeographically, the municipalities under study are included in the Mediterranean bioclimate, exhibiting a pluvio-seasonal oceanic bio-climatic region [38] with an average altitude ranging from 30 to 650 m ASL, mean annual precipitation of 660 mm, and a mean annual temperature of 14 °C. Temperature peaks and heatwave duration are major forest fire ignition drivers; data from http://portaldoclima.pt/en/ (accessed on 1 September 2020), indicates that on average, the yearly number of days with a temperature $\geq$35 °C is 12, where heatwave duration with temperatures $\geq$35 °C is 4 to 6 days. This region is severely affected by recurrent forest fires, where the main anthropic causes are associated with the misuse of fire, land abandonment, aging population, desertification, and the choice of forestry species (*Eucalyptus globulus*) influenced by economic yields.

The region, once prosperous with *Pinus pinaster* plantations, due to the last century's forestry policy, and speckled with olive orchards and lowland agriculture, is currently transformed into a mosaic of *Eucalyptus globulus* plantations, with remnants of *Pinus pinaster* and shrubland dominated by *Ulex* spp., *Erica* spp., *Pistacia* sp., *Myrtus* sp., and *Rubus* spp. where agriculture once thrived, confirming the abandonment trend that is further observed as we move northeastward. According to the data of the Official Administrative Chart of Portugal (DGT-2019), the three municipalities cover an area of approximately

95,823 ha, divided as follows: Tomar (35,120 ha), Ourém (41,666 ha), and Ferreira do Zêzere (19,037 ha) [36].

Figure 1. The geographical location of the study region.

2.2. Methods

The applied methodology derives from the previously published Modernized Dynamic Ignition Risk (MDIR) model [36] which correlates stressors, both static and dynamic, and resilience variables with land occupation type, returning potential rural fire ignition maps. The MDIR approach is driven by the InVEST Habitat Risk Assessment (HRA) model using the process described in Figure 2. The InVEST HRA was not primarily designed to evaluate fire ignitions; however, it utilizes a well-supported exposure-consequence framework that assesses spatial variation as a cumulative risk from multiple human activities across diverse conditions and land use patterns. The cumulative aspect of ignitions is addressed by various authors [39,40] who, in some cases, consider exposure-consequence [41] as important indicators of ignition risk in habitats strongly influenced by human activities.

The condition of a habitat is a key model determinant; as anthropogenic stressors continue to diversify and intensify, so too does the need for quick, clear, and repeatable ways of assessing risks to habitats, both now and under future management scenarios. These factors award HRA the necessary characteristics to adequately estimate ignition risk and deliver updateable risk maps. Considering that, this model explores the consequences of human activities through the assessment of cumulative risk; the ignition risk probability can be assessed in such a cumulative manner that exposure to risk is positively correlated with the number of stressors affecting a particular area.

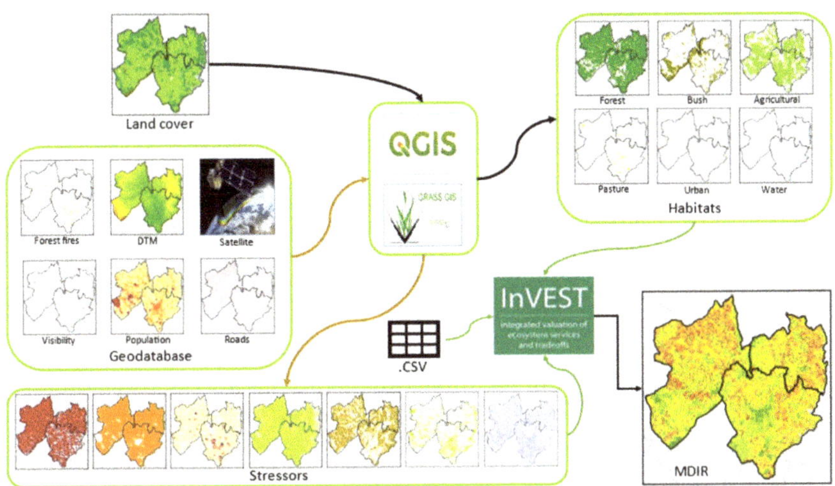

Figure 2. Modernized Dynamic Ignition Risk (MDIR) conceptual methodology.

The conceptual model is built using land use and occupation data (COS2018) obtained from the Directorate General of the Territory of Portugal (http://mapas.dgterritorio.pt/ DGT-ATOM-download/COS_Final/COS2018v1.0-NUT3/COS2018-V1-PT16I_Medio_Tejo. zip; accessed on 7 September 2020). Land use data is available with 9 classes of land occupation: artificialized territories, agriculture, pastures, agroforestry areas, forests, bush, uncovered bare soil or with little vegetation, wetlands, and surface water masses. The land use data was reclassified using the QGIS software to produce individual habitat maps, where some pairs of classes were joined, "artificialized territories" with "uncovered bare soil or with little vegetation," "agroforestry areas" with "forests," and "wetlands" with "surface water masses." These classes were grouped due to land occupation similarities, which in case of fire ignition, does not significantly influence results [14,20].

The model also considers a set of slow variables: Digital Terrain Model (MDT), population density, roads network, and viewshed.

The Digital Terrain Model (MDT) from Porto University Faculty of Sciences, (https: //www.fc.up.pt/pessoas/jagoncal/srtm/; accessed on 7 September 2020) was used to calculate the slope. The data was edited using the QGIS (2020) software slope function, followed by a reclassification into two classes, ≥ 10 and ≤ 10 degrees, where the layer from 0 to 10 degrees will be used as a stressor in the InVEST risk assessment model, since low slopes are considered ignition-prone areas according to the last 10-year records for this region. However, for fire spread, the authors refer to high slopes as favorable to fire spread [31].

Population density (inhabitants per km^2) was determined using the geographic and alphanumeric data from the 2011 census from the Portuguese National Statistics Institute (http://mapas.ine.pt/download/index2011.phtml, https://censos.ine.pt/xportal/xmain? xpid=CENSOS&xpgid=censos_lugar; accessed on 7 September 2020). QGIS software was used to delimitate the stressor layer of population density, which corresponds to areas below 250 inhabitants per km^2, and thus awarded an increased risk of ignition, due to desertification and elderly population, with little or no possibilities to maintain forested or agricultural spaces [42–44].

For the road network stressor, the municipal plans forest fire defense road network from the Portuguese Institute of Nature Conservation and Forests (https://fogos.icnf.pt/ infoPMDFCI/PMDFCI_PUBLICOlist.asp; accessed on 7 September 2020), was analyzed using the QGIS buffer function of 20 m from the road network, which was selected as a high risk, considering that most ignitions start in the vicinity of roads and attributed to negligent or criminal acts [43,45].

The viewshed QGIS map entered as a stressor in the InVEST model were the shadow areas, or areas not visible by at least three lookout posts/towers, corresponding to an additional risk. Early ignition detection improves early intervention and therefore, firefighting success. The viewshed analysis was calculated by applying the QGIS software plugin, using the MDT available (https://www.fc.up.pt/pessoas/jagoncal/srtm/; accessed on 7 September 2020), and the National Network of Lookout Posts made available by the Nature and Environment Protection Service of the Republican National Guard (SEPNA/GNR).

Fire ignition history is considered a dynamic stressor in the model obtained from the Nature and Environment Protection Service of the Republican National Guard (SEPNA/GNR) which, in Portugal, holds the responsibility to investigate and determine the ignition cause and starting point of fires. The stressor is determined with the QGIS Hotspot function, being assigned a radius of 1000 m at each point; only fire ignitions that occurred from the first day of the calendar year to the date of the satellite images were used for the calculation of the MDIR, hence the dynamic classification. This variable derives from the ignition trend of proximity to a known occurrence, attributed to both natural and anthropic causes [36].

The model also assumes another two dynamic variables, the Normalized Difference Vegetation Index (NDVI) and the Normalized Difference Water Index (S2-NDWI), evolving from the previously published MDIR model [36] which considered only one dynamic variable, NDVI. Indexes were calculated using 75 Multi Spectral Instrument (MSI) sensor Sentinel-2 satellite images from the months of May, June, July, August, and September for the years 2016 to 2020 (https://scihub.copernicus.eu/dhus/#/home, accessed on 1 November 2020). Downloaded study region satellite images, level 2A, were atmospherically, radiometrically, and geometrically corrected for the period under analysis, without the presence of clouds. Selected data values up to 0.4 [46] were considered for both indexes, taking into account the loss of chlorophyll and the absence of humidity as dynamic variables that influence the ignition of rural fires [47,48]. Sentinel-2 satellite imagery corresponding to bands 4 and 8, with 10 m resolution and band 12, with 20 m resolution, were analyzed using the QGIS Raster Calculator to produce chlorophyll NDVI (Equation (1)) and pre-fire vegetation humidity S2-NDWI (Equation (2)) maps (https://foodsecurity-tep.net/node/214 accessed on 1 November 2020).

$$\text{NDVI} = \frac{\text{NIR} - \text{RED}}{\text{NIR} + \text{RED}} \tag{1}$$

Theorem 1. *Normalized difference vegetation index (NDVI), Sentinel-2 Band 4 (RED), red region (Red: 650–680 nm, 10 m resolution); Band 8 (NIR) near infrared (NIR: 785–899 nm, 10 m resolution).*

$$\text{S2\_NDWI} = \frac{\text{NIR} - \text{SWIR2}}{\text{NIR} + \text{SWIR2}} \tag{2}$$

Theorem 2. *Normalized Difference Water Index (S2-NDWI) formulae using Sentinel-2 Band 8 (NIR) near infrared (NIR: 785–899 nm, 10 m resolution) and Band 12 (SWIR 2) Shortwave infrared (SWIR: ~1610–~2190 nm, 20 m resolution).*

Finally, the MDIR methodology involves the application of the Habitat Risk Assessment model (HRA) from InVEST (Integrated Valuation of Ecosystem Services and Tradeoffs) suite of models. The process involves grouping all produced data layers in a "host" folder, which will be subdivided into two other folders, one referring to the land occupation files representing the habitats, and the other where the different types of stressor maps are lodged; all files are used in vector format, which favors easy applicability and update. The main "host" folder will hold the vector files with the sub-regions of the study area, in vector format, and two CSV files, where the weights and characteristics of the relationship between habitats and stressors are assigned. Special attention should be given to the need for a direct match between the assigned names in the stressors data table and the stressor folder map, as they should match.

Tables were produced using Microsoft Excel and later saved as CSV files. The Habitat Stressor Info table (Appendix A, Figure A1) contains rows corresponding to the habitat's characteristics and stressors, respectively, with four columns:

- NAME—The name, that should be unique for each entry, and must match exactly those that appear in the "exposure_consequence_criteria.csv" file, never exceeding 8 characters.
- PATH—Corresponds to the path of the vector file of the input data. They can be absolute file paths.
- TYPE—Defines whether it corresponds to a habitat or a stressor.
- BUFFER STRESSOR—The desired buffer distance (in meters) to be used to expand the influence of a given stressor. It should be left blank for habitats, but should not be left blank for stressors or factors.

The file (exposure_consequence_criteria.csv) corresponds to the Score of Exposure and Consequence criteria that must also be in the "host" input folder. This file contains information on the impact of each stressor on each type of land use/occupation (i.e., exposure (B) and consequence (A) scores) for the different types of land occupation and stressors under analysis. The classification should be awarded an odd scale, which for this study, the scale varied from 1 to 5, since the fire risk maps are standardized in 5 classes.

(A) The Consequence criteria consist of the risk observed in certain areas being exposed to a stressor, i.e., the Consequences are determined by the sensitivity of a type of land occupation to a specific stressor and by the natural resilience that occupation holds to resist and recover from disturbances in general.

By default, the model includes two specific measures of sensitivity (frequency of disturbance, change in classification) and four measures of natural resilience (average growth, rotation rate, connectivity rate, natural recovery time). Each of these measures is described as:

- Average growth: corresponds to the average growth rates of the various species that make up a type of soil occupation. According to the soil occupation variables considered, values between 1 and 5 were assigned, where the maximum value corresponds to the highest average growth rates and consequent resilience.
- Rotation rate: consists of changing the type of occupation, or the natural or anthropic soil characteristics of the various species that make up the class of soil occupation. Species that suffer a high turnover are less sensitive to fire-related stressors and therefore, are given a higher resilience value.
- Connectivity rate: corresponds to the level of connection between the types of land occupation with similar characteristics. In this way, greater connectivity implies greater resilience; therefore, they are given a greater weight.
- Natural recovery time: habitats consisting of species that reach maturity earlier show a faster recovery rate after a disturbance than those that take longer to reach maturity. Consequently, greater weight is attributed to soil occupations where the predominant species rapidly reach maturity.
- Frequency of disturbance: corresponds to the frequency at which the soil occupation type is disturbed by a stressor, i.e., whether the type of land occupation is disturbed or not, with the occurrence of an event. High rates of disturbance imply greater sensitivity, and therefore, they are given a higher weight.
- Change in classification: the change in structure corresponds to the percentage of structural density change in a type of soil occupation when exposed to a given stressor. Types of land occupation that lose a high percentage of area when exposed to a certain stressor are highly sensitive, while those that lose little area are less sensitive, and as such, the latter are assigned a lower weight.

(B) Exposure criteria consists of the increasing risk to which certain areas are exposed when accumulating certain characteristics. The characteristics, from a temporal and spatial nature, consider that aspects such as management, overlap, and neighborhood contribute cumulatively to the aggravation of the risk factor.

- Management effectiveness: management can limit the negative impacts of stressors on the type of land occupation, so effective management reduces the likelihood of stress when compared to areas of land occupation where there is no management. As with other criteria, higher numbers represent higher exposure and, as such, less management effectiveness.
- Intensity of overlap: exposure depends not only on the overlap of the type of land occupation and stressors in space and time, but also on the cumulative effect of stressors.
- Neighborhood: exposure tends to increase when the types of land occupation in the vicinity are very similar; the opposite situation decreases exposure.

For each type of land occupation, it is necessary to have a classification and the assignment of a Data Quality (DQ) and a Weight. For each criterion it is necessary to specify whether it corresponds to a Consequence or Exposure. These criteria are characterized as:

- Rating—This is a measure of the impact of a criterion on a given type of land occupation in relation to the general ecosystem. The classification is an integer between 1 and 5, assigned by taking the published bibliography into consideration. These numbers can be updated as better information becomes available. A rating score of 0 will tell the model to ignore these specific criteria.
- DQ—This column represents the quality of the score data provided in the Rating column. Here, the model gives the user the ability to reduce the weight of less reliable data sources or to define particularly well-studied criteria. A low DQ indicates the best data quality, while a high DQ indicates limited data quality. In this study, the criterion of an intermediate value (3) was used.
- Weight—The weight criterion gives the possibility to determine the most important criteria for the system, regardless of data quality. A low weight matches more important criteria, while a high weight indicates less important criteria.
- E/C—This column indicates whether the criteria given are being applied to exposure or the consequence of the chosen risk equation. By default, all criteria in the Sensitivity or Resilience categories will be assigned to Consequence (C) in risk equations, and all criteria in the Exposure category will be assigned to Exposure (E) in the risk equation.

The InVEST HRA model considered the Multiplicative Risk calculation (Equation (3)), which, for the current study, considered the Cumulative Risk (Equation (4)) to the Ecosystem from Multiple Stressors as an integrative index of risk across all habitats in a grid cell (The Natural Capital Project, 2021) (Appendix A, Figure A1).

$$R_{ijkl} = E_{jkl} \cdot C_{jkl} \tag{3}$$

Theorem 3. *Multiplicative Risk calculation where risk to habitat "j" caused by stressor "k" in cell "l" is calculated as the product of the exposure and consequence scores.*

$$R_l = \sum_{j=1}^{J} R_{jl} \tag{4}$$

Theorem 4. *The cumulative risk for habitat "j" in cell "l" is the sum of all risk scores for each habitat. To assess the influence of multiple activities, the cumulative risk of all stressors was quantified for each habitat "l" as the sum of all risk scores for each combination of habitat and activity "j" as "R_jl", and annual and monthly results are presented on Figure 3.*

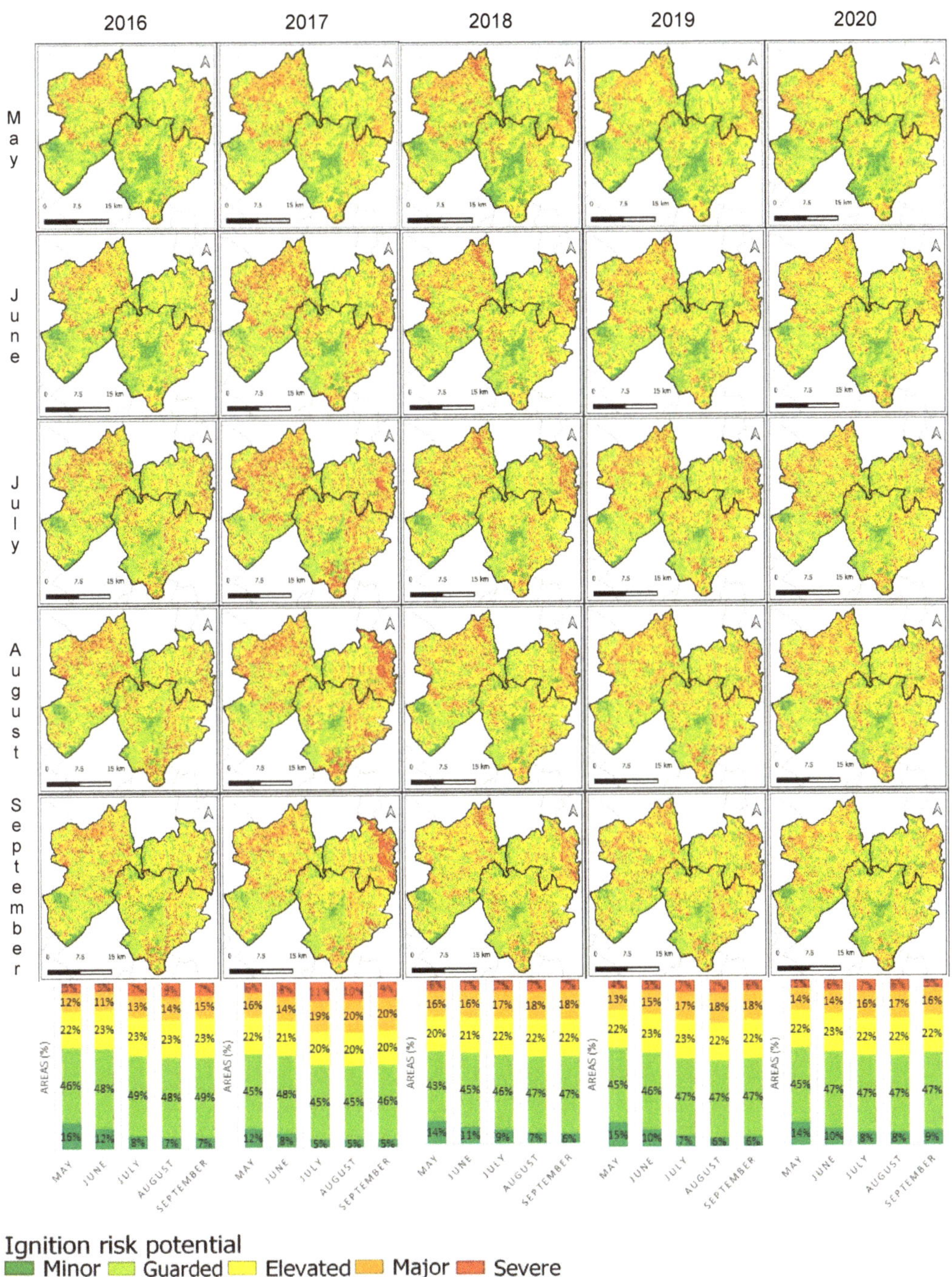

Figure 3. Ignition potential risk maps from the four critical fire risk months for the 2016–2020 period, followed by an analytic percentage graph of yearly representative areas of five risk classes.

The data ranks and variable valuation in the HRA model were attributed according to literature review and local regional characteristics, expressed as sensitivity and natural resilience measures. All attributed ranks were also discussed thoroughly by a transdisciplinary team of experts holding in-depth knowledge of the regional characteristics. The literature supported the choice of variables and ranked values defined for MDIR, forest road network [15,22,45,49], slope [20,26,50], population density [21,22,41], visualization basins [33], and fire history [5,51,52] combined with NDVI [4,30,53,54] and S2-NDWI [4,55] indices.

Since results are delivered as model maps with spatial risk distribution, the variable validation will undergo a sequential individual variable removal to determine the contribution of variables and interactions of the strategy.

3. Results and Discussion

The results obtained for the study region shown as potential ignition risk maps using the InVEST HRA model within the Modernized Dynamic Ignition Risk strategy (MDIR) are illustrated on Figure 3. The model connected seven different stressors: forest road network, slope, population density, visualization basins, and fire history, combined with NDVI and S2-NDWI indices. The time interval used in the analysis process comprises the months of May to September for the years 2016 to 2020. The results consider five risk classes, minor, guarded, elevated, major, and severe.

The ignition potential mapping (Figure 3) reveals that the area percentages in each risk class do not change significantly from year to year when considering homologous periods; however, they vary geographically from month to month and year to year, most noticeable for the years 2017 and 2018, when major fire events occurred. This may be attributed to the continuous ecological succession of recently fire-affected habitats, land use change, or natural succession processes. The stressors that contribute to the total gross areas class oscillations are mainly represented through the geographical dispersion of dynamic variables, namely, the NDVI and S2-NDWI together with the cumulative fire ignitions in each year, since weather conditions modify the response of vegetation to ignition probability. The vegetation index is sensitive to vegetation and water content, which is complementary with NDVI, where high values of S2-NDWI exhibit a high water content of the vegetation [4,56].

One of the observed trends in Figure 3 is the monthly decrease in lower risk class percentages associated with the increase in temperature and reduction of humidity, represented in the model by NDVI and S2-NDWI, and also correlating with regional weather conditions that cumulatively increase the higher risk categories. Another interesting fact is the percentage value of lower risk classes in the initial month of the critical season (May) whose values are associated with the reduction level in the very critical months of August and September, representing the difference between a harsher and calmer fire ignition season, which in turn is correlated with plant water content. Therefore, the amount of water in plants in the beginning of the season will dictate whether or not the year will exhibit recurrent ignitions and possibly larger fires, like those observed in 2017, a critical year for forest fires across the Portuguese territory [1,41].

Although the number of fire ignitions plays an active role in the MDIR model, it may also contribute to the success of model analysis, enhancing the ability to predict high ignition risk areas, particularly for areas exhibiting considerable anthropic modifications. Figure 4 exhibits the number of monthly ignitions for the years 2016–2020, 1025 in total, emphasizing the high number of ignitions in the month of July, as well as the exceptional year of 2017, in which the month of June observed a high incidence of 87 ignitions for this region, hence the importance of early season conditions.

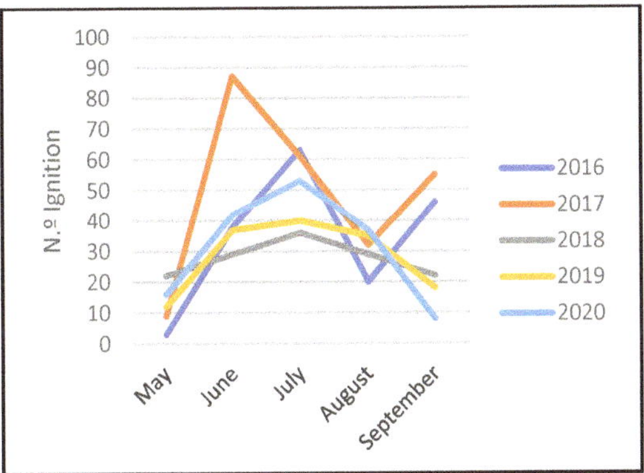

Figure 4. The number of monthly critical season ignitions for the years 2016–2020.

Considering the evolution of ignitions throughout the critical period of the five-year timeline (Figure 4), in general, the month of May exhibits the lowest number, with ignitions peaking during the month of July and decreasing during the month of August, concluding with a year-dependent variation in September. This trend may be interpreted as a result of plant water content and associated chlorophyll values; the sustainability during the critical period dictates a dryness threshold variability earlier or later within the critical season [55].

The variation during the month of September after 2017 may be attributed to a policy change in the Portuguese Civil Protection system, in which prior approval, by formal request through communication with the authorities, was required for slash-and-burn, and other forestry and agricultural practices; otherwise, the offense was punishable by law; hence, the reduction in ignitions is observed.

All in all, the 1025 ignitions observed between 2016 and 2020 are most common in forest habitats, making up 51% of the total; the ignitions occurring on agricultural habitats make up 35%, followed by the urban, bush, and pasture habitats exhibiting 9%, 3%, and 2% of the total, respectively (Figure 5).

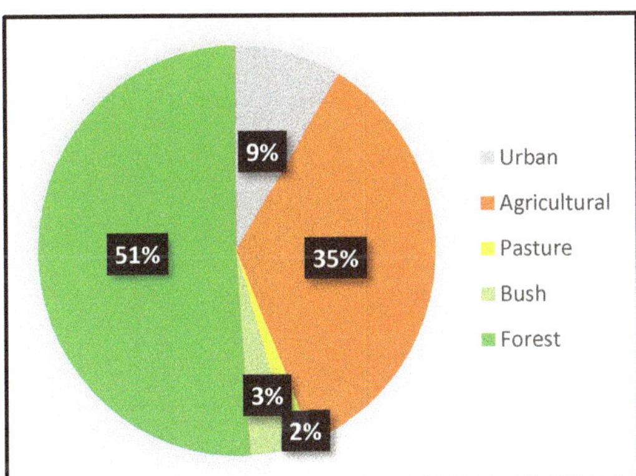

Figure 5. The percentage of ignitions per land use class.

The model's quality was analyzed using the number of ignitions that occurred since the last monthly satellite imagery update of NDVI and S2-NDWI. Pursuing this objective, the number of ignitions since the production of the initial MDIR until the new MDIR was calculated (monthly) and the percentage of ignitions that fell in each of the five risk classes was determined as a validation test (Figure 6).

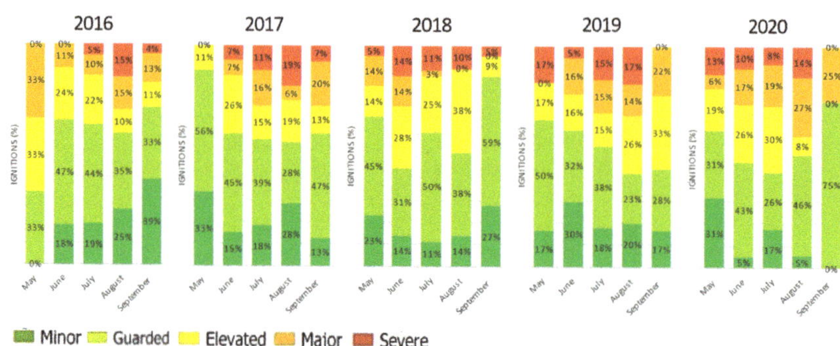

Figure 6. The monthly ignitions since the last MDIR model for the five risk classes during each year of the five-year study.

The monthly ignitions for the five-year analysis period regarding the MDIR model award a higher percentage of ignitions in the lower risk classes due to the high occurrence of these classes in the study area. Despite the lower representativity of the high-risk class areas, the number of ignitions follows a monthly trend, where a higher number of ignitions is observed in the months of June, July, and August, in accordance with the data in the literature [13,52].

The MDIR delivers a larger area of lower-risk classes, when compared to higher-risk classes (Figure 3); therefore, the representativity of ignitions is rather misleading regarding the number of ignitions occurring in the higher-risk classes. The relative representativeness of ignitions per area class was calculated using Equation (5), delivering a proportional percentage of occurrence per area class (Figure 7).

$$P_\% = \left(\left(\frac{rA}{rNi} \right) / At \right) * 100 \tag{5}$$

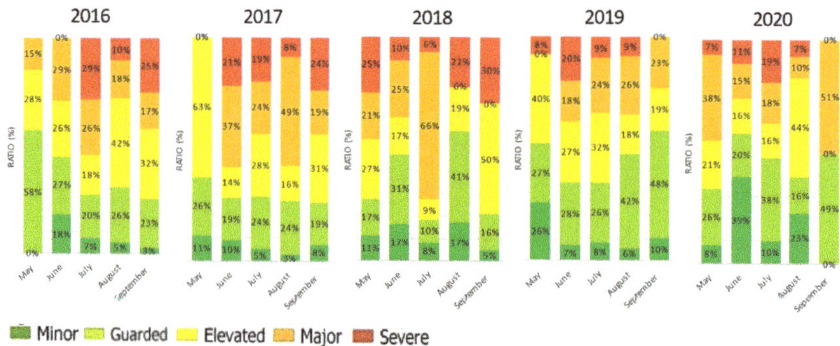

Figure 7. The rational representative percentage of ignitions per associated representative area.

Theorem 5. *The representative percentage of ignitions per associated representative area (P%) equals the relative area of each risk class (rA) divided by the relative number of ignitions in each class (rNi), divided by the total area of classes (At), and multiplied by 100.*

Using the class representative percentage (Equation (5)) allows a clearer interpretation of the success of the MDIR model (Figure 7), where the higher-risk classes "elevated," "major," and "severe" show a noticeable increase during the months of June, July, and

August, with over 50% of the ignitions occurring within these classes. The months of May and September, prone to weather variability, and concurring with the agricultural and forestry practices of slash-and-burn, may deliver a bias to the efficiency of the model.

Focusing on the representative area ignitions (Figure 7), the average model success, considering the higher-risk classes "elevated," "major," and "severe," is of 63%, 70%, 65%, 54%, and 55% for the years 2016, 2017, 2018, 2019, and 2020, respectively. The monthly success variability of each year can be attributed to weather conditions reflected in the model by the NDVI and S2-NDWI; the years evidencing most variability are 2019 and 2020, where additional policy measures may have played a relevant role. The total analysis of average success for the five-year period is 62%, which, if considered the preventive surveillance objective, should confer a significant reduction in the number of ignitions through early detection attributed to proximity monitoring [22].

The importance of the chosen variables was also analyzed by removing each individual variable/stressor from the MDIR model. The model's results for each individual run are presented in Appendix B, Figure A2. The averaged 2016–2020 comparative evaluation of variance from the final model's higher-risk classes "elevated," "major," and "severe" is expressed in Figure 8.

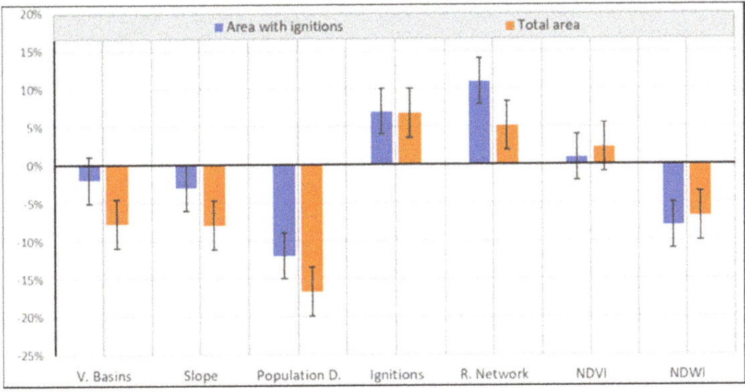

Figure 8. The area variance of the individual variable removal when compared with the MDIR model.

The variables "v. basins," "slope," "population d.," and "NDWI," when individually removed from the model, will reduce both the total area and the area with ignitions when compared with the model with the complete set of variables. "V. basins" and "slope" behave in a similar contributive way, where the total area input will be larger than that of the area with ignitions. Population density exhibits a larger contribution than the remaining variables, revealing the fragmented characteristics of the landscape with small, disperse, and numerous urban areas. The NDWI is the only variable where the area with ignitions increases more than the total area and whose contribution towards the model is rather positive. The contribution of NDWI towards the model's success derives from the fact that water content varies considerably throughout the season for this particular region where plants are well adapted to drought. The NDVI chlorophyll content alone may be misleading, whereas the cumulative characteristics of the model and the overlap of NDVI and NDWI will better assess the risk, hence the increase in the area with ignitions.

The considerable increase in the area with ignitions observed with the removal of "road networks," when compared with the modest increase in the total area, emphasizes the specificity of this variable in accounting for social ignition causes. Removing the "ignitions" proximity area from the model displays a balanced increase in both the total area and the area with ignitions, once again bringing to evidence the representativeness of this variable. Curiously, when removing "ignitions," all remaining variables behave in a similar unbalanced manner, either increasing or decreasing both the total area and the area with ignitions, once again revealing the disruptive influence of the social impact.

All the variables display a correlational trend between the total area and the area with ignitions, where area variance, as expected, influences the number of ignitions overlap. The geographic area representativeness of each variable certainly affects the model, a fact observed when "ignitions", "r. network," and "NDVI" variables are individually removed, increasing both the total and ignition areas, meaning that their presence will restrict the higher-risk areas, as expected from the model. The remaining variables, when individually removed, will decrease both areas, apparently indicating that their inclusion is not as favorable to the model, since their presence will increase the areas, most likely due to the higher geographic area representativeness, awarding areas with less representativity extra weight. Towards the current model's ambition of pinpointing specific high ignition risk areas for monitoring purposes, ultimately reducing these ignitions, results indicate that specific restricted geographic monitoring/intervention areas are of major importance, hence the representativeness towards model's success.

4. Conclusions

The integrated analysis of spatial, slow, and dynamic variables, shaped by the QGIS software in combination with the InVEST HRA model, allowed the production of geographically specific MDIR risk maps, which favor the implementation of a facilitated monitoring effort and consequent reduction of monitoring costs. Considering the General Public License (GPL) of all software's used in the production of MDIR, along with the detailed description of the process, means that this approach may be replicated by researchers worldwide. This may be of particular interest to developing countries with restricted or no access to licensed software and quality imagery [36].

Sentinel-2 satellite imagery is a valuable, free monitoring tool, with acceptable resolution, for assessing the variation in chlorophyll and water contents. This asset allows a dynamic approach to risk maps, enabling near-real estimate analysis. While NDVI and S2-NDWI are correlated, S2-NDWI positively influences NDVI; the opposite is not mandatory, and hence, the importance of including both as model variables. Furthermore, S2-NDWI grants the necessary balance with other variables in the model that are mostly socially driven.

The MDIR approach grants the identification of high ignition-risk areas, and can easily be adapted to current monitoring strategies. The use of this methodology confers further efficiency and efficacy in reducing fire ignitions, thus granting early critical season evaluation for determining if a particular year presents a higher ignition risk. The approach delivered adequate results in estimating ignitions, achieving, in several month, accuracy percentages of over 70%. For the current case study, the MDIR model, within its domain of applicability, returns a satisfactory range of accuracy consistent with the intended monitoring application.

Understanding that many anthropic factors are difficult to estimate, MDIR produced quality maps identifying areas of high ignition risk, delivering a 62% rate of success in predicting ignitions. From another angle, the importance of possessing a free, easily applicable approach such as MDIR is the possibility to study the impact of new policy implementation or an increased monitoring effort when analyzing the model's accuracy.

Despite the interesting results obtained for determining risk areas, the MDIR model would also benefit from the analysis of the ignition causes that gave rise to large fires in order to test other dimensions and capabilities of this model.

Author Contributions: Conceptualization, L.S., C.B. and V.L.; methodology, L.S. and V.L.; software, L.S. and V.L.; validation, L.S., V.L. and C.B.; formal analysis, L.S.; investigation, L.S.; resources, V.L.; data curation, V.L.; writing—original draft preparation, L.S.; writing—review and editing, L.S. and C.B.; supervision, L.S. All authors have read and agreed to the published version of the manuscript.

Funding: This research received no external funding.

Informed Consent Statement: Not applicable.

Data Availability Statement: All data used in the current study is under access via the public Internet and GEANT networks, with committed reliability and performance on https://sentinels.copernicus.eu/web/sentinel/home, accessed on 1 November 2020.

Acknowledgments: We would like to thank the Tomar Center for Nature Protection and the Republican National Guard for granting access to the fire ignition data and for the precious territorial and fire ignition-related knowledge shared.

Appendix A

HABITAT NAME	agricultural			water			florest			bush			pasture			urban			Criteria Type
HABITAT RESILIENCE ATTRIBUTES	Rating	DQ	Weight	Rating	DQ	Weight	Rating	DQ	Weight	Rating	DQ	Weight	Rating	DQ	Weight	Rating	DQ	Weight	E/C
growth rate	3	3	1	0	1	5	5	3	3	4	3	1	3	3	2	1	3	4	C
rotation rate	5	3	5	0	1	1	4	3	1	3	3	3	4	3	4	1	3	1	C
connectivity fee	4	3	2	0	1	1	5	3	1	3	3	3	2	3	5	2	3	3	C
natural recovery time	2	3	1	0	1	5	5	3	3	4	3	1	2	3	2	1	3	4	C
HABITAT STRESSOR OVERLAP PROPERTIES																			
Road	Rating	DQ	Weight	Rating	DQ	Weight	Rating	DQ	Weight	Rating	DQ	Weight	Rating	DQ	Weight	Rating	DQ	Weight	E/C
disturbance frequency	2	3	3	0	1	5	1	3	3	1	3	3	2	3	3	5	3	3	C
change in classification	2	3	3	0	1	5	1	3	3	5	3	3	2	3	3	5	3	3	C
management efficiency	3	3	2	0	1	5	5	3	1	4	3	2	3	3	2	1	3	3	E
intensity	2	3	3	0	1	5	5	3	3	3	3	3	4	3	3	1	3	3	E
neighborhood	2	3	2	0	1	5	3	3	2	1	3	2	3	3	2	5	3	3	E
Dpop	Rating	DQ	Weight	Rating	DQ	Weight	Rating	DQ	Weight	Rating	DQ	Weight	Rating	DQ	Weight	Rating	DQ	Weight	E/C
disturbance frequency	4	3	3	0	1	5	2	3	3	1	3	3	3	3	3	5	3	3	C
change in classification	2	3	3	0	1	5	5	3	3	1	3	3	2	3	3	5	3	3	C
management efficiency	3	3	2	0	1	5	5	3	1	4	3	2	3	3	2	1	3	3	E
intensity	2	3	3	0	1	5	5	3	3	3	3	3	4	3	3	1	3	3	E
neighborhood	2	3	2	0	1	5	3	3	2	1	3	2	3	3	2	5	3	3	E
Decli	Rating	DQ	Weight	Rating	DQ	Weight	Rating	DQ	Weight	Rating	DQ	Weight	Rating	DQ	Weight	Rating	DQ	Weight	E/C
disturbance frequency	1	3	3	0	1	5	5	3	3	4	3	3	3	3	3	2	3	3	C
change in classification	4	3	3	0	1	5	5	3	3	3	3	3	4	3	3	5	3	3	C
management efficiency	2	3	2	0	1	5	5	3	1	2	3	2	2	3	2	1	3	3	E
intensity	3	3	3	0	1	5	1	3	3	1	3	3	2	3	3	5	3	3	E
neighborhood	2	3	2	0	1	5	1	3	2	1	3	2	2	3	2	4	3	3	E
Visualizatiom	Rating	DQ	Weight	Rating	DQ	Weight	Rating	DQ	Weight	Rating	DQ	Weight	Rating	DQ	Weight	Rating	DQ	Weight	E/C
disturbance frequency	2	3	3	0	1	5	3	3	3	3	3	3	2	3	3	5	3	3	C
change in classification	3	3	3	0	1	5	5	3	3	1	3	3	3	3	3	5	3	3	C
management efficiency	3	3	2	0	1	5	3	3	1	2	3	2	3	3	2	1	3	3	E
intensity	5	3	3	0	1	5	3	3	3	3	3	3	4	3	3	1	3	3	E
neighborhood	3	3	2	0	1	5	3	3	2	3	3	2	3	3	2	3	3	3	E
Fire	Rating	DQ	Weight	Rating	DQ	Weight	Rating	DQ	Weight	Rating	DQ	Weight	Rating	DQ	Weight	Rating	DQ	Weight	E/C
disturbance frequency	1	3	3	0	1	5	1	3	3	1	3	3	3	3	3	4	3	3	C
change in classification	1	3	3	0	1	5	5	3	3	2	3	3	1	3	3	5	3	3	C
management efficiency	2	3	2	0	1	5	4	3	1	3	3	2	3	3	2	1	3	3	E

Figure A1. *Cont.*

	Rating	DQ	Weight	Rating	DQ	Weight	Rating	DQ	Weight	Rating	DQ	Weight	Rating	DQ	Weight	Rating	DQ	Weight	E/C
intensity	2	3	3	0	1	5	3	3	3	3	3	3	3	3	3	1	3	3	E
neighborhood	1	3	2	0	1	5	3	3	2	3	3	2	1	3	2	2	3	3	E
NDVI	Rating	DQ	Weight	Rating	DQ	Weight	Rating	DQ	Weight	Rating	DQ	Weight	Rating	DQ	Weight	Rating	DQ	Weight	E/C
disturbance frequency	4	3	3	0	1	5	2	3	3	1	3	3	3	3	3	5	3	3	C
change in classification	2	3	3	0	1	5	5	3	3	1	3	3	2	3	3	5	3	3	C
management efficiency	3	3	2	0	1	5	5	3	1	4	3	2	3	3	2	1	3	3	E
intensity	2	3	3	0	1	5	5	3	3	3	3	3	4	3	3	1	3	3	E
neighborhood	2	3	2	0	1	5	3	3	2	1	3	2	3	3	2	5	3	3	E
NDWI	Rating	DQ	Weight	Rating	DQ	Weight	Rating	DQ	Weight	Rating	DQ	Weight	Rating	DQ	Weight	Rating	DQ	Weight	E/C
disturbance frequency	4	3	3	0	1	5	2	3	3	1	3	3	3	3	3	5	3	3	C
change in classification	2	3	3	0	1	5	5	3	3	1	3	3	2	3	3	5	3	3	C
management efficiency	3	3	2	0	1	5	5	3	1	4	3	2	3	3	2	1	3	3	E
intensity	2	3	3	0	1	5	5	3	3	3	3	3	4	3	3	1	3	3	E
neighborhood	2	3	2	0	1	5	3	3	2	1	3	2	3	3	2	5	3	3	E

Figure A1. The Habitat Stressor Info Table.

Appendix B

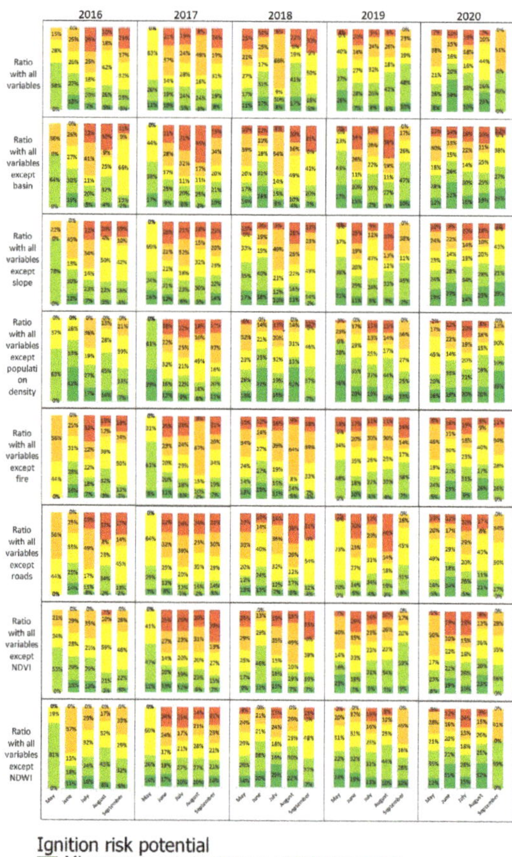

Figure A2. The comparative model results of the analysis of the variables influencing the model.

References

1. Couto, F.T.; Iakunin, M.; Salgado, R.; Pinto, P.; Viegas, T.; Pinty, J.P. Lightning modelling for the research of forest fire ignition in Portugal. *Atmos. Res.* **2020**, *242*, 104993. [CrossRef]
2. Kountouris, Y. Human activity, daylight saving time and wildfire occurrence. *Sci. Total Environ.* **2020**, *727*, 138044. [CrossRef]
3. Balch, J.K.; Bradley, B.A.; Abatzoglou, J.T.; Chelsea Nagy, R.; Fusco, E.J.; Mahood, A.L. Human-started wildfires expand the fire niche across the United States. *Proc. Natl. Acad. Sci. USA* **2017**, *114*, 2946–2951. [CrossRef]
4. García-Llamas, P.; Suárez-Seoane, S.; Taboada, A.; Fernández-Manso, A.; Quintano, C.; Fernández-García, V.; Fernández-Guisuraga, J.M.; Marcos, E.; Calvo, L. Environmental drivers of fire severity in extreme fire events that affect Mediterranean pine forest ecosystems. *For. Ecol. Manag.* **2019**, *433*, 24–32. [CrossRef]
5. Tessler, N.; Sapir, Y.; Wittenberg, L.; Greenbaum, N. Recovery of Mediterranean Vegetation after Recurrent Forest Fires: Insight from the 2010 Forest Fire on Mount Carmel, Israel. *L. Degrad. Dev.* **2016**, *27*, 1424–1431. [CrossRef]
6. Oliveira, S.; Oehler, F.; San-Miguel-Ayanz, J.; Camia, A.; Pereira, J.M.C. Modeling spatial patterns of fire occurrence in Mediterranean Europe using Multiple Regression and Random Forest. *For. Ecol. Manag.* **2012**, *275*, 117–129. [CrossRef]
7. Doerr, S.H.; Santín, C. Global trends in wildfire and its impacts: Perceptions versus realities in a changing world. *Philos. Trans. R. Soc. B Biol. Sci.* **2016**, *371*, 20150345. [CrossRef]
8. Alló, M.; Loureiro, M.L. Assessing preferences for wildfire prevention policies in Spain. *For. Policy Econ.* **2020**, *115*, 102145. [CrossRef]
9. Xu, W.; He, H.S.; Fraser, J.S.; Hawbaker, T.J.; Henne, P.D.; Duan, S.; Zhu, Z. Spatially explicit reconstruction of post-megafire forest recovery through landscape modeling. *Environ. Model. Softw.* **2020**, *134*, 104884. [CrossRef]
10. Radeloff, V.C.; Helmers, D.P.; Anu Kramer, H.; Mockrin, M.H.; Alexandre, P.M.; Bar-Massada, A.; Butsic, V.; Hawbaker, T.J.; Martinuzzi, S.; Syphard, A.D.; et al. Rapid growth of the US wildland-urban interface raises wildfire risk. *Proc. Natl. Acad. Sci. USA* **2018**, *115*, 3314–3319. [CrossRef]
11. Alcasena, F.J.; Salis, M.; Ager, A.A.; Castell, R.; Vega-García, C. Assessing wildland fire risk transmission to communities in northern spain. *Forests* **2017**, *8*, 30. [CrossRef]
12. Górriz-Mifsud, E.; Burns, M.; Marini Govigli, V. Civil society engaged in wildfires: Mediterranean forest fire volunteer groupings. *For. Policy Econ.* **2019**, *102*, 119–129. [CrossRef]
13. Parente, J.; Pereira, M.G. Structural fire risk: The case of Portugal. *Sci. Total Environ.* **2016**, *573*, 883–893. [CrossRef]
14. Fernandes, P.M. Fire-smart management of forest landscapes in the Mediterranean basin under global change. *Landsc. Urban Plan.* **2013**, *110*, 175–182. [CrossRef]
15. de Castro Galizia, L.F.; Rodrigues, M. Modeling the influence of eucalypt plantation on wildfire occurrence in the Brazilian savanna biome. *Forests* **2019**, *10*, 844. [CrossRef]
16. Molina, J.R.; Lora, A.; Prades, C.; Rodríguez y Silva, F. Roadside vegetation planning and conservation: New approach to prevent and mitigate wildfires based on fire ignition potential. *For. Ecol. Manag.* **2019**, *444*, 163–173. [CrossRef]
17. Quintano, C.; Fernández-Manso, A.; Stein, A.; Bijker, W. Estimation of area burned by forest fires in Mediterranean countries: A remote sensing data mining perspective. *For. Ecol. Manag.* **2011**, *262*, 1597–1607. [CrossRef]
18. Taylor, S.W.; Woolford, D.G.; Dean, C.B.; Martell, D.L. Wildfire prediction to inform fire management: Statistical science challenges. *Stat. Sci.* **2013**, *28*, 586–615. [CrossRef]
19. Jain, P.; Coogan, S.C.P.; Subramanian, S.G.; Crowley, M.; Taylor, S.; Flannigan, M.D. A review of machine learning applications in wildfire science and management. *Environ. Rev.* **2020**, *28*, 478–505. [CrossRef]
20. Carmo, M.; Moreira, F.; Casimiro, P.; Vaz, P. Land use and topography influences on wildfire occurrence in northern Portugal. *Landsc. Urban Plan.* **2011**, *100*, 169–176. [CrossRef]
21. Guo, F.; Su, Z.; Wang, G.; Sun, L.; Lin, F.; Liu, A. Wildfire ignition in the forests of southeast China: Identifying drivers and spatial distribution to predict wildfire likelihood. *Appl. Geogr.* **2016**, *66*, 12–21. [CrossRef]
22. Catry, F.X.; Rego, F.C.; Bação, F.L.; Moreira, F. Modeling and mapping wildfire ignition risk in Portugal. *Int. J. Wildl. Fire* **2009**, *18*, 921–931. [CrossRef]
23. Costafreda-Aumedes, S.; Comas, C.; Vega-Garcia, C. Human-caused fire occurrence modelling in perspective: A review. *Int. J. Wildl. Fire* **2017**, *26*, 983–998. [CrossRef]
24. Vilà-Vilardell, L.; Keeton, W.S.; Thom, D.; Gyeltshen, C.; Tshering, K.; Gratzer, G. Climate change effects on wildfire hazards in the wildland-urban-interface—Blue pine forests of Bhutan. *For. Ecol. Manag.* **2020**, *461*, 117927. [CrossRef]
25. D'Este, M.; Ganga, A.; Elia, M.; Lovreglio, R.; Giannico, V.; Spano, G.; Colangelo, G.; Lafortezza, R.; Sanesi, G. Modeling fire ignition probability and frequency using Hurdle models: A cross-regional study in Southern Europe. *Ecol. Process.* **2020**, *9*, 54. [CrossRef]
26. Rodrigues, M.; Costafreda-Aumedes, S.; Comas, C.; Vega-García, C. Spatial stratification of wildfire drivers towards enhanced definition of large-fire regime zoning and fire seasons. *Sci. Total Environ.* **2019**, *689*, 634–644. [CrossRef]
27. Chuvieco, E.; Congalton, R.G. Application of remote sensing and geographic information systems to forest fire hazard mapping. *Remote Sens. Environ.* **1989**, *29*, 147–159. [CrossRef]
28. Jaiswal, R.K.; Mukherjee, S.; Raju, K.D.; Saxena, R. Forest fire risk zone mapping from satellite imagery and GIS. *Int. J. Appl. Earth Obs. Geoinf.* **2002**, *4*, 1–10. [CrossRef]

29. Modugno, S.; Balzter, H.; Cole, B.; Borrelli, P. Mapping regional patterns of large forest fires in Wildland-Urban Interface areas in Europe. *J. Environ. Manag.* **2016**, *172*, 112–126. [CrossRef]

30. Hernandez-Leal, P.A.; Arbelo, M.; Gonzalez-Calvo, A. Fire risk assessment using satellite data. *Adv. Sp. Res.* **2006**, *37*, 741–746. [CrossRef]

31. Naderpour, M.; Rizeei, H.M.; Khakzad, N.; Pradhan, B. Forest fire induced Natech risk assessment: A survey of geospatial technologies. *Reliab. Eng. Syst. Saf.* **2019**, *191*, 106558. [CrossRef]

32. Matin, M.A.; Chitale, V.S.; Murthy, M.S.R.; Uddin, K.; Bajracharya, B.; Pradhan, S. Understanding forest fire patterns and risk in Nepal using remote sensing, geographic information system and historical fire data. *Int. J. Wildl. Fire* **2017**, *26*, 276–286. [CrossRef]

33. Phelps, N.; Woolford, D.G. Guidelines for effective evaluation and comparison of wildland fire occurrence prediction models. *Int. J. Wildl. Fire* **2021**, *30*, 225–240. [CrossRef]

34. Moritz, M.A.; Batllori, E.; Bradstock, R.A.; Gill, A.M.; Handmer, J.; Hessburg, P.F.; Leonard, J.; McCaffrey, S.; Odion, D.C.; Schoennagel, T.; et al. Learning to coexist with wildfire. *Nature* **2014**, *515*, 58–66. [CrossRef] [PubMed]

35. Gaglio, M.; Aschonitis, V.; Pieretti, L.; Santos, L.; Gissi, E.; Castaldelli, G.; Fano, E.A. Modelling past, present and future Ecosystem Services supply in a protected floodplain under land use and climate changes. *Ecol. Modell.* **2019**, *403*, 23–34. [CrossRef]

36. Santos, L.; Lopes, V.; Baptista, C. Modernized Forest Fire Risk Assessment Model Based on the Case Study of three Portuguese Municipalities Frequently Affected by Forest Fires. *Environ. Sci. Proc.* **2020**, *3*, 30. [CrossRef]

37. Nolè, L.; Pilogallo, A.; Saganeiti, L.; Bonifazi, A.; Santarsiero, V.; Santos, L.; Murgante, B. Land use change and habitat degradation: A case study from tomar (portugal). *Smart Innov. Syst. Technol.* **2021**, *178 SIST*, 1722–1731. [CrossRef]

38. Rivas-Martínez, S.; Rivas-Sáenz, S.; Penas-Merino, A. Worldwide bioclimatic classification system. *Glob. Geobot.* **2011**, *1*, 1–638.

39. Sari, F. Forest fire susceptibility mapping via multi-criteria decision analysis techniques for Mugla, Turkey: A comparative analysis of VIKOR and TOPSIS. *For. Ecol. Manag.* **2021**, *480*, 118644. [CrossRef]

40. Çolak, E.; Sunar, F. Evaluation of forest fire risk in the Mediterranean Turkish forests: A case study of Menderes region, Izmir. *Int. J. Disaster Risk Reduct.* **2020**, *45*, 101479. [CrossRef]

41. Oliveira, S.; Félix, F.; Nunes, A.; Lourenço, L.; Laneve, G.; Sebastián-López, A. Mapping wildfire vulnerability in Mediterranean Europe. Testing a stepwise approach for operational purposes. *J. Environ. Manag.* **2018**, *206*, 158–169. [CrossRef]

42. Chergui, B.; Pleguezuelos, J.M.; Fahd, S.; Santos, X. Modelling functional response of reptiles to fire in two Mediterranean forest types. *Sci. Total Environ.* **2020**, *732*, 139205. [CrossRef]

43. Navalho, I.; Alegria, C.; Quinta-Nova, L.; Fernandez, P. Integrated planning for landscape diversity enhancement, fire hazard mitigation and forest production regulation: A case study in central Portugal. *Land Use Policy* **2017**, *61*, 398–412. [CrossRef]

44. Chergui, B.; Fahd, S.; Santos, X.; Pausas, J.G. Socioeconomic Factors Drive Fire-Regime Variability in the Mediterranean Basin. *Ecosystems* **2018**, *21*, 619–628. [CrossRef]

45. Castillo Soto, M.E. The identification and assessment of areas at risk of forest fire using fuzzy methodology. *Appl. Geogr.* **2012**, *35*, 199–207. [CrossRef]

46. Viegas, D.X.; Almeida, M.; Raposo, J.; Oliveira, R.; Viegas, C.X. Ignition of Mediterranean Fuel Beds by Several Types of Firebrands. *Fire Technol.* **2014**, *50*, 61–77. [CrossRef]

47. Chrysafis, I.; Mallinis, G.; Tsakiri, M.; Patias, P. Evaluation of single-date and multi-seasonal spatial and spectral information of Sentinel-2 imagery to assess growing stock volume of a Mediterranean forest. *Int. J. Appl. Earth Obs. Geoinf.* **2019**, *77*, 1–14. [CrossRef]

48. Fernández-Manso, A.; Fernández-Manso, O.; Quintano, C. SENTINEL-2A red-edge spectral indices suitability for discriminating burn severity. *Int. J. Appl. Earth Obs. Geoinf.* **2016**, *50*, 170–175. [CrossRef]

49. Sevinc, V.; Kucuk, O.; Goltas, M. A Bayesian network model for prediction and analysis of possible forest fire causes. *For. Ecol. Manag.* **2020**, *457*, 117723. [CrossRef]

50. Mohammadi, F.; Bavaghar, M.P.; Shabanian, N. Forest Fire Risk Zone Modeling Using Logistic Regression and GIS: An Iranian Case Study. *Small-Scale For.* **2014**, *13*, 117–125. [CrossRef]

51. Bocken, N.M.P.; Geradts, T.H.J. Barriers and drivers to sustainable business model innovation: Organization design and dynamic capabilities. *Long Range Plann.* **2019**, *53*, 101950. [CrossRef]

52. Tessler, N.; Wittenberg, L.; Greenbaum, N. Vegetation cover and species richness after recurrent forest fires in the Eastern Mediterranean ecosystem of Mount Carmel, Israel. *Sci. Total Environ.* **2016**, *572*, 1395–1402. [CrossRef] [PubMed]

53. Peng, G.-X.; Li, J.; Chen, Y.-H.; Patah, N.A. A Forest Fire Risk Assessment Using ASTER Images in Peninsular Malaysia. *J. China Univ. Min. Technol.* **2007**, *17*, 232–237. [CrossRef]

54. Michael, Y.; Helman, D.; Glickman, O.; Gabay, D.; Brenner, S.; Lensky, I.M. Forecasting fire risk with machine learning and dynamic information derived from satellite vegetation index time-series. *Sci. Total Environ.* **2021**, *764*, 142844. [CrossRef] [PubMed]

55. Gao, B.C. NDWI—A normalized difference water index for remote sensing of vegetation liquid water from space. *Remote Sens. Environ.* **1996**, *58*, 257–266. [CrossRef]

56. Gao, B.-C. Naval Research Laboratory, 4555 Overlook Ave. *Remote Sens. Environ.* **1996**, *7212*, 257–266. [CrossRef]

MDPI
St. Alban-Anlage 66
4052 Basel
Switzerland
Tel. +41 61 683 77 34
Fax +41 61 302 89 18
www.mdpi.com

Forests Editorial Office
E-mail: forests@mdpi.com
www.mdpi.com/journal/forests